The

Chemical Engineering

Guide to Heat Transfer

Volume II

Equipment

Edited by

Kenneth J. McNaughton
and the Staff of Chemical Engineering

HEMISPHERE PUBLISHING CORPORATION
Washington   New York   London

McGraw-Hill Publications Co., New York, N.Y.

DISTRIBUTION OUTSIDE NORTH AMERICA
SPRINGER-VERLAG

Berlin   Heidelberg   New York   Tokyo

Copyright © 1986 by Chemical Engineering McGraw-Hill Pub. Co.
1221 Avenue of the Americas, New York, New York 10020

Printed in the United States of America.

Library of Congress Cataloging-in-Publication Data
Main entry under title:

The Chemical engineering guide to heat transfer.

   Includes bibliographies and index.
   Contents: v. 1. Plant principles—v. 2. Equipment.
   1. Heat—Transmission.   2. Heat exchangers.
I. McNaughton, Kenneth J.   II. Chemical engineering.
TJ260.C426   1985   660.2'8427   85-26987
ISBN 0-07-606939-7 (v. 1) Chemical Engineering
ISBN 0-07-606940-0 (v. 2) Chemical Engineering
ISBN 0-08116-465-0 (v. 1). Hemisphere
ISBN 0-89116-466-9 (v. 2) Hemisphere
DISTRIBUTED OUTSIDE NORTH AMERICA:
ISBN 3-540-16177-5 (v. 1) Springer-Verlag Berlin
ISBN 3-540-16178-3 (v. 2) Springer-Verlag Berlin

CONTENTS

PREFACE

When I was studying chemical engineering at college, I loved heat transfer. It was neat. Q = CAT. That's all I needed to know. And heat-in equals heat-out. Boy! If only the rest of life was so simple!

Some people say that engineers are drawn to science because they lack the skills to deal with people—that scientific types take comfort in being able to handle a field that responds according to inviolate laws, unlike their lawless and unpredictable fellow beings. Other observers, perhaps more charitable, suggest that those who inherit the skills to deal with a scientific universe may not give themselves the time to come to grips with the more elusive rules that attempt to explain human behavior.

I think both are fascinating fields worth pursuing. And who can say that the two won't come together? What with all the exciting developments in our understanding of molecular biology, surely we are overdue for some breakthroughs in psychology as well.

As it turns out, heat transfer isn't so simple anyway. But it is neat. And we are very fortunate to have developed our communication skills so well. In this book, for instance, we have accumulated ten years of practical wisdom from the pages of *Chemical Engineering,* on the subject of heat transfer, and the equipment that deals with it.

Here, in one volume, are the writings of fellow chemical engineers who also graduated with the basic understanding that Q = CAT and that heat-in equals heat-out. But they went on to specialize, and now they share with us their knowledge about the latest heat transfer equipment in today's chemical process industries.

This is where theory is put to the test: Boilers—how they work and how you can make them work better; Cooling—refrigeration, cooling towers, and chillers; Heating, insulation and winterizing; Condensers (with a calculator program); Dryers of all different sorts; and a pot pourri of other equipment, including fired heaters, fluidized beds, multistage evaporators and even solar ponds.

Life may not be so simple, but this book is going to make your life easier when it comes to heat transfer equipment. And so will the companion volume, which is geared to basic plant principles of heat transfer.

Section I
Boilers

Specifying and operating reliable waste-heat boilers

Boilers operating on waste heat from chemical processes are critical to many plants. Here is how to ensure that they will be troublefree.

Peter Hinchley, Imperial Chemical Industries Ltd.

☐ Many processes are economic only because they generate steam from their waste heat. A number of authors have described problems that have occurred in waste heat boilers and have emphasized the huge losses that resulted from some of these failures [1–4].

Although the engineer cannot alter the basic chemistry of processes, he can have an important influence on their efficiency, capital cost and long-term reliability. He can do this by influencing the way in which the heat recovery system is integrated into the process, and through the design and specification of that equipment.

Waste heat is usually present in processes in which the required chemical reactions take place at elevated temperatures. The heat is available in the process gases, and in the flue gas leaving the furnace that supplies the heat

required for endothermic reactions. The scale of a waste-heat recovery system varies over a wide range; a small acid-plant may produce 5 t/h (metric tons/hour) of steam at 15 atm and 300°C, whereas a large single-stream ammonia plant may produce up to 300 t/h of steam at up to 150 atm and 530°C. In some cases, the steam is exported to a factory, where it may be used for heating purposes only. In others, it may be used in large turbines that drive the main compressors, pumps and fans, with the exhaust being used both in the process and for space heating (thus making the plant self-sufficient in power).

Nevertheless, there are several features common to most plants that generate steam from waste heat and these are covered in the next section.

Originally published August 13, 1979

A typical waste-heat-boiler system

Fig. 1 shows the basic elements in a waste-heat-boiler installation. The water entering the unit has to pass through a water treatment plant into an on-plot treated-water storage tank, from which it is pumped into a deaerator. This deaerator is frequently preceded by an undeaerated-boiler-feedwater heater. The deaerated water, after treatment by the injection of chemicals, passes through boiler-feedwater pumps. These generate enough head to force the water through the boiler-feedwater heaters/economizers and into the steam drum. Steam is generated in waste-heat boilers, which are connected to the steam drum by downcomer and riser pipes.

There is a continuous circulation through these pipes; water flows from the steam drum into the boiler, and a mixture of steam and water leaves the boiler and enters the steam drum. This flow is either by forced circulation using boiler circulating pumps or by natural circulation (convection) using the driving head arising from the difference in density between the water in the downcomer pipes and the steam/water mixture in the riser pipes. The steam is separated from the water in the steam drum, passes through a superheater, and is then available for internal consumption or export.

In Fig. 1, arrows are used to indicate where heat is supplied, since it may come from the process gases or from flue gases, depending on the particular process and flowsheet.

Designing-in operability and reliability

At the conceptual-design stage of a project, it is important that there be a strong input from those who will subsequently be involved in the design and specification of equipment. The ideal arrangement is to have an informal multidisciplinary team work together. They can develop a flowsheet or plant design that satisfies the needs of the particular process, and at the same time takes into account the experience both of the engineering specialists and of personnel who have operated similar plants.

In the past, all too often a flowsheet was developed in isolation, for maximum heat recovery but without regard to engineering cost or complexity; then, steam conditions were selected purely to satisfy the heat and power requirements of this flowsheet. Little or no regard was paid to startup or part-load conditions. Nor was it recognized that actual plants do not behave exactly in accordance with theoretical prediction and that a small overestimate of the amount of heat recovered, when combined with a small underestimate of the power consumed, results in a major shortfall of the power for the plant.

As a result, many plants were difficult or impossible to start up, failed to achieve the design output rates, and were less efficient than expected. Naturally, changes were made to deal with these problems, but these proved costly in time and money.

An early decision to be made is the selection of the steam operating pressure and temperature, which can have a very important bearing on the reliability of the plant. The most severe steam operating-conditions on a process plant that are known to the author are 150 atm and 540°C. The steam temperature is limited by metallurgical considerations, since the steam could be heated to much higher temperatures by the waste-heat streams. Even though proven industrial turbines are available that will handle temperatures of 500 to 540°C, the boiler specialist can argue that the temperature should be reduced below 500°C. This is a level at which superheater tubes can still have a temperature margin, without the need to resort to austenitic steel (which brings with it the risk of chloride stress-corrosion cracking).

The process engineer can accept this advice and still avoid problems due to wetness in the low-pressure turbine by incorporating a reheater for the intermediate-pressure steam that reenters the turbine.

The selection of steam pressure is also important. There is no obvious limit in pressure for watertube boilers, but for many processes the author considers that firetube boilers are preferable (discussed later in the section, "Boilers heated by process gas").

In 1969, when the flowsheet was being developed for a large methanol plant, the technology of firetube boilers of the required size limited the pressure to 70 atm. The design team decided that a reliable boiler was a paramount need and the flowsheet was built around a steam pressure level of 70 atm. Pressures are no longer so limited, and if the decision were taken today the pressure could certainly be 100 atm and possibly higher.

The engineering specialist can influence the positioning of items of equipment in the flowstream, so as to ensure that reliable equipment can be designed and purchased. In this connection, an important consideration is the way the items interact when the plant is operating at a condition away from the design condition. Ideally, a computer model should be made of the proposed arrangement, to predict the performance of each item under all possible conditions including startup, various trip situations, part-load operation (each with the various heat-transfer areas overperforming and underperforming by, say, 10%). This technique may highlight potential problems that are so serious that the flowsheet needs to be rearranged. As an example, in the case of the flowsheet for an ammonia plant

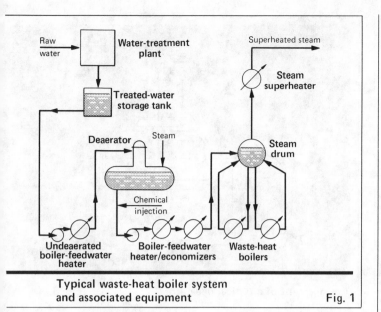

Typical waste-heat boiler system
and associated equipment   Fig. 1

eral application. It does not aim to detail special features or designs developed purely for one process.

The items are discussed in the order in which they occur in the process of producing steam, rather than in order of difficulty.

Boiler-feedwater treatment plant

The type of water treatment plant required depends on the quality of the raw water, the steam-generation pressure, and the heat flux at the steam-generation surface of the boilers. The subject is very complex, and beyond the scope of this article, but useful background is given by Strauss [5], and quantitative standards are laid down by the VGB* [6].

Nevertheless, the importance of water treatment cannot be stressed too highly, since a major proportion of boiler failures is linked with water problems. It is vital that the correct processes be selected and that the plant's management be provided at an early date with all information required for it to operate the plant correctly. Allowable working ranges should be agreed upon for key measurements such as pH, conductivity, and phosphate level, and provision should be made for the continuous monitoring of these properties.

The experience of many companies is that it is not uncommon for acid or alkali breakthrough to occur. A frequent cause, in demineralization plants, is the failure to carry out an adequate final rinse at the end of the regeneration cycle of any of the ion-exchange units.

There is some evidence to show that the risk of break-

*Vereinigung der Grosskraftwerksbetreiber eV, Essen, West Germany.

similar to that shown in Fig. 2, more than 14 operating conditions were studied.

Equipment reliability

Although flowsheets developed for different processes vary enormously, they frequently contain items of equipment common to most waste-heat boiler systems. The purpose of this section is to highlight areas of importance in those items of equipment that have gen-

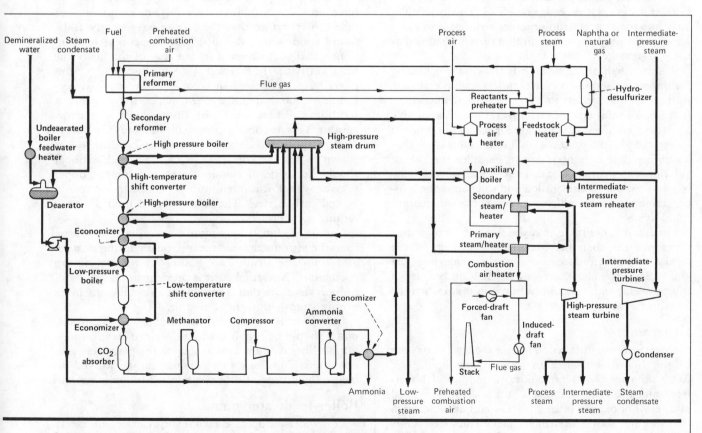

Steam-generation flow diagram for a typical ammonia plant   Fig. 2

through is greater with manually operated plants, and for this reason automated plants are preferred. In either case, an instrument should be installed in the treated-water line that continuously measures conductivity and actuates a trip should the reading exceed a preset level. This trip should dump the water to drain, or stop the transfer pump, thus ensuring that contaminated water never enters the treated-water tank. This tank should be large enough to allow time for locating and rectifying faults without shutting down the whole process.

Another common problem is the contamination of the return steam condensate, for example by cooling water, as a result of leakage in the vacuum condensers. Provision should be made for continuous monitoring of the conductivity of this and other return condensates, and automatic dump facilities should be provided.

Undeaerated-boiler-feedwater heaters

Heat exchangers are used on some plants to achieve maximum heat recovery by preheating the water fed to the deaerator, thus minimizing the use of low-pressure steam in the latter. Several features are of particular importance: The first is that care should be taken with the exchangers' sizing and with their control systems, in order to ensure that the feedwater does not reach too high a temperature. If the feedwater approaches the saturation temperature of the water in the deaerator, the normal pressure-control system of the deaerator will cut off or reduce its supply of steam and this will cause its performance to fall. If the deaerator is to operate at, say, 110°C, then the central system should bypass some of the heating gas, to ensure that the water entering the deaerator does not exceed, say, 85°C. This is particularly important at startup or under reduced-load operation, and these conditions must be carefully studied at the design stage.

Another item of importance is the choice of tubing for the exchanger. Carbon steel suffers serious corrosion pitting when subjected to oxygen at the temperatures that occur in this type of exchanger. Copper is suitable from the standpoint of waterside corrosion, but it might be attacked by the process gas. Stainless steel is another possibility, but it is necessary to consider the risk of chloride stress-corrosion cracking. Clearly, each case requires separate consideration and sometimes the solution might be a bimetallic, or a stress-corrosion-resistant stainless steel.

Finally, it is desirable that the heating fluid be at a lower pressure than the feedwater, if leakage of the former could introduce harmful chemicals into the boiler. Several boilers have suffered failures because of this problem, and it should be borne in mind when the flowsheet is being developed.

Deaerators

Deaerators are installed to reduce the level of oxygen in the boiler feedwater to a very low level, e.g., 0.005 ppm for a 100-bar boiler. A number of boiler-feedwater heaters/economizers and boilers have failed as a result of oxygen pitting [4]. Careful and thorough checking has shown that many proprietary deaerators fail to achieve their specified duties. In some cases, this has been due to insufficient steam for oxygen stripping

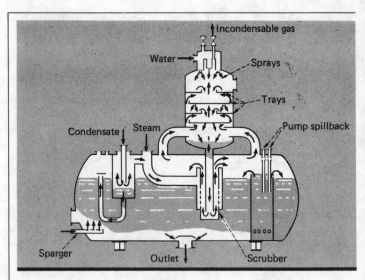

Arrangement of a typical deaerator   Fig. 3

because the water supplied to the deaerator has been too hot (as described above). In other cases, vacuum deaerators have been supplied, and air has been pulled into the deaerators or their extraction pumps. Experience has shown that it is extremely difficult to prevent this ingress, and for this reason pressure deaerators are to be preferred. On other occasions, even pressure deaerators have not worked until improved means have been adopted for steam-stripping oxygen from the water.

Leading suppliers of deaerators to power stations have provided similar designs for process plants but failed to recognize that the former have nearly 100% return condensate (with little raw makeup water containing oxygen), whereas in the case of the latter the situation is nearly reversed. When, subsequently, suppliers have recognized the problem, the required performance has sometimes been achieved by fitting a scrubbing section inside the storage part of the deaerator. This scrubber supplements the sprays, or sprays and trays, in the deaerator dome. A typical deaerator is shown in Fig. 3. The spray head, trays and scrubber should be made of stainless steel; but for the upper pressure-vessel, it is often more economic to use epoxy-coated carbon steel. The storage vessel itself may not require such protection.

Careful design can minimize the vibration arising from the introduction of "sparge" steam during startup, and from the steam that flashes off the hot return condensate. Nevertheless, it is prudent to design a stiff support structure that is capable of absorbing any loads arising from within the deaerator.

Clearly, care is required in selecting the deaerator, and it should be supplied with all-welded stainless-steel sample lines that are used as soon as the plant goes into operation. Thus, any shortcomings can be worked on before the pitting of other equipment becomes serious.

Boiler-feedwater pumps

Feedwater pumps, somewhat surprisingly, are rarely a cause of trouble. A number of manufacturers can supply reliable multistage pumps. The main stipulation

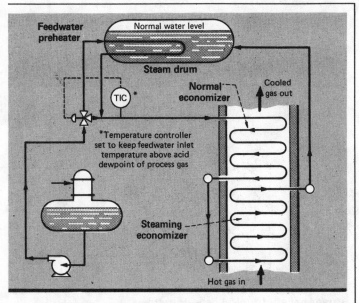

Arrangement of an economizer   Fig. 4

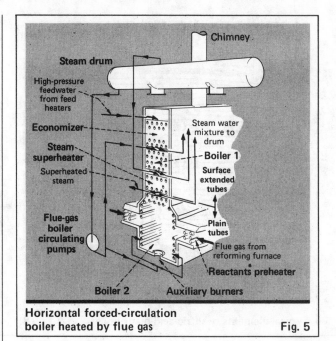

Horizontal forced-circulation
boiler heated by flue gas   Fig. 5

is that deaerator height and water inlet system are such as to avoid cavitation in the suction of the pump. Normally, at least two pumps are provided, and sometimes three.

Alternative sources of power are used; frequently one pump is motor-driven, with the other(s) driven by turbine(s) that are supplied with steam generated in the boilers they serve. The standby pump should start automatically if the flow, or pressure, in the feedwater line falls below a preset figure. Each pump must be fitted with a "leak-off" system, back to the deaerator, to ensure that the pump is guaranteed a minimum flowrate

(at low or no flow, a multistage high-head pump easily generates excessive pressure and heat, which would lead to rapid failure).

An important consideration for operational reliability is the stability of pumps running in parallel. Generally, this leads to a requirement that the pumps have a steep characteristic, e.g., a rise in head of 20%, from normal to minimum flow.

Boiler-feedwater heaters/economizers

These devices may take the form of a tube-in-shell exchanger in which the water is heated by process gas,

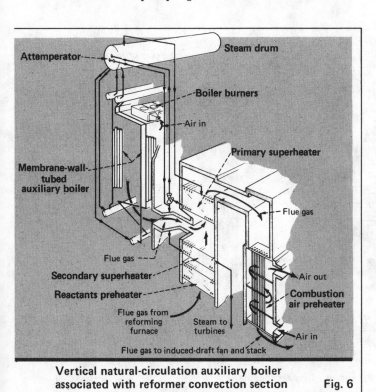

Vertical natural-circulation auxiliary boiler
associated with reformer convection section   Fig. 6

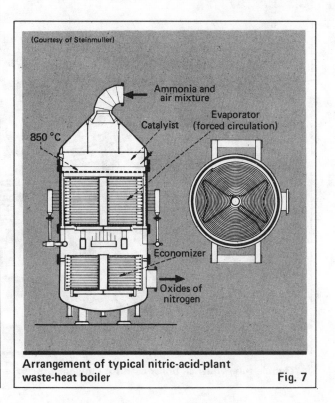

Arrangement of typical nitric-acid-plant
waste-heat boiler   Fig. 7

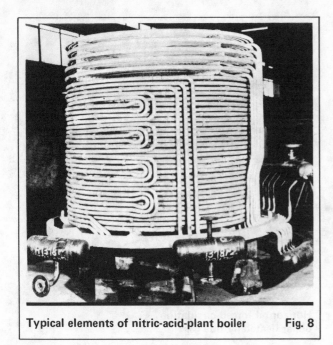

Typical elements of nitric-acid-plant boiler   **Fig. 8**

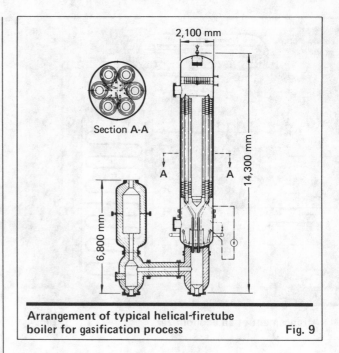

2,100 mm

Section A-A

14,300 mm

6,800 mm

Arrangement of typical helical-firetube boiler for gasification process   **Fig. 9**

or a tubular bundle over which flue gas or process gas passes. They are not a common cause of trouble unless corrosion takes place on the gas side (as a result of the tubewall temperature dropping below the acid dewpoint). This can be avoided at the design stage by careful selection of tube material or by introducing a small heat exchanger, which, by automatic control, ensures that the temperature of the water entering the boiler-feedwater heater is above the acid dewpoint.

For instance, on a sulfuric acid plant, it is possible to insert such an exchanger in the water space of the steam drum, and an automatic three-way valve can divert

sufficient water through this bundle so that the temperature of the water entering the boiler-feedwater heater is not less than, say, 130°C (Fig. 4).

The only general type of fault that can occur is for boiling to take place in an economizer not designed to generate steam. This can cause water hammer, vibration and general disturbance to the boiler system. At the process flowsheet stage, it is usually possible to avoid this risk by not aiming for too close an approach to saturation temperature in the economizer. If for any reason a close approach is considered desirable, a steaming economizer can be designed in which the

Gas outlet chamber
Riser header
Water/steam outlet
Oval headers
External tube
Internal tube
Downcomer distributor
Oval headers
Water inlet
Gas inlet chamber
Gas

"Double tube" quench boiler used for ethylene process, and diagram of its arrangement   **Fig. 10**

steam/water mixture is arranged to flow upward through the bundle and into the steam drum, thus avoiding any high pockets where the steam can separate and give rise to hydraulic disturbance (see Fig. 4).

Boilers

Boilers or evaporators can be broadly split into two categories: watertube, in which the water/steam mixture is inside the tubes, and firetube, in which the heating gas is inside the tubes. In either case, the boilers can be designed to operate on the forced or natural circulation modes described earlier. It is clear from the author's surveys [1,4] that they are potentially the most vulnerable items in waste-heat boiler systems.

Boilers heated by flue gas

When the heating fluid is flue gas, a watertube boiler is nearly always adopted. It is cheaper to keep the steam (which is at pressure) within small-diameter tubes and to contain the flue gas (which is normally slightly below atmospheric pressure) inside a large refractory-lined or water-cooled enclosure.

Two basic types of flue-gas boiler are common. The one usually adopted by process contractors, or by vendors of fired-heaters, consists of large horizontal tubes through which the boiler water is pumped (Fig. 5). The tubes are typically 3 or 4 in. nominal diameter, and usually have extended-surface fins to enhance the heat-transfer coefficient on the gas side (in cases where the boiler is subject to direct radiation, the first few rows of tubes are without fins).

The other type is that usually adopted when the

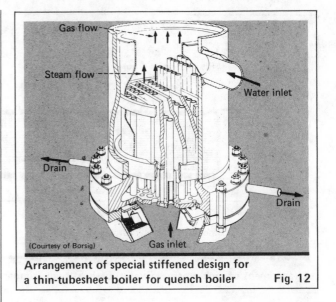

Arrangement of special stiffened design for a thin-tubesheet boiler for quench boiler   Fig. 12

equipment is supplied by a boiler vendor. Here, the evaporation generally takes place in relatively-small-diameter vertical tubes through which the steam/water mixture flows upward by natural circulation. The tubes are typically $1\frac{1}{2}$ in. nominal diameter and are often attached to each other by flat strips to provide a "membrane wall" construction (which avoids the need for refractory lining of the radiant chamber). The installation shown in Fig. 6 consists of an auxiliary boiler whose exhaust flue gases join those from the furnace before passing over a superheater and an air preheater.

Vertical natural-circulation boilers are intrinsically more reliable than forced-circulation horizontal boilers. Nevertheless, the vast majority of horizontal boilers are reliable but there is a potential risk either of dry-out, which can cause severe corrosion of the crown of the tubes if harmful chemicals are present in the boiler water, or of overheating, if the heat flux is high enough. The subject has been covered in detail by Hinchley [4].

Boilers heated by process gas

The design of boilers heated by process gas has attracted the interest of many clever engineers and there are many ingenious designs of boilers in operation on various processes [7-12]. The description and evaluation of these boilers is beyond the scope of this article. Nevertheless, it is appropriate to devote some space to these equipment items because they are potentially the most vulnerable in the whole system.

In some cases, the design of a boiler has evolved with the process and is virtually standard—for example the "flat pancake" watertube boilers used in nitric acid plants (Fig. 7 and 8). Such boilers have raised as much as 50 t/h of steam at up to 60 bar and 450°C. Boiler-makers new to this field have sometimes developed quite different designs but have experienced troubles because of a lack of knowledge of the process.

For example, the inclusion of insulation in the design has provided a site for the collection of humid oxides of nitrogen, which have turned into acid at plant shutdowns. This has caused corrosion of the inner liner, which has led to the ultimate failure of the shell and to

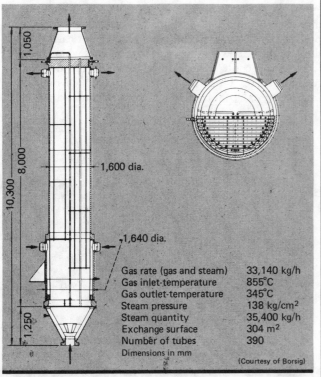

Gas rate (gas and steam)	33,140 kg/h
Gas inlet-temperature	855°C
Gas outlet-temperature	345°C
Steam pressure	138 kg/cm²
Steam quantity	35,400 kg/h
Exchange surface	304 m²
Number of tubes	390

Dimensions in mm

(Courtesy of Borsig)

Arrangement of special thin-tubesheet quench boiler for ethylene process   Fig. 11

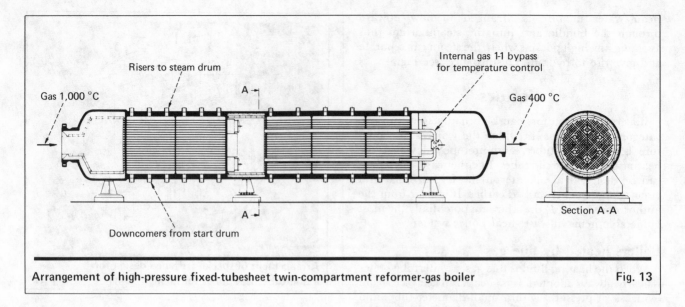

Arrangement of high-pressure fixed-tubesheet twin-compartment reformer-gas boiler   Fig. 13

corrosion of some evaporator tubes. Another example is the nitrate stress-corrosion cracking that has occurred when the shell has not been fully stress-relieved.

The need for a proper understanding of the impact of the process on the boiler design was underlined further in the case of some sulfuric acid plants. The steam pressure and the heating gas conditions were similar to those in a package firetube boiler, and the waste-heat boiler was designed and constructed to similar standards. However, minor leaks in tube-end welds—which probably would not have been noticed during the operation of a package boiler—resulted in a rapid shutdown of the whole plant.

The contractor's specialists in that process had failed to point out that a minor leak of water into the gas stream could not be tolerated because acid would be formed, which would rapidly corrode the equipment. Even worse, it had not been recognized that, because of the large heat capacity of the system and the need to purge it thoroughly, the minimum shutdown time to carry out the simplest repair was 10 days. Had these facts been understood at the design stage, the boiler would have been designed, fabricated and inspected to the stringent standards laid down for very-high-pressure waste-heat boilers.

In the oil gasification process, a design of boiler that has become well-established is the submerged helical-coil smoke-tube boiler, which is capable of cooling high-pressure process gases from temperatures as high as 1,500°C (Fig. 9).

In ethylene plants, there is a need to quench as quickly as possible the low-pressure high-temperature process gases leaving the cracking furnace. This is done in quench boilers, which operate at very high steam pressures (up to 100 atm) in order to maintain high tube-temperatures and thereby reduce the amount of deposition from the process gas. Two of the most common proprietary designs are the double-tube type described in Ref. 12 and shown in Fig. 10, and the special thin "ribbed tubesheet" design described in Ref. 10 and shown in Fig. 11 and 12.

The position is less clear with regard to the re-

formed-gas boilers on plants that produce town gas, methanol, ammonia and hydrogen. The importance of these boilers can be judged from the fact that they cool gas at pressures of around 30 bar from temperatures of up to 1,000°C and that the largest of these generate 233 t/h of steam at 140 bar in a single vessel. Here there is a clear divergence of opinion on the most preferable type. Some of the plants currently being built incorporate firetube boilers, whereas others use one of the many types of watertube boilers that are available.

The typical arrangement of a large, natural-circulation, firetube, reformed-gas boiler, generating steam at pressures in excess of 100 bar, is shown in Fig. 13. In order to reduce the stresses that arise due to the differential expansion between tube and shell, it is becoming increasingly common to build such boilers in two compartments. This has the added advantage of easing the duty on the valves that are required in cases where there is a need to control the temperature entering the next reaction stage. Thus, the bypass valve, which may be internal or external to the boiler, has to contend with a

Fabricated twin-compartment
reformer-gas boiler   Fig. 14

gas temperature of only 600 to 650°C instead of 1,000°C. Fig. 14 shows one of the largest boilers of this two-compartment design so far built.

The keys to success with this type of boiler are the adoption of a maximum heat-flux that is reasonable—having regard to the geometry and hydraulics of the boiler, the thorough understanding of the stresses in all parts of the boiler, the thermal protection of the inlet tubesheet, and the methods of attaching both the tube to the tubesheet, and the tubesheet to the shell and the channel.

Fig. 15 shows one successful design for dealing with the last three points, but various manufacturers have their own preferred ways of handling these problems. The author has a strong preference for the firetube-boiler design since it is far less sensitive to dirt or debris left in the system or released from the walls of the system during plant upsets. The presence of a 100 to 150-mm-thick layer of refractory inside the inlet channel results in a large "untubed" annulus on the shell side. Thus, a large amount of dirt or debris could accumulate in the base of the boiler without building up on the heat-transfer surface and without interfering with boiler circulation. This advantage disappears completely if the boiler is mounted vertically, and the author is strongly opposed to the use of vertical firetube boilers even if, as in some cases, they are attractive from the point of view of plant layout.

Of the watertube reformed-gas boilers that are built, the most common is the natural-circulation "bayonet-tube" design. This should not be taken as implying that it is the best boiler—it happens to be the design favored by the world's largest supplier of ammonia plants.

The principal features of this type of boiler are shown in Fig. 16 (left). Water from the steam drum enters the upper chamber and passes through the false tubesheet containing the bayonet tubes. At the bottom, it reverses and flows up through the annulus between the bayonet and scabbard tubes. The hot gas enters at the bottom of

the vessel and flows across baffles, leaving at the top.

The proponents of this design point out that the central downcomer tube is not heated and that the steam bubbles generated are required to flow only upward. This is basically a sound argument, although the details on some plants have been such that reverse circulation has occurred under certain circumstances and this has sometimes led to failure.

There are others who argue that the vertical U-tube boiler, Fig. 16 (center) used on many plants is equally satisfactory, particularly when circulation is ensured by the use of forced-circulation pumps. They also argue that the whole boiler duty can be achieved in a single exchanger, whereas in the other design the bayonet-tube boiler is followed by a firetube boiler. Several boilermakers have made two or three of their own designs of watertube reformed-gas boilers, incorporating features that have special merit but that often add to the cost and complexity of the units—Fig. 16 (right), 17 and 18.

Perhaps the one of most technical interest is the mono-tube boiler installed on a 1,000-t/d ammonia plant in France, shown in Fig. 19 and 20. This generates 137 t/h of steam at 152 bar in a single vessel. The installation is the most striking that the author has seen. The main reasons are that there is no large steam drum and associated structure, only a vertical separator that is little more than a pipe attached to the boiler; there are no risers or downcomers; there are no circulating pumps; the whole duty is handled inside one shell; and temperature control is achieved by an internal arrangement, thus avoiding hot-gas piping. Further details are given by Silberring [11].

The fact that this boiler has operated without trouble for five years does not imply that it is the most reliable watertube boiler—it is more a tribute to the quality of control of the plant and particularly of the water. Nevertheless, most of the watertube reformed-gas boilers known to the writer have low points (where any debris can collect) in the zones of highest heat flux, thus giving rise to the risk of corrosion or overheating of the type described in Ref. 4. Furthermore, they are all likely to fail if poor precommissioning or a plant upset causes a significant flow blockage at the inlet of one or more tubes. Notwithstanding these potential risks, some users have had reliable service from watertube boilers.

There is a significant point in favor of most designs of watertube boilers. Inadequate water-treatment can lead to the rapid failure of the tubes of either a watertube or a firetube boiler. In a watertube type, if a large number of tubes fail, a spare watertube bundle can be fitted in quite a short time, whereas a firetube boiler would require a very long time to retube or replace. Such major failures are rare, and if only a few tubes are affected they can be readily plugged off with either design.

Steam drums

The main purpose of a steam drum is to separate the steam from the steam/water mixtures that return from a boiler, but it also provides a holdup capacity of water to feed the boilers. In forced-circulation watertube boilers, the ratio of the quantity of water circulating to that

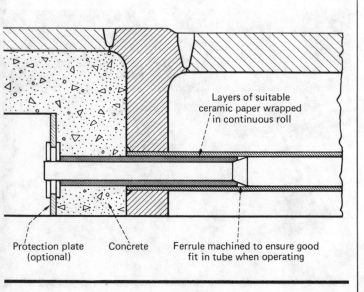

Layers of suitable ceramic paper wrapped in continuous roll

Protection plate (optional)   Concrete   Ferrule machined to ensure good fit in tube when operating

Some key features of firetube boiler   Fig. 15

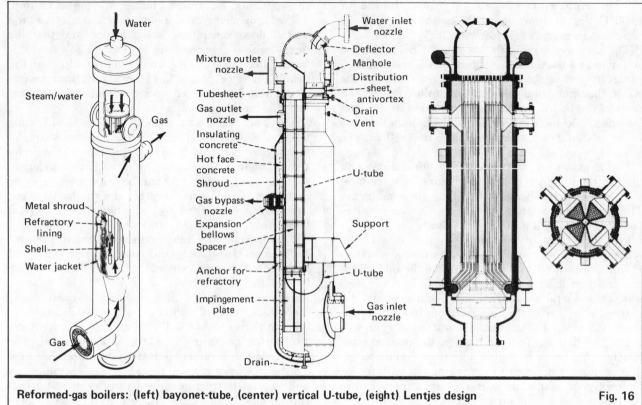

Reformed-gas boilers: (left) bayonet-tube, (center) vertical U-tube, (eight) Lentjes design   Fig. 16

which is converted into steam may be as low as 4 to 1, whereas in a natural-circulation firetube boiler it is normally between 10 to 1 and 20 to 1.

Thus, the primary stage of separation is the coarse one of removing the bulk of the water from the steam. The second stage is the removal of water droplets from the steam. This stage is of vital importance if the steam is subsequently to be superheated and used in turbines. The dissolved solids in the feedwater concentrate up to a hundredfold in the boiler water, depending on the amount of blowdown or purge from the system. As a result, any boiler water carried over in the steam contains a high level of dissolved solids.

The solids will be deposited in the superheater. Some solids will build up on the heat-transfer surfaces, resulting in overheating and even possibly creep rupture failure. Some may deposit in the turbine and cause a lowering of efficiency. The most serious event that can occur is "priming" of the steam drum when there is bulk carryover of water due to a plant upset or loss of control. Not only does this have a severe effect on the superheaters but it can lead to failure of the steam turbines.

A common design figure in many boiler specifications is that the steam drum shall have a separation efficiency of 99%. This has little meaning; it cannot be measured, and—even if it were achieved—it would be completely inadequate for most high-pressure boilers fitted with superheaters. It is better to state that the steam leaving the drum shall have a solids content of less than, say, 0.01 ppm of sodium.

An isokinetic sample point should be provided in the steam line from the drum. This should be connected to a continuous analyzer fitted with high-level alarms. Such a purity level usually can only be achieved in the sophisticated steam drums developed by leading boilermakers. Fig. 21 shows such a drum, in which the primary separation is carried out in a large number of small cyclones, and the final separation in packs of chevron pads. Knitted wiremesh pads are sometimes used and these are equally acceptable, but the author is very suspicious of those who claim that the guaranteed

Fabricated tube-bundle of Lentjes boiler   Fig. 17

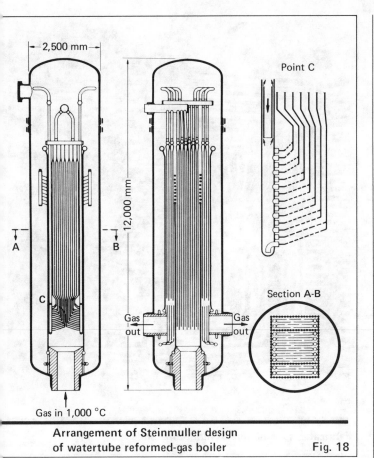

Arrangement of Steinmuller design of watertube reformed-gas boiler   Fig. 18

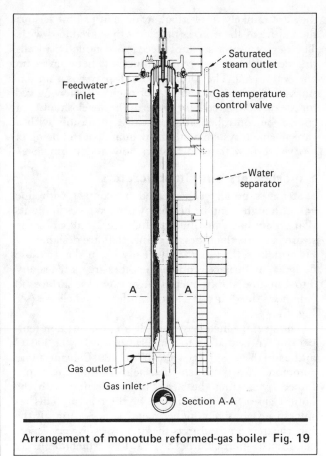

Arrangement of monotube reformed-gas boiler Fig. 19

purity can be achieved with simple baffles inside an ordinary pressure vessel.

Additional advantages accrue from the use of a specialist boilermaker's steam drum. He usually has developed special designs for the uniform distribution of chemicals for water treatment, for the uniform removal of blowdown water, and for the design of the downcomers to ensure that there is no underflow of the steam that could either upset the circulation of the system or cause cavitation in boiler circulation pumps. Nevertheless, experience shows that close liaison is necessary with the boilermaker to ensure that the standard design is modified where necessary to take account of any differences that arise in process waste-heat-boiler systems.

In one particular case, the boilermaker had selected the number of cyclones appropriate to the evaporation rate of his watertube boilers and had completely overlooked the fact that with the natural-circulation firetube boilers, the quantity of water returning with the steam was several times greater. Not only were the cyclones seriously overloaded but the chambers enclosing the riser outlets were damaged, nuts came loose, and carryover occurred on a massive scale.

Another problem that occurred and caused failures was reverse circulation in one or more boilers sharing a common steam drum. This had arisen because one of the boilers in the process was brought on load earlier than the others. The normal practice, of allowing all risers to enter a common equalized chamber, had been followed, and, as a consequence of this, the hydraulic head generated in the operating boiler had been suffi-

cient to initiate reverse circulation in the cold boiler.

This incident should not be taken as a reason for discontinuing the highly desirable practice of one common drum. The boilermaker, when made aware of the circumstances, can arrange the baffles to avoid the problem, albeit at the expense of some extra cyclones.

An important aspect of the steam drum concerns the maintenance of drum level. If the size of the steam drum is arrived at on the basis of providing adequate space for the steam purity equipment, then the holdup capacity between the normal drum level and that of the drum empty (i.e., exposed downcomer entry) may be only one or two minutes. This is probably inadequate when one bears in mind the many factors that demand the attention of the plant operators during the upset of a complex process plant.

In nitric acid plants, this is not a serious problem since the unit can be automatically tripped out before the level becomes dangerously low. However, in reforming plants, there is large heat capacity in the system, and flow through the reformers must continue in order to protect the catalyst. In these cases, more holdup capacity is desirable, and process contractors usually provide at least three minutes capacity—users that have suffered boiler failures caused by loss of level sometimes specify five minutes. The cost of this extra capacity is not insignificant for large plants.

An instrument for indicating water level, which has recently become available, is worthy of special mention, since it provides the operators in the control room with such a vivid indication of the steam drum level that they are unlikely to overlook it, no matter how serious the nature of the process problems they are faced with. The device uses a stack of electrical conductivity cells that detect the difference in conductivity between steam and water [13]. The control room display has a similar stack of 20-mm-square lights for each cell, glowing red for steam and green for water. The level should, of course, be controlled automatically, preferably with a three-element system, with the primary control being of feedwater flow, reset by steam flow and drum level.

Superheaters and attemperators

Steam is usually superheated by contact with flue gas, although it can be heated by process gases in plants that do not have any flue gas. In virtually all cases, the steam is inside the tubes. In plants that are designed by boilermakers, the tubes are typically $1\frac{1}{2}$ in. dia. In those designed by furnace vendors, the tubes are more usually 3 to 4 in. dia. and frequently have extended surfaces to enhance the outside heat-transfer coefficient (see Fig. 5).

The steam, which leaves the drum at saturation temperature, is heated up to temperatures as low as 300°C and as high as 540°C, with 460 to 480°C being more common. Clearly, the smallest heat-transfer area is obtained by heating the steam in counterflow with the hottest gases from the furnace or the reactor. This results in progressively higher metal temperatures in the superheaters, and subjects the final rows to very severe conditions—particularly the final two, which are likely to receive heat by radiation as well as by convection.

The tubes in such a superheater are vulnerable to creep-rupture failure due to overheating, if the gases are hotter than expected or if the steam flow is less than expected or if there are transient operational modes in which there is a mismatch between the flue-gas rate and the rate of steam production. In some designs, an attempt is made to minimize these effects by the insertion of one or two rows of evaporator tubes to shield the superheater.

The approach preferred by the author, for plants in which the risks described above are real, is shown in Fig. 22. The superheating duty is divided between a primary superheater and secondary superheater. The former, which heats the saturated steam from the drum, is in normal counterflow. The secondary superheater is in cocurrent flow with the hotter flue gas.

In cases where the computer simulations referred to earlier show that the steam temperature would still vary excessively under certain operating conditions, an attemperator can be installed to control the steam outlet temperature by cooling part of the steam from the primary superheater in the manner shown in Fig. 22. This is most easily accomplished in a drum attemperator, which consists of heat-exchanger bundles situated in the water space of the steam drum. A controlled amount of steam is cooled by evaporating water in the drum.

Notwithstanding the above precautions, it is the view of the author that for design purposes an additional margin of say 50°C should be provided on top of the predicted long-term metal temperature in order to cover unforeseen happenings, such as solids carryover

Fabricated tube-bundle of monotube boiler   **Fig. 20**

from the drum, stratification of hot gas, or some of the other happenings described in the foregoing section on steam drums. This margin is not sacrosanct and should be reviewed if, for example, it leads to a very large increase in cost by requiring a change from ferritic to austenitic steel.

This general approach to the design of superheaters should also be applied to any process heaters that are included in the flue-gas duct.

Combustion-air heaters

Brief mention must be made of combustion-air heaters, since they are a common feature of flue-gas streams of waste-heat-boiler systems. If the fuel is natural gas, then no problems should be expected, and the user can install an air preheater of the type shown in Fig. 6. For large plants, the capital cost of the air preheater can be reduced significantly by using the rotary type, but this saving has to be balanced against the slightly increased risk of unreliability brought about by the introduction of moving parts and the additional running costs arising from air leakage across the seals. Another alternative that warrants an economic evaluation is a design that uses internally and externally finned cast-iron tubes. The various alternatives are described, illustrated and discussed in Ref 15. A relatively new entry not covered in that review is a design based on the use of heat pipes.

If the fuel contains sulfur, more care is required. Glass tubing has been successfully used in the coldest passes of tubular air heaters. In the case of rotary air heaters, a

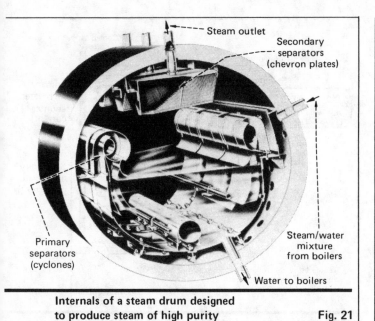

Steam outlet

Secondary
separators
(chevron plates)

Primary
separators
(cyclones)

Steam/water
mixture
from boilers

Water to boilers

**Internals of a steam drum designed
to produce steam of high purity**   Fig. 21

small steam-heated suction is often used to preheat the inlet air by the amount necessary to keep the main heater above the acid dewpoint.

Forced- and induced-draft fans

Induced-draft fans are installed on virtually all modern furnaces. In many cases, in order to achieve maximum heat recovery, the combustion air is preheated by the flue gas, and a forced-draft fan is required to overcome the pressure losses in the combustion-air preheater and in the burner windbox.

Many of the earlier plants incorporating such equipment had their outputs limited by the capacity of the forced- or induced-draft fans. The burners may have required more air to achieve their duty. Furthermore, that duty may have been underestimated. Consequently, the forced-draft fan could not supply that amount of air without being increased in size. The problem was often worse in the case of the induced-draft fan because the additional quantity of flue gas was increased still further by air leakage into the furnace and its convection section.

Clearly, it is desirable to introduce some margins, and the author advocates a figure of 10 to 20% on the forced-draft fan rate and 15 to 25% on the induced-draft fan rate. It is important when specifying the fan duty to ensure that this increased flow is achieved with pressure losses appropriate to the increased flow. This, of course, leads to a much larger driver, and it may be economic to supply a variable-speed drive, or inlet guide vanes, on the fans to minimize the power consumption should the plant not need the margins built into them.

Obviously, introduction of the fans could lead to a decrease in plant reliability. In the opinion of the author, reliability is best obtained by installing one high-quality fan rather than by duplicating ordinary fans.

Mechanical design and fabrication

It is inappropriate here to discuss the detailed mechanical design and fabrication of each of the many items of equipment in a complex waste-heat-boiler system but their importance should not be underestimated. The author [1,4] has highlighted costly failures that have arisen as a result of poor detailed design and bad workmanship. Frequently, companies use their best engineers to perform the conceptual and basic design and then pass on the execution of the contract without managing to convey the importance and requirement of the plant for special quality.

The client should take steps to ensure that this does not happen. Prior to the order, he or his main contractor should lay down procedures for submission for approval of detailed designs, welding procedures, quality-control schedules, manufacturing procedures, protection and shipping instructions. Also, he or the contractor should arrange to supplement the internal inspection by external inspection, as appropriate. These comments apply equally well to onsite fabrication, where there is probably even greater need for continuous scrutiny of the quality of workmanship including, in particular, the welding.

Precommissioning of
waste-heat-boiler systems

Objective

The objective of the precommissioning is to ensure that when the boiler system goes into service, it is free from debris, scale and corrosive products, and that all the metal surfaces are covered with a uniform tightly adhered layer of magnetite. It is desirable that the feedwater system be cleaned in a similar manner, to prevent the carrying of scale and debris into the steam-generation circuits.

Provision for cleaning at the design stage

The general approach to the cleaning processes should be set at a very early stage, so that the design of the equipment and piping that are to be cleaned incorporates the required piping connections, as well as access for inspection and physical removal of debris. All too often, this is not the case. Then, late and costly alterations are made to the piping, and undesirable compromises are accepted.

Piping connections should be provided and sized to achieve an optimum cleaning velocity over the heat-transfer surfaces (e.g., about 0.3 m/s for 5% hydrochloric acid solution). Where there are several boilers in parallel, provision may have to be made for cleaning each boiler separately.

The total system to be cleaned should be critically examined to establish where dirt and debris will collect during cleaning, and provision should be made for its physical removal. In the case of firetube boilers, the blowdown pipes in the bottom annulus can be attached to a removable pad. For some watertube boilers, a "mud pot" can be provided on the bottom of the distribution manifolds.

Cleanliness during construction

Equipment should be supplied in a clean dry condition, and in some cases shipping under a blanket of nitrogen is justified. Those involved in construction should be aware of the need to take special care over cleanliness. They should understand that if they were to leave behind only one foreign body it could, during operation, restrict the flow of water and cause a boiler failure (particularly if the boiler is of watertube design). Prefabricated sections of piping for the system should be cleaned mechanically to remove mill scale. Wire-brushing, even by power tools, is seldom a satisfactory method of descaling, but high-pressure water jets can remove a significant proportion of the scale. Alternatively, grit blasting can be used.

Cleanliness is an important aspect of inspection of equipment (1) prior to dispatch from the manufacturer's works, (2) after the equipment is received on site, and, (3) during inspection of the whole system prior to hand-over to the commissioning staff.

Planning the chemical cleaning procedure

A detailed plan of the cleaning procedure must be drawn up and agreed upon between the cleaning contractor and client, and there must be supervisors from both parties who know every detail of the programs. A careful check should be made to ensure that the pumps that the contractor proposes to use can achieve the rate corresponding to the optimum cleaning velocity, taking into account the resistance in the circuits. A suitable spare pump must always be available.

Cleaning operations

A typical procedure for the chemical cleaning of an important system operating under arduous conditions is outlined below.

Stage 1. Flush out to remove loose material. Use a water velocity equal to or greater than that which will occur in service.

Stage 2. Alkaline-degrease for 24 h at 95 to 100°C, using 1,000 to 1,500 ppm by weight of trisodium phosphate and a suitable detergent.

Stage 3. Acid-clean using 5% inhibited hydrochloric acid at 75 to 85°C. Continue until the total iron content in the circulating system has reached a definite plateau, typically 0.3 to 0.4 g/L.

Stage 4. Citric-acid flush using a 0.2% solution.

Stage 5. Passivate using ammoniated hydrazine solution at 95 to 100°C for 24 h.

Stage 6. Drain system and dry thoroughly.

In the case of the plant illustrated in Fig. 6, a special step was introduced between Stages 2 and 3. For nearly two days, water was circulated while tests were carried out to check, by contact flowmeter [14], that each of the membrane tubes in the auxiliary boiler had a similar flowrate and, particularly, that none of the tubes was blocked.

Storage

If the boiler system is not held under the correct conditions after cleaning, much of the passivation can be lost within a few days. The system can be stored wet (if filled with ammoniated hydrazine solution) or dry (using dehumidified air or nitrogen). The use of dehumidified air allows access to the system for further work or inspection, without costly, time-consuming purging and drying.

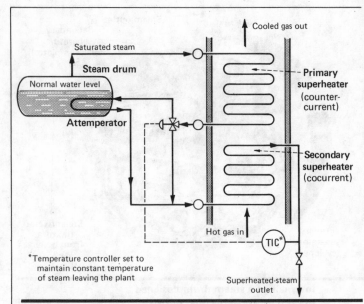

Superheater system that maintains constant outlet steam temperature; holds down metal temperatures   Fig. 22

Conclusions

The general approach and techniques described above have been developed as a result of dealing with the problems of earlier plants.

Striking success with large new plants demonstrates that high-temperature, high-pressure waste-heat-boiler systems can be incorporated into process plants without introducing significant risk of unreliability. These plants were designed taking into account the principal factors listed below as recommendations.

For the system as a whole:

Development of a process flowsheet should be done in close consultation with experienced engineering specialists who can influence the choice of steam conditions and the positioning of items of equipment in the flow-streams, in order to ensure that they are able to design and purchase reliable equipment.

A computer model of the system should be developed that can predict the duty of each item of equipment under startup, part-load and trip conditions, as well as under normal conditions.

The supplier of the equipment should be made aware of special aspects of particular problems, the chief of these being potential corrosion problems and the choice of materials of construction.

The supplier should be told of the need for high standards in detailed design, manufacture and quality control in the works and on site, and arrangements should be made for the monitoring of these matters to the extent judged appropriate for each particular case.

Special attention should be given to the precommissioning procedures, including mechanical cleaning and various chemical-treatment stages. The basic approach should be agreed upon at a very early stage, so that the design of the equipment and piping that are to be cleaned incorporates the required piping connections and access for inspection and the physical removal of debris.

The quality of the feedwater and return condensate

should be continuously monitored, and automatic trips should be installed in certain locations, in order to prevent contaminated water entering the treated-water storage tank or deaerator.

For individual items of equipment:

Undeaerated-boiler-feedwater heaters—Special care should be given to the choice of materials of construction, particularly with regard to the risk of corrosion on both sides of the tubes.

If possible, the heating fluid should be at a lower pressure than the feedwater, if leakage of the former could introduce harmful chemicals into the boiler.

Deaerators—The design should be very critically examined in view of the significant quantity of oxygen entering as a result of the high proportion of fresh makeup water in process plant systems.

The design of the system should ensure that, under all conditions, the amount of steam in the deaerator is sufficient for scrubbing out the oxygen.

The design of the deaerator should be such as to minimize the vibration arising from the introduction of heating steam and hot condensate.

Boiler-feedwater heaters/economizers—The risk of steaming should be assessed, and if this is considered likely, an appropriate design should be adopted.

The danger of "acid dewpoint" corrosion should be assessed, and if this is considered likely, a special design should be used.

Waste-heat boilers heated by flue gas—Where possible, vertical natural-circulation boilers should be used, and if horizontal boilers are essential, the design should be checked to ensure that "dryout" does not occur at the crown of the tubes.

Waste-heat boilers heated by process gas—Whenever possible, proprietary boilers should be selected that have evolved with the process and have become established as reliable. When this is not the case, the choice of type of boiler and of fabricator warrants most careful consideration, bearing in mind that the cost of failures has often exceeded the cost of the boiler. Horizontal natural-circulation firetube boilers have much to commend them, provided that the vendor is very knowledgeable in their design and fabrication. Nevertheless, a number of good designs of watertube boilers exist and have been used with success on plants where special care has been taken to remove debris from the system.

Steam drums—The sophisticated steam drums evolved by leading boilermakers should be used on systems with high-temperature superheaters, and severe standards should be enforced on the maximum level of contaminants permitted in the steam.

When several boilers share the same drum, its internal baffles should be so arranged as to ensure that reverse circulation is not initiated in any of the boilers.

Water storage capacity should be provided appropriate to the heat capacity of the equipment.

Superheaters—For severe-duty superheaters, the design should incorporate selected features that minimize the actual metal temperatures and maintain close control of steam outlet temperatures over all operating conditions.

The mechanical design should include an additional temperature margin to overcome unpredictable increases in metal temperatures.

Forced- and induced-draft fans—The duty specified should include a generous margin to ensure that the fans do not limit the plant output. Robust reliable machines should be selected.

Acknowledgements

The author wishes to acknowledge and thank the many people who, over the years, have contributed a great deal of information, help and advice on the matters discussed in this paper. They include numerous colleagues and designers, fabricators and operators from many parts of the world. Thanks are also due to vendors who provided several of the illustrations used in this article.

Illustration credits

The following companies supplied illustrative material for this article: Borsig, Fig. 11, 12; CCM-Sulzer, Fig. 19, 20; Foster Wheeler, Fig. 21; Oschutz, Fig. 8; Pullman Kellogg, Fig. 16 (left); S.H.G., Fig. 10; Steinmuller, Fig. 7, 9, 13, 14, 18; Struthers Wells, Fig. 16 (center).

References

1. Hinchley, P., Waste Heat Boilers in the Chemical Industry, paper presented at Institution of Mechanical Engineers Conference on "Energy Recovery in Process Plants," Jan. 29–31, 1975 and subsequently published in *Chem. Eng.*, Sept. 1, 1975, pp. 94–98.

2. Appl, P., and Frink, K., Troubles with Thin Tube Sheet Waste Heat Boilers, AIChE 1975 Ammonia Plant Safety, Vol. 18, pp. 113–121.

3. Sawyer, J. G., others, Causes of Shut Downs in Ammonia Plants, AIChE Safety in Air and Ammonia Plants, Vol. 14, pp. 61–68 and Vol. 16, pp. 4–9.

4. Hinchley, P., Waste Heat Boilers: Problems and Solutions, *Chem. Eng. Prog.*, Mar. 1977, pp. 90–96.

5. Strauss, S. D., Water Treatment, *Power*, June 1973.

6. Neue Richtlinien fur das Kesselspeisewasser und das Kesselwasser von Dampferzeugern, VGB Kraftwerkstechnik, No. 2–52, Apr. 1972, pp. 167–172.

7. Becker, J., Examples for the Design of Heat Exchangers in Chemical Plant, *Verfahrenstechnik*, Vol. 3, No. 8, pp. 335–350 (1969).

8. Kummel, V. J., Abhitze and Sonderkessel in der chemischen und petrochemischen Industrie, *Chem-Ing-Tech.*, Vol. 49 No. 6,5, pp. 475–479 (1977).

9. Capitaine, D., and Stoffels, Jentzsch, Der Einsatz von Abhitzekesseln und einige Konstructionsmerksmale, *Mitteilungen der VGB*, Vol. 49, No. 3, July 1969, pp. 165–173.

10. Deuse, K. H., Waste Heat Boilers in Large Plants, *Het Ingenieursblad*, 40E Jaargang (1971), No. 21.

11. Silberring, L., Waste Heat Recovery in Ammonia Plants, CPE—Heat Transfer Survey 1969, and Energetische Probleme der Ammoniak—Erzeugung, *Chem-Ing-Tech*, Vol. 43, January 1971, No. 12, pp. 711–720.

12. Knulle, H. R., Problems with Exchangers in Ethylene Plants, *Chem. Eng. Prog.*, Vol. 68, No. 7, pp. 53–56 (1972).

13. Hasler, E. F., and Martin, R. E., Hydrastep: a 'Fail Operative' Gauge System to replace Visual Boiler Water Level Gauges, *Measurement and Control*, Dec. 1971, pp. 366–371.

14. Roughton, J. E., Detection of Boiler Tube Blockages. Experience with CERL Contract Flowmeter, *CEBG Digest*, May, 1973, pp. 7–10.

15. Berman, H. L., Fired Heaters—IV, How to reduce your fuel bill, *Chem. Eng.*, Sept. 11, 1978, pp. 165–169.

The author

Peter Hinchley works in Project Management for the Agricultural Div., Imperial Chemical Industries Ltd., P.O. Box 6, Billingham, Cleveland TS23 1LD, England. For almost ten years he was manager of the Furnace & Boiler section, where he had extensive experience in design, specification and purchasing of waste-heat boilers. He holds a first-class Honors degree in mechanical engineering from Sheffield University and is a Member of the Institution of Mechanical Engineers.

Retrofitting coal-fired boilers: economics for process plants

After investigating the economics of installing coal-fired boilers in existing plants, this article offers guidance on conditions for profitable investments. It also examines the consequences of tighter pollution regulations and the maturing of fluidized-bed technology.

C. Thomas Breuer, Arthur D. Little, Inc.

☐ New coal-fired boilers appear economically attractive as an alternative to oil- and gas-fired boilers in many U.S. chemical process plants. Uncertainties in federal natural-gas policies and in future supplies and prices of fuel oil can also make coal the preferred fuel for meeting large, steady demands for process steam.

About 60% of the fuel consumed in the chemical process industries (CPI) is estimated to be burned in boilers. Coal and miscellaneous fuels (principally, process wastes) amount to about 27%. Oil and natural gas account for the remainder. Thus, there is significant potential for displacing oil and natural gas with coal.

Basis of the analysis

It is assumed in this analysis that existing gas-fired-boiler capacity is adequate to fill all of a plant's steam demands. It is further assumed that the gas-fired boiler will be kept in working condition, capable of filling in for the coal-fired boiler and meeting demand for steam that exceeds the latter's capacity.

The focus is on CPI plants in which steam demand is high enough for the installation of a coal-fired boiler having a nameplate firing rate of between 75 and 250 million Btu/h. This size range is chosen for two reasons:

1. The steam supply economics are transitional. There are many plants in which steam demand may be most economically supplied by one of three types of coal-fired boilers that will be evaluated in this article. But there are others for which oil or gas will continue to be the most economical fuel.

The economics of coal are neither compelling (as is often the case with boilers fired with more than 250 million Btu/h of fuel) nor prohibitive (as is generally the case with boilers fired with less than 75 million Btu/h). Within this transition range, enlightening comparisons can be made among the different boiler technologies, and an examination of the comparisons is possible with regard to regional differences in prospective energy prices.

2. State regulations on emissions of air pollutants apply to boilers in this size range. Although these regula-

tions vary, they are generally less strict than the federal New Source Performance Standards applicable to larger boilers. The implications of possible federal air pollution regulations on boilers in this size range will be considered, as well as the full commercialization of the fluidized-bed boiler, with its potential for superior pollution control.

Financial and tax parameters

All costs in this analysis are in 1982 constant dollars. Therefore, only real changes in fuel and other operating and maintenance costs over the life of the system need be of concern. Unless otherwise specified, the values of the financial parameters are:

Useful life—20 yr; the consequences of variations in return on equity are described.

Debt/equity ratio—30/70; this is typical for large U.S. companies.

Real interest rate—3.5%; this is appropriate for a constant-dollar analysis.

Property tax rate—2%.

Corporate tax rate—50%; this includes federal and state rates.

Depreciation life—5 yr; by the accelerated cost recovery system.

Inflation rate—8%; this is required only for calculating tax depreciation, because the analysis otherwise is based on constant dollars.

Investment tax credit—10%.

Texas fuel prices are used for the base-case projection, and Alabama and Illinois prices for the alternative projections (Table I). These 1982 constant-dollar prices are based on U.S. Dept. of Energy projections, allowing for the discounted effects of real-price increases through the year 2005.

Options in coal-fired boilers

Summarized in Table II are the key characteristics of the principal commercially available coal-fired boilers. The following options are considered:

■ Stoker-fired boiler burning compliance coal.

Originally published September 17, 1984

Energy prices in 1982 constant dollars, f.o.b. process plant   Table I

Plant site	Coal, \$/million Btu				Natural gas, \$/million Btu	Electricity, ¢/kWh
	Meets state SO$_2$ regulations		Meets 1971 federal New Source Performance Standards			
	Crushed	Double-size	Crushed	Double size		
Texas	2.76	2.86	2.76	2.86	4.64	6.2
Alabama	2.59	2.69	3.04	3.14	6.47	4.43
Illinois	2.27	2.37	2.92	3.02	6.53	4.43

Source: U.S. Dept. of Energy and Arthur D. Little, Inc.

- Pulverized-coal-fired boiler burning compliance coal.
- Pulverized-coal-fired boiler burning high-sulfur coal, with the fluegas desulfurized.
- Atmospheric fluidized-bed boiler burning high-sulfur coal.

All these systems can limit particle emissions to 0.03 lb/million Btu.

Stoker-fired—This is the choice for many small-to-moderate-capacity coal-fired boilers. At the small end, its capital cost is lower than that for any of the other options. In many respects, it is the least complex to operate, in part because little, or no, coal processing need be done onsite. Because coal pulverizing is not involved, the stoker boiler is operationally reliable and consumes electrical power at a relatively low rate. An additional advantage is that such a boiler can burn coarse refuse.

(Because only a limited number of plants can take advantage of this opportunity, the burning of such fuels is not considered in this analysis.)

The stoker is not chosen more often than it is principally because of three factors:

1. It cannot burn small-particle coal very well. It must be fed double-sized coal—i.e., coal crushed a specified maximum size, then screened to reject smaller sizes. The rejected coal must be sold, probably at a discount on a million-Btu basis. All these costs must be allocated to the double-sized coal, making it higher-priced than the simple run-of-the-mine coal that can be burned in pulverized-coal-fired boilers.

2. As Table II shows, its efficiency is lower than that of the other boilers considered. This is due to its relatively high-rate loss of unburned carbon and its greater requirements for excess air.

3. Although its capital cost is the lowest (Table II), it cannot be scaled up readily, its natural size being limited to a steam capacity of 400 – 500 million Btu/h.

Pulverized-coal-fired—This boiler is highly efficient and can burn run-of-the-mine coal. It has been successfully scaled up to steam capacities of over 10 billion Btu/h. Compared to the stoker, its disadvantages are higher capital cost and higher electric-power consumption. Furthermore, to make it capable of burning solid wastes, additional fuel-handling and combustion equipment must usually be installed at the bottom of the furnace. There, refuse can be burned similarly to the way it is burned in the stoker-fired boiler.

Pulverized-coal-fired with fluegas desulfurization—This option is the same as the preceding one, except that it includes limestone-slurry fluegas desulfurization. The addition substantially hikes capital and electric-power costs and reduces overall efficiency. This option's advantage is greatest when regulations require 70 – 90% sulfur removal. Because this regulatory scenario is not evaluated, this option will not be considered.

Fluidized-bed—This system's boiler efficiency is comparable to that of the pulverized-coal boiler burning run-of-the-mine coal. Its capital cost is comparable to electric-power costs, and higher than those of the other

Operating characteristics and capital costs of coal-fired boilers   Table II

Characteristic	Boiler type			
	Spreader-stoker	Pulverized-coal-fired	Pulverized-coal-fired with fluegas desulfurization	Fluidized-bed
Efficiency, %	83.7	87.5	87.1	87.5
Fuels burned	Double-sized compliance coal* and coarse solid refuse	Run-of-mine compliance coal	Run-of-mine high-sulfur coal	Run-of-mine high-sulfur coal and solid refuse
Electric-power requirement, 150-million-Btu/h steam rating, kW	252	415	631	610
SO$_2$ control	No	No	Yes	Yes
Total capital required for a 150-Btu/h steam facility, \$million	11.5	12.5	About 16	About 12.5

*Compliance coals, by definition, can be burned in compliance with applicable SO$_2$ control regulations without additional sulfur control. High-sulfur coals, by definition, have a sulfur content too high to be considered compliance coals.
Source: Arthur D. Little, Inc., and U.S. Dept. of Energy [1,2]

boilers without fluegas desulfurization. Because sulfur dioxide emission is controlled by a sorbent bed inside the boiler, boiler efficiency is higher, and power consumption and capital cost lower, than for the pulverized-coal boiler with fluegas desulfurization. It shares the fired boiler's advantage of being able to burn solid refuse.

The principal disadvantage is that the technology is immature. The largest fluidized-bed boiler installed so far generates only approximately 400 million Btu/h of steam. About 100 units have been built or ordered worldwide; thus, although the operating experience accumulated to date is limited, a track record should be thoroughly established over the next few years.

On the basis of available coal supplies and state regulations on air pollution, the fluidized-bed boiler would not offer meaningful cost savings in coal burning at many locations. Therefore, only the pulverized-coal and stoker boilers and current pollution regulations will be considered in the initial analysis of steam costs. The fluidized-bed boiler will be evaluated afterwards as an option that may be employed in complying with tighter future regulations.

Economic comparison of the options

Total annual costs for owning and operating stoker and pulverized-coal boilers at an existing process plant in Texas are compared in Table III. Boiler rating, annual capacity factor and rate of return on equity (12% in constant dollars) are specified. The comparison is based on resulting average steam costs.

The steam costs projected for both boilers are virtually identical. A comparison with other boilers having capacity factors between 20% and 80%, and with Illinois and Alabama energy prices, also yields inconclusive results. In fact, total steam costs on an annual average for the two options remain within 5% in all cases. Therefore, the choice must be based on other factors—e.g., fuel supply, and familiarity with operation and maintenance.

The fuel-supply consideration that may favor the stoker boiler is the current or future availability of one or more solid-waste fuels, which it burns more readily. On the opposite side, stoker (i.e., double-sized) coal is not available in many localities. In such a case, a significant premium would have to be paid to obtain a reliable supply.

Familiarity with operation and maintenance may favor either type of boiler. Substituting an oil or gas type for a coal-fired boiler adds significantly to operating and maintenance burdens. Consequently, many buyers opt for a certain type based on experience with it.

Coal vs. gas

Although the economics of the stoker and pulverized-coal boilers are similar, the economics of both differ markedly from those of natural-gas boilers. Capital, operating and maintenance costs and electricity consumption are all much lower for gas boilers.

Indeed, in the cases considered, capital cost is not attributed to the gas boiler because it is already in place. However, the cost of fuel (the predominant variable cost of steam production) is substantially higher for gas firing. Because of these economic differences, there is a

Factors that limited CPI coal consumption

Coal consumption in the CPI actually declined from 1974 through 1980. In addition to site-specific factors noted elsewhere, further penetration by coal as a CPI fuel was held back by:

■ Conservation—During this period, total CPI production increased at an average rate of about 4.4%/yr. However, energy consumption declined, due to improved energy efficiency. Coal consumption fell at an average rate of 0.9%/yr, while oil and gas usage dropped 1.8%/yr. Thus, higher production could have been supported by existing boilers.

■ Energy prices and policies—Uncertainty, due to such nonmarket factors as OPEC decisions, federal price regulation, and possible federal prohibitions on burning some fuels, had clouded projections of fuel availability and prices. Consequently, many investment decisions were based on the assumed continuation of existing fuel prices.

■ Capital requirements—In the face of steady, or even declining, steam demand, installing a coal-fired boiler simply to cut fuel costs called for a major investment that would not add to plant output.

■ Coal's extra demands—Switching to coal places additional burdens on plant personnel. Fuel procurement, operation and maintenance, and technical support requirements are much more complex.

niche for solid-fuel boilers, and another for gas boilers.

To simplify the analysis, the gas boiler is compared with only one class of coal boiler for a particular boiler rating: (1) a stoker boiler for systems rated to generate steam equal to, or less than, 100 million Btu/h; and (2) a pulverized-coal boiler for systems rated to produce steam at more than 100 million Btu/h.

Given the similar economics of the two types of boiler, this is a reasonable approach. Both boilers are assumed to be subject to current air-pollution regulations. This allows burning the less expensive high-sulfur coal appropriate to each region, as shown in Table I.

In Table IV, the annual cost of owning and operating a new stoker boiler rated to generate steam at 75 million Btu/h is compared with the cost of continuing to operate an existing gas boiler at the same capacity. On the basis of a moderate 12% constant-dollar rate of return on equity, installing a coal boiler would not be economically attractive at the Texas plant, even if its annual capacity factor were as high as 70%. An unrealistic factor of over 80% would be necessary. However, the economics for a coal boiler would be very attractive for an Alabama plant.

The differing results in the two cases can be explained almost entirely by the difference in the projected fuel costs for the two states. Natural gas is expected to remain inexpensive in Texas, but coal to be expensive (based on shipping it from the Rocky Mountains and the Northern Great Plains). On the other hand, coal is expected to be slightly cheaper in Alabama than in Texas, and natural gas substantially more expensive.

A similar analysis is presented in Table IV for a pulverized-coal boiler rated at 150 million Btu/h of

Boiler-system annual costs—Texas site*		Table III
	Boiler type	
Cost component†	Spreader-stoker	Pulverized-coal-fired
Capital charges (12% constant-dollar return on equity)	1,364	1,482
Property taxes and insurance	229	249
Electricity	96	158
Limestone	—	—
Solid-waste disposal	82	79
Other operating and maintenance costs	760	791
Coal	3,145	2,903
Total annual cost	5,676	5,662
Cost per million Btu of steam, $	6.17	6.15

*Basis: 150-million-Btu/h steam capacity, 70% capacity factor and allowed SO_2 emission of 1.2 lb/million Btu of fuel.
†Costs are in thousands of 1982 constant dollars, except as noted.
Source: Arthur D. Little, Inc., and U.S. Dept. of Energy [2]

steam and operated at a 70% capacity factor. At this scale, the coal boiler generates steam at a total cost nearly equal to that produced from natural gas in Texas. In Alabama, coal's cost advantage is even more pronounced. Again, the difference for the two states can be explained almost entirely in terms of prospective fuel costs.

Thus, identifying the niches defined by boiler rating and capacity factor for coal should begin with analyses such as those presented in Table IV. However, such comparisons of total cost imply analytical accuracy and general applicability of the analysis that are inappropriate. Given analytical uncertainties and site-by-site variations, a more appropriate method of generic cost comparison is presented below.

Uncertainties in analyses

The sources of substantial uncertainty and site-to-site variation include:

1. Remaining useful life. This varies from site to site, and generally cannot be predicted with certainty for a particular site.

2. Site-specific construction requirements. These will increase boiler costs above the level projected in the article (which is appropriate for a highly favorable site). Also, these can seldom be projected with accuracy in making the preliminary analysis of coal-boiler economic feasibility.

3. Real rate of return required. This can vary with owner and time.

4. Delivered fuel prices. These cannot be predicted with accuracy over the life of the boiler.

In the face of these, and other, site-specific uncertainties and variations, it is useful to compare coal vs. gas in terms of ranges of difference in steam costs. These ranges are defined as the percentage difference between projected steam costs based on the continued burning of gas exclusively vs. coal firing. The ranges are mapped, as functions of capacity and capacity factor, in Fig. 1 for Alabama and Texas energy prices.

The Alabama case indicates that coal boilers having capacity factors and steam ratings falling in Zone A offer 20%, or larger, steam cost savings. In such cases, coal boilers will almost certainly be economically attractive, overcoming all but the most severe site-specific disadvantages (short remaining life, environmental restrictions, inadequate plan area, etc.).

Coal boilers falling in Zone B (0 to 20% steam cost savings) may be attractive. Those in Zone C probably will not be attractive unless a solid process waste is available for co-firing.

Comparison with the Texas case shows that almost all coal boilers fall into Zone C (possibly attractive with solid-waste co-firing). This is not surprising, again reflecting the importance of the coal-vs.-gas cost differential ($1.88/million Btu in Texas vs. $3.88 in Alabama).

Analysis based on site steam supply

After the preliminary analysis indicates that a coal boiler could be economical, the next step is the complex one of selecting the most appropriate boiler rating. Consider the case of an Illinois plant facing the prospective fuel costs given in Table I. Determining the capacity of the boiler that should be installed necessitates con-

Comparative steam costs— coal-fired boiler vs. existing gas-fired boiler								Table IV
	Stoker-fired boiler*				**Pulverized-coal-fired boiler†**			
State	Texas		Alabama		Texas		Alabama	
Boiler fuel	Coal	Natural gas	Coal	Natural gas	Coal	Natural gas	Coal	Natural gas
Capital charges (12% constant-dollar return on equity)	824	—	824	—	1,482	—	1,482	—
Property taxes and insurance	139		139		249		249	
Electricity	48	32	34	23	158	64	113	46
Solid-waste disposal	40	—	40	—	79	—	79	—
Other operating and maintenance costs	549	283	549	283	791	348	791	348
Fuel	1,572	2,483	1,471	3,462	2,903	4,966	2,725	6,924
Total annual costs	3,172	2,798	3,065	3,768	5,662	5,378	5,439	7,318
Cost per million Btu of steam, $	6.89	6.08	6.66	8.19	6.15	5.85	5.91	7.96

Source: Arthur D. Little, Inc., and U.S. Dept. of Energy [2]
*75 million Btu/h
†150 million Btu/h

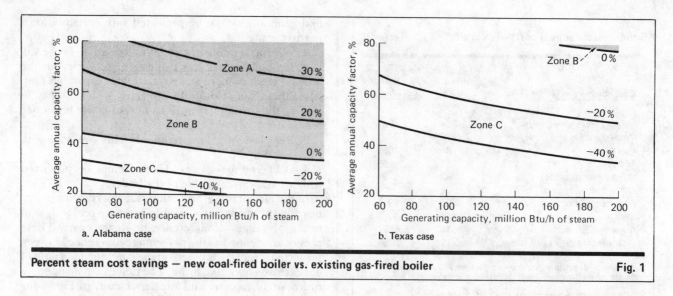

a. Alabama case

b. Texas case

Percent steam cost savings — new coal-fired boiler vs. existing gas-fired boiler   Fig. 1

verting the annual steam-demand pattern into a load-duration curve.

(An idealized, simplified load-duration curve is shown in Fig. 2 for example purposes only.)

In Fig. 2, Zone X represents nearly constant steam demand based on favorable ambient conditions and continuous process operation; Zone Y primarily reflects seasonal variations in steam demand, such as for heating buildings, additional preheating of boiler feedwater and feedstocks, and transient process conditions; and Zone Z denotes steam demand during partial and total process shutdowns.

A coal boiler may be very appropriate for a plant with a steam demand such as that shown in Fig. 2. It could fulfill substantially all the Zone X demand, and might be sized to serve Zone Y demand. Boiler maintenance could be performed during the Zone Z time. Such a load-duration curve suggests three possible approaches to boiler sizing:

1. Service baseload steam demand, generating steam at 100 million Btu/h.

2. Meet total steam demand, 200 million Btu/h.

3. Satisfy an intermediate level of demand, with exact boiler capacity based on achieving a specified hurdle rate of return on equity.

Choosing from among these alternatives, and sizing the boiler if the last is selected, requires estimating the rate of return that will be earned for all three cases. Simply comparing the earned rate of return on each of the three types of boilers as a whole is not sufficient. Instead, determining the boiler size entails selecting the boiler type whose return will meet, or exceed, the corporate hurdle rate on the entire investment and which earns precisely this hurdle rate on the last increment of steam capacity. What follows is an example of appropriate incremental investment analysis.

Steps in the financial analysis

First, confirm that all the boilers in the 100 to 200 million-Btu/h size range will earn an aggregate return on equity that exceeds the corporate hurdle rate (here, assumed to be 12%). Then, evaluate the incremental rate of return on equity, considering the incremental

capital charges that the facility could support, based on the fuel cost saving attributable to the last capacity increment.

These fixed capital charges include actual return on equity, corporate income taxes (after allowance for applicable tax benefits), and debt service incurred to partially finance the investment.

The fixed-charge rate corresponding to these total charges (expressed as a percentage of the total capital investment) can then be calculated as a function of rate of return on equity (based on the financial and tax parameters given previously). For a hurdle equity rate of 12%, the total fixed-charge rate would be 11.9%.

Constant-dollar capital charge rate is plotted in Fig. 3 as a function of constant-dollar rate of return on equity. The fixed-charge rate that corresponds to any hurdle equity rate can be identified by using the curve shown in the figure.

Next, develop an expression for the incremental fixed charge that can be supported by the fuel cost saving produced by incremental boiler capacity as a function of total boiler capacity.

Such an expression is: Supportable incremental fixed-

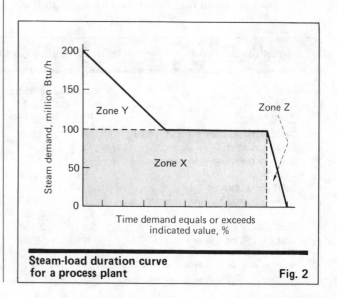

Steam-load duration curve for a process plant   Fig. 2

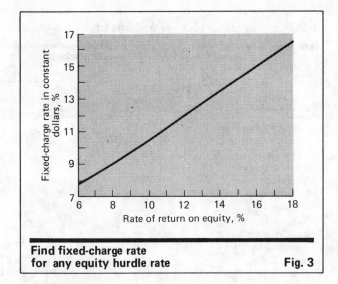

**Find fixed-charge rate
for any equity hurdle rate**   Fig. 3

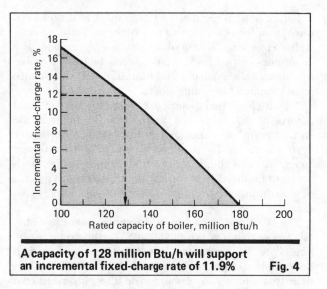

**A capacity of 128 million Btu/h will support
an incremental fixed-charge rate of 11.9%**   Fig. 4

charge rate = d(allowable capital charges) ÷ d(capital cost) = [d(allowable capital charges)/d(boiler capacity)] ÷ [d(capital cost)/d(boiler capacity)].

However, capital charges are, by definition, "allowable" if they equal net annual saving attributable to the coal boiler, exclusive of capital charges. Substituting yields: Supportable incremental fixed-charge rate = [d(net annual saving excluding capital charges)/d(boiler capacity)] ÷ [d(capital cost)/d(boiler capacity)].

This incremental approach, calling for differentiation, is necessary because unit capital cost and annual saving vary with boiler capacity in a complex manner—both incremental capital cost and annual saving per unit capacity decline as boiler rating increases. Incremental capital cost per unit capacity declines with economics of scale. Annual saving per incremental unit of capacity also declines, because each capacity increment is used for fewer hours per year than the preceding one.

The annual net saving (exclusive of capital charges) is equal to the value of the annual natural-gas saving minus all of the following: value of the coal burned annually; annual cost of ash disposal; the difference in annual cost of electricity for running the coal boiler and the gas boiler; the difference in labor cost for operating coal vs. gas boiler; the difference in maintenance cost for coal vs. gas boiler; and property taxes and insurance for the coal boiler.

Each of the foregoing, as well as the capital cost itself, can be expressed as a function of the coal boiler steam-generating capacity, S (in million Btu/h of steam), for a boiler fulfilling the steam demand depicted in Fig. 2. The functional forms and parameter values used are based on information obtained from equipment vendors, field experience and publications. Note that the annual quantity of coal-fired steam generated depends on three factors:

1. The number of hours in the average year—8,766.
2. The average quantity of steam generated per hour. It is assumed that the boiler operates at maximum capacity or serves maximum steam demand (whichever is smaller) during the part of the year when Zones X and Y of Fig. 2 represent the steam demand, and that it is out of service for planned maintenance when steam demand

drops, as in Zone Z. This complex function of S is valid for values of S falling between 100 and 200 million Btu/h of steam, because of the declining increments of steam demand served by additional increments of capacity.

3. A boiler operability factor, which accounts for occasions when the boiler is inoperable.

Substituting values for the foregoing factors into the supportable incremental fixed-charge rate equation and carrying out the implied differentiation with respect to boiler capacity, S, yields: Supportable incremental fixed-charge rate =
$$\{[30.2(1-S/200)-(5.73/S^{0.7})] \div [155/S^{0.48}+114/S^{0.18}]\} - 0.05.$$

The resulting supportable incremental fixed-charge rates as a function of rated coal-boiler capacity, S, are shown in Fig. 4. Note that an incremental fixed-charge rate of 11.9% (which corresponds to a 12% return on equity) is supportable up to a capacity of 128 million Btu/h.

Again, site-specific uncertainties and variations render this level of accuracy questionable. However, the basic conclusion is valid: Only modest incremental investments above the level required for the baseload (100 million Btu/h) boiler can be justified. Continued use of natural gas to meet seasonal and daily load swings would appear to be more economical than sizing the coal boiler to meet them.

Such conclusions were found to be valid in numerous analyses of actual process-plant steam supplies. There are cases in which only a coal boiler to meet maximum demand was evaluated, and rejected as uneconomical, although it could have been readily demonstrated that a smaller-baseload coal boiler would have been quite economical.

Fluidized-bed boiler and SO₂ regulations

Current federal regulations on the control of sulfur dioxide emissions apply to boilers rated to fire 250 million Btu/h, or more, of fuel. They do not apply to boilers as small as those considered in this article. Today, only state regulations pertain to such boilers. These regulations generally permit burning less-expensive, higher-sulfur coals.

Future federal regulations may require new boilers to burn compliance coal, or to take other measures to limit sulfur emissions. If compliance coal is mandated, the economics of coal firing would clearly be less favorable in states such as Alabama and Illinois, where compliance coals command premium prices.

The fluidized-bed boiler offers less-expensive internal control of SO_2 emissions, and may make possible the consumption of noncompliance coals despite tighter regulations. Total annual cost is compared with average steam cost in Table V for two boiler types—pulverized-coal and fluidized-bed—operating in Alabama. The cost of the continued operation of a natural-gas boiler is also given.

The basis of comparison is the same as that in Table IV, except that it is assumed that SO_2 emission must be limited to 1.2 lb/million Btu. With the fluidized-bed boiler able to burn coal that costs 45¢/million Btu less than that which can be burned in the pulverized-coal boiler, the fluidized-bed boiler's additional operating costs appear to be justified, based on a 12% constant-dollar return on equity.

The key to the choice between the fluidized-bed and pulverized-coal boilers is this differential between the costs of the coal that can be burned. As shown in Table IV, the fluidized-bed boiler's advantage is modest, based on the 45¢/million-Btu cost differential predicted for Alabama.

This advantage could be expected to be more substantive in Illinois, for which a differential of 65¢/million Btu is predicted. Where compliance coals are the least expensive coals that are available, the advantage would vanish.

However, note again that the cost differentials are quite small compared to the overall costs projected for steam. As with the choice between the stoker and pulverized-coal boilers, the decision whether to select a high-sulfur-coal-burning fluidized-bed boiler or a low-sulfur-coal-burning pulverized-coal boiler may be made primarily on grounds other than such a simply projected comparison of steam costs.

An important factor is likely to be whether the fluidized-bed boiler has been sufficiently proven operationally in comparison with the pulverized-coal boiler. Also critical is whether there will be an opportunity in the future to burn wastes in the fluidized-bed boiler, which is better suited to this role than the stoker boiler, especially if the moisture content of the wastes is high.

Finally, a detailed comparison of the operating characteristics of each boiler should be made with requirements posed by the steam demand to be served. For example, if rapid loading or long-term operation at reduced load, or both, is required, these factors should be considered.

Also note that the inclusion of the fluidized-bed boiler as an option does not significantly alter the coal-vs.-gas decision. The pulverized-coal boiler burning compliance coal and the fluidized-bed boiler burning higher-sulfur coal are similarly competitive with the continued burning of natural gas: they offer substantial operating (principally fuel) cost savings that must offset higher capital costs.

As noted previously, the case considered here gives no

| Boiler-system annual costs for Alabama site* | | | Table V |

Costs in thousands of 1982 constant dollars			
	Boiler type		
Cost component 70% capacity factor	Pulverized-coal without desulfurization	Fluidized-bed	Natural-gas
Capital charges (12% constant-dollar return on equity	1,482	1,482	
Property taxes and insurance	249	249	
Electricity	113	166	46
Limestone	—	69	
Solid-waste disposal	79	148	
Other operating and maintenance costs	791	791	348
Coal	3,198	2,724	6,924
Total annual costs	5,912	5,629	7,318
Cost per million Btu of steam, $	6.42	6.11	7.95
40% capacity factor			
Total annual costs	4,459	4,297	4,431
Cost per million Btu of steam, $	8.48	8.17	8.42

*Steam-generating capacity of 150 million Btu/h.
Allowed SO_2 emission of 1.2 lb/million Btu of fuel.
Source: Arthur D. Little, Inc., and U.S. Dept. of Energy [2]

credit to fluidized-bed boilers for their very substantial advantages in burning wastes and mixes of different fuels, in NO_x control, etc. Analysis of cases that did allow credit for such advantages would show fluidized-bed boilers to be the system of choice in still more industrial installations. Also, fluidized-bed combustion is an evolving technology: incremental improvements and development of second-generation systems should broaden its appeal further.

References

1. "Cogeneration Technology Alternatives Study, Final Report," Vol. 4—Heat Sources, Balance of Plant, and Auxiliary Systems, prepared by Power Systems Div. of United Technologies Corp. for National Aeronautics Space Admin., Contract DEN3-30, for the U.S. Dept. of Energy, January 1980.
2. "Cogeneration Technology Alternatives Study," Vol. 6, Part 1—Coal-Fired Noncogeneration Process Boiler, Section A, prepared by General Electric Co. for the National Aeronautics Space Admin., Contract DEN3-31, for the U.S. Dept. of Energy, May 1980.

The author

C. Thomas Breuer is a staff member of the Chemical and Metallurgical Engineering Section of Arthur D. Little, Inc. (Acorn Park, Cambridge, MA 02140; tel.: 617-864-5770). He has been extensively involved in the field of coal use, for both electric utilities and industrial plants, throughout the world. Prior to joining Arthur D. Little, he worked for Georgia Power Co. A member of Sigma Xi and a licensed engineer in Massachusetts, he holds a Ph.D. in Energy Technology Assessment from the Massachusetts Institute of Technology and a B.S. in physics from Emory University.

Nebraska Boiler Co., Inc.

Packaged boilers for the CPI

Unless you specify carefully, the boiler you buy may have a shorter-than-expected life and be less efficient than foreseen.

Charles W. Hawk, Jr., Olin Corp.

☐ The environment in the chemical process industries (CPI) today demands that the specifiers of packaged boilers consider both the potential for corrosion and energy efficiency.

Packaged boilers are commonly used in the CPI because they are less costly than field-erected boilers, and their size range meets the average plant's needs. Most packaged boilers are standardized for common pressures, temperatures, capacities, fuels, etc. However, owing to today's CPI plant environment, it is not unusual for the standard packaged boiler to have a shorter-than-expected life, and to be less energy efficient than anticipated.

Hence, the engineer should carefully consider boiler specifications prior to the purchase of the boiler. The additional cost of extra corrosion protection and energy efficiency will be returned in extended boiler life and in energy savings. This article will consider those areas to which special attention should be paid: boiler controls, mechanical design and energy efficiency.

Packaged boilers are commonly available with capacities up to about 120,000 lb/h of steam (3,500 hp). Typically, the pressure vessel, burner controls, refrac-

tory and exterior insulation are assembled as a unit and tested at the manufacturer's plant. Packaged boilers can be shipped by truck or rail to the jobsite. The fuel lines, supply piping, and electrical equipment are then installed, to make the boiler operational.

Packaged boilers may be either of the fire-tube or water-tube type. This article will primarily address water-tube boilers that are gas- or liquid-fired.

Burner controls

CPI environments often accelerate the corrosion rate of burner controls, resulting in downtime, nuisance burner-trips, and significantly shorter equipment life. Burner control systems are designed to prevent continued operation of a boiler when a hazardous condition exists. Particularly, they must prevent explosions that could cause injuries or damage the boiler.

In addition, the controls assist the operator in starting and stopping the burners and the fuel equipment. They must also prevent damage to burners and fuel equipment during normal operation, and must avoid false trips of fuel equipment (when an unsafe condition does *not* exist).

The typical packaged boiler, for services such as hospitals, colleges, or other applications in a limited corrosive environment, might have the following features:

■ Light-duty design control panel with ungasketed panel doors, and containing lighted window covers and bezels.

■ Tight layout inside the control panel, to conserve space and minimize panel size.

Originally published May 16, 1983

■ Working relays used with printed circuit boards, probably having gold contacts.

■ Electrical circuitry using microswitches.

■ Soldered connecting wiring.

■ Wiring neither numbered nor color coded (although terminal strips would be numbered).

■ Timers of a single-cam-operated type.

■ Field-located devices, such as high- and low-pressure switches of a NEMA Type 1 (not water- or dustproof).

Corrosion-resistant controls

In the CPI environment, whether the packaged boiler is located outside, or in an enclosure, corrosion-resistant design is very desirable. In general, equipment having a NEMA Type 4 (or 4X) enclosure is best. Type 4 implies water- and dustproof, while Type 4X includes a nonmetallic material for the enclosure.

In addition, a freestanding panel eliminates any problems associated with space limitations. The panel layout should allow easy access for maintenance and wiring changes; particularly, consideration should be given to room for future component additions. Finally, consider using an instrument-air purge on the panel, to ensure that corrosive gases do not accelerate deterioration of components.

The components, whether panel or field mounted, should have corrosion-resistant features and be heavy duty. For example, relays should be of a conventional machine-tool type. Also, pushbuttons should be oiltight NEMA 4X types, and timers should be heavy duty.

All wiring in the panel should be done with 18AWG-THHN wire, color coded and numbered; outside the panel, the wire should be 14AWG-THHN type. (Of course, flame-detection devices will require shielded wiring.) The terminal strips should be numbered, and all wiring connections should be of the screw-type, not soldered.

Field devices such as pressure switches, selector switches and power lights should have NEMA Type 4 enclosures. The annunciator for alarm conditions should preferably be of the first-out type* and have hermetically sealed relays. Other field-mounted devices— e.g., pressure gages, feedwater and pressure controls— should be water- and dustproof, if possible.

Mechanical design

The mechanical design of the CPI packaged boiler can greatly extend boiler life. Some drum connections for packaged units are $\frac{1}{2}$- or $\frac{3}{4}$-in. size. However a 1-in. drum connection is preferred because the increased diameter not only provides better corrosion protection but also a more substantial structure to prevent accidental bending at the connection.

Air dampers, control valves, and other moving linkages associated with the boiler should contain grease fittings, if practical. If such fittings are not possible, use a good coating of grease at contact points. Regular lubrication will minimize corrosion of contact points.

When a boiler is fired with fuel oil more than half the time, consider a rear boiler panel of stainless steel or other corrosion-resistant material. Sulfur gases and condensation often accelerate the corrosion of the typical packaged boiler carbon-steel rear boiler panel.

Boiler-tube life

A typical packaged boiler might have a maximum input at rated load of 60,000 to 70,000 Btu/ft^3 of furnace volume. This particular design criterion implies that the minimum furnace volume is used to generate the maximum boiler output—i.e., the cheapest boiler to produce the most steam. As a result, the boiler tubes are "worked" at their maximum potential when the boiler is being fired at this maximum input. During the life of the boiler, cycling due to process changes, shutdowns, etc., will cause the boiler tubes to be worked to their fullest.

A specification of maximum input at rated load of 50,000 Btu/ft^3 of furnace volume will result in extended life of the boiler tubes. Of course, at initial purchase, there will be some additional costs associated with the added furnace volume. But the life of the boiler will be extended considerably.

Firing with byproduct gas

A byproduct gas, such as hydrogen, may often be fired in a packaged boiler. Remember that trace impurities sometimes exist in such gas. For example, traces of caustic could result in premature corrosion of the copper components normally found in the pressure switches.

In addition, consider the tube and drum arrangements, if byproduct gas is to be fired. Packaged boilers are available in "A," "D" and "O" configurations. The "D" type is often designated, particularly one without baffles, so as to minimize dead space in the boiler. In this type, the steam drum is located longitudinally along one side, and the water drum is directly below, near the bottom. (Dead spaces in the boiler can result in byproduct gases being trapped after the flame has been extinguished.)

Spare-parts standardization

Minimizing the spare parts for a packaged boiler is also important, because of the need for replacement due to corrosion. Where there are other boilers in the plant, considerable maintenance-cost savings can result by standardizing on parts from a single manufacturer, so that they will be interchangeable and familiar to maintenance personnel. Once the boiler specification has been put together, it should be reviewed with plant maintenance and warehouse personnel to ensure that spare parts will be minimized.

The same applies to control items that are also used in the plant process itself. For example, a low-excess-air burner might use an infrared scanner, and infrared scanners may be also used elsewhere in the plant. So may control valves and relief valves. It may be more expensive to specify similar controls for the boiler, but money will be saved on maintenance in the long run.

Energy efficiency

Obviously, the boiler should be energy efficient. Depending upon the load factor of the boiler, a relatively

---

*A first-out alarm remains flashing even after other alarms are activated by a boiler shutdown. Hence an operator can troubleshoot the first condition without having to check all the others before restarting the boiler.

small expenditure at the time of initial purchase will yield significant returns when calculated over the boiler's entire life.

One of the most common reasons for poor efficiency is excess combustion air. Theoretically, in order to burn all of the fuel, a specific (stoichiometric) quantity of air is required. When more than this theoretical amount is used, energy is wasted in heating the excess air. However, to guarantee complete combustion and prevent smoke formation, even the most efficient boilers must operate with some excess air. About 10 to 20% excess air offers a good compromise between safety, efficiency and pollution control. Excess air is regulated by the proper adjustment of combustion controls, coupled with flue-gas analysis.

Low-excess-air burners are now available that can yield significant energy savings. Due to its mechanical design, a low-excess-air burner can use much less air than the normal packaged boiler. The term "low excess air" is applied to burners that require only 5% excess air or less. Another benefit of the low-excess-air burner is that its use can result in lower stack-emissions and lower NO_x design capability.

Combustion control systems

Combustion control systems for packaged boilers range from the simple to the very sophisticated. It is a mistake to automatically assume that there is an economic payout of energy savings from investing in the more-sophisticated systems. It is not unusual for such a system to cost $60,000, or more, while saving only a few thousand dollars.

A combustion-control system generally serves two functions: setting the fuel/air ratio, and maintaining load. Generally speaking, there are two approaches to achieving combustion control: using positioning control (the simplified system) and using metering control (more sophisticated).

A positioning-control system can be either on/off, high/low/off, or modulating positioning. In the modulating-positioning control system, the air and fuel control-valve settings are continuously varied. Both settings are determined by the steam load on the boiler. The typical CPI packaged boiler can use positioning control via mechanical linkages, or control rods, which are connected to the fuel valves and air damper.

The more-sophisticated metering system can be either semi-metering or full-metering. For the full-metering system, one input (either fuel or air) is controlled according to load. The flow of the other input is measured and controlled according to the measured flow of the first input. Normally, this sophisticated system can only be justified when there are simultaneous multiple fuel firings, variable fuel properties, or use of multiple burners.

Automatic oxygen-trim control is a popular method, based on the measured oxygen levels in the exhaust, for trimming the level of air fed to the boiler. It can be applied to either the positioning system or the metering system.

However, it is not unusual to discover that, by analyzing the distribution of the operating time versus the load ranges, oxygen-trim control does not have an energy-savings payout. It is a common mistake to retrofit existing systems with oxygen-trim control based on a false analysis of savings. The false justification for such control results from comparing a poorly tuned existing control system with the proposed oxygen-trim system. Often, one can take the existing control system—whatever shape it is in—and achieve significant savings by reducing the excess air by proper tuning of the system load characteristics. Such savings can be maintained by regular monitoring of the boiler efficiency. Properly, one should compare the proposed oxygen-trim control system to the best that the present control system can achieve.

Using economizers

The typical CPI packaged boiler can achieve an annual savings of 3% by installing an economizer that preheats the boiler feedwater by using waste heat from the flue gas. In general, a 1% increase in efficiency can be obtained by decreasing the flue-gas temperature by 40°F, thereby raising the boiler feedwater temperature by 10°F.

An economizer is justifiable only if there is inadequate heat-transfer surface in the boiler to remove sufficient heat given off by the flame. However, due to the typical sizing of the packaged boiler, it is not unusual to find that an economizer is justifiable. Remember that with conventional economizers, there is a minimum flue-gas temperature required to avoid corrosion. This temperature depends upon the fuel used and the economizer design. For natural gas, the minimum temperature is generally 250°F; for No. 2 fuel oil, 275°F; and for high-sulfur oils, 320°F.

By atomizing fuel oil with air, rather than with steam, an average annual energy saving of 1% can be achieved, because more energy is required to produce steam than to produce compressed air. Steam atomization usually requires 1% of the energy in the fuel, versus a fraction of a percent for air.

Summary

The CPI packaged boiler often has a shorter-than-expected life and is less efficient than anticipated. Proper boiler specifications, written prior to purchase, can overcome these deficiencies. Implementation of the corrosion resistance and energy recommendations given in this article will ensure extended boiler life and good energy efficiency.

The author

Charles W. Hawk, Jr., is a project manager of Olin Corp.'s Southeast Regional Engineering group, P.O. Box 248, Charleston, TN 37310; (615) 336-4360. He has been responsible for engineering on several capital projects at Olin's chlor/alkali plants. He was previously a maintenance engineer at Olin's Augusta, Ga., plant. His prior experience was with Procter & Gamble and Martin Marietta. He holds a B.S. in mechanical engineering from the University of Tennessee, has had one patent issued and has several pending, and is a registered professional engineer in Georgia, Alabama and Tennessee.

Boiler heat recovery

To minimize fuel costs, heat can be recovered from both flue gases and boiler blowdown.

William G. Moran and *Guillermo H. Hoyos,**
Engineering Experiment Station,
Georgia Institute of Technology

☐ Boiler flue gases are rejected to the stack at temperatures at least 100 to 150°F higher than the temperature of the generated steam. Obviously, recovering a portion of this heat will result in higher boiler efficiencies and reduced fuel consumption.

Heat recovery can be accomplished by using either an economizer to heat the water feedstream or an air preheater for the combustion air. Normally, adding an economizer is preferable to installing an air preheater on an existing boiler, although air preheaters should be given careful consideration in new installations.

Economizers are available that can be economically retrofitted to boilers as small as 100 hp (3,450 lb/h steam produced).

Fig. 1 can be used to estimate the amount of heat that can be recovered from flue gases. Two main assumptions have been made in developing this graph:

1. The boiler operates close to optimum excess-air levels. (It does not make sense to use an expensive heat-recovery system to correct for inefficiencies caused by improper boiler tuneup.)

2. The lowest temperature to which the flue gases can be cooled depends on the type of fuel used: 250°F for natural gas, 300°F for coal and low-sulfur-content fuel oils, and 350°F for high-sulfur-content fuel oils.

These limits are set by the flue-gas dewpoint, or by cold-end corrosion, or heat driving-force considerations.

Example

A boiler generates 45,000 lb/h of 150-psig steam by burning a No. 2 fuel oil that has a 1% sulfur content. Some of the condensate is returned to the boiler and mixed with fresh water to yield a 117°F boiler feed. The stack temperature is measured at 550°F.

Determine the annual savings (assuming 8,400 h/yr boiler operation) that will be achieved by installing an economizer in the stack.

Assume that fuel energy costs $3/million Btu.

*This material has been prepared by staff members of Georgia Tech's Industrial Energy Extension Service, which is a continuing energy-conservation program funded by the State of Georgia Office of Energy Resources.

Originally published December 3, 1979

Babcock & Wilcox Ltd.

From steam tables, the following heat values are available:

For 150°F saturated steam   1,195.50 Btu/lb
For 117°F feedwater   84.97 Btu/lb

The boiler heat output is calculated as follows:

$$\text{Output} = 45{,}000(1{,}195.50 - 84.97)$$
$$= 50 \text{ million Btu/lb.}$$

Using the curve for low-sulfur-content oils, the heat recovered that corresponds to a stack temperature of

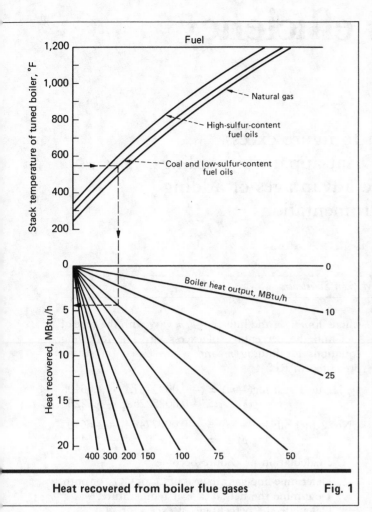

Heat recovered from boiler flue gases   **Fig. 1**

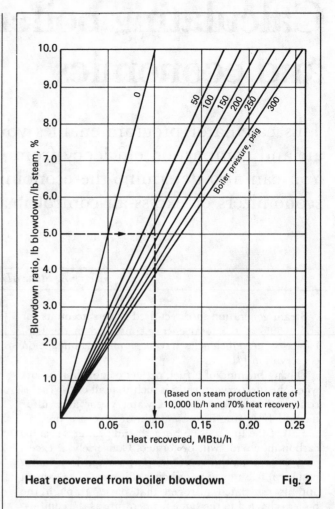

Heat recovered from boiler blowdown   **Fig. 2**

550°F and a boiler duty of 50 million Btu/h can be read from the graph as 4.3 million Btu/h. The annual savings are:

4.3 MBtu/h × $3/MBtu × 8,400 h/yr = $108,000/yr

Recovering heat from boiler blowdown

Heat can also be recovered from boiler blowdown, by preheating the boiler makeup water through use of a heat exchanger. This is done most conveniently in continuous-blowdown systems.

Example

In a plant where the cost of steam is $2.50/MBtu, 1,250 lb/h are continuously purged in order to avoid buildup of solids in the boiler tubes. Determine the yearly savings (assume 8,760 h of operation per year) if the blowdown is passed through a heat exchanger to raise the temperature of the makeup water. The boiler generates 125-psig steam at a rate of 25,000 lb/h.

First calculate the blowdown ratio:

$$\text{Blowdown ratio} = \frac{1{,}250 \text{ lb/h blowdown}}{25{,}000 \text{ lb/h steam}} = 5\%$$

From Fig. 2, the heat recovered, corresponding to 5% blowdown ratio and 125-psig boiler pressure, is 0.100 MBtu/h.

Since Fig. 2 is based on a steam production rate of 10,000 lb/h, the actual yearly savings for this plant are:

Savings:

$$= 0.100 \frac{\text{MBtu}}{\text{h}} \times \frac{25{,}000 \text{ lb/h}}{10{,}000 \text{ lb/h}} \times 8{,}760 \frac{\text{h}}{\text{yr}} \times \frac{\$2.5}{\text{MBtu}}$$

$$= \$5{,}500$$

The values obtained from Fig. 2 are based on a makeup water temperature of 70°F and a heat recovery of 70%.

Suggested actions

Determine stack temperature after boiler has been carefully tuned up. Then determine minimum temperature to which stack gases can be cooled and study economics of installing an economizer or air preheater. Also study savings from installing a heat exchanger on blowdown.

The authors

William G. Moran is a research engineer at the Engineering Experiment Station, Georgia Institute of Technology, Atlanta, GA 30332. His present fields of interest include industrial and commercial energy conservation, alternative energy sources, and research and development. He holds a B.S. from the University of Massachusetts and an M.S. from Rensselaer Polytechnic Institute, both in mechanical engineering, and is a registered professional engineer in Georgia.
Guillermo H. Hoyos was a research engineer with Georgia Tech's Engineering Experiment Station when these materials were prepared. He holds B.S. and M.S. degrees in chemical engineering from Georgia Institute of Technology. He now resides in Bogota, Colombia.

Calculating boiler efficiency and economics

This calculator program enables you to figure excess air and combustion efficiency from Orsat-apparatus readings. You can also determine the economic advantages of adding economizers or excess-air-control instrumentation.

Terry A. Stoa, ADM Corn Sweeteners

Steam generation in direct-fired boilers accounts for about 50% of the total energy consumed in the U.S. Hence, boiler efficiencies have a significant impact on conservation [1].

During burning of a fuel, perfect combustion occurs when the fuel/oxygen ratio is such that all of the fuel is converted to carbon dioxide, water vapor and sulfur dioxide. If an insufficient amount of oxygen is present, not all of the fuel will be burned, and products such as carbon monoxide will be formed. Conversely, if excess oxygen is present, it serves no purpose and is in fact a major contributor to poor boiler efficiency. The oxygen and its associated nitrogen that passes through the boiler is heated to the same temperature as the combustion products. This heating uses energy that would otherwise be available to produce steam.

Boiler efficiency is the ratio of heat output (steam and losses) to the heat input (fuel, feedwater, combustion air). Flue-gas analysis and stack-temperature measurements can be used to monitor efficiency.

The percentage of excess combustion-air is determined by analyzing the boiler exit gases for oxygen or carbon dioxide (or both). Assuming that the gases consist solely of O_2, CO_2 and N_2, the following equation can be used:

$$A_x = \frac{\%O_2}{0.266(100 - \%O_2 - \%CO_2) - \%O_2} \times 100$$

where $\%O_2$ and $\%CO_2$ are found by an Orsat-type device [2].

Equations have been developed using just $\%O_2$, e.g.:

$$A_x = \frac{a \times \%O_2}{1 - 0.0476 \times \%O_2}$$

where Factor a is characteristic of the fuel being burned [3]. Based on curves presented in Ref. 4:

a, natural gas $= 4.5557 - (0.026942 \times \%O_2)$
a, No. 2 fuel oil $= 4.43562 + (0.010208 \times \%O_2)$

Boiler efficiency and net flue-gas temperature* follow a linear relationship:

$$E = 1/m \times (T - b)$$

where slope $1/m$ and intercept b/m vary with the type of fuel and the percentage of excess air. The following equations for finding m and b are based on curves presented in Ref. 4:

Natural gas: $\log(-m) = -0.0025767A_x + 1.66403$
$\log(b) = -0.0025225A_x + 3.6226$

No. 2 fuel oil: $\log(-m) = -0.0027746A_x + 1.66792$
$\log(b) = -0.0027073A_x + 3.6432$

The calculation procedure becomes:

1. Determine flue-gas analysis of O_2 or CO_2, or both.
2. Determine the net stack exit temperature, T.
3. Calculate the percentage of excess air, A_x.
4. Calculate m and b.
5. Calculate boiler efficiency, E.

The calculated efficiency does not account for radiation or carbon losses. It is a measure of stack heat losses.

Once the efficiency is found, steam costs can be determined by the equation:

$$C_s = \frac{1,000 \times C_f \times H}{E}$$

This cost accounts only for the fuel portion. For a more accurate figure one must include chemical treatment costs, electric costs, labor costs, etc.

Efficiency calculations provide a sound basis for evaluating conservation projects such as installation of economizers and excess-air controls. Potential dollar savings can be based on either constant steam outputs or present fuel costs, as seen in Fig. 1 and 2.

Similarly, efficiency improvements at constant fuel input will result in capacity increases, as shown in Fig. 3.

A computer program was written to perform the described calculations, using a Texas Instruments TI-59 programmable pocket calculator. The calculator eliminates the need for charts, tables and nomographs, while providing the user with fast, dependable results. The

*Net flue-gas temperature is the difference between ambient temperature and the stack temperature measured after the last heat-transfer surface of the boiler.

Originally published July 16, 1979

Program for TI-59 calculator determines boiler efficiency and percentage of excess air. Inputs required are flue-gas temperature
Table I continues

Step	Key	Code	Step	Key	Code	Step	Key	Code	Step	Key	Code	Step	Key	Code	Step	Key	Code
000	76	LBL [1]	062	42	STO	124	04	4	186	00	0	248	02	2	310	06	6
001	11	A	063	15	15	125	02	2	187	00	0	249	02	2	311	02	2
002	42	STO	064	00	0	126	07	7	188	00	0	250	01	1	312	02	2
003	00	00	065	91	R/S	127	01	1	189	00	0	251	03	3	313	06	6
004	91	R/S	066	76	LBL [6]	128	07	7	190	00	0	252	69	OP	314	95	=
005	76	LBL [2]	067	17	B'	129	69	OP	191	69	OP	253	02	02	315	22	INV
006	12	B	068	22	INV	130	01	01	192	02	02	254	03	3	316	28	LOG
007	42	STO	069	86	STF	131	03	3	193	69	OP	255	06	6	317	42	STO
008	02	02	070	00	00	132	05	5	194	05	05	256	00	0	318	08	08
009	91	R/S	071	02	2	133	06	6	195	25	CLR [10]	257	00	0	319	76	LBL [14]
010	76	LBL [3]	072	02	2	134	02	2	196	69	OP	258	00	0	320	30	TAN
011	14	D	073	08	8	135	00	0	197	00	00	259	00	0	321	94	+/-
012	42	STO	074	08	8	136	00	0	198	53	(	260	00	0	322	85	+
013	03	03	075	02	2	137	00	0	199	43	RCL	261	00	0	323	43	RCL
014	91	R/S	076	42	STO	138	00	0	200	11	11	262	00	0	324	02	02
015	76	LBL [4]	077	05	05	139	00	0	201	65	×	263	00	0	325	95	=
016	19	D'	078	93	.	140	00	0	202	43	RCL	264	69	OP	326	55	÷
017	42	STO	079	00	0	141	69	OP	203	00	00	265	03	03	327	43	RCL
018	01	01	080	02	2	142	02	02	204	85	+	266	69	OP	328	07	07
019	91	R/S	081	06	6	143	69	OP	205	43	RCL	267	05	05	329	95	=
020	76	LBL [5]	082	09	9	144	05	05	206	12	12	268	98	ADV [12]	330	42	STO
021	16	A'	083	04	4	145	25	CLR [8]	207	54	)	269	93	.	331	09	09
022	86	STF	084	02	2	146	69	OP	208	65	×	270	00	0	332	06	6 [15]
023	00	00	085	94	+/-	147	00	00	209	43	RCL	271	00	0	333	01	1
024	01	1	086	42	STO	148	01	1	210	00	00	272	02	2	334	00	0
025	09	9	087	11	11	149	06	6	211	55	÷	273	05	5	335	00	0
026	04	4	088	04	4	150	01	1	212	53	(	274	07	7	336	03	3
027	05	5	089	93	.	151	03	3	213	01	1	275	06	6	337	02	2
028	08	8	090	05	5	152	03	3	214	75	-	276	07	7	338	00	0
029	42	STO	091	05	5	153	07	7	215	93	.	277	94	+/-	339	03	3
030	05	05	092	05	5	154	69	OP	216	00	0	278	65	×	340	69	OP
031	93	.	093	07	7	155	01	01	217	04	4	279	43	RCL	341	04	04
032	00	0	094	42	STO	156	01	1	218	07	7	280	06	06	342	43	RCL
033	01	1	095	12	12	157	07	7	219	06	6	281	85	+	343	00	00
034	00	0	096	93	.	158	06	6	220	65	×	282	01	1	344	69	OP [16]
035	02	2	097	01	1	159	02	2	221	43	RCL	283	93	.	345	06	06
036	00	0	098	07	7	160	00	0	222	00	00	284	06	6	346	03	3
037	08	8	099	42	STO	161	00	0	223	54	)	285	06	6	347	07	7
038	42	STO	100	13	13	162	00	0	224	95	=	286	04	4	348	05	5
039	11	11	101	01	1	163	00	0	225	42	STO [11]	287	00	0	349	07	7
040	04	4	102	93	.	164	00	0	226	06	06	288	03	3	350	06	6
041	93	.	103	09	9	165	00	0	227	66	PAU	289	95	=	351	05	5
042	04	4	104	42	STO	166	69	OP	228	66	PAU	290	22	INV	352	02	2
043	03	3	105	14	14	167	02	02	229	66	PAU	291	28	LOG	353	01	1
044	05	5	106	01	1	168	69	OP	230	87	IFF	292	94	+/-	354	69	OP
045	06	6	107	05	5	169	05	05	231	00	00	293	42	STO	355	04	04
046	02	2	108	93	.	170	25	CLR [9]	232	39	COS	294	07	07	356	43	RCL
047	42	STO	109	05	5	171	69	OP	233	68	NOP	295	93	. [13]	357	02	02
048	12	12	110	42	STO	172	00	00	234	02	2	296	00	0	358	69	OP [17]
049	93	.	111	15	15	173	02	2	235	01	1	297	00	0	359	06	06
050	01	1	112	00	0	174	07	7	236	04	4	298	02	2	360	06	6
051	05	5	113	91	R/S	175	03	3	237	01	1	299	05	5	361	01	1
052	42	STO	114	76	LBL [7]	176	02	2	238	01	1	300	02	2	362	00	0
053	13	13	115	15	E	177	01	1	239	07	7	301	02	2	363	00	0
054	93	.	116	25	CLR	178	03	3	240	69	OP	302	05	5	364	04	4
055	08	8	117	69	OP	179	69	OP	241	01	01	303	94	+/-	365	04	4
056	42	STO	118	00	00	180	01	01	242	02	2	304	65	×	366	03	3
057	14	14	119	01	1	181	01	1	243	07	7	305	43	RCL	367	06	6
058	01	1	120	04	4	182	06	6	244	06	6	306	06	06	368	69	OP
059	04	4	121	03	3	183	06	6	245	02	2	307	85	+	369	04	04
060	93	.	122	02	2	184	02	2	246	00	0	308	03	3	370	43	RCL
061	02	2	123	02	2	185	00	0	247	00	0	309	93	.	371	06	06

and the percentage of oxygen in flue gases (as determined by Orsat analysis).   **Table I continued**

Step	Key	Code		Step	Key	Code		Step	Key	Code		Step	Key	Code		Step	Key	Code		Step	Key	Code
372	69	OP	18	434	00	00		496	66	PAU		524	03	3		552	01	1		580	00	0
373	06	06		435	93	.		497	66	PAU		525	00	0		553	01	1		581	00	0
374	06	6		436	00	0		498	65	×		526	00	0		554	07	7		582	69	OP
375	01	1		437	00	0		499	93	.		527	00	0		555	02	2		583	03	03
376	01	1		438	02	2		500	01	1		528	01	1		556	07	7		584	02	2
377	07	7		439	07	7		501	55	÷		529	04	4		557	00	0		585	07	7
378	02	2		440	07	7		502	43	RCL		530	69	OP		558	00	0		586	01	1
379	01	1		441	04	4		503	09	09		531	03	03		559	69	OP		587	04	4
380	02	2		442	06	6		504	95	=		532	03	3		560	01	01		588	03	3
381	01	1		443	94	+/-		505	42	STO		533	07	7		561	01	1		589	06	6
382	69	OP		444	65	×		506	10	10		534	04	4		562	05	5		590	06	6
383	04	04		445	43	RCL		507	25	CLR	23	535	01	1		563	03	3		591	02	2
384	43	RCL		446	06	06		508	69	OP		536	03	3		564	02	2		592	00	0
385	09	09		447	85	+		509	00	00		537	06	6		565	03	3		593	00	0
386	69	OP		448	01	1		510	01	1		538	06	6		566	06	6		594	69	OP
387	06	06		449	93	.		511	05	5		539	02	2		567	03	3'		595	04	04
388	25	CLR		450	06	6		512	03	3		540	00	0		568	07	7		596	69	OP
389	69	OP		451	06	6		513	02	2		541	00	0		569	06	6		597	05	05
390	00	00		452	07	7		514	03	3		542	69	OP		570	03	3		598	43	RCL
391	98	ADV		453	09	9		515	06	6		543	04	04		571	69	OP		599	10	10
392	43	RCL		454	02	2		516	03	3		544	69	OP		572	02	02		600	58	FIX
393	09	09		455	95	=		517	07	7		545	05	05		573	02	2		601	03	03
394	91	R/S		456	22	INV		518	69	OP		546	43	RCL		574	00	0		602	99	PRT
395	76	LBL	19	457	28	LOG		519	02	02		547	03	03		575	01	1		603	58	FIX
396	39	COS		458	94	+/-		520	06	6		548	99	PRT		576	00	0		604	09	09
397	02	2		459	42	STO		521	03	3		549	02	2	24	577	01	1		605	98	ADV
398	01	1		460	07	07		522	03	3		550	01	1		578	00	0		606	98	ADV
399	04	4		461	93	.	21	523	00	0		551	04	4		579	01	1		607	98	ADV
400	01	1		462	00	0														608	91	R/S
401	01	1		463	00	0																
402	07	7		464	02	2																
403	69	OP		465	07	7																
404	01	01		466	00	0																
405	02	2		467	07	7																
406	07	7		468	03	3																
407	06	6		469	94	+/-																
408	02	2		470	65	×																
409	00	0		471	43	RCL																
410	00	0		472	06	06																
411	05	5		473	85	+																
412	01	1		474	03	3																
413	00	0		475	93	.																
414	03	3		476	06	6																
415	69	OP		477	04	4																
416	02	02		478	03	3																
417	00	0		479	02	2																
418	00	0		480	95	=																
419	03	3		481	22	INV																
420	02	2		482	28	LOG																
421	02	2		483	42	STO																
422	04	4		484	08	08																
423	02	2		485	61	GTO	22															
424	07	7		486	30	TAN																
425	00	0		487	68	NOP																
426	00	0		488	76	LBL																
427	69	OP		489	10	E'																
428	03	03		490	43	RCL																
429	69	OP		491	03	03																
430	05	05		492	65	×																
431	98	ADV	20	493	43	RCL																
432	25	CLR		494	01	01																
433	69	OP		495	66	PAU																

Comments

1. % O_2	9. Print "LOAD:"	17. Print " _____ % XS"
2. T	10. Calculate A_x	18. Print " _____ % EFF"
3. C_f	11. Print "FUEL: GAS"	19. Print "FUEL: #2 OIL"
4. H	12. Calculate m	20. Calculate m
5. No. 2 fuel oil	13. Calculate b	21. Calculate b
6. Natural gas	14. Calculate E	22. Calculate C_s
7. Print "BOILER:"	15. Print " _____ % O_2"	23. Print "COST/MM BTU:"
8. Print "DATE:"	16. Print " _____ T, °F"	24. Print "FUEL COST/1,000 LB"

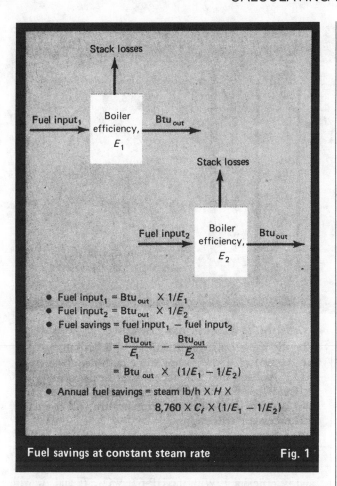

- Fuel input$_1$ = Btu$_{out}$ × 1/E_1
- Fuel input$_2$ = Btu$_{out}$ × 1/E_2
- Fuel savings = fuel input$_1$ − fuel input$_2$

$$= \frac{Btu_{out}}{E_1} - \frac{Btu_{out}}{E_2}$$

$$= Btu_{out} \times (1/E_1 - 1/E_2)$$

- Annual fuel savings = steam lb/h × H ×

$$8,760 \times C_f \times (1/E_1 - 1/E_2)$$

Fuel savings at constant steam rate   Fig. 1

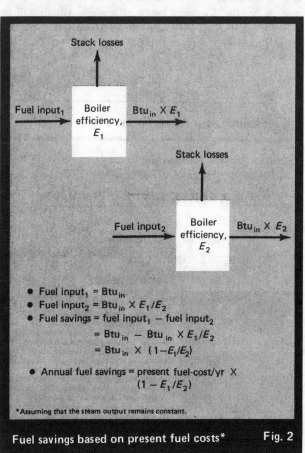

- Fuel input$_1$ = Btu$_{in}$
- Fuel input$_2$ = Btu$_{in}$ × E_1/E_2
- Fuel savings = fuel input$_1$ − fuel input$_2$

$$= Btu_{in} - Btu_{in} \times E_1/E_2$$

$$= Btu_{in} \times (1 - E_1/E_2)$$

- Annual fuel savings = present fuel-cost/yr ×

$$(1 - E_1/E_2)$$

*Assuming that the steam output remains constant.

Fuel savings based on present fuel costs*   Fig. 2

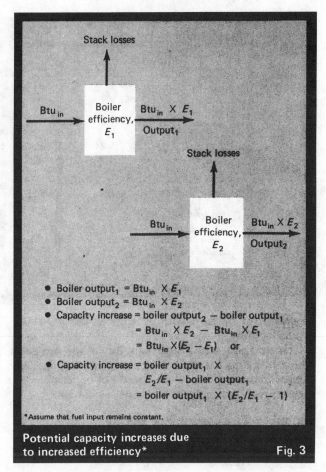

- Boiler output$_1$ = Btu$_{in}$ × E_1
- Boiler output$_2$ = Btu$_{in}$ × E_2
- Capacity increase = boiler output$_2$ − boiler output$_1$

$$= Btu_{in} \times E_2 - Btu_{in} \times E_1$$

$$= Btu_{in} \times (E_2 - E_1) \quad \text{or}$$

- Capacity increase = boiler output$_1$ ×

$$E_2/E_1 - \text{boiler output}_1$$

$$= \text{boiler output}_1 \times (E_2/E_1 - 1)$$

*Assume that fuel input remains constant.

Potential capacity increases due to increased efficiency*   Fig. 3

Nomenclature

a	Fuel factor
A_s	Percentage of excess combustion air
b	Intercept × m
C_f	Fuel cost, \$/10^6 Btu
C_s	Steam cost, \$/1,000 lb
CO_2	Flue-gas carbon dioxide, %
E	Boiler efficiency, %
H	Steam enthalpy, Btu/lb of steam
m	Slope^{-1}
O_2	Flue-gas oxygen, %
T	Net flue-gas temperature, °F

program can be used with or without the PC-100A printer option.

The program and instructions for its use are presented in Tables I and II. Note that variables may be entered in any order, and they need not be reentered if they do not change for subsequent calculations.

Example 1: Economizer saving

Boiler analysis without an economizer is found to be: 5%O_2 in the exit gases, where net flue temperature is 550°F. The economizer vendor claims that a 200°F reduction can be realized.

What is the annual saving if C_f = \$2.79/10^6 Btu

User instruction for loading and running program that calculates boiler efficiency and % excess air.   Table II

Step	Procedure	Enter	Press		Display
1	Partition calculator	2	2nd OP	17	799.19
2	Enter program, sides 1–4	1			1
		2			2
		3			3
		4			4
3	Enter variables in any order:				
	a. %O$_2$	%O$_2$		A	O$_2$
	b. T, net flue-gas temperature	T		B	T
	c. C_f, fuel cost $/10^6 Btu*	C_f		D	C_f
	d. H, steam enthalpy Btu/lb steam*	H	2nd	D'	H
4	Enter fuel type:				
	a. No. 2 fuel oil		2nd	A'	0
	b. Natural gas		2nd	B'	0
5	Calculate efficiency			E	A_x flashes 3 times, stops showing E
6	Calculate C_s, steam cost/1,000 lb		2nd	E'	H flashes 3 times, stops showing C_s

*These variables are required only for steam cost calculations.

(No. 2 fuel oil), and the capacity of the boiler is 100,000 lb/h steam ($H = 1,160$ Btu/lb)?

Procedure: 1. Find E_1.
          2. Find E_2.
          3. Calculate saving.

Find E_1	Find E_2
BOILER! Existing	BOILER! With economizer
DATE!	DATE!
LOAD!	LOAD!
FUEL! *2 OIL	FUEL! *2 OIL
5. % O2	5. % O2
550. T, 'F	350. T, 'F
29.44002625 % XS	29.44002625 % XS
80.6386846 %EFF	85.82422822 %EFF

Solution: Annual fuel saving
$= $ steam lb/h $\times H \times 8,760 \times C_f \times (1/E_1 - 1/E_2)$

$$= \frac{100,000 \times 1,160 \times 8,760 \times 2.79 \times 0.0749}{10^6}$$

$$= \$212,348$$

Equations for coal and #6 fuel oil

The above program can be modified to accommodate two more fuels by changing the appropriate steps in sections 5, 6 and 10 of the program. For low-sulfur coal, use the following:

$$a = 4.4477 + 0.025446 \times \%O_2$$
$$m = -(0.0226225 + 0.00015719A_x)^{-1}$$
$$b = 96.8609 (0.0226225 + 0.00015719A_x)^{-1}$$

For #6 fuel oil, use the following:
$$a = 4.3957 + 0.7078/\%O_2$$

$$\log(-m) = -0.0027A_x + 1.6544$$

$$\log(b) = -0.0027A_x + 3.6345$$

For HP-67/97 users

The HP version of the program operates in a very similar manner to the TI version, except that there is no alphabetic printout with the former. Table III provides a listing of the HP program, and Table IV gives user instructions. Table V lists the constants that must be stored prior to running the program. Note: At step 4, the HP-97 prints %O$_2$, T, A_x, and % efficiency. With the HP-67, %O$_2$, T, and A_x will appear briefly, followed by % efficiency, which will remain in the display. At step 5, the HP-97 prints C_f and C_s. The HP-67 flashes C_f, followed by C_s.

	User instructions for HP version			Table IV
Step	Procedure	Enter	Press	Display
1	Load program			
2	Enter variables in any order:			
	a) % O$_2$	%O$_2$	A	%O$_2$
	b) T, Net flue-gas temperature, °F	T	B	T
	c) C_f, Fuel cost, $/10^6$ Btu*	C_f	D	C_f
	d) H, Steam enthalpy Btu/lb steam*	H	d	H
3	Enter fuel type (select one only):			
	a) No. 2 fuel oil, or		a	0
	b) Natural gas		b	1
4	Calculate efficiency, E		E	See te
5	Calculate C_s, steam cost/1,000 lb		e	See te

*These variables are required only for steam cost calculations.

Listing of boiler efficiency program—HP version   **Table III**

Step	Key	Code	Step	Key	Code	Step	Key	Code	Step	Key	Code	Step	Key	Code
001	*LBL0	21 00	046	CLX	-51	091	1	01	136	PRTX	-14	181	.	-62
002	P⇄S	16-51	047	RCL1	36 01	092	.	-62	137	RCL9	36 09	182	6	06
003	F1?	16 23 01	048	X=Y?	16-33	093	6	06	138	PRTX	-14	183	4	04
004	GTO1	22 01	049	GTO0	22 00	094	6	06	139	CLX	-51	184	3	03
005	0	00	050	1	01	095	4	04	140	SPC	16-11	185	2	02
006	R/S	51	051	R/S	51	096	0	00	141	RCL9	36 09	186	+	-55
007	*LBL1	21 01	052	*LBLE	21 15	097	3	03	142	R/S	51	187	10^x	16 33
008	1	01	053	RCLA	36 11	098	+	-55	143	*LBLC	21 13	188	STO8	35 08
009	R/S	51	054	ENT↑	-21	099	10^x	16 33	144	CLX	-51	189	GTOc	22 16 13
010	*LBLA	21 11	055	.	-62	100	CHS	-22	145	.	-62	190	*LBLe	21 16 15
011	STOA	35 11	056	0	00	101	STO7	35 07	146	0	00	191	RCLC	36 13
012	R/S	51	057	4	04	102	.	-62	147	0	00	192	ENT↑	-21
013	*LBLB	21 12	058	7	07	103	0	00	148	2	02	193	RCLD	36 14
014	STOB	35 12	059	6	06	104	0	00	149	7	07	194	X	-35
015	R/S	51	060	X	-35	105	2	02	150	7	07	195	SPC	16-11
016	*LBLD	21 14	061	CHS	-22	106	5	05	151	4	04	196	SPC	16-11
017	STOC	35 13	062	1	01	107	2	02	152	6	06	197	SPC	16-11
018	R/S	51	063	+	-55	108	2	02	153	CHS	-22	198	.	-62
019	*LBLd	21 16 14	064	STO6	35 06	109	5	05	154	ENT↑	-21	199	1	01
020	STOD	35 14	065	RCL2	36 02	110	CHS	-22	155	RCL6	36 06	200	X	-35
021	R/S	51	066	ENT↑	-21	111	ENT↑	-21	156	X	-35	201	RCL9	36 09
022	*LBLa	21 16 11	067	RCLA	36 11	112	RCL6	36 06	157	1	01	202	÷	-24
023	SF0	16 21 00	068	X	-35	113	X	-35	158	.	-62	203	STO0	35 00
024	CF1	16 22 01	069	RCL3	36 03	114	3	03	159	6	06	204	CLX	-51
025	2	02	070	+	-55	115	.	-62	160	6	06	205	RCLC	36 13
026	2	02	071	RCLA	36 11	116	6	06	161	7	07	206	PRTX	-14
027	8	08	072	X	-35	117	2	02	162	9	09	207	F1?	16 23 01
028	8	08	073	RCL6	36 06	118	2	02	163	2	02	208	P⇄S	16-51
029	2	02	074	÷	-24	119	6	06	164	+	-55	209	RCL0	36 00
030	X⇄Y	-41	075	STO6	35 06	120	+	-55	165	10^x	16 33	210	DSP3	-63 03
031	CLX	-51	076	F0?	16 23 00	121	10^x	16 33	166	CHS	-22	211	PRTX	-14
032	RCL1	36 01	077	GTOC	22 13	122	STO8	35 08	167	STO7	35 07	212	DSP9	-63 09
033	X=Y?	16-33	078	SPC	16-11	123	*LBLc	21 16 13	168	.	-62	213	SPC	16-11
034	GTO0	22 00	079	.	-62	124	CHS	-22	169	0	00	214	SPC	16-11
035	0	00	080	0	00	125	ENT↑	-21	170	0	00	215	SPC	16-11
036	R/S	51	081	0	00	126	RCLB	36 12	171	2	02	216	RTN	24
037	*LBLb	21 16 12	082	2	02	127	+	-55	172	7	07	217	R/S	51
038	CF0	16 22 00	083	5	05	128	RCL7	36 07	173	0	00			
039	SF1	16 21 01	084	7	07	129	÷	-24	174	7	07			
040	1	01	085	6	06	130	STO9	35 09	175	3	03			
041	9	09	086	7	07	131	RCLA	36 11	176	CHS	-22			
042	4	04	087	CHS	-22	132	PRTX	-14	177	ENT↑	-21			
043	5	05	088	ENT↑	-21	133	RCLB	36 12	178	RCL6	36 06			
044	8	08	089	RCL6	36 06	134	PRTX	-14	179	X	-35			
045	X⇄Y	-41	090	X	-35	135	RCL6	36 06	180	3	03			

Data storage locations—HP version   **Table V**

Storage area		Value
Primary	1	19458
	2	0.010208
	3	4.43562
	6	A_x*
	9	% Eff*
	A	%O_2
	B	T, (°F)
Secondary	1	22882
	2	−0.026942
	3	4.5557
	6	A_x*
	9	% Eff*

*Indicates values calculated by the program.

References

1. Fundamentals of Boiler Efficiency, Lubetext D250, The Exxon Corp., 1976.
2. Sisson, Bill, Combustion Calculations for Operators, *Chem. Eng.*, June 10, 1974, p. 106.
3. Shinskey, F. G., "Energy Conservation Through Control," Academic Press, New York, 1978.
4. Schmidt, Charles M., Finding Efficiencies of Stoker-Fired Boilers, in "The 1977 Energy Management Guidebook," by the editors of *Power* magazine. McGraw-Hill, Inc., p. 85.

The author

Terry A. Stoa is Plant Engineer for ADM Corn Sweeteners, Inc., a division of Archer Daniels Midland Co. P.O. Box 1470, Decatur, IL 62526. His responsibilities at the wet-corn-milling facility include plant expansion, energy conservation, process control, and providing technical assistance to the production group. He also has had experience in soybean solvent-extraction and synthetic-fiber spinning. He received a bachelor's degree in chemical engineering from the University of North Dakota and is a member of AIChE.

Boiler circulation during chemical cleaning

Present practices for inducing boiler circulation by heating can cause problems. Here is why, plus some hints for doing a better cleaning job.

Barry O. Shorthouse, Consultant

☐ During chemical cleaning of a large boiler, using chelants, thermal cycling is frequently applied to stimulate circulation. But there may be low flowrates in various sections of the water wall.

The water-tube temperatures, fluid velocities, densities and concentration show considerable variation. The tube velocity controls the residence time of any particular fluid element and the supply of unreacted chelant. Because the solvation process is reversible and concentration-dependent on the chelant, and the coordination compounds are temperature-sensitive, a fairly precise control of temperature and heat input is important. Otherwise, material dissolved in one place may possibly be precipitated in another.

Using thermal cycling

Typically, at the start of a cycle, the steam-drum metal temperature is approximately 260°F. By means of a heat lance (or some other auxiliary heat input), the temperature is raised to about 300°F, and sometimes as high as 320°F. By successive heatings and coolings, some form of circulation should occur that will enable unreacted chelant to contact iron-oxide deposits.

However, certain restraints exist. The temperature should not exceed 320°F because of thermal degradation of both fresh chelant and the iron complex. The concentration of free complexing agent should not fall below a minimum value, normally 1–2%, to avoid a reverse reaction referred to as "dropping out" or precipitating.

Water-tube velocity

A simplified analysis shows some predictable and some surprising consequences of this procedure. The time taken for heating is typically 2.5 h, and the time for cooling from 320°F to 260°F is about 2 h. These times may be longer, but are seldom shorter.

We may write $Q = mC_p\,\Delta T$ or: $\dot{q} = \dot{m}C_p\,\Delta T$

where $\dot{q}$ = rate of heat input; $\dot{m}$ = mass flow of water; C_p = heat capacity; and ΔT = temperature rise.

Assembling the components:

$$\dot{q} = \dot{m}C_p\,\Delta T;\ m = \rho V\ \text{or}\ \dot{m} = \rho_{av}\cdot\dot{V}\ \text{and}\ \dot{V} = uA$$

so: $\dot{m} = u\rho_{av}\cdot A$ and $\dot{q} = u\rho_{av}A\cdot\Delta T$ where $C_p = 1$

Originally published August 22, 1983

36

Further, $\Delta P = Ku^2 = \sum\left(\dfrac{1}{2}\rho u^2 + 4\left(\dfrac{R}{\rho u^2}\right)\dfrac{l}{d}\rho u^2\right)$

$$= \sum \rho u^2\left(\dfrac{n}{2} + 4\left(\dfrac{R}{\rho u^2}\right)\dfrac{l}{d}\right)$$

also $\Delta P = hg\,\Delta\rho$

Combining these equations:

$$\dfrac{h\cdot g}{u\cdot\rho_{av}A}\dfrac{\Delta\rho}{\Delta T} = \dfrac{Ku^2}{\dot{q}},\ \text{or}\ u^3 = \dfrac{hg\dot{q}}{K\rho_{av}A}\cdot\dfrac{\Delta\rho}{\Delta T}$$

From steam tables:

$$\dfrac{\Delta\rho}{\Delta T} = 0.03204\ \text{lb/(ft}^3)(°F);\ \rho_{av} = 57.58\ \text{lb/ft}^3$$

K, A and h depend on the geometry of the plant (Fig. 1).

For a typical 1-million-lb/h boiler, the holdup of water is approximately 360,000 lb, and the effective thermal mass of the metal containment, tubes, downcomers, headers, steam drums, etc., approximately doubles this value.

Therefore $\dot{q}$, in Btu/s,

$$= \dfrac{720,000 \times 1 \times (320 - 260)}{2.5 \times 3,600} = 4,800\ \text{Btu/s}$$

or 2,400 Btu/s for water alone

This compares with a rate of over 300,000 Btu/s in the radiant section alone during normal operation.

The heat transferred into or out of the water can only go via the water-wall tubes. As the temperature in the tubes rises, the liquid density falls, and a density gradient induces circulation between the water in the walls and the rest of the system. The density gradient can be expressed as:

$$\Delta P = hg\,\Delta\rho$$

where h = height of boiler; g = gravitational constant; $\Delta\rho$ = density difference between the fluid in the water walls and the downcomer steam drum, etc.; and ΔP = pressure difference between hot and cold sides.

The thermal lift, or Grashof force, is resisted by the fluid forces of viscosity and kinetic-head loss. It may be simply expressed as:

$$\Delta P = Ku^2$$

where K = a constant; and u = mean fluid velocity.

Nomenclature

A	Area of flow, ft^2
C_p	Heat capacity, Btu/(lb)(°F)
g	Gravitational constant, 32.2 ft/sec^2
h	Height of boiler, ft
K	Constant, lb/ft^3
m	Mass flow of water lb/s
n	Number of velocity heads lost, dimensionless
ΔP	Pressure differential between fluid in water walls and downcomer steam drum, etc. = $hg\Delta\rho$, $lb/ft\ s^2$
q	Quantity of heat input, Btu
$\dot{q}$	Rate of heat input, Btu/s
R	Resistance to flow per unit area of pipe surface = $lb/(ft)(s)^2$
$R/\rho u^2$	Friction factor, dimensionless
ΔT	Temperature rise, °F
u	Mean fluid velocity, ft/s
v	Volume of flow, ft^3/s
$\Delta\rho$	Density difference between hot and cold sides, lb/ft^3

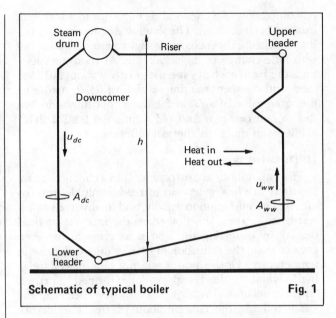

Schematic of typical boiler   **Fig. 1**

For a 10^6-lb/h boiler, the physical data are: $h = 112$ ft; Downcomer internal diameter = 4.12 in. = 0.3433 ft; Water-wall-tube internal diameter = 2.44 in. = 0.2033 ft; Average downcomer length = 155 ft; Average water-wall-tube length = 186 ft; Average downcomer bends, exits, entry loss in velocity heads = 1.5; Average water-wall loss in velocity heads = 6.0.

$$\Delta P = \sum \left[57.58u_{dc}^2\left(\frac{1.5}{2} + 4(0.002)\frac{155}{0.3433}\right)\right] +$$
$$\left[57.58u_{ww}^2\left(\frac{6}{2} + 4(0.002)\frac{186}{0.2033}\right)\right]$$
$$= 251.1u_{dc}^2 + 594.1u_{ww}^2$$

Now $u_{dc} \cdot A_{dc} = u_{ww} \cdot A_{ww}$

$$\therefore u_{dc}^2 = \left(\frac{20.66}{10.55}\right)^2 u_{ww}^2 = 3.81u_{ww}^2$$

where: A_{dc} = flow area of downcomer = 10.55 ft^2; and A_{ww} = flow area of water wall = 20.66 ft^2, so that:

$$\Delta P = 3.81 \cdot 251.1u_{ww}^2 + 594.1u_{ww}^2 = 1,551u_{ww}^2$$

Substituting values:

$$u_{ww}^3 = \frac{112 \cdot 32.2 \cdot 2,400 \cdot 0.3204}{1,551 \cdot 57.58 \cdot 20.60}; \quad u_{vw} = 0.532 \text{ ft/s}$$

$$\text{Reynolds number} = \frac{ud\rho}{\mu} = \frac{0.532 \times 0.2033 \times 57.58}{0.3 \times 0.672 \times 10^{-3}}$$
$$\doteq 30,000$$

This suggests fully turbulent flow, a volume flowrate of 10.4 ft^3/s through the water walls, a residence time in the tubes of 6 min, and a circulation time of 10 min.

Variation in flowrate

However, there are problems. The circulation velocity depends upon the value of K, which depends on both the geometry of the system and the density change with temperature. Large variations exist in downcomer lengths, number of bends, and header-exit and -entry losses in any particular section of a boiler.

A greater variation is seen with water-wall tubes, particularly those making up the nose, screen and roof of the boiler. Choosing more-appropriate geometric values for, say, individual front- and rear-wall tubes, and assuming equal energy inputs per unit area, more-realistic values are 1.13 ft/s and 0.19 ft/s, almost an order of magnitude difference between individual tubes. (Assuming equal heat input per unit area is challenged later.)

The analysis so far has concerned the cycle while the temperature is rising. When cooling commences, the fluid in the water wall is cooled, becomes more dense, and descends. Flow is reversed. The process is not instantaneous—not only has the momentum of the fluid to be dissipated, but the structure and the water have a thermal inertia. The situation is complicated owing to uncertain heat-transfer coefficients, and so on, but estimates calculated (and actually measured) suggest values of 5 to 10 min before significant flow reversal is established.

There is an inherent danger in measuring the steam drum metal-temperature only, as this is an average temperature of drum and contents and not a direct measure of the fluid temperature in a particular water-wall tube.

A better indication is obtained from measurements taken at roof-tube, riser or upper-header locations. Once flow has been established, the temperature differential over the length of the tubes may be about 5°F; but at the outset of a cycle or a flow reversal, the tube water-temperature may differ from that of the steam drum by 50°F or more.

Density change with composition

It has been shown that to establish a flow velocity of, say, 0.5 ft/s, with a given resistance to flow in the water tubes, requires a density difference of 0.124 lb/ft^3, established by thermal input. However, the density of the fluid is also determined by its composition.

Fresh chelant solution is charged into the boiler and,

on completion of a chemical cleaning, up to 4,000 lb of iron has been solvated. The solution density changes primarily because the molecular weight almost doubles, but with little change in molar volume. At room temperature, the fresh chelant's specific gravity is about l.0175 as used, and the spent solution's is about 1.0332. Assuming the same value of $d\rho/dt$ as for water, not unreasonable, the $d\rho$ between new and old solution is 0.9072 lb/ft^3, about seven times the thermal difference.

Implications

The implications are surprising. If in a chemical cleaning there are heavy deposits of readily soluble iron oxides that go into solution rapidly, and the hydrodynamic resistance restricts the flow, then the increase in fluid density by dissolving the iron is as great as, or even greater than, the reduction in density due to the rise in temperature. Flow upwards will not take place, and a hydrodynamic "dead leg" may be formed. Further, "spent" chelant still retains 2% or so free chelant in the water wall. If the concentration of free chelant approaches zero, the solvated iron or copper may be thrown out of solution—a so-called "dropping out." The possibility exists of the density increase due to solvation overcoming the weak circulating force caused by a modest temperature increase. Further, should the heat input not be uniform throughout the furnace chamber, then the thermal circulating force may be weak and the flow nonexistent. Confirmation of low flow is shown in downcomer-tube surface-temperatures during some cleanings. Water-wall tubes may even act as secondary downcomers in certain circumstances.

The net result is that circulation throughout the boiler during some chemical cleaning is uncertain, and the problem of ensuring circulation during an air blow is even more complex. Iron and copper may be solvated from one part of the boiler and, due to long residence times and a local depletion of chelant, redeposited in another part. The variability in the burden of iron and copper removed in successive cleanings may in some part be accounted for by being dependent on the heating pattern and the flow conditions induced. If the solvation rate is purely chemically controlled, then adequate concentration of free chelant must be maintained to thwart reversibility. If the reaction is diffusion-controlled, then

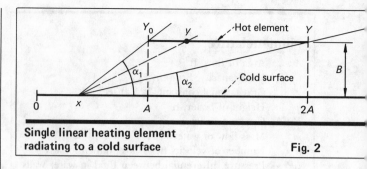

Single linear heating element radiating to a cold surface   **Fig. 2**

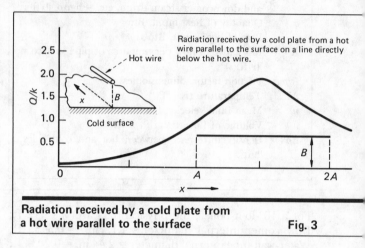

Radiation received by a cold plate from a hot wire parallel to the surface on a line directly below the hot wire.

Radiation received by a cold plate from a hot wire parallel to the surface   **Fig. 3**

the flowrate, Reynolds number and boundary-layer thickness become dominant features.

Thermal flux patterns

During a chemical cleaning operation, the rate at which heat is added to the boiler system is less than one-fiftieth of the heat input during the normal operation. Several methods are used to supply the heat—such as use of an auxiliary oil or gas lance or a very brief firing of one of the coal mills. Furnace gas circulation patterns may not be fully established and it is highly probable that these patterns would not be similar to those in normal operation. To estimate the water circulation, it is also necessary to know the thermal inputs to various sections of the water walls.

A simple method can be used to estimate these inputs.

Data for problem concerning energy from a hot plate			Table I
x	(Q/k)	z	$(Q/k)_z$
0.0	0.0022	0.0	0.075
0.5	0.041	0.5	0.241
1.0	0.091	1.0	0.894
1.5	0.273	1.8	1.414
2.0	0.970	2.0	0.894
2.5	1.655	2.5	0.241
3.0	1.788	3.0	0.0757
3.5	1.655	3.5	0.0318
4.0	0.970	4.0	0.0162
4.5	0.273	4.5	0.009
5.0	0.091	5.0	0.005

Product $(Q/k)_x$ by $(Q/k)_z$							Table II
	$z \longrightarrow$						
		0.0	0.5	1.0	1.5	2.0	2.5	3.0
x	0.0	0.001	0.005	0.019	0.031	0.019	0.005	0.001
	0.5	0.003	0.009	0.036	0.057	0.036	0.009	0.003
	1.0	0.006	0.021	0.081	0.128	0.081	0.021	0.006
	1.5	0.020	0.065	0.244	0.386	0.244	0.065	0.020
	2.0	0.072	0.233	0.867	1.371	0.867	0.233	0.072
	2.5	0.124	0.398	1.479	2.340	1.479	0.398	0.124
	3.0	0.134	0.430	1.598	2.528	1.598	0.430	0.134
	3.5	0.124	0.398	1.479	2.340	1.479	0.398	0.124
	4.0	0.012	0.233	0.867	1.371	0.867	0.233	0.072
	4.5	0.020	0.065	0.244	0.386	0.244	0.065	0.020
	5.0	0.006	0.021	0.081	0.128	0.081	0.021	0.006

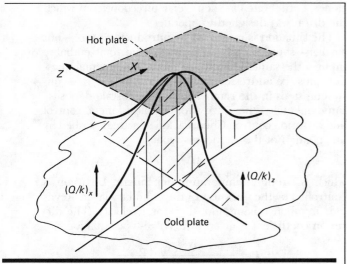

Radiation received by a cold plate from a hot plate parallel to the surface   **Fig. 4**

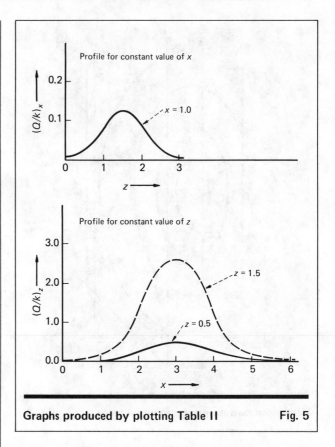

Graphs produced by plotting Table II   **Fig. 5**

The analysis, when compared with, say, the FURDEC program of the CEGB (the U.K.'s Central Electricity Generating Board) is not sophisticated (even being primitive), but has the advantages of being easy to apply, being quickly executed, and providing insight to those who are not furnace experts. The results of such an analysis lead to conclusions that are partly qualitative, but are not at variance with more-complex programs. This provides a conceptual framework that eases the understanding of more-precise approaches.

Consider Fig. 2, which depicts a single linear heating-element radiating thermally to a cold surface.

$$\text{Energy received at Point } x = dq = k \sin \alpha$$

where k = some constant that embraces emissivity, temperature difference, etc.

As y goes from Y_0 to Y, α goes from α_1 to α_e

$$\therefore dq = Q = k \int_{\alpha_1}^{\alpha_2} \sin \alpha \, d\alpha = k(\cos \alpha_2 - \cos \alpha_1)$$

but $\cos \alpha_2 = \dfrac{(2A - x)}{\sqrt{B^2 + (2A - x)^2}}$

and $\cos \alpha_1 = \dfrac{(A - x)}{\sqrt{B^2 + (A - x)^2}}$

or $Q = k\left[\dfrac{(2A - x)}{\sqrt{B^2 + (2A - x)^2}} - \dfrac{(A - x)}{\sqrt{B^2 + (A - x)^2}}\right]$

A plot of Q/k against x for a given separation of B is given in Fig. 3.

If a hot plate replaces the wire, then a similar analysis—at right angles to the original axis—yields expressions of the same form. The product of these two expressions yields the total energy received by the plate. Fig. 4 clarifies the point.

The data are given in Table I with the values:

$$A_x = 2 \qquad B_x = 0.5 \text{ for the } x \text{ direction}$$
$$A_z = 1 \qquad B_z = 0.5 \text{ for the } y \text{ direction}$$

It should be noted that Q/k is solely an accommodation coefficient, which is a measure of the diminished radiation received due to geometric position. It is akin to burn marks produced by some inefficient toasters.

Table II compiles the product of these factors; from this, a contour map may be drawn, which shows lines of constant thermal flux. The values give rise to a series of concentric contours of elliptical shape, with the geometric center coincident with that of the hot plate. The gradient of the contours is not constant, as may be seen from a plot of $(Q/k)_x$ at constant z and $(Q/k)_z$ at constant x. Fig. 5 shows typical values for such a system.

Let us return now to the partial firing of a boiler during a chemical cleaning operation, and consider the flame—or combustion zone—produced by an oil lance. The flame is luminous, reasonably well defined, and occupies a volume contained by a space 5 ft × 8ft × 12 ft. It would present an area of approximately 5 ft × 8 ft at about 12 ft distant. The geometry is obviously not as well defined as the hot plate discussed above. However, the basic principle is identical; the energy emitted from this source is subject to the same geometric considerations as that of the hot plate.

Suppose a contour map of equal flux intensities were projected onto the surface of the wall, using the same computational methods as with the hot plate and the cold surface. This is shown in Fig. 6. Consider three water-wall tubes AA', BB', and CC'. Tube AA', while rising through the boiler, will meet an increasing thermal flux until it reaches the maximum value near the center of the thermal-flux map. It will have the maximum heat input of the three tubes.

Tube BB', although exposed for a longer period, does not experience the same high thermal intensities. Tube

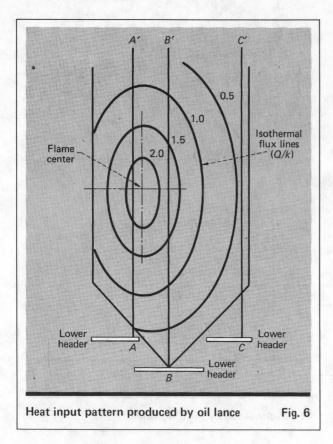

Heat input pattern produced by oil lance   **Fig. 6**

CC', with the shortest exposure path, encounters the lowest intensities.

The total quantity of heat received by a tube is the integral of the thermal flux and the length of travel through the boiler.

If the water enters the headers at the same temperature, the mean temperature-rise will not be the same for all three tubes. The more asymmetric the firing pattern, the greater the exit water-temperature difference. It has been shown that the mean velocity in a water-wall tube is:

$$u^3 = \frac{h \cdot g \cdot \dot{q}}{K \cdot \rho_{av}} \cdot \frac{\Delta \rho}{A \, \Delta T}$$

where h = height of furnace tube; K = flow resistance constant; ρ_{av} = average fluid density; A = area for flow; g = gravitational constant; $\Delta \rho$ = density change with temperature; and $\dot{q}$ = heat input rate.

Lower values of $\dot{q}$ will reduce the circulation velocity in the tubes. An incomplete axial mix in the steam drum will create greater flow divergence. There will be a contribution of convective heat transfer at the roof and screen tubes but the tube spacing and pitch are arranged for considerably higher gas velocities and volumes. The pressure drop will be minimal and will not redistribute the combustion gases significantly to the cooler tubes. Basically, the energy required for pumping fluid around any particular part of the circuit is obtained from the heat transferred into the water. Low heat input means low circulation rate.

During the reverse of the cycle—that is, when the furnace is cooled with ambient air—flow is reversed by extracting heat from the furnace walls. It is arguable that the resistance to flow will be considerably greater as the liquid negotiates the steam-drum furniture (e.g., cy-

clones, baffles, etc.) in the reverse direction from which the drum was designed to operate.

The situation is improved by introducing more—but smaller—sources of heat, provided the total enthalpy input is the same. However, a more fundamental question must be addressed, mainly, what are the rate-controlling steps in the removal of iron and scale deposits? Fundamentally, there are two, chemical kinetic control and diffusion control. Chemical kinetics are described by an equation of the Arrhenius type:

$$k = Ae^{-E/RT}$$

which is strongly temperature-dependent. Diffusion is controlled by the thickness of the boundary layer, Reynolds number, mixing and, therefore, velocity. The diffusion coefficient is given by a Wilke-type equation:

$$D_{AB} = \frac{7.4 \times 10^{-8}(X_B M_B)^{1/2}T}{\mu V_A^{0.6}}$$

which is more moderately temperature-sensitive. Laboratory experiments show that in a bomb reaction vessel, both temperature and stirring rates influence the solvation of iron-oxide test samples. Thus, oxide removal will be controlled by firing patterns and will lead to inconsistent removal of oxide, unless the reactions are allowed to go to completion.

Remedies

Power-plant engineers and cleaning vendors are urged to clean boilers as quickly as possible; little tolerance is shown to procedures that prolong the outage time. Some actions will help in doing a good cleaning job:

■ The use of more, but smaller, sources of thermal input, positioned as low as possible within the furnace, and sized to the thermal path that that particular section of water wall has to traverse (Fig. 6).

■ The temperature of each section of water wall should be monitored and the heat input to that section controlled by that temperature. This may require greater heating inputs for sections with more resistance to flow, due to either hydrodynamic or density forces.

■ Samples for chemical analysis should be taken from the headers of sections suspected of restricted flow. The conventional steam-drum sample should not be the sole indicator of the progress of a chemical cleaning.

■ If possible, auxiliary pumping should be used to assist circulation in the most difficult sections. In general, forced-circulation chemical cleanings present fewer problems than do natural-circulation ones.

The author

Barry O. Shorthouse, 1313 Hollywood Drive, Monroe, MI 48161, is a consultant to utilities on steam raising, alternative fuel systems and chemical control. He studied chemistry at Southampton and Loughborough Universities, and chemical engineering at Imperial College, London, and Cambridge University, Cambridge, England, from which he received a Ph.D. He has taught in universities in both the U.S. and U.K. and is a member of AIChE, the Institution of Chemical Engineers (London), the Engineering Soc. of Detroit and the Royal Institute of Chemistry, London.

Venting requirements for deaerating heaters

Here is how boiler-water deaerators work, plus ways of determining how much steam will be vented and how to save energy.

Arthur C. Knox, Jr., HELEX Div., A. C. Knox, Inc.

☐ Deaerating heaters use steam stripping to remove dissolved gases from boiler feedwater. (If the gases are not removed, boiler corrosion results.) Steam is continuously bled off from the deaerator to scavenge noncondensable gases, a technique that also wastes energy. This article addresses the ways of minimizing energy losses.

There are four basic procedures for treating boiler feedwater to remove corrosive gases: 1) chemical, 2) physical, 3) physical and chemical, and 4) deionizing. The most cost-effective method for most boilers is the combination of physical and chemical treatment (deionized water is used in utility-sized boilers).

Chemical treatment

Dissolved oxygen can be removed chemically from boiler feedwater. This method is used primarily in small commercial boilers—generally for boilers of less than 300 horsepower capacity (10,000 lb/h of steam) operating on an intermittent basis. Chemical treatment removes oxygen only, not carbon dioxide. This carbon dioxide, most of which is created by the thermal decomposition of bicarbonates and carbonates, results in corrosion of steel through the action of dissolved CO_2 (carbonic acid).

The first of the chemical reagents used to scavenge oxygen was sodium sulfite, 7.88 parts of which react with 1 part of oxygen (by weight).

$$2 Na_2SO_3 + O_2 \longrightarrow 2 Na_2SO_4$$

In practice, 10 parts of sodium sulfite are added to assure removal of the oxygen. This type of treatment adds to the dissolved-solids content of the boiler water, thus requiring additional blowdown to maintain a safe operating level of dissolved solids.

Another drawback is possible thermal breakdown of sodium sulfite into sulfur dioxide and hydrogen sulfide gases. Hence, sodium sulfite is not recommended by one water-treatment-chemical manufacturer, in high-pressure boilers above 1,800 psi, and not recommended above 650 psi by another manufacturer. However, sodium sulfite decomposition can take place in boilers of medium pressure due to "hot spots" in the radiant section of the boiler, and also when thick scale formation causes high skin temperatures sufficient to decompose the sulfite. For this reason, it is important not to use too great an excess of sulfite. With the advent of its use as a rocket propellant, hydrazine became available as a scavenger for oxygen. One part, by weight, of hydrazine reacts with one part of oxygen.

$$N_2H_4 + O_2 \longrightarrow N_2 + 2 H_2O$$

Hydrazine has an advantage over sulfite in that its use results in the formation of an inert gas instead of a solid. Generally, hydrazine is preferred in boiler operations at pressures above 1,000 psi, if the possibility of sulfite breakdown is to be avoided.

Manufacturers of chemical oxygen scavengers also provide these chemicals with added catalysts that help speed up the reaction rate with oxygen. Heavy-metal salts such as cobalt sulfate and cobalt chloride are added to improve reaction speed.

Physical treatment

To remove dissolved gases for boiler feedwater, particularly for industrial boilers operating continuously, steam stripping (a physical treatment), supplemented by

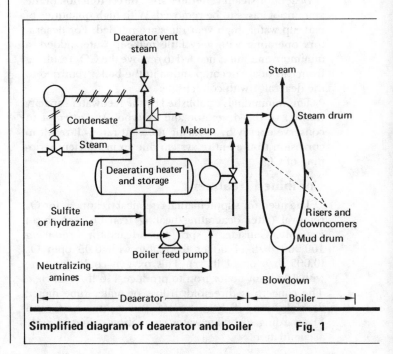

Simplified diagram of deaerator and boiler **Fig. 1**

Originally published January 23, 1984

chemical treatment, usually is employed. Deaerating heaters, commonly called "deaerators" are designed to strip the condensate return and makeup water (used to replace both blowdown and steam that is not returned as condensate).

These deaerators consist of a horizontal or vertical tank that acts as a feedwater reservoir, and a vertical section—the actual deaerator (see Fig. 1). This deaerator section contains spray nozzles located at the top, through which the makeup water is sprayed into rising steam (injected for heating and purging).

The most efficient deaerator from the viewpoint of degassing ability is the spray-tray type, which contains a number of contacting trays (that may be of various designs). Another type employs packing instead of trays. The least efficient type contains sprays only—neither packing nor trays—although manufacturers have developed spray improvements that produce results approaching tray-type oxygen removal.

Live steam passes into the deaerator to raise the water temperature and thereby raise the operating pressure of the deaerator. Generally a pressure of 5 psig is used in industrial deaerators, but others work at 15 psig and some at atmospheric pressure. As steam rises through the deaerator, countercurrent to the downflowing water, it strips out the noncondensable gases.

An excess of steam—beyond that required to supply sensible heat—is used to convey the heavy corrosive gases upward, and eventually to the atmosphere. The continuous venting of steam and gases is accomplished by a manually controlled valve atop the deaerator.

The water spray is the most cost-effective section because the heating is by direct contact and is capable of removing 75% to 95% (by weight) of the dissolved gases. The remaining 25% to 5% cannot be totally removed because the steam and condensate lower the pH of the makeup, which shifts the carbonate/bicarbonate equilibrium, resulting in the locking of a significant part of the free CO_2 into bicarbonate ions.

Deaerator steam vent rates are a direct function of the amount of gases to be removed. With high quantities of makeup water, high vent rates are needed. For deaerators operating with very little makeup water added, a minimum amount is needed to remove the CO_2 resulting from carbonate decomposition in the boiler returned to the deaerator with condensate.

This minimum is established by the feedwater capacity (rating) of the deaerator and its type. Any attempt to conserve energy by reducing the vent rate will result in corrosion to the entire system due to insufficient stripping of CO_2.

Combined treatment

The need for supplemental chemical treatment for O_2 removal (after deaerating/heating) can be seen in an example. Consider a spray-type deaerator producing 100,000 lb/h of boiler feedwater with 0.05 ppm O_2 (0.005 lb/h or 44 lb/yr). This oxygen could combine with 102 lb of boiler iron to produce 146 lb/yr of rust. The concern is threefold: 1) loss of boiler tube metal; 2) reduction of heat transfer owing to scale; and 3) retarding of boiler water circulation owing to scale accumulation.

Non-condensable gases present in deaerator vent streams	
Gas	**Source**
Oxygen	Air dissolved in makeup water
	Air dissolved in returned condensate
Nitrogen	Air dissolved in makeup water
	Air dissolved in returned condensate
	Reaction product of hydrazine and oxygen
Carbon dioxide	Carbon dioxide dissolved in makeup water
	Condensate returned
	Reaction product in thermal breakdown of bicarbonates and carbonates
Ammonia	Breakdown of organic nitrogenous compounds in makeup water [3]
	Reaction product in breakdown of hydrazine in presence of excess oxygen requirements [4]
Hydrogen	Reaction product from iron and carbonic acid when pH drops to 5.9 and below
Methane	Reaction of atomic hydrogen with iron carbide in steel
Sulfur dioxide and hydrogen sulfide	Thermal breakdown of sodium sulfite in boiler

Gases

Of the total of eight different gases that may be present in boiler feedwater, oxygen, nitrogen and carbon dioxide constitute the major portion and are always present. For the others, see the table (above).

Hydrogen gas is generated in the system when the water pH drops to 5.9 and below. At this pH, iron and carbonic acid react to form ferrous bicarbonate and hydrogen. At elevated temperatures, the hydrogen-ion concentration in water increases to 20 to 30 times that at atmospheric pressure.

Atomic hydrogen can react with the iron carbide in steel to form methane—a reaction occurring above 430°F [5,6].

O_2 and CO_2 can each react with the steel in boiler systems, and together the mixture is more corrosive than the sum of the two—it can even attack stainless steel and Monel [7]. CO_2 forms carbonic acid in water, which reacts with steel to form soluble ferrous bicarbonate, which is responsible for fictitiously high pH readings of condensate, and frequently fools operators into believing that the condensate is not severely corrosive.

Carbon dioxide

All of the bicarbonates and about 80% of the carbonates decompose thermally [4]. When the carbonate decomposes it yields CO_2 and either the metal or alkali oxide, which in turn hydrolyzes to form the alkaline hydroxide.

Thus the boiler is a producer of alkalinity. It increases the pH of the boiler feedwater from 7 to about 10. But the CO_2 in the steam results in an acid condensate, by forming carbonic acid. The acid condensate, returned to the deaerator, neutralizes the makeup water, reducing its

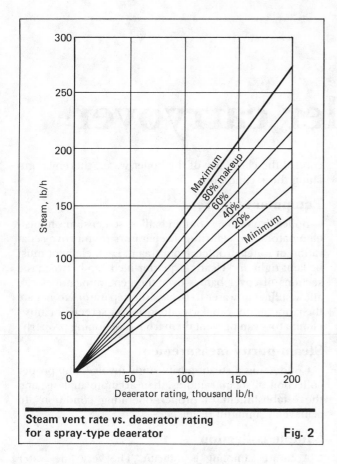

Steam vent rate vs. deaerator rating for a spray-type deaerator **Fig. 2**

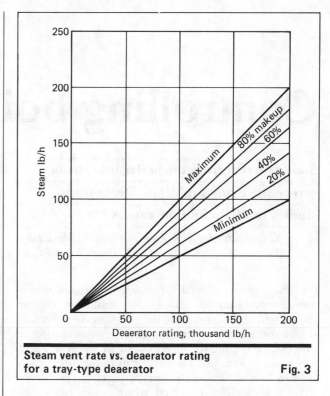

Steam vent rate vs. deaerator rating for a tray-type deaerator **Fig. 3**

pH from about 7.8 to 7.0. Trays in the deaerator will remove the free CO_2.

Neutralizing amines can be injected into the boiler feedwater, ahead of the boiler. These amines, which include cyclohexylamine, benzylamine and morpholine, volatilize in the boiler and, upon condensation, neutralize the carbonic acid, raising the pH above 7.0 to 8.0 or 8.3. These chemicals are added in direct proportion to the amount of free carbon dioxide present or generated in the boiler. Insufficient steam stripping in the deaerator to remove free CO_2 results in extra use of amines.

Deaerator vent steam—average cases

In practice, deaerator vent-steam rates range from 0.05 to 0.14% of the deaerator rating. The actual amount is dependent on: the type of deaerator (which determines efficiency of gas removal); amount of makeup water; properties of makeup water; and condition of condensate return.

To estimate the amount of steam venting from spray- and tray-type deaerators (the two most common types), Figs. 2 and 3 illustrate the amounts for deaerators rated up to 200,000 lb/h. In the case of a 100,000-lb/h unit of the spray type, the steam vent rate varies from a minimum of 70 to a maximum of 140 lb/h. For a tray type of the same size, the rate ranges from 50 to 100 lb/h.

Energy losses

Steam venting from deaerating heaters represents a significant heat loss. It may range from a high of 0.8% down to 0.3% of the energy in the heated feedwater.

Until recently, there has been no practical means of

recovering energy in the vent steam. Industry has not accepted the shell-and-tube heat exchanger for deaerator vent condensing because its configuration, size, weight, and thermal and hydraulic characteristics are not compatible with the deaerator heat and water balance. However, a new type of heat exchanger has become available for recovering energy from the deaerator vent steam, by using it to preheat makeup water [9].

References

1. Korten, E. C., The ASME Boiler and Pressure Code, *TAPPI*, Vol. 37, No. 6, June 1954, p. 242.
2. Othmer, D. F., *Ind. Eng. Chem.*, Vol. 26, p. 576 (1929).
3. Reid, William T., "External Corrosion and Deposits in Boilers and Gas Turbines," Elsevier Pub. Co., Amsterdam, 1971.
4. "Betz Handbook of Industrial Water Conditioning," 7th ed., Betz Laboratories, Inc., Trevose, Pa., 1976.
5. Obrecht, Malvern R., Steam and Condensate Return Line Corrosion: Its Causes and Cures, *Heat./Piping/Air Cond.*, July 1964, p. 117.
6. Tresden, R. S., Guarding Against Hydrogen Embrittlement, *Chem. Eng.*, June 29, 1981, p. 105.
7. Monroe, E. S., Jr., Effects of CO_2 in Steam Systems, *Chem. Eng.*, Mar. 23, 1981, p. 209.
8. Monroe, E. S., Jr., Condensate Recovery Systems, *Chem. Eng.*, June 13, 1983, p. 119, Fig. 6.
9. HELEX Div. of A. C. Knox, Inc., Cincinnati, Ohio.

The author

Arthur C. Knox is the principal of A. C. Knox, Inc., Consulting Engineers, 525 Purcel Ave., P.O. Box 5029, Cincinnati, OH 45205; phone: (513) 921-5028. A graduate of the University of Cincinnati, he is is a licensed professional engineer in a number of states. He has 25 years of experience in industrial utilities and energy systems, holds patents on heat exchangers, and has authored articles on pollution control, fuels and combustion.

Controlling boiler carryover

Solids carried from the boiler into steam can create many problems. Steam-purity studies can help find the causes.

R. C. Andrade, J. A. Gates and *J. W. McCarthy,*
Drew Chemical Corp.

☐ The presence of small quantities of inorganic salts in water carried from the boiler to the steam can increase the potential for corrosion (or erosion-corrosion) in steam condensate systems or cause contamination in processes requiring the direct use of steam. With steam for power generation, carryover can cause superheater failures, loss of turbine efficiency, and related problems.

With a carefully designed steam-purity study, the causes of carryover can be pinpointed, then eliminated.

Causes of carryover

Carryover into the steam of substances present in boiler water is caused both by entrainment of small droplets of boiler water in the steam leaving the drum and by volatilization of salts that are dissolved in the steam. Mechanical entrainment, which can occur in all steam generators, can be minimized through mechanical or operational changes, so this type of carryover is the most common target of steam-purity studies. (Volatile carryover, while of major concern in high-pressure installations, cannot be prevented by mechanical or operational modifications, and will not be discussed further.)

Mechanical entrainment can be divided into three categories: priming, foaming, and equipment failure.

Priming

Priming usually results from a sudden reduction in boiler pressure caused by a rapid increase in the steam load. This causes steam bubbles to form throughout the mass of water in the steam drum, flooding the separators or dry pipe. Priming may also result from excessively high water levels. Priming results in a violent "throwing" of large slugs of boiler water into the steam. The problem can usually be minimized by changes in operation.

Foaming

Foaming is the buildup of bubbles on the water surface in the steam drum. This reduces the steam release space, and, by various mechanisms, causes mechanical entrainment. Foaming is almost always the result of improper chemical conditions in the boiler water, including alkalinity, suspended solids, dissolved solids, and organic surfactants and detergents. Your water-treatment-chemical supplier should provide a program to control the chemistry of the boiler water, thereby regulating these conditions.

Equipment failure

Boiler drums may contain baffles, screens, mesh mist-eliminators and centrifugal separators to improve separation of water droplets from steam. Each element must be kept tight and clean. A quarter-inch gap between the sections of cover baffles over the generating tubes can allow sufficient water to bypass the separators to negate their operation. Similarly, deposits on screens or mist-eliminators can prevent them from functioning properly.

Steam-purity measurement

Conducting a steam-purity study involves the proper collection of steam samples, their accurate analysis, and the establishment of proper operating conditions to achieve meaningful results.

Sample collection

Accurate sampling is difficult. The very fine water droplets in saturated steam remain dispersed uniformly for only several tube diameters in the direction of flow; then they agglomerate and run along the tube walls.

Sampling superheated steam is even more difficult. Particulate matter and substances soluble in the steam can deposit on the sampling-line walls, effectively depleting the sample of some important components.

The basis of the problem is that a two-phase, and sometimes a three-phase, vapor—liquid—solid dynamic system must be sampled to reflect the absolute composition of the total mass. The special sampling nozzle designs specified by ASTM and ASME are recommended. However, installation of such nozzles generally requires that the boiler be shut down first.

Often, a carryover problem occurs suddenly while the boiler is being operated at high capacity, when a shutdown is impossible. The plant's engineers may then opt to use a standard steam takeoff. Such samples should be taken from a point as close as possible to the boiler, and the results interpreted for their relative values only.

Sample analysis

The following methods have been used for the online analysis of steam samples:
- Ion exchange.
- Conductivity.
- Flame photometry.
- Specific-ion electrodes.
- Radioactive tracers.

Of these, the specific-ion electrode method for measuring the sodium content of the steam is considered to offer the most benefits. This method relies on the ratio

Originally published December 26, 1983

of sodium to dissolved solids being the same in the steam as in the boiler water from which it came. From experience, we know that the solids-to-sodium ratio is usually in the range of 2.8–3.0 to one. Furthermore, the sodium content in the saturated steam can be directly related to the potential for deposit formation in both turbines and superheaters (Table I).

The specific-ion electrode offers high sensitivity (0.1 ppb Na) and low lag time. The lag time that does occur may tend to shave off the highest peaks, but still clearly shows meaningful deviations. The unit is portable, and easy to operate and maintain.

Conducting the tests

A sufficient amount of baseline data is necessary to evaluate a potential carryover situation. This usually involves the collection and analysis of steam samples at various loads, with the boiler in a base-loaded (no load swings) condition.

With the base line established, samples should be collected and analyzed with the boiler operating under normal loading conditions. It is important that boiler-water chemical conditions be kept within their recommended limits during this phase of testing. Since most sodium analyzers incorporate continuous strip-chart re-

corders, data can easily be collected for 12 to 24 hours (or longer). These data can then be compared with the baseline study to determine any variations.

If the problem cannot be resolved during these initial tests, alternative conditions must be established at different boiler loads, such as:

- High and low water levels in the steam drum.
- Rapid load swings (usually accomplished with the boiler on manual control).
- Overconcentration of the boiler water (excessive conductivity).

Parameters should be varied individually, and detailed records kept. In addition to records of data pertinent to the above variables, such items as a fuel change, gas pressure drop, feedwater temperature change, change in condensate-return flow or source, etc., should be included. Reference samples of the boiler water should be taken at a number of points during the test, especially before and after any change that would have a significant effect on boiler-water chemistry.

Implementing these procedures will assure the most accurate test results possible. Additionally, they will enable keeping a close watch on steam purity and avoiding the downtime and lost efficiency caused by carryover.

The following are examples of the use of steam studies to determine the basic causes of carryover.

Study in 300-psig refinery boilers

In a Southwestern refinery, the boilers had been in a clean condition for several years, except that the superheaters required acid cleaning at each annual inspection. Steam-purity studies had never been carried out to determine the cause of carryover.

Sample points were determined and stainless-steel sample tubing and coolers were installed at ground level. Although the control-board reading indicated a 50%

Relation of sodium levels to boiler deposition			Table I
Sodium, ppb	Total solids, ppm	Turbine deposition	Superheater deposition
3.3	0.01	No	No
3.3-33	0.01-0.1	Possible	No
33-333	0.1-1.0	Yes	Possible
333	1.0	Yes	Yes

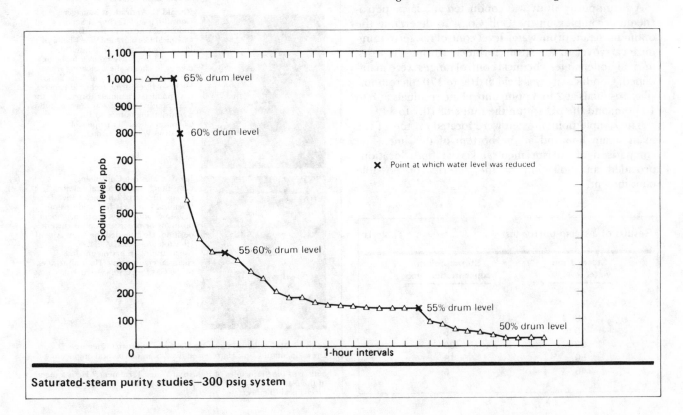

Saturated-steam purity studies—300 psig system

water level in the steam drum, the onsite gage reading varied from 60 to 80%. The high water-level was suspected of contributing to the carryover problem.

To help control carryover, the boiler-water total alkalinity had been controlled within the recommended range of 450 to 650 ppm (but generally at 450 ppm) by adjusting the blowdown rate. Despite alkalinity control, deposition was occurring in the superheater sections.

The sample line was heavily purged in preparation for the steam-purity test, the sodium analyzer was calibrated, and operation was stabilized. Measured sodium levels quickly pegged out at 1,000 ppb. Testing was continued for several hours without any reduction in sodium level. (The initial tests were performed without any change in plant operations.)

The second phase of testing involved step reductions from the initial observed steam-drum level of 65% down to 50% in 5% decrements, with sodium levels being monitored after operations stabilized at each drum level. The plotted data (p. 52) show the dramatic reduction of sodium from a concentration of 1,000 ppb to 30 ppb.

Since the only operating variable that had been changed was the steam-drum water level, this factor was considered responsible for the high sodium level in the saturated steam. By maintaining the level at 50% the carryover problem had been solved.

The boiler-plant operator and instrument technician worked together to calibrate the instrument-board level meters. Later, it was found that the blowdown rate could be decreased somewhat to maintain the total alkalinity at the upper end of the recommended range without increasing impurity carryover in the saturated steam. A subsequent boiler inspection confirmed the absence of deposition in the superheaters.

Study of petrochemical complex's boilers

A steam-purity study was conducted at a large petrochemical complex on the Gulf Coast to determine the optimum steam-drum water-level control range, to minimize carryover to the steam turbines. During the tests, normal boiler-water-chemical control ranges were maintained. Conductivity was held at 160 to 170 micromhos, silica residuals at 2 to 3 ppm, phosphate residuals at 8 to 10 ppm, and the pH within the range of 10.7 to 11.0.

The sample points tested were located on top of the main steam line and at the bottom of this line. The comparison of measurements taken at both locations provided an indication of the severity of possible entrainment.

During the initial tests, the water level was held at 42%, with a constant steam rate of 140,000 lb/h. The sample obtained from the top of the line had a sodium level of 0.6 ppb, whereas the one from the bottom had a level of 4.0 ppb.

As testing continued, the boiler feedwater pumps were switched; causing a momentary change in the system's hydraulic stability. The feedwater rate to the boiler sustained a sudden drop, followed by a large upward surge. As a result, the water level rose from 40 to 48%, with an increase in the sodium reading to 110 ppb.

Operating conditions were again stabilized at a water level of 42%, with a corresponding sodium concentration of 4.0 ppb. The water level was then slowly increased and stabilized at each level. These adjustments were made over a period of several hours to ensure that the sodium analyzer was recording a constant sodium concentration at each water level. The results are summarized in Table II.

After completing these tests, we recommended that the steam-drum water level be held within the range of 42 to 48%, with a target level of 45%. This mode of operation would minimize carryover, while avoiding the annoyance of frequent low-water-level alarms. During the tests, is was noted that the control-board level readings and the actual drum's gage-glass level corresponded within plus or minus 0.5 to 1%.

This steam-purity study enabled us to identify two critical operational factors for limiting carryover. Potential problems have been avoided by maintaining the water level below 48%, and by minimizing the occurrence of sudden water-level surges.

The authors

Ronald C. Andrade is Manager, Consulting and Technical Services, Industrial Chemicals Div., Drew Chemical Corp., One Drew Chemical Plaza, Boonton, NJ 07005, telephone: (201) 263-7600. He holds a bachelor's degree in chemical engineering from the College of the City of New York. He has written many technical papers, has conducted seminars throughout the U.S., South America and Europe, and is a contributing author to Drew Chemical's text, "Principles of Industrial Water Treatment."

Jay A. Gates is a Consultant, Consulting and Technical Services Group, Industrial Chemicals Div., Drew Chemical Corp. He has held several field-sales and field-sales-management positions before assuming his current position. He attended the University of Houston and Texas A&M, concentrating in petroleum and chemical engineering. He is a member of the Gas Processors Assn.

John W. McCarthy is Marketing Manager, Industrial Chemicals Div., Drew Chemical Corp. He received his B.S. in mechanical engineering from the New Jersey Institute of Technology, and is a major contributing author to the "Principles of Industrial Water Treatment." He is a member of the National Assn. of Corrosion Engineers and the American Soc. of Mechanical Engineers.

Results of a steam-purity test	Table II

Steam-drum water level, %	Steam sodium concentration, ppb
42	4
45	4
48	4
50	15
52	30
50	18
45	6

Section II
Cooling

Selecting refrigerants for process systems

Here is a rundown on the refrigerants generally used in
the chemical process industries, together with the
factors that should be taken into account when trying
to determine the best one for a particular process.

Howard W. Sibley, Carrier Group, United Technologies

☐ In selecting industrial refrigeration cycles and systems, the end-user frequently considers for refrigerant
selection only those fluids with which he or she is most
familiar. This is particularly true of installations where
plant gas, such as propane, is readily available.

This article discusses the major factors to be evaluated in order to optimize both first cost and operating
cost.

Refrigerant types

The term "refrigerant" covers a wide variety of organic and inorganic compounds that share a common
property of absorbing heat at low temperature and
pressure, and rejecting it at high temperature and pressure—usually with a change of phase. Refrigerants fall
into three general categories of compounds:

1. Saturated and unsaturated aliphatic hydrocarbons, e.g., propane and propylene.
2. Aliphatic halogenated hydrocarbons, e.g., dichlorodifluoromethane.
3. Inorganic gases, e.g., air, CO_2, SO_2, NH_2, and Cl_2.
For the saturated and unsaturated aliphatic hydrocarbons and aliphatic halogenated hydrocarbons, a
numerical coding (naming) system has been developed
that describes the refrigerants' molecular structure by
the use of a general formula having the form ABCD, in
which:

A = the number of double bonds
B = (the number of carbon atoms) − 1
C = (the number of hydrogen atoms) + 1
D = the number of fluorine atoms

Originally published May 16, 1983

49

For example, the refrigerant dichlorodifluoromethane, CCl_2F_2, is designated R12 (R = refrigerant), computed thus:

A = 0
B = 1 − 1 = 0
C = 0 + 1 = 1
D = 2

Dropping the two initial zeroes, we have R12.

Similarly, the unsaturated hydrocarbon propylene, $CH_3CH=CH_2$, is designated R1270, computed as follows:

A = 1
B = 3 − 1 = 2
C = 6 + 1 = 7
D = 0

and the saturated hydrocarbon propane, C_3H_8, is designated R290:

A = 0
B = 3 − 1 = 2
C = 8 + 1 = 9
D = 0

Inorganic refrigerants are handled somewhat differently. They are assigned three-digit numbers, the first of which is 7, with the following two numbers being the molecular weight of the gas. Hence, NH_3 (molecular weight 17) is coded 717 and SO_2 (molecular weight 64) is coded 764.

One of the commonly used halocarbons, R500, is an azeotrope of the mixture R12 and R152a. Since an azeotrope is a mixture whose vapor composition is identical to its liquid composition, the mixture boils and condenses at constant composition throughout the refrigeration cycle. Azeotropes have been assigned a three-digit code starting with 5. The second and third

digits reflect the chronological order of discovery. Table I lists common refrigerants in use worldwide.

Parameters for optimum operation

In general, the selection process takes into account the following physical and thermodynamic parameters of the refrigerant:
- Temperature-pressure characteristics in the evaporator, condenser and compressor.
- Latent heat of vaporization.
- Specific volume of vapor.
- Specific heat of liquid.
- Molecular weight.
- Adiabatic head.
- Power.

A refrigeration cycle has low- and high-side pressure boundaries—represented in Fig. 1 by P_s and P_c, respectively. The design working pressures for the refrigeration machine are established by the temperature-pressure characteristics of the refrigerant, and the low- and high-side system requirements.

The low side is typically the process load that requires cooling, and the high side a heat-rejection sink such as a cooling tower. The process temperature and available high-side heat-sink temperatures interact with the heat-transfer characteristics of the refrigerant to establish the working temperature differences, ΔT_c and ΔT_s, as illustrated in Fig. 1. These set the refrigerant condensing and evaporating temperatures (t_c) and (t_s). The corresponding saturated pressures establish the material-selection criteria for the condenser, evaporator and compressor.

The pressure-temperature relationship for a refrigerant may be expressed by the Clapeyron equation:

$$\frac{dP}{dT} = \frac{\Delta H_{fg}}{T(V_v - V_l)} \tag{1}$$

which shows the rate of change of the vapor pressure

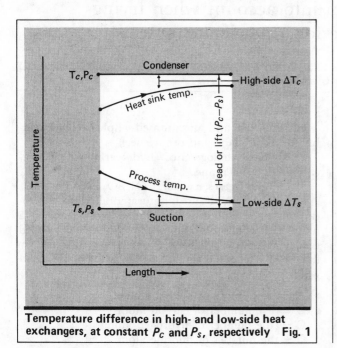

Temperature difference in high- and low-side heat exchangers, at constant P_c and P_s, respectively   Fig. 1

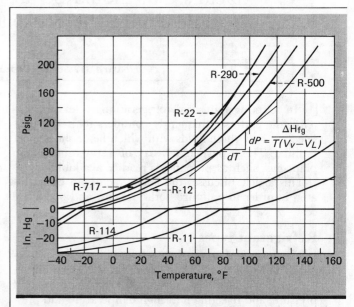

Saturated-refrigerant temperature-pressure diagram   Fig. 2

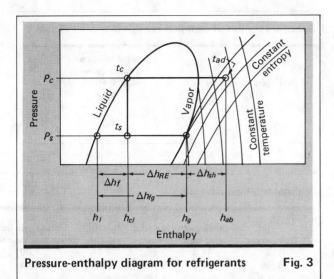

Pressure-enthalpy diagram for refrigerants   Fig. 3

Nomenclature

C_p	Specific heat
C	Mean specific heat of condensate between t_c and t_s
D	Impeller diameter, in.
g	Gravitational constant, 32.2 ft/s²
h, H, h	Specific enthalpy
$\Delta H_{RE}, \Delta h_{RE}$	Latent heat corrected for adiabatic flashing, Btu/lb
$\Delta H_{fg}, \Delta h_{fg}$	Latent heat of vaporization Btu/lb
HP	Horsepower
$K = C_p/C_v$	Isentropic exponent
M	Molecular weight, lb/lb-mol
N	Impeller rotating speed, rpm
N	Number of stages of compression
P	Pressure, psia
P_c	Pressure in condenser, psia
P_s	Pressure in evaporator, psia
Q	Heat transferred in cooler load
Q_e	Heat transferred in evaporator
t	Condensing temperature, °F
t	Evaporator temperature, °F
T	Temperature, °F
T_c	Condensing temperature, °F
T_s	Evaporator temperature, °F
U	Impeller tip speed, ft/s
U_o	Acoustic velocity, ft/s
V	Volume, ft³
V_r	Volumetric flowrate, ft/s
V_s	Specific volume, lb/ft³
W_{ad}	Adiabatic head, ft
W_R	Mass flowrate, corrected for adiabatic flash gas, lb/min
W	Mass flowrate based on heat of vaporization
W_f	Flash gas, lb/min
Z_d	Compressability factor at discharge
Z_s	Compressability factor at suction
μ	Head coefficient, dimensionless

Subscripts

l	Liquid
v	Vapor

with temperature, dP/dT, in terms of the latent heat of vaporization, ΔH_{fg}, volume of the vapor, V_v, and volume of the liquid, V_l, at temperature T and at a pressure equal to the vapor pressure. In this equation, ΔH_{fg} and V are in molal units, and temperature and pressure are in absolute units.

The Clapeyron equation shows the interrelationship

Common refrigerants in use worldwide   Table I

Refrigerant number	Chemical name	Chemical formula	Molecular weight	Normal boiling point, °F
170	Ethane	CH_3CH_3	30	-127.5
290	Propane	$CH_3CH_2CH_3$	44	-44.2
717	Ammonia	NH_3	17	-28.0
718	Water	H_2O	18	212
729	Air	—	29	-318
744	Carbon dioxide	CO_2	44	-109
764	Sulfur dioxide	SO_2	64	14.0
1150	Ethylene	$CH_2=CH_2$	28	-155
1270	Propylene	$CH_3CH=CH_2$	42.1	-53.7
11	Trichloromono-fluoromethane	CCl_3F	137.4	74.8
12	Dichlorodifluoro-ethane	CCl_2F_2	120.9	-21.6
113	Trichlorotrifluoro-ethane	CCl_2FCClF_2	187.4	117.6
114	Dichlorotetrafluoro-ethane	$CClF_2CClF_2$	170.9	38.4
13Bl	Monobromotrifluoro methane	$CBrF_3$	148.9	-72.0
22	Monochlorodifluoro-methane	$CHClF_2$	86.5	-41.4
30	Methylene chloride-	CH_2Cl_2	84.9	105.2
40	Methyl chloride	CH_3Cl	50.5	-10.8
50	Methane	CH_4	16.0	-259
152A	Difluoroethane	CH_3CHF_2	66	-12.4
500	Refrigerants 12/152a	$CCl_2F_2/$ CH_3CHF_2	99.29	-28.0

Mass flowrate advantage based on latent heat   Table II

Refrigerant number	Latent heat of vaporization at -20°F	Mass flowrate advantage relative to R12
R12	70.87	1.00
R290	175.47	2.47
R717	583.60	8.23

of four key refrigerant parameters and applies to an absolute-pressure-temperature plot, but is shown in Fig. 2 as illustration of its physical significance.

Referring to Fig. 3, the latent heat of vaporization in the Clapeyron equation is the difference in enthalpy between the refrigerant vapor (h_g) and liquid (h_l) at constant suction pressure, and is defined as:

$$\Delta H_{fg} = h_g - h_l \qquad (2)$$

ΔH_{fg} represents the maximum heat transferable to the refrigerant during its change in phase from liquid to

Specific heats of common refrigerants	Table III

Refrigerant	Liquid specific heat at 100°F
R12	0.24
R290	0.6727
R717	1.158

Specific volumes of common refrigerants	Table V

Refrigerant number	Specific volume of saturated vapor at −20°F, ft3/lb
R12	2.44
R290	3.95
R717	14.68

Corrected latent-heat values adjusted for specific heat and adiabatic cooling			Table IV

Refrigerant	H_{fg}	H_{re}	Mass flowrate advantage relative to R12
R12	71	44.0	1.00
R290	175	102.7	2.33
R717	583	450	10.2

Volumetric-flow advantage				Table VI

Refrigerant	Corrected latent-heat	Specific heat	Specific volume	Volumetric-flow advantage over R12
R12	44.0	0.24	2.44	1.00
R290	102.7	0.6727	3.95	1.44
R717	450	1.158	14.68	1.72

vapor at constant suction pressure and is the maximum theoretical refrigeration effect.

The mass flowrate of refrigerant, W_r, associated with this evaporation would be:

$$W_r = \frac{Q_e}{\Delta H_{fg}} \qquad (3)$$

where Q_e = evaporator heat transferred from the load; this implies that refrigerants with high latent heats will require lower mass flowrates per ton of refrigeration. On this basis, R290 and R717, for example, would appear to have significant advantage over R12, as indicated in Table II.

However, as refrigerant condensate passes from the condenser at Point t_c to the cooler at Point t_s, some of the condensate adiabatically flashes to cool the remaining liquid to evaporator conditions. This flash gas reduces the available refrigerant for cooling and adds to the total refrigerant weight flow according to the following equation:

$$W_R = \frac{Q_e}{\Delta H_{re}} = W_r + W_f \qquad (4)$$

where Q_e = heat transferred in evaporator; W_r = refrigerant flow from evaporator load; and W_f = flash gas.

ΔH_{re}, in Eq. (4), is a corrected latent-heat value that adjusts the refrigerant flow calculation to account for non-useful flash gas effect, and is referred to as the effective heat of vaporization. It is calculated as follows (see Fig. 3):

$$\Delta H_{re} = \Delta H_{fg} - \Delta H_f \qquad (5)$$

ΔH_f is the fraction of latent heat expended to produce non-useful adiabatic cooling, and is a function of liquid specific heat:

$$\Delta H_f = C_p(t_c - t_s) \qquad (6)$$

where C_p = mean specific heat of the condensate between condensing temperature, t_c, and evaporator temperature, t_s.

Refrigerants with low liquid specific-heats produce less flash gas and less refrigerant flow than those with high specific heats. Table IV shows adjusted latent heat values from Table III for specific heat and adiabatic cooling. Note that R190 shows slightly less advantage after correcting for its high specific heat; on the other hand R717 shows greater advantage over R12 because its very high latent heat more than offsets its high specific-heat effect.

The next factor in this analysis is specific volume, because the size of heat exchangers and compressors will be directly related to the volume of the gas being handled. The volumetric flowrate is related to the mass flowrate by the relationship:

$$V_r = \frac{Q_e}{\Delta H_{re}} V_s \qquad (7)$$

where V_s is the vapor specific-volume.

Table V shows specific volumes of the three various refrigerants.

If we now construct a table showing volumetric flow advantages of refrigerants, taking into consideration specific heat, effective latent heat and specific volume, as in Eq. (7), we will obtain the result shown in Table VI.

The apparent advantages of R190 and R717 over R12—shown in Table II, based on specific heat alone—nearly disappear after factoring in the low liquid specific-heat and low specific-volume advantages of halocarbon refrigerants. In the selection process, the volumetric flowrate is typically kept as low as practical to limit compressor equipment size—therefore, a low specific-volume and low liquid specific-heat are desirable. On the other hand, the latent heat of vaporization should be high to limit volumetric flow, according to Eq. (7). Generally, the selection process will be a compromise between conflicting desirable properties such as these.

The next factor to consider in the selection process is molecular weight. In Fig. 2, the pressure differential, $P = P_c - P_s$ is referred to as the "lift" or "head," and

Adiabatic head of various refrigerants			Table VII
Temperature, °F		Adiabatic head, W_{ad}, ft		
t_s	t_c	R12	R290	R717
−20	100	12,210	31,630	127,130

Theoretical horsepower per ton of refrigeration requirements for common refrigerants			Table VIII
Temperature, °F		Theoretical power, HP/ton		
t_s	t_c	R12	R290	R717
−20	100	1.681	1.866	1.711

defines the static pressure-difference between cooler and condenser. There is a second head referred to as energy head, which is the work input in ft-lb required to "lift" a pound of gas from P_s to P_c. The energy head is called the "adiabatic head," and is given by (see Fig. 2):

$$W_{ad} = 778(h_{ad} - h_g) = 778(\Delta h_{sh}) \qquad (8)$$

The adiabatic head, from a thermodynamic standpoint, is the difference in vapor enthalpy between the discharge at t_{ad} and suction t_{se}. For an adiabatic process, the head may be expressed by:

$$W_{ad} = \frac{Z_s + Z_d}{2}\left(\frac{1{,}545}{M_w}\right)T_s\left[\frac{\left(\frac{P_2}{P_1}\right)^{(K-1)/K} - 1}{(K-1)/K}\right] \qquad (9)$$

which shows that the adiabatic head is inversely proportional to molecular weight; therefore, high-molecular-weight refrigerants produce lower adiabatic heads. Table VII shows that halocarbon refrigerants have significantly lower adiabatic heads than other refrigerants, due to their high molecular weight, under the same load condition.

What significance does low adiabatic head have on equipment? For a given centrifugal compressor design, the number of compression stages is approximated by:

$$N_{st} = \frac{W_{ad}g}{\mu U^2} \qquad (10)$$

and the impeller diameter is given by:

$$N = 153.3\frac{W_{ad}}{D} \qquad (11)$$

Therefore, use of low-adiabatic-head refrigerants results in few stages of compression, smaller impeller diameters, or lower compressor speed—all of which result in smaller compressor size.

Molecular weight vs. compressor size

The following example illustrates the effect of the refrigerant's molecular weight on compressor size:

Basis for comparison—R12 and R290 at −20°F suction and 100°F condensing.

Comparison 1—Assume impeller tip velocity = 90% of acoustic velocity, head coefficient is 0.5 for both cases, and the compressor speed is 10,000 rpm.

Acoustic velocity =

$$U_o = 39.3\sqrt{\frac{KgTZ}{M_w}}$$

$$U_o\,(R290) = 39.3\sqrt{\frac{1.13 \times 32.2 \times 440 \times 0.95}{44}}$$

$$= 731 \text{ ft/s}, \; U = 658 \text{ ft/s}$$

$$U_o\,(R12) = 39.3\sqrt{\frac{1.12 \times 32.2 \times 440 \times 1.0}{120.9}}$$

$$= 450 \text{ ft/s}, \; U = 405 \text{ ft/s}$$

$$N_{ST}\,(R290) = \frac{W_{ad}g}{\mu U^2} = \frac{31{,}630 \times 32.2}{(0.5)(658)^2}$$

$$= 4.7 \cong 5 \text{ stages}$$

$$N_{ST}\,(R12) = \frac{W_{ad}g}{\mu U^2} = \frac{12{,}210 \times 32.2}{(0.5)(405)^2}$$

$$= 4.81 \cong 5 \text{ stages}$$

$$D_{R290} = 229\,\frac{U}{N} = \frac{229(658)}{10{,}000} = 15.1 \text{ in.}$$

$$D_{R12} = 229\,\frac{U}{N} = \frac{229(405)}{10{,}000} = 9.3 \text{ in.}$$

Comparison 2—Assume impeller tip velocity = 400 ft/s

$$N_{ST,(R290)} = \frac{31{,}630 \times 32.2}{(0.5)(400)^2} \cong 13 \text{ stages}$$

$$N_{ST,(R12)} = \frac{12{,}210 \times 32.2}{(0.5)(400)^2} \cong 5 \text{ stages}$$

To drive a refrigeration machine, the horsepower/ton of refrigeration required, is given by:

$$HP/\text{ton} = 6.06 \times 10^{-3} \times W_{ad}/\Delta H_{re} \qquad (12)$$

Since it has already been established that the W_{ad} for higher-molecular-weight refrigerants is lower, it holds that the HP/ton is similarly lower, as shown in Table VIII.

Halocarbons have lower power requirements per ton of refrigeration than do the other refrigerants. The following example shows how low adiabatic head and HP/ton influence compressor operating costs. We will compare R290 (propane) and R12 (dichlorodifluoromethane).

Basis of comparison—1,000 tons of refrigeration at −20°F suction temperature, 100°F condensing temperature, and power costing $.04/kWh.

Adiabatic head (W_{ad}):

For R290   (W_{ad}) = 31,360 ft
For R12   (W_{ad}) = 12,210 ft

Energy consumption (HP/ton):

For R290   1.87 HP/ton × 1,000 tons = 1,866 HP

For R12   1.68 *HP*/ton × 1,000 tons = 1,681 *HP*

11% less *HP* required for R12

Annual operating cost

R290 $445,440

R12   401,288

Savings using R12 $ 44,152

Current trends in refrigerant use

Use of halocarbon refrigerants, in terms of refrigeration duty, may be listed in five categories:

1. Water chilling.
2. Chilling of secondary coolants: brines, inhibited glycol, hydrocarbons.
3. Refrigerant condensing.
4. Process vapor condensing.
5. Heat reclaiming systems.

All five categories can apply to chemical-process-industries requirements. Chilled-water duty typically falls in the range of 40°F suction and 105°F condensing temperatures (such as reactor cooling in polyethylene processing). Secondary coolants apply to low-temperature applications, where calcium chloride, methanol, ethylene glycol, propylene glycol, etc., are used as secondary refrigerants for low-temperature process cooling. Refrigerant condensing applies to systems in which the condensed fluid in the tubeside of the chiller is distributed to the system as a liquid, but returns as a mixed liquid and gas.

In the latter case, the refrigeration machine serves as a condenser for the secondary refrigerant. The most common application of this system is ammonia condensing. In process-vapor condensing, the cooler of the refrigeration machine is used for condensing chemicals from the vapor phase (such as chlorine, HCl, and hydrocarbon mixtures) and for dewpoint control of natural gas in a partial condensing mode.

Heat reclaiming applications for refrigerants are becoming more common as cost of energy increases and energy sources are less available. Although heat reclaiming is predominantly applied to residential and large-building systems, in recent years industrial applications for heat reclaiming have developed, particularly using R22 for reciprocating compressors and R114 for centrifugals.

In Europe, for example, utilities that provide district heating are considering extracting heat from seawater, and then boosting the temperature to 194°F by refrigeration-heat reclaiming. R114 is particularly suited for the high stage of this application because it requires fewer stages of compression, and has a low volumetric flowrate and good superheat characteristics.

Refrigerant for a specific application

As previously discussed, selection of a refrigerant involves consideration of such parameters as specific volume, molecular weight, and operating pressures. Generally, the best refrigerant for a system is that which requires the smallest compressor for operation at design conditions.

Low-pressure refrigerants, such as R11, are preferred for smaller refrigeration tonnages (usually below 400

tons). Despite the high specific volume, the equipment is smaller and less costly because it is designed for low-pressure standards.

For larger tonnages, the use of higher-pressure, lower-specific-volume refrigerants reduces the physical size requirement of the compressor and gas spaces, resulting in lower first cost, lighter machine weight and lower installation cost. In higher tonnage ranges of applications, R114, R12, R500 and R22 are in common use for industrial process machines.

The following list describes the typical (but not all-inclusive) applications of common refrigerants:

R11—Used in applications requiring cooling loads below 400 tons, and process chilled-temperature levels of 40°F suction and 105°F condensing.

R114—Used for process refrigeration loads up to 2,000 tons at 40°F suction, 105°F condensing, as well as to 100 tons at −40°F.

R12—Used for process refrigeration loads up to 5,500 tons at 40°F suction, and 105°F condensing levels; 450 tons at −50°F suction; and 2,600 tons at 0°F suction. By adding compressor stages, suction temperatures between −50°F and −100°F are obtainable.

R500—Used for process refrigeration loads up to 6,500 tons at 40°F suction and 105°F condensing; 650 tons at −30°F suction; and 2,100 tons at 0°F suction. R500 can be used in the low stage of a staged low-temperature application.

R22—Used for loads up to 10,000 tons at 40°F suction, 105°F condensing; and 3,000 tons at suction temperatures near 0°F, and lower.

In general, one- and two-stage compressors are used for process cooling at 40°F suction and 105°F condensing. Three- and four-stage compressors can achieve suction temperatures down to −10°F and −40°F, respectively. It is possible to interconnect compressor casings with a hermetic sleeve and coupling in so-called "drive through" fashion. This assembly has only one seal and one driver set of gear and motor. Temperature to −100°F can be achieved in this manner.

Conclusions

Based on the key considerations that should be given to the selection of refrigerants for an industrial process, the halocarbon refrigerants are shown to have considerable advantage over other types in terms of equipment first-cost and operating-cost.

The author

Howard W. Sibley is manager of Materials Engineering (chemical engineering and metallurgy) for the Commercial Div.—Applied Products, Carrier Group, United Technologies, P.O. Box 4808, Carrier Parkway, Syracuse, NY 13211, telephone: (315) 432-6000. He holds a B.S. in chemical engineering from the University of Maine, is a member of the American Soc. of Heating, Refrigeration and Air Conditioning Engineers Technical Committee 3.6 on Corrosion and Water Treatment, and holds U.S. and foreign patents on conventional and solar-powered absorption refrigeration equipment.

Part XVI:
Costs of refrigeration systems

Refrigeration systems are mainly used to control volatile organic vapors from storage and distribution facilities. Cost graphs are presented for surface-condensing and direct-contact-condensing systems.

William M. Vatavuk, U.S. Environmental Protection Agency, and Robert B. Neveril, Gard, Inc.

☐ Refrigeration systems are normally coupled with surface condensers, absorbers or adsorbers to remove and recover volatile vapors from inlet exhaust streams. Of the several configurations available, two have been used to recover volatile organic vapors: surface-condensing and contact-condensing systems.

Part I of this series, Oct. 6, 1980, p. 165, presented parameters and a procedure for sizing entire air-pollution control systems; Part II, Nov. 3, p. 157, dealt with factors for estimating capital costs and annualized operating costs. Parts III through XV provided data for estimating the size and cost of the following equipment that make up air-pollution control systems: Part III, Dec. 1, p. 111—capture hoods; Part IV, Dec. 29, 1981, p. 71—ductwork; Part V, Jan. 26, 1981, p. 127—gas conditioners; Part VI, Mar. 23, p. 223—dust-removal and water-handling auxiliary equipment; Part VII, May 18, p. 171—fans and accessories; Part VIII, June 15, p. 129—exhaust stacks; Part IX, Sept. 7, p. 139—electrostatic precipitators; Part X, Nov. 30, p. 93—venturi scrubbers; Part XI, Mar. 22, 1982, p. 153—baghouses; Part XII, July 12, p. 129—incinerators; Part XIII, Oct. 4, p. 135—gas absorbers; Part XIV, Jan. 24, 1983, p. 131—carbon adsorbers; and Part XV, Feb. 21, p. 89—flares. In these previous articles, all costs have been updated to December 1977, except when otherwise noted.
For information about the authors, see the Feb. 21, 1983 issue, p. 90.

Components of the two systems

The surface-condensing system typically consists of a dehumidifier (to remove moisture), a recuperative heat exchanger (to precool the stream), and a vapor condenser (to remove the captured vapor). Coupled to the dehumidifier and vapor condenser are refrigeration units, which circulate coolant to the equipment (Fig. 1). Note in Fig. 1 that the condenser is served by a *cascade* refrigeration unit—i.e., two or more refrigerators linked

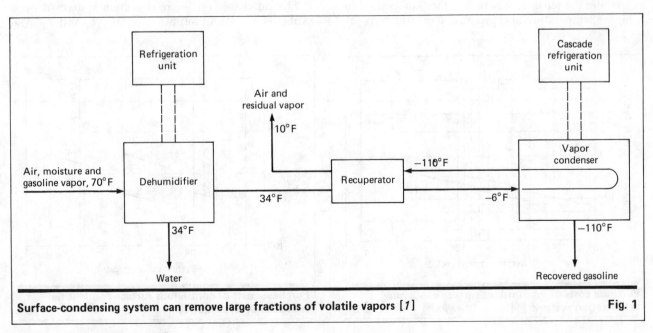

Surface-condensing system can remove large fractions of volatile vapors [1] Fig. 1

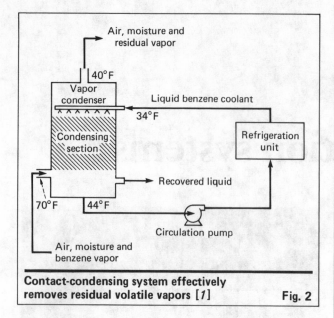

Contact-condensing system effectively removes residual volatile vapors [1]   Fig. 2

ant rejects its sensible and latent heat in a condenser (at temperatures ranging from 0 to −40°F) and is vaporized when it passes through an expansion valve, completing the cycle. (For more details on refrigeration principles, see Ref. 2 and 3.)

Effectiveness and efficiency

The effectiveness of a refrigeration unit is measured by the amount of heat it removes, usually expressed in "tons" (1 ton = 288,000 Btu/d):

$$Q_r = Q_a + Q_v + Q_c \qquad (1)$$

Here, $Q_a = m_a C_a(T_1 - T_2)$ = sensible-heat cooling of air, Btu/h; $Q_v = m_{v1} C_v(T_1 - T_2)$ = sensible-heat cooling of inlet volatile vapors, Btu/h; $Q_c = m_c \Delta H_v$ = heat evolved by condensing volatile vapors, Btu/h; m_a, m_{v1} and m_c = masses of air, inlet volatile vapors and condensate, respectively, lb/h; C_a and C_v = mean heat capacities of air and volatile vapors, respectively, between T_1 and T_2, Btu/(lb) (°F); ΔH_v = latent heat of volatile vapors at T_2, Btu/lb; and T_1 and T_2 = inlet and outlet temperatures, respectively, of the gas stream, °F.

Condensate mass is calculated from:

$$m_c = 0.00763 \, (M_v G/T_1)[(y_1 - y_2)/(1 - y_2)] \quad (2)$$

Here, M_v = molecular weight of volatile vapors, lb/lb-mol; G = volume of inlet gas stream (system capacity), gal/d; and y_1 and y_2 = inlet and outlet saturated concentrations of volatile vapors by volume—which are functions of the vapor pressures of the volatile vapors at T_1 and T_2.

Eq. (2) is based on two assumptions: negligible water-vapor content, and constant-pressure operation at 1 atm. These are reasonable assumptions for typical vapor-recovery applications.

The control efficiency, E, of a refrigeration system is simply the quotient of m_c and m_{v1}, the condensate and inlet-vapor mass flowrates, respectively:

$$E = m_c/m_{v1} = (1/y_1)[(y_1 - y_2)/(1 - y_2)] \qquad (3)$$

Costs of surface-condensing systems

The purchase costs of refrigeration systems depend explicitly on G, and implicitly on M_v, T_1 and T_2. Also

to produce coolant below −100°F. Such low temperatures are often needed to remove large fractions of volatile organic vapors from the inlet stream.

The contact-condensing system does not produce such low temperatures. In it, the inlet vapors enter a packed column, to be contacted by a circulating organic liquid, such as benzene. The vapors are absorbed by the liquid, and the stream of air, moisture and residual vapor exits the top of the column (Fig. 2).

The temperature of the contact-condensing system must be kept above 32°F to prevent ice formation. Alone, the contact-condensing system is usually less efficient than the surface-condensing one in removing volatile vapors. However, installed downstream, it removes residual vapors not trapped by the latter.

Although new techniques (such as air-refrigeration cycle) have been experimented with, the mainstay of refrigeration units is still the basic mechanical-compression cycle. Here, heat absorbed from the gas stream evaporates the refrigerant, which is then compressed to a higher temperature and pressure. Next, the refriger-

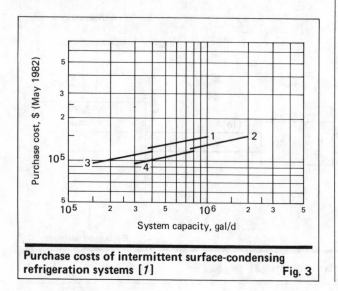

Purchase costs of intermittent surface-condensing refrigeration systems [1]   Fig. 3

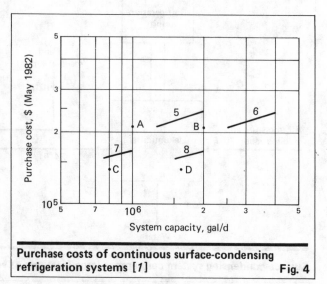

Purchase costs of continuous surface-condensing refrigeration systems [1]   Fig. 4

Cost parameters for surface-condensing refrigeration systems			
Fig. 3 and 4 curve number	Capacity range, 1,000 gal/d	Parameter a	Parameter b
1	375-1,000	11,000	0.187
2	750-1,500	9,620	0.187
3	150-400	7,990	0.208
4	300-800	6,920	0.208
5	1,250-2,000	2,310	0.322
6	2,500-4,000	1,850	0.322
7	750-1,000	8,540	0.215
8	1,500-2,000	7,360	0.215

influencing costs are the mode of operation—continuous or intermittent—and certain design conditions, such as whether the system includes a chiller precooler for dehumidifying the gas stream.

Costs in May 1982 dollars of intermittently operated surface-condensing systems—those that must be shut down for one hour daily for defrosting—are plotted in Fig. 3. Here, Curves 1 and 2 cover systems that include chiller precoolers, and Curves 3 and 4 those that do not. Additionally, Curves 1 and 3 and Curves 2 and 4 correspond, respectively, to outlet volatile-vapor concentrations of 0.33 and 0.67 lb/1,000 gal of liquid transferred from the source [4].

(Note: one gallon of transferred liquid displaces one gallon of vapor and air mixture; for gasoline vapors at 70°F and 1 atm, these are, respectively, equivalent to 1.45% and 2.9% by volume.)

Costs of continuously operated surface-condensing systems are plotted in Fig. 4.

Curves 7 and 8 and Points C and D give costs of systems of one set of compressors, each set with a split condensing coil (while one part of the coil is condensing, the other is defrosting). Curves 5 and 6 and Points A and B provide costs of systems of two sets of compressors, each set having dual, full-capacity coils.

Curves 5 and 7 and Points A and C correspond to an outlet volatile-vapor concentration of 0.33 lb/1,000 gal of transferred liquid; and Curves 6 and 8 and Points B and D to 0.67 lb/1,000. Additionally, Points A through D represent costs of systems without precoolers.

Each system comes with a volatile-vapor analyzer, condenser chart recorder, and recovered-product meter.

The systems with precoolers do not include the cost of the heat-transfer fluid (usually ethylene glycol). The quantity of fluid required varies with system capacity; a typical system needs 150 lb. This would add only about $50 to the purchased cost of the equipment, based on May 1982 prices [5].

Costs of surface-condensing systems can be approximated via Eq. (4):

$$C_s = aG^b \qquad (4)$$

Here, C_s = purchase cost in May 1982 dollars, and a, b = parameters. Values for a and b are listed in the table.

Costs of direct-condensing systems

Fig. 5 displays costs of direct-contact condensing systems in terms of equipment size in tonnage. Eq. (5) fits the data in Fig. 5:

$$C_d = 8.12 + 5.69T \qquad (5)$$

Here, C_d = purchase cost of contact-condensing system in November 1979 dollars, and T = cooling capacity in tons of refrigeration.

Operating costs

Refrigeration systems require little operating or maintenance labor—usually about 1/2 labor-hours/shift for each [6]. However, electricity costs (for running compressors, circulation pumps, etc.) can be significant. In the case of the surface-condensing system, average electricity usage ranges from 4.1–4.3 kWh/1,000 gal of inlet flow, with 4.15 kWh being typical [4]. That of the contact-condensing system is less, approximately 2 kW/ton in the range 35–40°F [1].

The value of the vapor condensate recovered should be subtracted from the operating and maintenance costs and capital charges. In some instances, this recovery credit could exceed these costs. To compute the annual credit, multiply the condensate mass, m_c, by the value of the condensate and the number of hours that the system operates annually.

Acknowledgment

The authors are grateful to Robert J. Honegger of GATX Terminals Corp., who provided valuable material for this article.

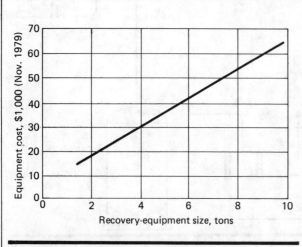

Purchase costs of direct-contact-condensing refrigeration systems [1]   **Fig. 5**

References

1. Honegger, R. J., "Environmental Control Systems: Refrigeration Methods of Vapor Recovery," Gard, Inc., Niles, Ill., Dec. 21, 1979.
2. Smith, J. M., and Van Ness, H. C., "Introduction to Chemical Engineering Thermodynamics," 2nd ed., McGraw-Hill, Inc., New York, 1959.
3. Mehra, Y. R., Refrigeration Systems for Low-Temperature Processes, *Chem. Eng.*, July 12, 1982, p. 94.
4. Price data from Edwards Engineering Corp., Pompton Plains, N.J., 1982.
5. *Chemical Marketing Reporter*, May 17, 1982, p. 56.
6. Vatavuk, W. M., and Neveril, R. B, Estimating Costs of Air-Pollution Control Systems, Part II: Factors for Estimating Capital and Operating Costs, *Chem. Eng.*, Nov. 3, 1980, p. 157.

Trouble-shooting compression refrigeration systems

Comparing operating conditions to design values quickly locates the source of problems that can arise in refrigeration systems. Here are procedures for performing such analyses and the corrective actions to be taken.

Kenneth J. Vargas, Gulf Canada Resources Inc.

☐ An efficient way to troubleshoot a compression refrigeration system is by using a pressure-enthalpy chart for the refrigerant in the system [1,3].

The design conditions for the refrigeration system being studied can be superimposed on the operating conditions on such a chart. Thus, we can compare the operating cycle to the design values.

We will consider here the smaller refrigeration systems (up to 20 hp) and the larger units. The reason for this division is that some problems such as an undercharge or overcharge of refrigerant in the system, while common to the smaller systems, are not often encountered in the larger ones.

Small refrigeration systems

The most effective method for detecting faults in the smaller units is to determine the suction and discharge pressures at the compressor. Knowing the pressure, we

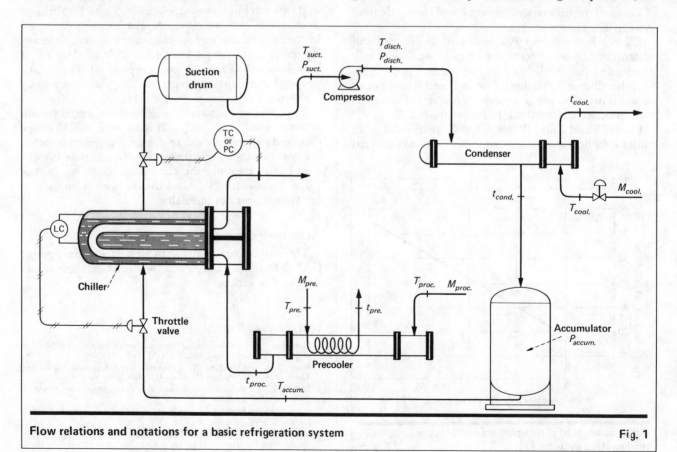

Flow relations and notations for a basic refrigeration system

Fig. 1

Originally published March 22, 1982

can find the temperature automatically from the refrigerant charts [1,3].

Usually, these smaller units have provisions for hooking up service equipment such as valves and gages in order to add refrigerant. These service points are located before and after the compressor.

Some of the most common problems, their causes and remedies for smaller refrigeration units are listed in Table I. It is probably best to construct a pressure-enthalpy chart for the operating conditions, and compare these with the design values. Then, by using Table I, we can determine the problems. Care must be taken with the smaller units to avoid an overcharge or undercharge of refrigerant.

Large refrigeration systems

In order to simplify the analysis of a refrigeration system for troubleshooting, we will use the notation shown in Fig. 1. Thus, when we refer to a pressure, temperature or flow, we will use its symbol. Subcoolers, economizers, etc. will be neglected in order to clarify the troubleshooting procedures—we will consider only a basic refrigeration system (Fig. 1).

The first step in troubleshooting large systems is to draw a pressure-enthalpy diagram of the operating conditions. A comparison of the data in this diagram to the design conditions will give the clues to the problems. These problems can be categorized as:

- Fouling of precooler.
- Impurities in the refrigerant.
- Fouling of the chiller.
- Malfunctions of the throttling valves.
- Fouling or loss of capacity of the condenser.
- Malfunctions of the compressor.

We will examine each of these problems and analyze their respective symptoms.

Fouling of precooler

If the precooler becomes fouled, the chiller will not be able to cool the process fluid, $M_{Proc.}$, to the desired temperature. This fouling could occur on the side of the process fluid or of the cooling-medium fluid, $M_{Pre.}$.

From the heat-exchanger design sheets, we obtain the duty, q, and the inlet and outlet temperatures for both the shellside and tubeside of the exchanger. We can now calculate the heat-transfer coefficient, U, from:

$$q = UA\,[f(\mathrm{LMTD})] \qquad (1)$$

where q = duty, Btu/h; A = heat-exchanger (i.e., tube bundle) surface area, ft²; U = heat-transfer coefficient, Btu/(h)(ft²)(°F); f = heat-exchanger correction factor, dimensionless (this factor depends on heat-exchanger characteristics). For countercurrent flow $f = 1.0$; $(LMTD)$ = log mean temperature difference, °F.

The log mean temperature difference is the approach of the ΔTs between the shellside and tubeside. For countercurrent flow (as shown by the sketch) we calculate U from:

$$(LMTD) = \frac{(T_{hot} - t_{cold}) - (T_{cold} - t_{hot})}{\ln\left(\dfrac{T_{hot} - t_{cold}}{T_{cold} - t_{hot}}\right)} \qquad (2)$$

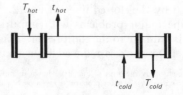

Trouble-shooting small refrigeration units	Table I

Symptom: High discharge pressure

Cause	Remedy
Ambient conditions too hot for air or water.	Change location of unit to obtain cooler air or water for condenser.
Insufficient air across condenser.	Check for obstructions on condenser coils or faulty fan operation.
Refrigeration overcharge.	Purge.
Air in refrigerant system. This gives an extreme increase in head pressure.	Purge air at highest point of condensing unit (air is lighter than refrigerant).
Fouled condenser.	Clean or replace condenser.
No water.	Open condenser water valve.

Symptom: Low discharge pressure

Cause	Remedy
Shortage of refrigerant.	Find and fix leak. Recharge refrigerant reservoir.
Malfunction of compressor's suction or discharge valves.	Clean or replace valves or seats. Overhaul compressor.

Sumptom: Low suction pressure

Cause	Remedy
Insufficient process fluid in the chiller.	Check superheat at chiller outlet. Increase the flow of process fluid to chiller by eliminating obstructions, etc.
Poor refrigerant flow.	Check for restrictions in refrigerant flow in line or at expansion valve. Clean out line or expansion valve.
Short refrigerant charge.	Find refrigerant leak (characterized by hissing at expansion valve because high-pressure vapor rather than liquid passes through valve to evaporator). Refill.
Low process temperature	No problem (low head pressure will be observed as well).

Symptom: High suction pressure

Cause	Remedy
High loading of evaporator coil.	Too hot a load for refrigerant capacity. Upgrade unit.
System operates at low superheat.	Expansion valve has to be adjusted or compressor damage may occur.

Symptom: Process side of chiller temperature too high

Cause	Remedy
Refrigerant shortage	Fix leak. Refill with refrigerant.
Expansion valve plugged, or restriction at refrigerant suction.	Change or repair valve or line.
Misadjustment of expansion valve.	Lower superheat setting for expansion valve.
Unit too small.	Debottleneck and upgrade refrigeration capacity.
Fouled chiller.	Clean.

Note: For an exchanger where there is a change of state and a temperature change, the (*LMTD*) is an average for each state.

After calculating the appropriate numerical values, we determine *U* from:

$$U = q/Af(LMTD) \qquad (3)$$

By comparing the value of *U* for the operating unit as calculated from Eq. (3) to the design value for *U*, fouling can be detected. The ratio of the calculated value of *U* to the design value of *U* should be in the range of 0.90 to 1.00. If this is not the case, fouling is the most probable cause. A problem similar to fouling is the drop in velocity of the fluid through the tubeside.

The first thing to do to defoul a heat exchanger is to determine whether the tubeside or shellside is fouled. In any service, one side of the exchanger is more prone to fouling than the other. The side that is suspect should be cleaned first. If the *U* value does not significantly improve after the exchanger is put back into operation, the other side should also be cleaned.

We will cover fouling and loss of fluid velocities in greater detail when we review fouling of the chiller. This will give us a better appreciation of the actual conditions causing decreases in the values of the overall heat-transfer coefficient, *U*.

Impurities in the refrigerant

On occasion, it may be desirable to change the physical properties of a refrigerant by adding lighter- or heavier-molecular-weight components to an existing refrigerant. The incremental returns, however, from such a procedure are slight compared to the problems that may arise.

An impure refrigerant undergoes changes in almost all of its pressure/temperature characteristics. Hence, the pressure-enthalpy charts have to be drawn for the specific composition of the refrigerant. Commonly, impurities in a refrigerant disrupt the refrigeration cycle. Substances such as air in fluorinated-hydrocarbon refrigerant or ethane in propane can significantly affect the shape of the enthalpy curve of the refrigeration cycle.

The first thing to do is to plot a pressure-enthalpy diagram for the refrigeration system, as shown in Fig. 2* by the solid lines. Label each line in the cycle. To define the cycle, we need the suction pressure and temperature, $P_{suct.}$ and $T_{suct.}$, respectively; discharge pressure and temperature, $P_{disch.}$ and $T_{disch.}$; condenser temperature, $t_{cond.}$; and accumulator pressure, $P_{accum.}$.

Let us next consider the situation where impurities are present in the refrigerant. The refrigeration cycle is now defined by the following:

$T_{suct.} = 40°F, T_{disch.} = 180°F, t_{cond.} = 90°F$

$P_{suct.} = 40$ psia, $P_{disch.} = 300$ psia, $P_{accum.} = 250$ psia

The plot for the cycle containing an impure refrigerant is shown by the dotted lines in Fig. 2.

One of the major problems with impurities in the refrigerant is damage to the compressor. This may arise if the suction pressure is reduced by a change in the

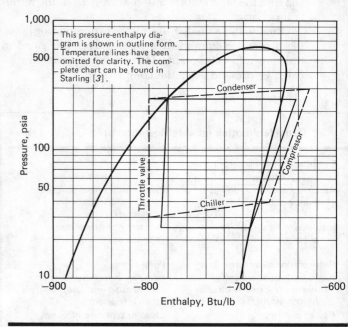

How the impurities affect the propane refrigeration cycle   **Fig. 2**

refrigerant composition or by impurities containing water. Water forms ice and hydrates[†] and has other unwanted side effects in hydrocarbon refrigerants.

It is better to operate the refrigeration unit at design conditions, and not get into unpleasant situations later on. If we observe a cycle such as that shown by the dotted lines in Fig. 2, it may not necessarily point to refrigerant impurities. The following items should be checked:

■ Pressure/temperature gages on compressor suction and discharge.

■ Chiller level is within range.

■ Condenser/accumulator has an adequate amount of liquid in it, as indicated by level.

To confirm the presence of impurities in the refrigerant, we take a sample of the refrigerant, or purge the system of lighter substances at the top of the condenser/accumulator.

Chiller fouling

Chiller fouling is indicated in refrigeration systems when process chilling is lost with little or no change in the temperature and pressure conditions at the compressor suction or discharge. As before, compare the pressure/enthalpy chart with the design values. If no significant change in the overall cycle is evident and loss of refrigeration capacity is present, the problem is in the chiller.

All problems with chiller capacity manifest themselves with a decrease in the heat-transfer coefficient, *U*. As with any heat exchanger, Eq. (1) will model the situation. A decrease in *U* can be due to several factors. The most common is fouling. Loss in *U* can also arise from a loss in fluid velocity on the process side of the chiller or

---

*The pressure-enthalpy diagram of Fig. 2 is shown in outline form. The detailed pressure-enthalpy chart for propane is given by Starling [3].

†Hydrates are hydrocarbon-and-water mixtures that freeze at fairly high temperatures.

from level-control problems. First, let us consider fouling in the chiller:

Fouling—As the chiller becomes fouled, U decreases as does the heat duty q. However, these are not proportional, and U is usually more pronounced—the ($LMTD$) restores equilibrium by increasing. Consider the conditions shown in Fig. 3 for a chiller; compare the nonfouled and fouled cases to enable us to understand what is happening.

Nonfouled chiller—Apply Eq. (1) and let $f = 1.0$ and consider that the heat-transfer area, A, remains fixed. Then:

$$U = q/A(LMTD) \qquad (3)$$

Next, substitute into Eq. (2) to find:

$$(LMTD) = \frac{(30 - 0) - (10 - 0)}{\ln(30/10)} = \frac{20}{1.10} \simeq 18.2°F$$

Now, find the duty q. Care must be taken if there is a change of state in the chiller. The energy–mass balance around the chiller is:

$$q = M_{gas}H_{gas}^n + M_{liq.}H_{liq.}^n - [m_{gas}H_{gas}^{n'} + m_{liq.}^{n'}H_{liq.}^n]$$

where M = mass flowrate in, lb/h; m = mass flowrate out, lb/h; H^n = enthalpy at $n°F$, Btu/lb.

To proceed with our example, let the hydrocarbon streams be:

Stream	lb/h	psia	Molecular weight
M_{gas}	10,000	1,000	20
$M_{liq.}$	0		
m_{gas}	8,000	1,000	17.55
$m_{liq.}$	2,000		30

From tables of enthalpies for hydrocarbon, we find:

$$H_{gas}^{30°} = 152 \text{ Btu/lb}$$
$$H_{gas}^{10°} = 150 \text{ Btu/lb}$$
$$H_{liq.}^{10°} = -35 \text{ Btu/lb}$$

We find the heat duty, q, to be:

$$q = 10,000(152) - [8,000(150) + 2,000(-35)]$$
$$q = 390,000 \text{ Btu/h}$$

By substituting the appropriate values into Eq. (3), calculate the overall heat-transfer coefficient in terms of the surface area, A, as:

$$U = \frac{390,000}{A(18.2)} \simeq \frac{21,429}{A} \text{ Btu/(h)(ft}^2\text{)(°F)}$$

Fouled chiller—We begin the analysis for this case by calculating the ($LMTD$) for the fouled chiller:

$$(LMTD) = \frac{(30 - 0) - (20 - 0)}{\ln(30/20)} = 24.7°F$$

Because of poor heat transfer across the tubes due to fouling, less process gas will condense in the tubes. For illustrative purposes, let us assume that 1,500 lb/h condenses, vs. 2,000 lb/h for the nonfouled case. The heat duty for the fouled chiller becomes:

$$q = 10,000(152) - [8,500(155) + 1,500(-30)]$$
$$q = 247,500 \text{ Btu/h}$$

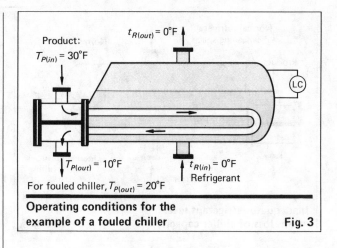

Operating conditions for the example of a fouled chiller   **Fig. 3**

The heat-transfer coefficient, U_F, for the fouled exchanger becomes:

$$U_F = 247,500/A(24.7) \simeq 10,020/A$$

Hence, the ratio $U_F/U = (10,020/A)/(21,429/A)$, or $U_F \simeq (1/2)U$.

To understand how fouling comes about, let us calculate U empirically from the following equation:

$$\frac{1}{U} = \frac{1}{h_o} + r_o + r_w + r_i\left(\frac{A_o}{A_i}\right) + \frac{1}{h_i}\left(\frac{A_o}{A_i}\right) \qquad (4)$$

where h_o and h_i are the outside and inside heat-transfer coefficients, respectively, Btu/(h)(ft^2)(°F); r_o and r_i are the outside and inside scale factors, and r_w is the transfer factor for the metal wall, (h)(ft^2)(°F)/Btu; and A_o and A_i are the outside and inside area of the tubes, ft^2.

The inside and outside heat-transfer coefficients, h_i and h_o, are a function of diameter, velocity, viscosity, specific heat, thermal conductivity and length. The variables for this relationship change according to whether the flow is laminar ($N_{Re} \leqslant 2,100$) or turbulent ($N_{Re} > 2,100$). Most chiller processes are designed for turbulent flow through the tubes because this results in better heat exchange.

Let us assume that fouling is on the process side (i.e., inside the tubes) and that the h_i and h_o do not change (only viscosity or velocity can change them). Assume also that for a clean chiller the following values apply:

$$h_o = 300 \text{ Btu/(h)(ft}^2\text{)(°F)}$$
$$h_i = 100 \text{ Btu/(h)(ft}^2\text{)(°F)}$$
$$r_w = \text{Negligible}$$
$$r_o = 0.0015 \text{ (h)(ft}^2\text{)(°F)/Btu}$$
$$r_i = 0.001 \text{ (h)(ft}^2\text{)(°F)/Btu}$$

We then substitute into Eq. (4), letting the value $A_o/A_i \simeq 1$, in order to evaluate U for the clean chiller:

$$\frac{1}{U} = \frac{1}{300} + 0.0015 + 0.001 + \frac{1}{100} = 0.0158$$
$$U = 1/0.0158 = 63.3 \text{ Btu/(h)(ft}^2\text{)(°F)}$$

For the fouled chiller, compute the fouling factor, r_i, by establishing U_F. We had previously calculated $U_F = 10,020/A$. If $A = 338.5 \text{ ft}^2$, then:

$$U_F = 10,020/338.5 = 29.6 \text{ Btu/(h)(ft}^2\text{)(°F)}$$

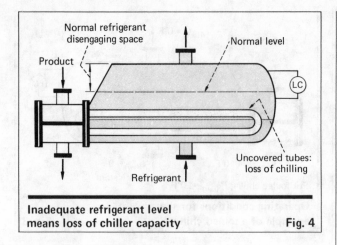

Inadequate refrigerant level
means loss of chiller capacity   **Fig. 4**

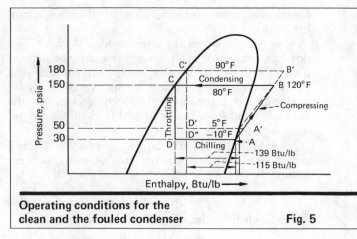

Operating conditions for the
clean and the fouled condenser   **Fig. 5**

We can now substitute U_F into Eq. (4) in order to compute the r_i for the fouled chiller:

$$(1/29.6) = .0033 + 0.0015 + r_i + 0.01$$
$$r_i = 0.019$$

Hence, the fouling factor went from 0.001 for the clean chiller to 0.019 for the fouled chiller.

If fouling is not the problem with the chiller, then loss of velocity through the chiller can be the cause, or change in viscosity, or level-control effects on the refrigerant side.

Loss of velocity—(we observe a similar problem with an increase in viscosity). Let the new lower velocity $V_N = \frac{1}{2}V_{Design}$. Since the heat-transfer coefficient, h_i or h_o, is a function of diameter, velocity, viscosity, specific heat, thermal conductivity, length, we can focus on velocity and write:

$$h_{iD} = V_D^n f(X) \qquad (5)$$

where h_{iD} is the heat-transfer coefficient at design conditions, V_D^n is velocity at design conditions, and $f(X)$ represents the remaining variables affecting the heat-transfer coefficient. The exponent, n, in Eq. (5) has values of $0.33 \leqslant n \leqslant 0.8$. For turbulent flow, $n = 0.8$. Hence, we can relate the new heat-transfer coefficient, h_{iN}, to the lower velocity as:

$$h_{iN} = (\tfrac{1}{2}V_D)^{0.8} f(X)$$
$$h_{in} = 0.57(V_D)^{0.8} f(X)$$

Substituting the appropriate values for this example into Eq. (4) and using $h_{iN} = 0.57 h_{iD}$, we get:

$$\frac{1}{U} = \frac{1}{300} + 0.0015 + 0.001 + \frac{1}{(100 \times 0.57)} = 0.0233$$
$$U = 1/0.0233 = 42.8 \text{ Btu/(h)(ft}^2)(°F)$$

Hence, the overall heat-transfer coefficient has decreased from 63.3 to 42.8.

Level control—Not maintaining an adequate level in the chiller can result in loss of refrigeration capacity (Fig. 4), as some tubes will not be immersed in refrigerant. Heat removal is most efficient when the refrigerant is liquid. As the refrigerant evaporates, it takes energy in the form of heat from the refrigerant liquid. In turn, this liquid takes heat from the tubes and the process liquid on the inside of the tubes. Therefore, all chiller tubes on the shellside must be covered with refrigerant.

Throttle valves

Problems arising with the throttling valve result in too much or too little refrigerant being fed to the chiller. If too little refrigerant is fed, a loss in refrigeration capacity occurs in a manner similar to that described for level control. If too much refrigerant is fed to the chiller, liquid can build up in the compressor suction drum and eventually get into the compressor suction line. Refrigerant flashing now occurs in places other than the chiller, and refrigeration capacity is reduced.

On occasion, throttle-valve problems can be caused by false or low level in the refrigeration accumulator. Thus, very little liquid and much refrigerant vapor is throttled, and refrigeration capacity is lost.

If the throttle valve is always in the open position (just refrigerant vapor passing through), it is due to the absence of refrigerant liquid in the accumulator. Check the refrigerant level—especially for false level indication due to plugs or icing/hydrate formation.

Fouled condenser

To identify a fouled condenser or loss of condensing capacity, plot the temperature/pressure conditions of the refrigeration cycle. Then, compare the condenser temperature to its design value. A higher condensing temperature indicates condenser problems.

Water condenser—The pressure-enthalpy diagram for a condenser in nonfouled (i.e., clean) and fouled conditions is shown in Fig. 5. On comparing the nonfouled (ABCD in Fig. 5) and fouled (A'B'C'D') cycles, we find the condenser's critical conditions to be:

	Nonfouled	Fouled
Compressor suction:		
Pressure, psia	30	50
Temperature, °F	−10	+5
Compressor discharge:		
Pressure, psia	150	180
Temperature, °F	120	140
Condenser exit:		
Pressure, psia	150	180
Temperature, °F	80	90
Enthalpy difference, ΔH, Btu/lb	139	115

Let us analyze what is happening to the refrigeration cycle during condensing and throttling/chilling on the pressure-enthalpy diagram of Fig. 5.

Condensing: If the accumulator is operating normally, all of the refrigerant in the condenser will con-

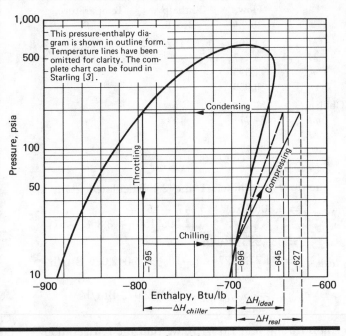

This pressure-enthalpy diagram is shown in outline form. Temperature lines have been omitted for clarity. The complete chart can be found in Starling [3].

Operating conditions of problem for propane refrigeration cycle   **Fig. 6**

dense. However, fouling lowers the heat-transfer coefficient, U (as in the case of the chiller, r_o or r_i gets larger due to the fouling film on the shellside or tubeside of the condenser).

As in the chiller case, the duty, q, decreases somewhat due to loss in heat exchange:

$$q_{water} = m_w C_P \Delta T \qquad (6)$$

where m_w = mass flowrate of water, lb/h; C_p = specific heat of water = 1, Btu/(lb)(°F); and ΔT = temperature difference of water, °F.

As ΔT decreases and all other variables remain constant, q decreases. Eq. (3) relates q and U. Since q and U have decreased, the $(LMTD)$ must increase to restore equilibrium. An increase in the $(LMTD)$ means an increase in condensing temperature, and the refrigerant is now on line B′C′ on Fig. 5.

The Point C′ is the new refrigerant saturated liquid boundary. Initially, throttling will take place along line C′D″. However, the enthalpy available has decreased from 139 to 115 Btu/lb. To compensate, more refrigerant has to be fed to the chiller (mass flowrate compensates for energy loss). The pressure in the chiller increases and the new refrigeration cycle is A′B′C′D. Only subcooling could decrease C′ to C in order to achieve a ΔH of 139 Btu/lb.

Air condenser—Loss of capacity in an air condenser occurs in the same manner as for a water condenser but for different reasons. The basic reason is a decreased airflow across the tubes, which may be caused by:

■ Fouled finned-tubes, caused by debris.
■ Clogged inlet screens.
■ Incorrect action of the air louvers.
■ Improper pitch of the fan blades.
■ Wrong speed of rotation for the fan.
■ Excessive clearance of the blade tips.

Loss of air around the condenser tubes affects U due to a decrease in air velocity, but a reduction in the mass flow of air causes the exit air temperature to rise. Hence, the $(LMTD)$ increases, and the same abnormal refrigeration cycle as in the water condenser takes place (i.e., similar to the cycle A′B′C′D′ of Fig. 5).

Compressor problems

In many instances when a complex problem with a refrigeration system cannot be identified, the compressor is blamed. Definite assurance that the driver and not the compressor causes the problem can save money and downtime in tearing down the compressor.

Compressor efficiency and horsepower consumption are the two key items for evaluating compressor and driver operation. The easiest way to find compressor efficiency and horsepower draw is to plot them on a pressure-enthalpy diagram.

Let us assume the following conditions for propane refrigerant: suction pressure and temperature are 18 psia and −20°F, respectively; discharge pressure and suction are 189 psia and 165°F. To prepare diagram:

1. Plot the complete refrigeration cycle (Fig. 6).
2. Notice that the actual compression line does not follow the isentropic line. In Fig. 6, the true isentropic line is shown dashed.
3. Consider that an accurate compression line must include the pressure drop at both suction and discharge. (On Fig. 6, this is difficult to do.) Typical suction and discharge pressure drops are: 3 to 5 psi for reciprocating compressors, and 1 psi for all other types.

To determine the compressor adiabatic efficiency, η_C, solve:

$$\eta_C = \frac{\Delta H_I}{\Delta H_R} \qquad (7)$$

where ΔH_I is the ideal change in enthalpy between suction and discharge conditions, plotted along the isentropic line (dashed line in Fig. 6), and ΔH_R is the difference in enthalpies between suction and discharge at actual compressor conditions (solid line in Fig. 6).

Note: An accurate compression line should account for suction/discharge pressure losses. A 5-psi drop should suffice for the pressure line.

For our example, the efficiency becomes:

$$\eta_C = \frac{-645 - (-696)}{-627 - (-696)} = \frac{51}{69} = 0.74$$

Typical compressor adiabatic efficiencies are:

Reciprocating	0.85 to 0.95
Screw	0.80 to 0.85
Centrifugal	0.70 to 0.75

If our compressor is a centrifugal, it is running at its correct efficiency. When compressor efficiency is low, we should look for:

■ Open unloading pockets, or leaking internal valves. Such leaks are serious because the discharge temperature gets extremely high, and the system may burn out. The compression line in this case diverges excessively from the isentropic line, and ΔH_R gets very high.

■ Extreme ambient conditions.

■ Cylinder wear on reciprocating compressors, or wear on labyrinth seals or wheels on centrifugals.

■ Faulty bypass valves on suction and discharge.

■ Compressor driver problems. To verify that the driver is operating correctly, we must determine the ideal horsepower, obtain the design horsepower and calculate the horsepower draw ratio for the compressor from the following:

$$(HP)_{d.r.} = (HP)_I/(HP)_D \qquad (8)$$

where $(HP)_{d.r.}$ is the horsepower draw ratio, $(HP)_I$ is ideal horsepower (i.e., horsepower drawn during compressor operation), and $(HP)_D$ is the nameplate (i.e., design) horsepower.

We determine the ideal horsepower for:

a. Electric-motor drivers, from:

$$(HP)_I = EI\,\eta_m/746 \qquad (9)$$

where E is voltage, V; I is current, A; η_m is motor efficiency; and 746 W = 1 hp.

b. Other drivers (steam turbines, etc.), from:

$$(HP)_I = M_R\,\Delta H_C/2{,}545 \qquad (10)$$

where M_R is the refrigerant circulation rate, lb/h; ΔH_C is the compressor enthalpy rise, Btu/lb; and 2,545 Btu/h = 1 hp.

The compressor enthalpy, ΔH_C, is found from Fig. 6 where $\Delta H_{chiller} = \Delta H_C$:

$$\Delta H_C = -696 - (-795) = 99 \text{ Btu/lb}$$

To find the refrigeration circulation rate, M_R, might be more difficult depending on whether there is a flowmeter on the line. Be sure that the flowmeter is working correctly and is placed to record single-phase flow.

Let us now consider another way for determining the refrigeration rate. The heat taken up by the refrigerant in the chiller is equal to the heat removed from the process, or:

$$q_{process} = q_{refrig.}$$

To determine the heat removed from the process, we must evaluate:

$$q_{process} = \sum_{i=1}^{k} [M_{gas,i} \times H^n_{gas,i} + M_{liq,i} \times H^n_{liq,i} -$$
$$(m_{gas,i} \times H^{n'}_{liq,i} + m_{liq,i} \times H^{n'}_{liq,i})] \quad (11)$$

where M_i or m_i is the mass flowrate into (M) or out of (m) the chiller in liquid or gaseous phase for chiller i; H^n_i is the enthalpy of the process fluid at $n°$F for chiller i; and k is the k-th chiller (i.e., the summation is for all chillers in service).

To determine the heat taken up by refrigerant, we evaluate:

$$q_{refrig.} = M_R \Delta H_C \qquad (12)$$

where M_R is the refrigerant circulation rate, lb/h, and ΔH_C is the compressor enthalpy rise, Btu/lb.

Since our earlier example had a $\Delta H_C = 99$ Btu/lb, we can write an equation for the refrigerant circulation rate in terms of ΔH_C as:

$$M_R = q_{refrig.}/99 = q_{process}/99 \qquad (13)$$

Let us work out an example for two chillers operating under the following conditions:

Chiller 1:

	Flow, lb/h	Enthalpy, Btu/lb	Temperature, °F
M_{gas}	4,806	152	60
$M_{liq.}$	5,447	−25	60
m_{gas}	1,395	150	0
$m_{liq.}$	8,858	−57	0

Chiller 2:

	Flow, lb/h	Enthalpy, Btu/lb	Temperature, °F
M_{gas}	34,773	127	39
$M_{liq.}$	—	—	—
m_{gas}	34,773	−18	25
$m_{liq.}$	—	—	—

The duty for Chillers 1 and 2 is calculated from:

$$q_{Chiller\,1} = 4{,}806(152) + 5{,}447(-25) -$$
$$[1{,}395(150) + 8{,}858(-57)]$$
$$= 889{,}993 \text{ Btu/h}$$

$$q_{Chiller\,2} = 34{,}773[127 - (-18)] = 5{,}042{,}085 \text{ Btu/h}$$

$$q_{process} = \sum_{i=1}^{2} q_{Chiller\,i} = 5{,}932{,}078 \text{ Btu/h}$$

Substituting into Eq. (13), we find the refrigeration recirculation rate, M_R, to be:

$$M_R = 5{,}932{,}078/99 \simeq 59{,}920 \text{ lb/h}$$

Using this value in Eq. (10), we establish the ideal horsepower for this example as:

$$(HP)_I = 59{,}920(51)/2{,}545 \simeq 1{,}201 \text{ hp}$$

If the driver for this installation is designed for 1,500 hp, we find the horsepower draw ratio from Eq. (8) as:

$$(HP)_{d.r.} = 1{,}201/1{,}500 \simeq 0.80$$

Normally, the $(HP)_{d.r.}$ is between 0.75 and 0.85. Hence, for this example, the compressor driver is operating correctly.

References

1. "Engineering Data Book," 9th ed., Gas Processors Suppliers Assn. [formerly Natural Gas Processors Suppliers Assn.], Tulsa, Okla., 1972.
2. Canjar, L., and Manning, F., "Thermodynamic Properties & Reduced Correlations for Gases," Gulf Publishing, Houston, 1967.
3. Starling, K. E., "Fluid Thermodynamic Properties for Light Petroleum Systems," Gulf Publishing, Houston, 1973.
4. Young, V. W., and Young, G. A., "Elementary Engineering Thermodynamics," McGraw-Hill, New York, 1941.
5. "Gas Technology Notes," Canadian Natural Gas Processors Assn., Calgary, Alta., Canada, 1975.

The author

Kenneth J. Vargas is an operations engineer at the Rimbey Operations of Gulf Canada Resources Inc., P.O. Box 530, Rimbey, AB T0C 2J0, Canada. A graduate of the U.S. Air Force Academy with a B.Sc. in mathematics and mechanics, he has previously worked in process and maintenance engineering for Eldorado Nuclear Ltd. Canada and Du Pont Mexico.

Refrigeration systems for low-temperature processes

Understanding the thermodynamics of the vapor-compression cycle provides the basis for practical design of actual refrigeration units. Here is detailed information for designing single-stage, multistage and cascaded systems.

Yuv R. Mehra, El Paso Hydrocarbons Co.

☐ Refrigeration systems are common in processes related to the petroleum-refining, petrochemical and chemical industries. The selection of a refrigerant is generally based upon its availability and cooling range, and previous experience with it.

For instance, in an olefins plant, pure ethylene and propylene are readily available; whereas in a natural-gas processing plant, ethane and propane are at hand. Propane or propylene may not be suitable in an ammonia plant because of the risk of contamination, while ammonia may very well serve the purpose. Fluorocarbons have been used extensively because of their non-flammable characteristics.

Due to their inherent properties, a variety of refrigerants as listed in Table I are used quite economically over a wide range of cooling temperatures.

All types of compressors—reciprocating, screw and centrifugal—are used for refrigeration services. The theory of refrigeration can be applied to any compressor, but side loads are usually considered only in centrifugal compressors. Hence, we will confine our discussion to centrifugal machines.

The refrigeration effect can be achieved by using one of the following cycles: (a) vapor compression (reversed Carnot), (b) expansion (reversed Brayton), (c) absorption, and (d) steam jet (water-vapor compression). All of these cycles have been used successfully in industrial refrigeration, but the majority of installations use vapor compression. Therefore, we shall discuss only the reversed Carnot cycle.

Thermodynamics of cycles

A Carnot cycle is composed of two isothermal and two isentropic processes, as represented on the temperature-entropy (*T-S*) and the pressure-enthalpy (*P-H*) diagrams in Fig. 1a. Here, Process 1-2 represents expan-

sion or pressure reduction; Process 2-3 represents heat rejection at constant temperature; Process 3-4 represents compression or increase in pressure; and Process 4-1 completes the cycle by heat addition at constant temperature. To carry out a Carnot cycle, it is necessary to have an ideal fluid.

However, real cycles, both direct and reverse, operate with fluids that undergo phase changes during the cycles. It is important to recognize that real processes transfer heat at essentially constant pressure instead of constant temperature.

A vapor-compression or a reversed Carnot cycle includes the same processes that occur during the expansion of a fluid in the direct Carnot cycle but in a reversed order. This cycle can be represented on *T-S* and *P-H* diagrams, as shown in Fig. 1b.

Vapor-compression cycle

A vapor-compression cycle can also be represented by hooking up equipment in the sequence shown in Fig. 2a. In order to illustrate the processes involved in this refrigeration cycle, let us consider each step:

Expansion process—The expansion process, Point 1 to Point 2 (1-2) in Fig. 2a, can also be referred to as an isenthalpic process. In a refrigeration cycle, it can be accomplished by flashing the liquid refrigerant through a control or expansion valve. The process can be represented on a *P-H* diagram, as in Fig. 2b.

Every refrigerant has its own *P-H* diagram that represents all thermodynamic properties. From a refrigeration standpoint, the envelope formed by the bubble-point curve and the dew-point curve joining each other at the critical point is very important.

The area left of the bubble-point curve represents subcooled liquid refrigerant; the area between the bubble-point and dew-point curves represents the presence

Originally published July 12, 1982

65

of vapor and liquid refrigerant, while the area to the right of the dew-point curve is superheated vapor. Saturated liquid exists along the bubble-point curve, while saturated vapor is present along the dew-point curve.

The starting point in a refrigeration cycle is the availability of liquid refrigerant. As will become apparent later, this liquid in most cases is at its saturation pressure at a given temperature. Therefore, Point 1 (the starting point) is located on the bubble-point curve and is at its saturation pressure of P_1, psia, at an enthalpy of h_{L1}, Btu/lb. In an expansion process, the pressure is reduced by flashing the liquid through a control valve to a pressure P_2, psia. The lower pressure (P_2) is a function of the desired refrigeration temperature T_2, °F, as determined by the vapor-pressure curve. For pure refrigerants, the saturation pressure and temperature lines under the envelope are the same, and run horizontally across the bubble-point and dew-point curves.

At the desired refrigeration temperature that corresponds here to P_2, the enthalpy of saturated liquid is h_{L2}, while the corresponding saturated vapor enthalpy is H_{V2}. Since the expansion process (1-2) occurs across the control valve and no energy has been exchanged, the enthalpy at the outlet of the control valve is the same as at the inlet, h_{L1}. This process is represented in Fig. 2b by a vertical line between Points 1 and 2.

Since Point 2 is inside the envelope, both vapor and liquid coexist. In order to determine the amount of vapor formed in the expansion process, let X be the fraction of liquid at low pressure P_2 with enthalpy h_{L2}. Therefore, the fraction of vapor formed during the expansion process with an enthalpy H_{V2} is $(1 - X)$. Hence, we may write equations for the heat balance, the fraction of liquid formed, and the fraction of vapor formed as:

$$(X)h_{L2} + (1 - X)H_{V2} = h_{L1} \tag{1}$$

$$X = \frac{H_{V2} - h_{L1}}{H_{V2} - h_{L2}} = \frac{a}{b} \tag{2}$$

$$(1 - X) = \frac{h_{L1} - h_{L2}}{H_{V2} - h_{L2}} = \frac{(b - a)}{b} \tag{3}$$

Evaporation process—This portion of the cycle (2-3) absorbs heat by the evaporation of liquid refrigerant through its latent heat. As shown in Fig. 2c, this process is completed at constant pressure and temperature. The

vapor formed in the expansion process (1-2) does not provide any refrigeration. Physically, the evaporation takes place in a heat exchanger—sometimes referred to as an evaporator or a chiller. The refrigeration is provided by the cold liquid, and its refrigeration effect can be defined as:

$$X(H_{V2} - h_{L2}) = (H_{V2} - h_{L1}), \text{ Btu/lb} \tag{4}$$

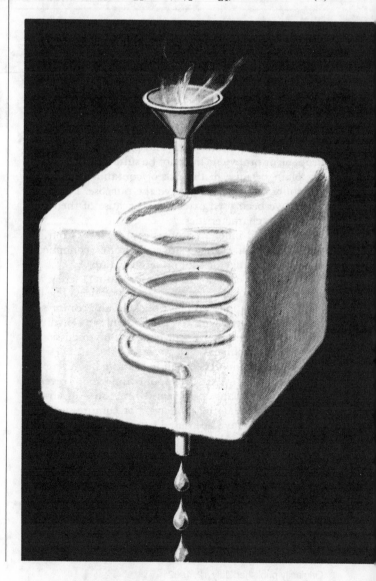

Cooling ranges for refrigerants	Table I

Refrigerant	Temperature range, °F
Methane	−200 to −300
Ethane and ethylene	−75 to −175
Propane and propylene	40 to −50
Butanes	60 to 10
Ammonia	80 to −25
Refrigerant-12	80 to −20

Polytropic efficiency of centrifugal compressors	Table II

Normal inlet flow range, ft³/min	Nominal polytropic efficiency, η_p
500— 8,000	0.76
6,000— 23,000	0.77
20,000— 35,000	0.77
30,000— 58,000	0.77
50,000— 85,000	0.78
75,000—130,000	0.78
110,000—160,000	0.78
140,000—190,000	0.78

Source: Ref. 1

The refrigeration effect (or refrigeration capacity) refers to the total amount of heat absorbed in the chiller, and is generally expressed as "tons of refrigeration," or Btu/unit time. A ton of refrigeration equals 12,000 Btu/h, or 200 Btu/min. To determine the refrigerant flow (m, lb/h) required through the evaporator, we divide the refrigeration duty (Q_{Ref}, Btu/h) by the refrigeration effect ($H_{V2} - h_{L1}$), Btu/lb, or:

$$m = Q_{Ref}/(H_{V2} - h_{L1}) \qquad (5)$$

Compression process—The refrigerant vapors leave the chiller or evaporator at its saturation pressure P_3. The corresponding refrigeration temperature is T_3 (since $T_2 = T_3$) at an enthalpy of H_{V2}. The entropy at this point is S_3, Btu/(lb)(°F). These vapors are compressed isentropically to pressure P_1 along Line 3-4' (Fig. 2d) having an entropy S_3.

The adiabatic work, W_{ad}, for compressing the refrigerant from P_2 to P_1 is given by:

$$W_{ad} = (H'_{V1} - H_{V2})m, \text{ Btu/h} \qquad (6)$$

where m is flow of refrigerant through the compressor, lb/h; and ($H'_{V1} - H_{V2}$) is the adiabatic head, ΔH_{ad}.

We determine H'_{V1} from refrigerant properties at P_1 and an entropy of S_3. Since the refrigerant is not an ideal fluid and since the compressors for such services do not operate ideally, adiabatic efficiency, η_{ad}, has been defined to compensate for the inefficiencies of the compression process. Therefore, the actual work of compression, W, can be calculated from:

$$W = \frac{W_{ad}}{\eta_{ad}} = \frac{m(H'_{V1} - H_{V2})}{\eta_{ad}}$$
$$= m(H_{V1} - H_{V2}), \text{ Btu/h} \qquad (7)$$

The enthalpy at discharge is given by:

$$H_{V1} = \frac{\Delta H_{ad}}{\eta_{ad}} + H_{V2} \qquad (7a)$$

The work of compression can also be converted to horsepower, and expressed as gas horsepower (*GHP*), or:

$$(GHP) = \left(\frac{\Delta H_{ad}}{\eta_{ad}}\right)\left(\frac{m}{2,544.5}\right) \qquad (7b)$$

where 2,544.5 Btu/h = 1 hp. For most refrigerants, charts and tables of thermodynamic properties are readily available [1,2].

Eq. (6) is very convenient for determining compression work. In order to use Eq. (6), we should know η_{ad}. Since commercial compressors have standardized frame sizes, compressor manufacturers provide the nominal polytropic efficiency, η_p, for their equipment. Table II lists the nominal polytropic efficiency for one compressor manufacturer's line [1]. Knowing the inlet flow to the compressor at P_3 and T_3, the corresponding polytropic efficiency can be determined; and from this polytropic efficiency, the adiabatic efficiency can be obtained from:

$$\eta_{ad} = \frac{(0.77 + 0.16x)}{(1 + 0.357x)} + 1.333x^{0.16}(\eta_p - 0.77) \qquad (8)$$

where:
$$x = [(P_1/P_2)^{(K-1)/K} - 1] \qquad (9)$$

and K = heat capacity ratio, C_p/C_v, of the gas. Heat capacity ratios for several common refrigerants are shown in Table III.

Condensation process—The superheated refrigerant

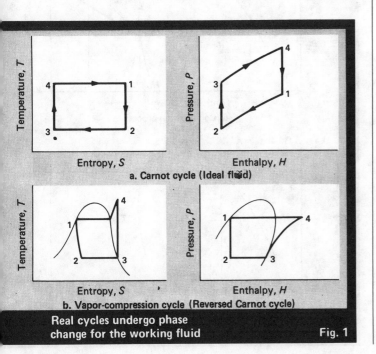

a. Carnot cycle (Ideal fluid)

b. Vapor-compression cycle (Reversed Carnot cycle)

Real cycles undergo phase change for the working fluid   Fig. 1

Heat-capacity ratios for common refrigerants Table III	
Refrigerant	**K^***
Methane	1.31
Ethane	1.19
Ethylene	1.24
Propane	1.13
Propylene	1.15
n-Butane	1.09
Isobutane	1.10
Refrigerant-12	1.12
Ammonia	1.31

*$K = C_p/C_v$, heat-capacity ratio at 60°F and 1 atm.

leaving the compressor at P_1 and T_4 (Point 4 in Fig. 2e) is cooled at constant pressure until its temperature reaches the dew point T_1, and refrigerant vapors begin to condense at constant temperature.

Under the condensation process, all heat added to the refrigerant during evaporation and compression must be removed so that the cycle can be completed by reaching Point 1 (the starting point) on the P-H diagram, as shown in Fig. 2e or 2b.

By adding the refrigeration effect to the heat of compression, we calculate the condensing duty, Q_{cd}, from:

$$Q_{cd} = m(H_{V1} - h_{L1}) = $$
$$m[(H_{V2} - H_{L1}) + (H_{V1} - H_{V2})] \quad (10)$$

It is important to note that the condensing pressure of the refrigerant is a function of the cooling medium available—air, cooling water or another refrigerant. The cooling medium removes the heat input, Q_{cd}, from the refrigeration cycle. In other words, the required discharge pressure of the compressor is established by the cooling medium. If a compressor cannot reach condensing pressure, condensation will not occur; the compressor will surge and the refrigeration unit will shut down.

Sometimes the saturated liquid is subcooled in the condenser to eliminate flash gas during the expansion process (1-2) of the cycle. This helps to reduce the circulation rate of refrigerant, m, as determined from Eq. (2), (3), (4) and (5), where X becomes 1 and $(1 - X)$ equals zero.

Actual refrigeration system

Thus far in our discussion, we have ignored the effects of pressure drop in the piping and heat exchangers associated with the refrigeration cycle. A single-stage refrigeration system with inclusion of pressure drops is shown in Fig. 3a. In this system:

$$P_1' = P_1 + \Delta P_1$$
$$P_2' = P_2 - \Delta P_2$$

where ΔP_1 = pressure drop between compressor discharge nozzle and inlet nozzle of the receiver (typically the range is between 5 and 10 psi), and ΔP_2 = pressure drop between the outlet of evaporator or chiller and inlet nozzle of the compressor (typically 1.5 psi).

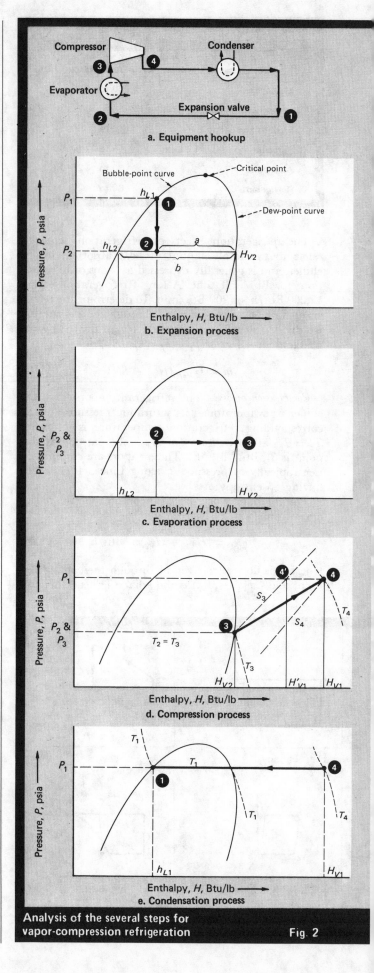

a. Equipment hookup

b. Expansion process

c. Evaporation process

d. Compression process

e. Condensation process

Analysis of the several steps for vapor-compression refrigeration   Fig. 2

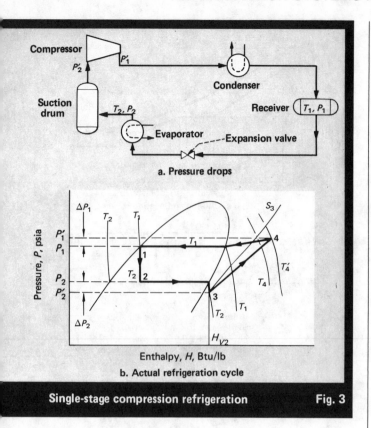

a. Pressure drops

b. Actual refrigeration cycle

Single-stage compression refrigeration   Fig. 3

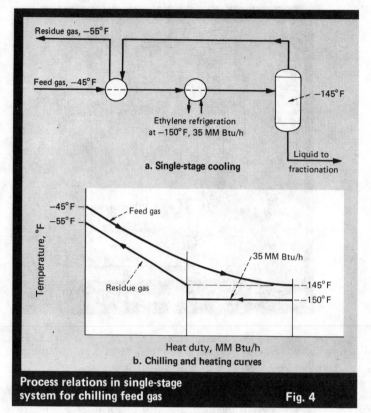

a. Single-stage cooling

b. Chilling and heating curves

Process relations in single-stage system for chilling feed gas   Fig. 4

The compression ratio, r, across the compressor is:

$$r = (P_1'/P_2')^{1/N} \qquad (11)$$

where N = number of compression stages.

For a single-stage machine, the compression ratio then becomes: $r = P_1'/P_2'$.

The actual refrigeration cycle on a P-H diagram is shown in Fig. 3b. Note that compression, Process (3-4), does not start at the dew-point curve, and that the discharge pressure of the compressor is greater than the condensing pressure. The vapors entering the compressor, Point 3, are slightly superheated, and the isentropic line is slightly to the right in Fig. 3b. This would result in a compressor discharge temperature T_4' greater than T_4 if pressure drops were not included. Also, T_3 at Point 3 is slightly cooler than refrigeration temperature T_2. It is important to note that P_1 is a function of the condensing medium, and P_2 is a function of the required refrigeration temperature.

Single-stage system

Let us consider a process, as shown in Fig. 4a, where the feed-gas stream should be cooled to $-145\,°F$ to maximize liquids recovery before rejecting uncondensible gases.

In order to minimize energy consumption, it is advantageous to recover cryogen from the residue gas by cooling the feed gas. The chilling and heating curves for the feed and residue-gas streams are plotted (Fig. 4b) to determine whether temperature crossovers occur, and the required refrigeration level. The final feed-gas temperature of $-145\,°F$ can be easily achieved by a $-150\,°F$ refrigeration level.

If ethylene refrigerant is available, we obtain its saturation pressure at $-150\,°F$ as 17.15 psia from the vapor-pressure curve for ethylene [2,3]. A typical pressure drop around the suction side of a compressor is 1.5 psi. Hence, we find a compressor inlet pressure of 15.65 psia (i.e., $17.15 - 1.5$). Since this inlet pressure is greater than 14.7 psia (atmospheric pressure), it is quite safe to use ethylene refrigerant for this service.

The single-stage ethylene refrigeration cycle is shown in Fig. 5. In order to complete the refrigeration cycle, it is assumed that propylene refrigerant will be available at $-50\,°F$ so that ethylene can be condensed at $-45\,°F$. The enthalpy data in Fig. 5 for ethylene refrigerant, necessary to determine the refrigerant flowrates, are readily available [2,3].

We will now evaluate for this single-stage cycle the refrigerant flowrate, m; adiabatic work, ΔH_{ad}; compression ratio, r; inlet volumetric flowrate, V_f; discharge enthalpy, H_{V1}; and gas horsepower, (GHP). We begin by substituting appropriate values into Eq. (5) to find m:

$$m = \frac{35 \times 10^6}{(1,016.3 - 876.05)} = 249,554 \text{ lb/h}$$

From Starling [2], we obtain the ethylene entropy at a compressor inlet pressure of 15.65 psia and an enthalpy of 1,016.3 Btu/lb as 1.6783 Btu/(lb)(°R). Therefore, the isentropic enthalpy at a compressor discharge pressure of 194.75 psia is 1,085.25 Btu/lb.

Substituting into Eq. (6) yields the adiabatic head:

$$\Delta H_{ad} = 1,085.25 - 1,016.3 = 68.95 \text{ Btu/lb}$$

From Table III, we find the heat-capacity ratio, K, for ethylene as 1.24. And from Starling [2], we find the specific volume of ethylene as 7.327 ft³/lb at 15.65 psia

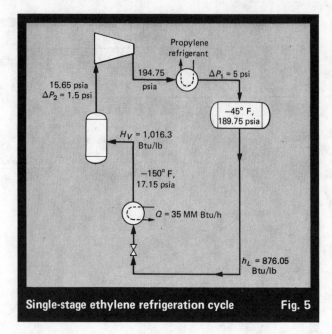

Single-stage ethylene refrigeration cycle   Fig. 5

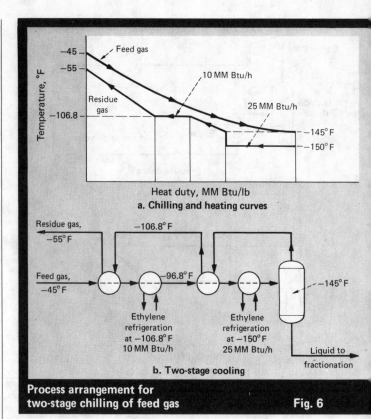

a. Chilling and heating curves

b. Two-stage cooling

Process arrangement for
two-stage chilling of feed gas   Fig. 6

and 1,016.3 Btu/lb. Therefore, the inlet volumetric flowrate becomes:

$$V_f = (249,554 \times 7.327)/60 = 30,474 \text{ ft}^3/\text{min}$$

The polytropic efficiency is 0.77 for inlet volumetric flowrates ranging from 20,000 to 35,000 ft³/min. Making the necessary substitutions into Eq. (8), we determine the adiabatic efficiency, η_{ad}, to be 0.7125. Substituting into Eq. (7a) yields the discharge enthalpy:

$$H_{V1} = (68.95/0.7125) + 1,016.3 = 1,113.07 \text{ Btu/lb}$$

From Eq. (7b), we determine the compression power:

$$(GHP) = \frac{(68.95)(249,554)}{(0.7125)(2,544.5)} = 9,492 \text{ hp}$$

The gas horsepower, (GHP), does not include losses due to mechanical seals and gears. The brake horsepower, (BHP), is defined as:

$$(BHP) = (GHP) + Losses$$

For our discussion, we will not address these losses because they depend upon the type of seals, operating speeds, compressor designs, etc. Typically, these losses range from 50 hp to 150 hp.

From Eq. (10), we establish the condenser duty:

$$Q_{cd} = (1,113.07 - 876.05)(249,554)$$
$$= 59.15 \times 10^6 \text{ Btu/h}$$

Two-stage system

Process chilling curves may indicate energy savings. Let us consider the same heating and cooling duties as for the single-stage system. For the projected two-stage system, the heating and cooling curves are shown in Fig. 6a. By splitting the residue-gas heating curve, the ethylene refrigeration duty at −150°F could be reduced from 35MM Btu/h* to 25MM Btu/h, while the remaining 10MM Btu/h can be provided at −106.8°F. The second level of refrigerant was determined by using

*MM stands for million, i.e., 10⁶.

equal compression ratio between the two stages. The new schematic for the process is shown in Fig. 6b. Here, the process feed gas is still chilled from −45° to −145°F. The two-stage refrigeration system can be represented by Fig. 7.

In order to determine the interstage refrigeration level for a two-stage system, we must determine the compression ratio per stage from Eq. (11):

$$r = (194.75/15.65)^{1/2} = 3.53$$

The interstage pressure (i.e., the pressure between the first and second stages) is 15.65 × 3.53, or 55.21 psia. Hence, the pressure at the second-stage chiller equals 55.21 + 1.5, or 56.71 psia. From the vapor-pressure curve for ethylene [2,3] the refrigeration temperature is equivalent to −106.8°F.

Substituting into Eq. (5), we find the refrigerant flowrate, m, through each chiller:

$$m_1 = \frac{25 \times 10^6}{(1,016.3 - 837.63)} = 139,924 \text{ lb/h}$$

$$m_2 = \frac{10 \times 10^6}{(1,024.71 - 876.05)} = 67,268 \text{ lb/h}$$

where m_1 is flowrate through first stage, and m_2 through second stage.

Liquid flow to the first-stage chiller (139,924 lb/h) is provided by flashing the liquid refrigerant from the refrigerant receiver at −45°F, and bypassing the second-stage chiller.

In order to determine the flow of liquid refrigerant from the receiver, let us consider the heat and material balances shown in Fig. 8. Here, let X lb/h denote the refrigerant bypassing the second-stage chiller, which

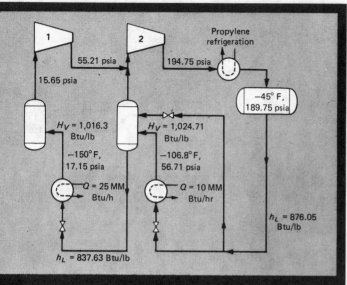

Two-stage ethylene refrigeration system   **Fig. 7**

produces 67,268 lb/h of refrigerant vapor at $-106.8°F$. These vapors flow through the second-stage suction drum, and leave overhead. The liquid required from the second-stage flash drum for the first-stage chiller comes from the quantity X.

By material balance (see Fig. 8), we find the vapors leaving the second-stage suction drum as: $X + 67,268 - 139,924$, or $(X - 72,656)$ lb/h. By heat balance around the suction drum, we can determine the amount of liquid, X, required for the second-stage suction drum from:

$$(X - 72,656)(1,024.71) + (139,924)(837.63) =$$
$$X(876.05) + (67,268)(1,024.71)$$
$$X = 176,085 \text{ lb/h}$$

In order to calculate adiabatic work for the first stage,

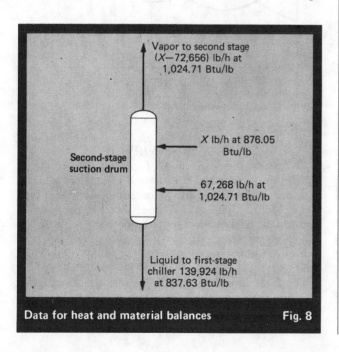

Data for heat and material balances   **Fig. 8**

we need the isentropic enthalpy at 55.21 psia. From Starling [2], we find the first-stage inlet entropy as 1.6783 Btu/(lb)(°R), and the corresponding isentropic enthalpy at 55.21 psia as 1,045.32 Btu/lb. Substituting into Eq. (6), we find:

$$\Delta H_{ad} = 1,045.32 - 1,016.3 = 29.02 \text{ Btu/lb}$$

For ethylene refrigerant, $r = 3.53$, $k = 1.24$. From Starling [2], we obtain the specific volume for ethylene at 15.65 psia and 1,016.3 Btu/lb as 7.327 ft³/lb. Therefore, the inlet volumetric flowrate for the first stage is:

$$V_{f,1s} = (139,924 \times 7.327)/60 = 17,087 \text{ ft}^3/\text{min}$$

The polytropic efficiency, η_p, is 0.77 for flowrates ranging from 6,000 to 23,000 ft³/min. And from Eq. (8), we calculate the adiabatic efficiency, η_{ad}, as 0.74.

The required compression power for the first stage is obained by using Eq. (7b):

$$(GHP)_{1s} = \frac{(29.02)(139,924)}{(0.74)(2,544.5)} = 2,157 \text{ hp}$$

Using Eq. (7a), we determine the first-stage discharge enthalpy as:

$$H_{V1,1s} = (29.02/0.74) + 1,016.3 = 1,055.52 \text{ Btu/lb}$$

A material balance around the second compression stage yields the second-stage vapor flow:

$$m_{v,2s} = 176,085 - 72,656 + 139,924 = 243,353 \text{ lb/h}$$

A heat balance around the second compression stage yields the second-stage inlet enthalpy:

$$H_{V2,2s} = \frac{(1,055.52)(139,924) + (1,024.71)(103,429)}{243,353}$$
$$H_{V2,2s} = 1,042.43 \text{ Btu/lb}$$

From Starling [2], we find inlet entropy at 55.21 psia and 1,042.43 Btu/lb as 1.6711 Btu/(lb)(°R), and the isentropic enthalpy at 194.75 psia as 1,081.35 Btu/lb.

Substituting into Eq. (6), we obtain the adiabatic head across the second stage as:

$$\Delta H_{ad} = 1,081.35 - 1,042.43 = 38.92 \text{ Btu/lb}$$

The required compression power for the second stage is determined from Eq. (7b):

$$(GHP)_{2s} = \frac{(38.92)(243,353)}{(0.74)(2,544.5)} = 5,030 \text{ hp}$$

The total compression horsepower, $(GHP)_N$, is:

$$(GHP)_N = \sum_{i=1}^{N}(GHP)_i \qquad (12)$$

where N is the number of stages.

Hence, the compression horsepower required for the two-stage system becomes:

$$(GHP)_2 = 2,157 + 5,030 = 7,187 \text{ hp}$$

Using Eq. (7a), we now calculate the second-stage discharge enthalpy:

$$H_{V1,2s} = (38.92/0.74) + 1,042.43 = 1,095.02 \text{ Btu/lb}$$

Substituting into Eq. (10) yields the condenser duty:

$$Q_{cd} = (1,095.02 - 876.05)(243,353 \times 10^{-6})$$
$$Q_{cd} = 53.29\text{MM Btu/h}$$

Three-stage system

In order to conserve energy, an alternative process scheme may be considered in which liquids are separated at $-78°$, $-112°$ and $-145°$F instead of having all of the liquids formed at $-145°$F. This helps to reduce the total refrigeration duty from 35MM Btu/h to 32MM Btu/h because liquids condensed at warmer temperatures may not be subcooled to $-145°$F. A heating and chilling curve is shown in Fig. 9a.

By performing an analysis similar to that for a two-stage system, the total compression horsepower and condenser-duty requirements for a three-stage system can be determined. For a three-stage ethylene refrigeration system having the duties shown in Fig. 9, we find the total horsepower for the ethylene compressor and the condenser duty to be:

$$(GHP)_3 = 5,477 \text{ hp}; \quad Q_{cd} = 45.94\text{MM Btu/h}$$

Effect of interstages

To illustrate the effect of interstages, let us consider the results for the three ethylene systems:

Stages, N	1	2	3
$(GHP)_N$, hp	9,492	7,187	5,477
Q_{cd}, MM Btu/h	59.15	53.29	45.94

There is a definite reduction in horsepower and condenser-duty requirements. For these refrigeration systems, the condenser duty becomes the refrigeration load for the propylene or propane refrigerant. In summary, if the compression horsepower for ethylene refrigeration can be reduced by shifting refrigerant load from low levels to warmer levels, the cascaded refrigerant will also have lower compression-power requirements.

For a propane refrigeration system, Table IV illustrates the effect of interstages, without using the refrigeration at intermediate levels. It is clear from these data that energy consumption is reduced as the number of stages is increased. However, it is important to recognize that the installation cost of such refrigeration systems also increases as the number of stages. The optimum overall cost will be a function of the specific system, and has to be worked out for a set of criteria.

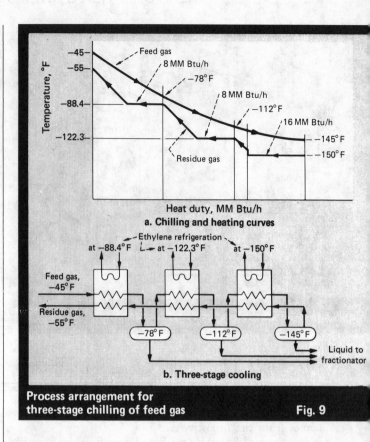

a. Chilling and heating curves

b. Three-stage cooling

Process arrangement for three-stage chilling of feed gas   Fig. 9

Based upon experience with several refrigeration systems, the following correlation has been developed as a general rule for determining the optimum number of refrigeration levels:

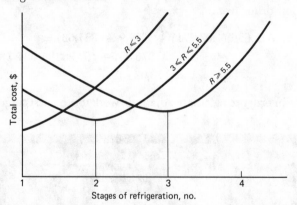

where R, the compression ratio, is the condensing pressure divided by the low-level suction pressure (psia).

Effect of condensing medium

The availability and choice of a medium for condensing the refrigerant has a profound effect on the compression horsepower and condensing-duty requirements. Mehra [3] illustrated the effect of the condensing medium on refrigeration requirements for single-, two- and three-stage systems. Let us consider a single-stage propylene refrigerant system, as summarized in Table V.

The table illustrates that the colder the condensing medium, the lower the refrigeration requirements. For most Gulf Coast locations, a condensing temperature of $100°$F is common. If a comparable system is located

Effect of interstages in a propane refrigeration system			Table IV
Stages, N	1	2	3
Refrigeration duty, MM Btu/h	1.0	1.0	1.0
Refrigeration temperature, °F	−40	−40	−40
Refrigerant condensing temperature, °F	100	100	100
Compression requirements, (GHP)	292	236	224
Reduction in (GHP), %	Base	23.7	30.4
Condenser duty, MM Btu/h	1.743	1.60	1.575
Reduction in condenser duty, %	Base	8.9	10.7

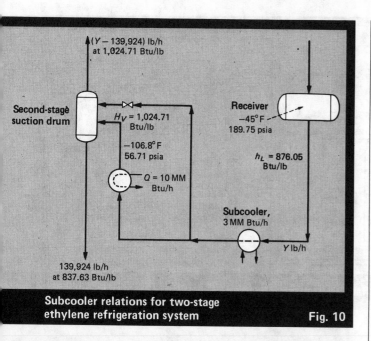

Subcooler relations for two-stage ethylene refrigeration system   Fig. 10

Effect of condensing temperature in a single-stage propylene refrigeration system   Table V

Condensing temperature, °F	60	80	100	120	140
Refrigeration duty, MM Btu/h	1.0	1.0	1.0	1.0	1.0
Refrigeration temperature, °F	−50	−50	−50	−50	−50
Compression requirement, (GHP)	211	267	333	428.5	554
Change in (GHP), %	−57.8	−24.7	Base	28.7	66.4
Condenser duty, MM Btu/h	1.54	1.675	1.84	2.085	2.42
Change in condenser duty, %	−19.5	−9.9	Base	13.3	31.5

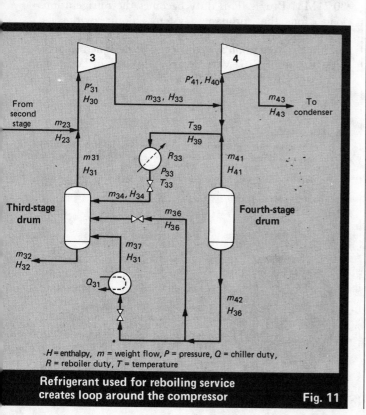

H = enthalpy, m = weight flow, P = pressure, Q = chiller duty, R = reboiler duty, T = temperature

Refrigerant used for reboiling service creates loop around the compressor   Fig. 11

where colder ambient temperatures prevail, a significant amount of energy could be saved. Table V also indicates to a certain extent the changes between summer and winter conditions as well as between day and night operations.

Refrigerant subcooling

Subcooling liquid refrigerants is very common in refrigeration systems. Subcooling the refrigerant reduces the energy requirements. It is carried out when an auxiliary source of cryogen is readily available, and the source stream needs to be heated. Subcooling can be accomplished by simply installing a heat exchanger on the appropriate refrigerant and process streams so as to make best use of available cryogen.

Let us consider installing a 3MM-Btu/h subcooler on liquid ethylene refrigerant from the receiver at −45°F in the example for the earlier two-stage ethylene refrigeration system. The second stage of this system and related data are shown in Fig. 10.

Let Y, lb/h, be the refrigerant flowrate to the subcooler, and $(Y − 139,924)$ the amount of vapor leaving the second-stage suction drum. We determine Y by performing the energy balance around the second stage:

$$876.05Y + (10 \times 10^6) = 837.63(139,924) +$$
$$1,024.71(Y − 139,924) + (3 \times 10^6)$$
$$Y = 223,174 \text{ lb/h}$$

When we compare this value of Y to the value of X for the earlier two-stage system without a subcooler, we find a decrease in the flowrate of 20,179 lb/h, i.e., $(243,353 − 223,174)$. The lower flowrate means reduced compression horsepower and condenser duty.

The enthalpy of the refrigerant leaving the subcooler becomes:

$$876.05 − [(3 \times 10^6)/223,174] = 862.61 \text{ Btu/lb}$$

From Starling [2], we find the temperature of −65.9°F, which corresponds to the enthalpy of 862.61 Btu/lb.

The flowrate of refrigerant through the second-stage chiller becomes:

$$\frac{10 \times 10^6}{(1,024.71 − 862.61)} = 61,690 \text{ lb/h}$$

As a result of subcooling, the flow of refrigerant through the second-stage chiller has been reduced from 67,268 lb/h to 61,690 lb/h. This subcooling not only saves energy but also reduces the size of piping and equipment, which now handle a lesser amount of circulating refrigerant.

Refrigerant for reboiling

Refrigerants have been successfully used for reboiling services wherever applicable conditions exist. Reboiling is similar in concept to subcooling—heat is taken out of the refrigeration cycle.

In reboiling service, the refrigerant changes phase from vapor to liquid, and is essentially at constant temperature and pressure. In subcooling service, the temperature of the liquid refrigerant is lowered before flashing. The liquid produced in a reboiler service is flashed to the next available stage to produce useful

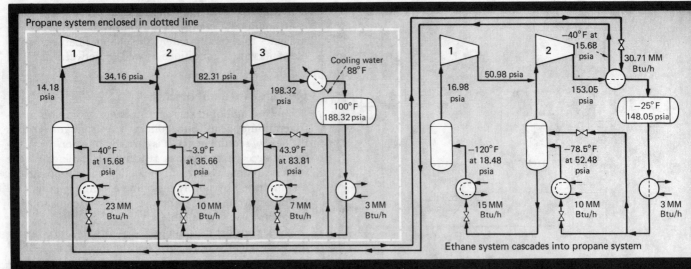

Ethane system rejects heat to cooling water in cascaded ethane-to-propane refrigeration system   **Fig. 12**

refrigeration. The refrigerant condensing pressure is a function of the reboiling temperature plus an appropriate temperature difference as a driving force.

When refrigerant is used for reboiling service and the liquid is flashed to the next stage, it creates a loop around the compressor, which has to be solved over several iterations. To illustrate the mechanics, let us consider, the reboiler arrangement in Fig. 11. The outline for the iteration is:

Step 1—Assume $H_{39} = H_{41}$.

Step 2—Determine m_{34} from $R_{33}/(H_{39} - H_{34})$. H_{34} is known for saturated liquid.

Step 3—Determine m_{31}.

Step 4—Determine m_{33} and H_{33}.

Step 5—If $m_{34} < m_{41}$, proceed to calculations for the next stage by determining:

$$H_{40} = \frac{m_{33}H_{33} + (m_{41} - m_{34})H_{41}}{m_{33} + m_{41} - m_{34}}$$

Step 6—If $m_{34} > m_{41}$, then:

$$H_{39} = \frac{m_{41}H_{41} + (m_{34} - m_{41})H_{33}}{m_{34}}$$

In this case, vapors will be withdrawn from the compressor's fourth-stage inlet, and $m_{43} < m_{33}$.

Step 7—Repeat Steps 2 through 6 until:

$$(H_{39})_{n-\text{th calc.}} = (H_{39})_{(n-1)\text{th calc.}}$$

Refrigerant cascading

In the cascading of refrigerants, warmer refrigerants condense cooler ones. Based on the low-temperature requirements of a process, a refrigerant that is capable of providing the desired cold temperature is selected.

For example, the lowest attainable temperature from ethylene refrigerant is −150°F (for a positive compressor-suction pressure), whereas the lowest temperature level for propylene is −50°F (for a similar positive pressure). The warmest temperature level to condense ethylene is its critical temperature of about 50°F. This temperature requires unusually high compression

ratios—making an ethylene compressor for such service complicated and uneconomical.

In a refrigeration cycle, energy is transferred from lower to higher temperature levels economically by using water or ambient air as the ultimate heat sink. In order to condense ethane at 50°F, a heat sink at about 45°F is necessary. Since most cooling-water temperatures are higher than 45°F, propane refrigerant can be cascaded with ethane refrigerant to transfer energy from the ethane system into the cooling water.

An example of a cascaded system is shown in Fig. 12, where an ethane system cascades into a propane system. The condenser duty for the ethane system is 30.71MM Btu/h. This duty becomes the refrigeration load for the propane system along with its 23MM Btu/h refrigeration at −40°F. Therefore, the propane refrigeration system has to be designed to provide a total of 53.71MM Btu/h at −40°F in addition to 10MM Btu/h at −3.9°F and 7MM Btu/h at 43.9°F.

References

1. Elliott Multistage Compressors, Bull. P-25A, Elliott Co., Jeanette, Pa., 1975.
2. Starling, K. E., "Fluid Thermodynamic Properties for Light Petroleum Systems," Gulf Publishing, Houston, 1973.
3. Mehra, Y. R., Refrigerant properties of Ethylene, Propylene, Ethane and Propane, *Chem. Eng.*, Dec. 18, 1978, p. 97; Jan. 15, 1979, p. 131; Feb. 12, 1979, p. 95; Mar. 26, 1979, p. 165.

The author

Yuv R. Mehra is manager, projects and technology, El Paso Hydrocarbons Co., P.O. Box 3986, Odessa, TX 79760; telephone: (915) 333-7530. He is responsible for new natural-gas-liquids extraction plants. Previously, he was in process engineering with El Paso Products Co., and several other firms. He is the author of several technical publications and holds a U.S. patent. He has an M.S. in chemical engineering from the University of California at Los Angeles, is a registered professional engineer in California and Texas, and is a member of AIChE, National Soc. of Professional Engineers, and the Texas Soc. of Professional Engineers.

Sizing dual-suction risers in liquid overfeed refrigeration systems

By using large- and small-diameter risers in parallel, one can avoid problems with two-phase flow over widely varying flowrates.

Donald K. Miller, York Div., Borg-Warner Corp.

In refrigeration systems, liquid overfeed (sometimes called liquid recirculation) describes the pumping of an excess of liquid refrigerant—at essentially evaporator temperature—to a heat-transfer device that serves as the evaporator.

Fig. 1 shows typical applications for a liquid overfeed refrigeration system in a chemical plant. The figure depicts evaporator coils or plates that are immersed in a liquid or gas that requires cooling or condensation, heat exchangers, cold storage of products by use of finned coils, and mechanized processing with ambient cooling. There is a vertical rise of return lines, where necessary, to accommodate the orientations of equipment.

Liquid refrigerant from the refrigeration system condenser flashes through an expansion device to the accumulator-pump receiver. By flashing a portion of the condenser liquid to gas, at the lower pressure in the accumulator, a cold liquid refrigerant is produced.

If an excess of this liquid is pumped to the evaporator(s), some of it will evaporate and form vapor; the remainder will stay liquid. The amount of vapor formed depends on the latent heat of the liquid and the refrigeration load of the evaporators. The refrigerant vapor must be returned to the accumulator—via the suction risers—together with the excess pumped liquid. At the accumulator-pump receiver, the vapor joins the flash gas from the expansion process and passes through the gas and liquid separator; thence back to the compressor. The excess liquid returned to the accumulator is recirculated, with additional cold liquid from the accumulator, via the pumping system. Since both liquid and vapor are trying to go through the evaporator and the suction line(s) at the same time, a two-phase flow situation exists.

Advantages of liquid overfeed systems

The considerable agitation of the excess liquid through the evaporator passages thoroughly wets the entire evaporator surface. This gives efficient use of surface with, theoretically, a zero-superheat condition on leaving the evaporator. Therefore, high heat-fluxes can be realized per unit area of evaporator surface.

With proper circuiting and geometry of the evaporator passages, the pressure drop of the two-phase flow through the evaporator can be limited to reasonable values—1° to 2.5°F equivalent. This promotes efficient use of evaporator surface and reduces the power consumption of the refrigeration plant as compared with a brine recirculation system that has a brine temperature range plus a significant temperature penalty at the evaporator. It also eliminates the brine cooler. Depending on the external coefficient of heat transfer for the evaporator (external coefficient controlling), liquid overfeed systems with the same evaporator temperature and the resulting larger LMTD (log mean temperature difference) may also reduce the size of the evaporator cooling coils needed.

The excess refrigerant liquid pumped to the evaporators and returned through the suction line(s) is effective in continuously flushing the inside of the evaporator and piping surfaces of any compressor-oil accumulations. The amount of the excess liquid flow at full load is determined by both the evaporator configuration and the orientation of liquid inlet and outlet connections. Generally for up-flow orientation, the evaporator will, in effect, operate flooded at all times, and as little as 25% excess can be used. The overfeed ratio—ratio of lb/min liquid-refrigerant pumping rate per lb/min of refrigerant evaporated (based on the refrigerant's latent heat at the average evaporator temperature)—would be 1.25 in this case. For evaporator surface geometries and orientations that make wetting more difficult, overfeed ratios of 2.5 to 3.0 are quite common. For downfeed where the evaporator is free-draining to an accumulator or large header, and particularly for large-diameter evaporator passages where wetting would be difficult, overfeed ratios of 5.0 to 8.0 are not unusual.

Dual pumps are usually used for backup, and both can be activated for pulldown load conditions. Excessively high overfeed ratios will cause excessive pressure drops and temperature penalties in the evaporator and the suction line(s). The minimum feed for the lowest

Originally published September 24, 1979

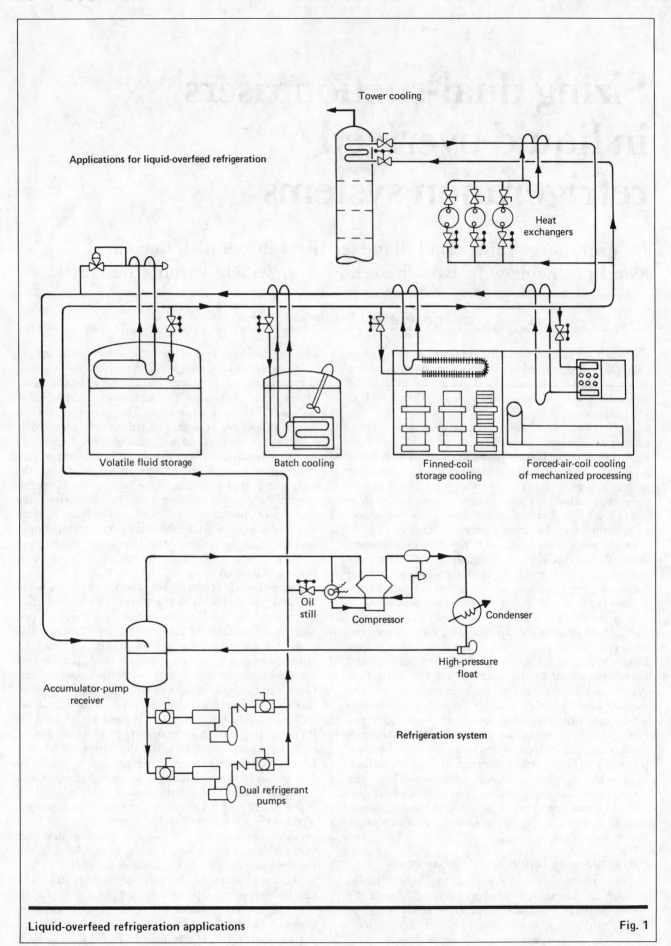

Applications for liquid-overfeed refrigeration

Tower cooling

Heat exchangers

Volatile fluid storage

Batch cooling

Finned-coil storage cooling

Forced-air-coil cooling of mechanized processing

Oil still

Compressor

Condenser

High-pressure float

Accumulator-pump receiver

Dual refrigerant pumps

Refrigeration system

Liquid-overfeed refrigeration applications

Fig. 1

Static-head penalties with a 12-ft vertical column of refrigerant liquid   Table I

	Refrigerant type				
	R-12	R-22	R-502	R-717	R-290
Evaporator temperature, °F	-22	-22	-22	-22	-22
Top temperature, °F	-52.5	-37	-35	-30.3	-27.8
Static-head penalty, °F	30.5	15	13	8.3	5.8
Capacity penalty, %	53.4	27.5	20.5	14.7	7.4
Power penalty, % kW/ton	47.4	21.5	13.6	10.0	6

LMTD, the proper cooling rate and temperature, and the highest practicable suction pressure at the compressor with adequate oil return, is the optimum performance point for the system.

The required pumping rate for a liquid-overfeed ratio of 3.0 is 15% to 25% of that for brine recirculation when a halocarbon refrigerant is used, and as small as 6% of the brine rate with ammonia as a refrigerant.

As the load is reduced, the pumping rate of the liquid is usually held constant at the full-load rate. Therefore a system selected for 3.0 overfeed ratio at full load could have an overfeed ratio of 15.0 for a 20% load condition. Obviously the two-phase pressure drop will vary, not only because of the changing gas flow but also because of the changing ratio of liquid to gas flowrates. Because of the reduced flow of gas at part load, and therefore a reduced velocity, the proper distribution of refrigerant is aided by maintaining the liquid flowrate constant.

Comparing different systems, one finds that liquid-overfeed refrigeration systems 1) can reduce power consumption versus either a direct-expansion system that was converted to overfeed, or a brine system, 2) will continuously remove oil from the evaporators, and thus maintain steady, efficient operation, 3) can reduce evaporator-surface requirements and eliminate brine-cooler requirements, and 4) will inherently increase flexibility and simplify operation and maintenance, if properly designed. On the other hand, liquid-overfeed systems 5) require the greatest amount of refrigerant, and 6) require greater attention to the prevention of pump cavitation—by liberal sizing of pump suction lines and by provision of adequate NPSH (net positive suction head)—and some attention to incrementations and timing of loading mechanisms for compressors.

The refrigerant can be any volatile substance compatible with the materials available, and can be used at any temperature level practical and consistent with the fluid properties and system components. It can be used for moving heat between process vessels, and refrigerating almost any item or substance, whether remote or close to the refrigeration plant, provided the piping is properly sized to reduce both two-phase pressure drop and static-head penalty to optimum values.

Static-head penalty for vertical lifts

A volatile substance exerts a vapor pressure that varies with temperature. If one divides 144 by the den-

Representative data for a batch-cooling refrigeration load   Table II

Operating mode	Refrigeration, tons*	Temperature at evaporator outlet,°F	Liquid to evaporator, gal/min	Approximate temperature of accumulator liquid,°F
Pulldown	16	-15	5	-21
Holding, maximum load	4.1	-22	2.5	-28
Holding, minimum load	2.0	-22	2.5	-28

*One ton equals 200 Btu/min heat-removal rate.

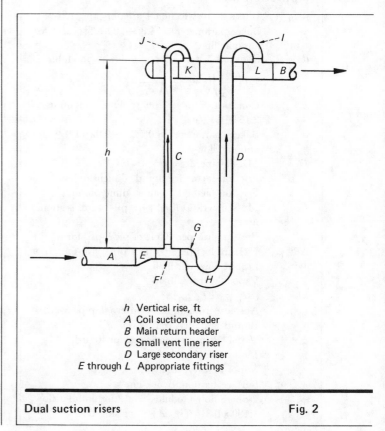

h Vertical rise, ft
A Coil suction header
B Main return header
C Small vent line riser
D Large secondary riser
E through L Appropriate fittings

Dual suction risers   Fig. 2

Equivalent lengths of piping components in circuits F-K and F-G-L   Table III

	Nominal pipe size, in.					
	1	1½	2	2½	3	4
C, Straight pipe, ft	12	12	12	12	12	12
2 Short-radius elbows, equiv. ft	2.4	5.2	7.8	8.4	10.6	14.4
2 Tees, branch, equiv. ft	10.4	16.8	21.0	26.0	32.0	44.0
Total line C, equiv. ft	24.8	34.0	40.8	46.4	54.6	70.4
C, 3° F ΔP equiv.*, psi/100 ft	4.073	2.97	2.48	2.18	1.85	1.44
D, Straight pipe, ft			12	12	12	12
1 Straight-run tee, equiv. ft			2.5	2.9	3.6	4.5
1 tee, branch, equiv. ft			10.5	13.0	16.0	22.0
5 Short-radius elbows, equiv. ft			17.0	21.0	26.5	36.0
Total line, D, ft			42.0	48.9	58.1	74.5
D, 3° F ΔP equiv.,* psi/100 ft			2.41	2.07	1.74	1.36

$$*\text{Example} \quad -22° F \quad 14.56 \text{ psia}$$
$$-25° F \quad 13.55 \text{ psia}$$
$$\Delta T, 3° F \quad 1.01 \text{ psi } \Delta P$$

$$\frac{1.01}{0.248} = 4.073 \text{ psi/100 ft for } C, \text{ size 1 in.}$$

Nomenclature

A or A_x	Internal cross-sectional area of pipe, ft^2
a	Baker parameter = $4.8 - 0.3125 (12D)$ for annular flow
B_x	Constant for 2-phase-flow modulus = $531 \left(\frac{W_l}{W_v}\right)\left(\frac{\sqrt{\rho_l \rho_v}}{\rho_l^{2/3}}\right)\left(\frac{\mu_l^{1/3}}{\sigma_l}\right)$
B_y	Constant for 2-phase-flow modulus = $2.16 W_v/A \sqrt{\rho_l \rho_v}$
b	Baker parameter = $0.343 - 0.021(12D)$ for annular flow
D	I.D. of pipe, ft
f_l	Friction factor for liquid, dimensionless
f_v	Friction factor for vapor, dimensionless
G	Mass flowrate, lb/(h)(ft^2 pipe cross-sectional area)
ΔP_{l-100}	Pressure loss per 100 ft of pipe, psi, for liquid
ΔP_{v-100}	Pressure loss per 100 ft of pipe, psi, for vapor
W_l	Liquid flowrate, lb/h
W_g	Vapor flowrate, lb/h
x	$(\Delta P_{l-100}/\Delta P_{g-100})^{1/2}$
ε	Absolute roughness of internal pipe wall, ft
μ_l	Liquid viscosity, centipoise, or lb/ft-h
μ_v	Vapor viscosity, centipoise, or lb/ft-h
ρ_l	Liquid density, lb/ft^3
ρ_v	Vapor density, lb/ft^3
σ	Surface tension, dynes/cm
ϕ	2-phase flow modulus = ax^b for annular flow; $1{,}190 \times 0.815/(W_l/A)^{0.5}$ for slug flow

sity (lb/ft^3) of the liquid at a given temperature and pressure, one obtains the feet of liquid column per psi pressure-change equivalent. With refrigerants, this pressure change is important because it changes the temperature level of evaporator operation, which in turn affects the efficiency of system operation.

For an upflow evaporator-coil bank that either is immersed in a process tank fluid or is at the top of a volatile-chemical storage tank with vertical return lines, proper consideration must be given to the static-head temperature penalty *for all load conditions*. Otherwise inadequate control of temperatures may be obtained. At high loads with a large percentage of the vertical piping volume occupied by flowing vapor (for, say, a dispersed or mist flow condition), the static-head penalty will not be too severe. However, for a low-load condition where there can be essentially a full column of liquid with only occasional slugs of vapor, or with slight bubble flow, the full static head of liquid could exert a temperature penalty on the evaporator that would be clearly excessive.

Typical values for refrigerants

For an equilibrium condition (no subcooling and no inerts) a column of liquid refrigerant in a vertical return line will vary in density and temperature depending on the height. Typical refrigerants might include R-12, R-22, R-502, R-717 (ammonia), R-290 (propane) or, as in many cases, mixed hydrocarbons. If there is, say, a 12-ft column height of pure liquid, and a saturated temperature-pressure condition of $-22°F$ is desired at

Sizing Line *C* to handle the minimum holding load   Table IV

	Minimum load, small riser (−22°F)		
Item	1-in. Sched. 40	1½-in. Sched. 40	2-in. Sched. 40
Tons, evaporator	2	2	2
Density liq., lb/ft^3	92.89	92.89	92.89
Lb/min, liq.	25.419	25.419	25.419
Density vapor, lb/ft^3	0.39173	0.39173	0.39173
Lb/min, vapor	5.627	5.627	5.627
I.D., ft	0.8740	0.13420	0.17220
A_x, ft^2	0.00600	0.01414	0.02330
μ_l, centipoise	0.3750	0.3750	0.3750
μ_v, centipoise	0.01015	0.01015	0.01015
μ_l, lb/ft-hr	0.9075	0.9075	0.9075
μ_v, lb/ft-hr	0.024563	0.024563	0.024563
B_x	78.71	78.71	78.71
B_y	20,149	8,549.8	5,118.6
Type flow	Annular	Annular	Slug
DG/μ, vapor	200,220	130,452	101,584
ϵ	0.00015	0.00015	0.00015
ϵ/D	0.001716	0.001118	0.00087
f, vapor	0.0231	0.0222	0.0218
Velocity, ft/s (vapor)	39.9	16.98	10.275
$\Delta P/100$ ft, ft fluid	653.37	74.06	20.754
$\Delta P/100$ ft, psi (vapor)	1.78	0.2015	0.0565
DG/μ, liq.	24,481	15,950	12,421
f, liq.	0.0283	0.0296	0.0302
Velocity, ft/s (liq.)	0.76	0.3225	0.1957
$\Delta P/100$ ft, ft fluid	0.2905	0.03562	0.010434
$\Delta P/100$ ft, psi (liq.)	0.1874	0.02298	0.00673
$x = \sqrt{\dfrac{\Delta P_{l-100}}{\Delta P_{v-100}}}$ =	0.3245	0.3377	0.345
a (annular) =	4.4722	4.29688	4.15406
b (annular) =	0.32097	0.30919	0.29959
$\phi = ax^b$ =	3.116	3.0717	
$\phi = \dfrac{1{,}190\,x^{0.815}}{(W_l/A)^{0.5}}$ =			1.954
2-Phase ΔP, psi/100 ft			
$\phi^2 \cdot \Delta P_v$	17.283	1.901	0.216 ΔP
−22°F, psia	14.56	14.56	
C 2-Phase ΔP, psi	4.29	0.65	Note: Slug flow gives excessive static head
K Pressure, psia	10.27	13.91	
K Temperature, °F	−36.2	−23.9	
Temperature penalty, °F	14.2	1.9	> 30.0

Note: A size 1¼-in. line would give approx: $\left(\dfrac{1.61}{1.38}\right)^5 \times 1.901 = 4.11$ psi/100 ft (N.G.)

the evaporator, one can integrate the temperature-pressure-density variations for each refrigerant and calculate the required temperature level at the top of the liquid column. Then one could determine the capacity and power penalty requirements of a two-stage compressor system to handle the refrigeration load as compared to a situation where there was no more than the normally-allowed 3°F suction riser penalty. The results are shown in Table I.

Obviously, the static-head penalty has to be reduced by some means. Also, the two-phase pressure drop must be controlled within a reasonable amount.

Two-phase pressure drop

A convenient method of calculating pressure drop substitutes an equivalent length of straight pipe or tubing for each fitting or valve. Then one determines the total pressure drops of the various straight lengths. For single-phase flow with either liquid or gas, the Moody data [2] are appropriate.

For two-phase flow there can be at least seven patterns of flow, as defined by Kern [4].

Many academic methods have been proposed for determining two-phase pressure drop. The Lockhart-Martinelli [3] method will generally provide conservative sizing (or oversizing) of lines larger than about 1-in. nominal pipe size. For a method that gives relatively fast answers that are not overly conservative, and correlate with some experience on actual refrigeration systems, the reader is referred to the Baker map and the procedure of Kern [4,8] for the sizing of single pipes. For small increments of temperature change, the evaluation of pressure drop is essentially isothermal and will not lead to significant error.

Single riser sizing

A single vertical riser sized for minimum pressure-drop at full flow of vapor (full refrigeration load) would produce an excessive liquid-refrigerant static-head penalty on the refrigeration system at greatly reduced loads. Also, at greatly reduced loads and low vapor velocities in the single large line, the probability of adequate oil return to the refrigeration plant is diminished.

On the other hand, a single riser sized for high vapor-velocity at greatly reduced load would be counterproductive. It would assure dispersed or annular flow (so that static-head penalty would be eliminated and so that there would be adequate oil return), but it would result in an excessive pressure-loss penalty at high loads.

Why dual risers are needed

What is really needed is a piping network that permits optimum velocities and pressure drops, as well as effective oil return, at all load conditions. These features are provided by a trapped large line and a smaller vapor-bleed line upstream of the trap, such as depicted in Fig. 2. Both lines conduct flow at maximum load. At a greatly reduced load, the large trapped line D is sealed by trapped liquids, and vapor flows through the smaller line C.

This concept has been used for many years to move oil up hot gas-discharge lines or suction lines that un-dergo wide variation in loading, when conventional single-phase refrigerant flow is employed. The actual sizing method practiced varies with the individual; the literature is somewhat vague and subject to interpretation. Hitherto, much was left to the judgment and experience of the individual designer.

However, a more scientific approach is essential if the method is to be applied with confidence to liquid-overfeed systems using a wide range of refrigerants and other fluids. This article has been written to achieve that end.

In the following example, much use is made of Ref. [1 – 6]. A further refinement, that may prove quite a challenge, could be pursued using adaptations from the work of DeGance and Atherton [7] particularly for the system piping. However, quite satisfactory results for the sizing of dual suction risers can be obtained by using the concept proposed in this example.

Sizing for minimum load

A batch-cooling load followed by a tapering holding load might be as shown in Table II.

This particular plant uses R-12 refrigerant; the vertical rise, h, of Fig. 2 is 12 ft, a maximum of 3°F temperature-penalty of the riser was permitted, and the piping is Schedule 40 black steel. The header K-L-B (Fig. 2) was sized for about 3°F temperature-penalty back to the accumulator. It was assumed that the liquid pump would deliver the liquid to the evaporator at essentially the evaporator outlet-temperature level, because of added pump-motor heat and heat pickup through insulation, with negligible pressure-drop in the evaporator itself. Therefore, the lb/min of vapor is based on the latent heat at the evaporator outlet.

At −15°F, the pressure of R-12 is 17.14 psia; density of liquid is 92.19 lb/ft³; sp. vol. of saturated vapor is 2.1923 ft³/lb; latent heat is 70.3514 Btu/lb; viscosity of liquid is 0.8833 lb/ft-h, or 0.365 centipoise; viscosity of vapor is 0.024926 lb/ft-h, or 0.0103 centipoise; and the surface tension of the liquid is 16 dynes/cm.

At −22°F, the pressure of R-12 is 14.56 psia; density of liquid is 92.89 lb/ft³; sp. vol. of saturated vapor is 2.5528 ft³/lb; latent heat is 71.081 Btu/lb; viscosity of liquid is 0.9075 lb/ft-h, or 0.375 centipoise, viscosity of vapor is 0.024563 lb/ft-h, or 0.01015 centipoise, and the surface tension is 16.5 dynes/cm.

Holding at −22°F

At maximum: 4.1 tons:

$$2.5 \times \frac{92.89}{7.48} = 31.046 \text{ lb/min liquid feed}$$

$$\frac{4.1 \times 200}{71.081} = 11.536 \text{ lb/min evaporated and}$$
returned as vapor via vertical risers

$$31.046 − 11.536 =$$
19.51 lb/min liquid returned via vertical risers

At minimum: 2 tons:

$$\frac{2 \times 200}{71.081} = 5.627 \text{ lb/min evaporated and}$$
returned as vapor via vertical risers

$$31.046 − 5.627 =$$
25.419 lb/min liquid returned via vertical risers

Determining flow and pressure drop in each "pair" of risers   Table V

	Pair I		Pair II	
	C 1½ in. (small)	D 2½ in. (large)	C 1½ in. (small)	D 3 in. (large)
Temperature, °F	-15	-15	-15	-15
Lb/min vapor	12.362	33.124	8.350	37.136
Lb/min liquid	4.386	11.752	2.962	13.176
ρ_l, lb/ft^3	92.19	92.19	92.19	92.19
ρ_v, lb/ft^3	0.4562	0.4562	0.4562	0.4562
μ_l, centipoise	0.365	0.365	0.365	0.365
σ_l, dynes/cm	16	16	16	16
W_l/W_v	0.355	0.355	0.355	0.355
B_x	2.676	2.676	2.676	2.676
A_x, ft^2	0.01414	0.03322	0.01414	0.05130
I.D., ft	0.1342	0.2057	0.1342	0.2557
W_v/A	52,456	59,827	35,431	43,434
$2.16/\sqrt{\rho_l \rho_v}$	0.3331	0.3331	0.3331	0.3331
B_y	17,472	19,927	11,801	14,467
Flow	Annular	Annular	Annular	Annular
I.D., in.	1.61	2.469	1.61	3.068
a	4.2969	4.0284	4.2969	3.8413
b	0.30919	0.29115	0.30919	0.27857
μ_l, lb/ft-hr	0.8833	0.8833	0.8833	0.8833
μ_v, lb/ft-hr	0.024926	0.024926	0.024926	0.024926
ϵ/D	0.0011177	0.000729	0.001117	0.000587
DG/μ, vapor	282,417	493,715	190,760	445,561
f_v	0.021	0.019	0.0212	0.0181
Vapor velocity, ft/s	31.94	36.43	21.57	26.45
ΔP/100 ft, ft vapor	247.9	190.35	114.1	76.9
ΔP/100 ft, psi	0.785	0.603	0.362	0.244
DG/μ, liq.	2,828	4,943	1,910	4,461
f_l	0.0222 to 0.0445	0.0382	0.033	0.0395
Liq. velocity, ft/s	0.05608	0.06396	0.03787	0.04643
ΔP/100 ft, ft liq.	0.000808 to 0.001619	0.00118	0.0005476	0.0005171
Liq. ΔP/100 ft, psi	0.000517 to 0.001037	0.0007555	0.0003506	0.00033105
$x = \sqrt{\dfrac{\Delta P_{l-100}}{\Delta P_{v-100}}} =$	0.02566 to 0.03635	0.0354	0.03112	0.03683
$\phi = ax^b =$	1.385 to 1.542	1.523	1.4697	1.5313
2-Phase ΔP, psi/100 ft $\phi^2 \cdot \Delta P_{v-100} =$	1.506 to 1.87	1.399	0.782	0.572
2-Phase ΔP, total psi	0.512 to 0.636	0.684	0.266	0.332
P at -15°F, psia	17.14	17.14	17.14	17.14
P at K and L	16.50	16.46	16.87	16.81
T at K and L, °F	-16.7	-16.8	-15.7	-15.9
Temperature penalty, °F	1.7	1.8	0.7	0.9

C = 1½ in.
D = 2½ in.
Adequate

C = 1½ in.
D = 3 in.
Has ~ 1°F less penalty

Pulldown

At 16 tons:

$$5 \times \frac{92.19}{7.48} = 61.624 \text{ lb/min liquid feed}$$

For the purpose of this analysis it is assumed that the turbulent condition of the flow causes equal liquid-vapor ratios in lines C and D. Once flow is established through D, at start of pulldown, D continues to conduct vapor and liquid flow even at lower loads until liquid flow is interrupted or the momentum of liquid and vapor is significantly reduced to that existing near the minimum holding load.

There is an equivalent-length difference between circuits F-K and F-G-L, and this should be considered. The solution to the problem of sizing C and D involves trial assumptions of flows in each until the 2-phase pressure-drops in each circuit balance. To simplify the calculations, the assumptions shown in Table III were made.

The first task is to size Line C to handle the minimum holding load under an annular or dispersed-flow condition, which would assure lack of static-head penalty. To do this, the Baker map [4, 8] should be examined. Moody's friction-factor curves must be referred to [2]. The calculations are shown in Table IV. It is therefore determined that a nominal size $1\frac{1}{2}$-in. Schedule 40 pipe for Circuit C is optimum for the minimum holding load.

Now, at a pulldown load of 16 tons, determine the size of the Line D. Assume equal liquid-vapor ratio for C and D. An approximate relationship of flows in C and D is determined by the equation:

$$\text{lb/min, } C = \frac{\text{lb/min total vapor or liquid}}{1 + \left[\left(\frac{\text{I.D.}D}{\text{I.D.}C}\right)^{2.73} \times \left(\frac{\text{equiv. length } C}{\text{equiv. length } D}\right)^{0.5}\right]}$$

(Total vapor = 45.486 lb/min and total liquid = 16.138 lb/min)

With $1\frac{1}{2}$ in. Schedule 40 at 1.610″ I.D. and 34.0 ft equivalent length for Line C, then evaluate the quantity of vapor and liquid in Lines C and D for two sizes of D.

Pipe size, in.	I.D., in.	Equiv. length, ft	Vapor Lb/min, C	Vapor Lb/min, D	Liquid Lb/min, C	Liquid Lb/min, D
$2\frac{1}{2}$	2.469	48.9(D)	12.362	33.124	4.386	11.752
3	3.068	58.1(D)	8.350	37.136	2.962	13.176

The type of flow must be determined for each "pair" of risers in each circuit and appropriate equations used for the pressure drop. These calculations are shown in Table V.

Conclusion

One can use a $1\frac{1}{2}$-in. pipe for C and $2\frac{1}{2}$-in. for D, although a $1\frac{1}{2}$-in. for C and 3-in. for D would have the lowest penalty.

With $1\frac{1}{2}$-in. and 3-in. risers, the two extremes of loading produce 1.9°F to 0.8°F penalty on the evaporator. At the beginning of pulldown, Riser D begins to conduct vapor and liquid and continues to do so at lower pulldown loads, until some time during the holding mode when the single riser, C, carries all the liquid and vapor. If this occurs at maximum holding load, the temperature penalty by calculation is 4.4°F maximum. It is difficult to calculate precisely just when Riser C "takes over" completely, because so much depends on the sustained momentum of the fluids. If loads gradually increase from a minimum, the point at which D begins to conduct flow would be at a higher load than where D ceases to conduct on falling load. The relative physical location of parts F and G influence the "cut-in" and "cut-out" points of Line D. G should be immediately adjacent to F.

Summary

Fig. 1 shows a few applications that use the flexibility and efficient operation made possible with liquid-overfeed refrigeration systems. By using dual risers, excessive static-head or excessive two-phase pressure-drop penalties are eliminated for all situations from maximum to minimum load. Also, close control of product temperatures and effective oil return are realized.

The conceptual approach described here for sizing the dual risers for these systems has been successful in assuring minimum temperature penalty, optimum oil return and maximum potential for operating-cost savings in refrigeration systems.

References

1. American Soc. of Heating, Refrigerating and Air Conditioning Engineers, "ASHRAE Handbook of Fundamentals," Chap. 4, 16, 32, ASHRAE, New York, N.Y., 1977.
2. Moody, Lewis F., Friction Factors for Pipe Flow, *Trans. ASME (Am. Soc. Mech. Eng.*, pp. 671–684 (1944).
3. Lockhart, R. W., and Martinelli, R. C., Proposed Correlation of Data for Isothermal Two-Phase, Two-Component Flow in Pipes, *Chem. Eng. Prog.*, Vol. 45, No. 1, pp. 39–48.
4. Kern, Robert, How to Size Process Piping for Two-Phase Flow, *Hydrocarbon Process.*, Oct. 1969, pp. 105–116.
5. Crane Co., "Flow of Fluids Through Valves, Fittings, and Pipe," Crane Technical Paper No. 410.
6. American Soc. of Heating, Refrigerating and Air Conditioning Engineers, "ASHRAE Handbook of Systems," Chap. 25 and 26, ASHRAE, New York, 1976.
7. DeGance, Anthony E., and Atherton, Robert W., Chemical Engineering Aspects of Two-Phase Flow, Parts 1–8, *Chem. Eng.*, Mar. 23, Apr. 20, May 4, July 13, Aug. 10, Oct. 5, Nov. 2, 1970 and Feb. 22, 1971.
8. Kern, Robert, Piping Design for Two-Phase Flow, *Chem. Eng.*, June 23, 1975, pp. 145–151.

The author

Donald K. Miller is Chief Engineer, Absorption and Reciprocating Compressors Development, York Div. of Borg-Warner Corp., P.O. Box 1592, York, PA 17405, where he is involved in guiding development, manufacture, application and construction of refrigeration components and systems. He holds a BSChE from the University of Missouri, is a registered professional engineer in Pennsylvania, and is a member of the American Soc. of Heating, Refrigeration and Air Conditioning Engineers and Vice-Chairman of the Air-Conditioning and Refrigeration Institute's Reciprocating Liquid-Chilling Packages Engineering Committee.

Variable fan speeds cut cooling-tower operating costs

Regulating the air flow through cooling towers by controlling fan speed via a.c. adjustable-speed drives can significantly reduce power costs, lessen maintenance expenses and extend equipment life.

James D. Johnson, General Electric Co.

☐ With electricity at 6¢/kWh, a typical mechanical-draft cooling tower with a 75-hp fan in continuous operation, driven by an a.c. induction motor, generates a power bill of nearly $32,000/yr (75 hp × 0.746 kW/hp × 8,760 h/yr × $0.06/kWh × 1/0.92 efficiency = $31,964/yr).

Because fan hp rating is based on required air flow for worst-case wet- and dry-bulb temperatures (which occur only a small part of a year), air flow—hence, power cost—can be reduced, over 50% in many cases, while cooling is still maintained.

Fig. 1 shows the relationship between air flow (ft^3/min) and power for centrifugal fans. Note that the required driving hp declines rapidly as air flow is reduced. For centrifugal fans, the fan law states that flow is proportional to fan speed, and power required is proportional to speed cubed. Therefore, slower fan speeds result in large reductions in power requirements.

Savings from adjusting fan speed

Adjustable-speed operation of cooling-tower fans has been made practical by recently developed inverter

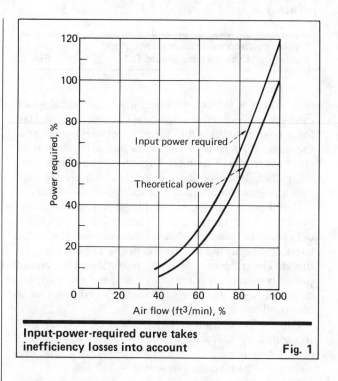

Input-power-required curve takes inefficiency losses into account **Fig. 1**

power units, which have made variable-speed operation of standard induction motors reliable and cost-effective. Such operation can result in significant savings in electrical power costs, reduced water wind-drift loss, longer mechanical-equipment life, and less noise and vibration.

One method of accurately evaluating the possible en-

Operating data for southern Ohio cooling tower, with 75-hp fan motor

Operating time, h	Average ambient temperature, °F	Wet-bulb temperature, °F	Required air flow, ft^3/min	Fan motor size, hp	Motor rating, kW	Power usage, kWh
22	97	76	275,728	73.8	55.0	1,210
378	92	74	256,818	59.6	44.5	16,821
1,837	77	68	214,359	34.7	25.9	47,578
2,380	62	57	171,135	17.6	13.2	31,416
1,973	47	43	140,891	10.0	7.5	14,798
1,706	32	30	122,985	6.6	4.9	8,359
464	13	12	106,222	4.3	3.2	1,485
8,760						121,667

Originally published August 8, 1983

83

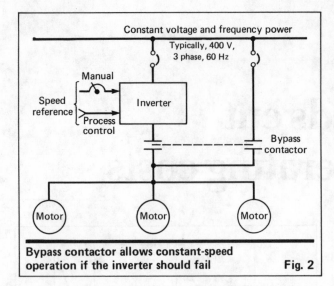

Constant voltage and frequency power

Typically, 400 V, 3 phase, 60 Hz

Manual

Speed reference

Inverter

Process control

Bypass contactor

Motor   Motor   Motor

Bypass contactor allows constant-speed operation if the inverter should fail   **Fig. 2**

ergy saving is to start with actual weather data, which can be secured from the National Climatic Center. Data for a typical middle Ohio River Valley site are given in the table. By means of such data and a psychrometric chart (as in the "Chemical Engineers' Handbook," 5th ed., p. **20**-6), the cooling-tower air flows required for various ambient conditions can be calculated (see table).

For these air flows, the required fan motor horsepower and kW ratings are determined. For the various operating periods, the kWh consumptions are calculated. These are totaled to obtain the annual kWh demand. The difference between this total and the annual power required for continuous full-speed motor operation represents the energy saving obtained from variable-speed operation.

For the annual power requirement of 121,667 kWh for adjustable-speed operation (from the table):

Adjustable-speed kWh = $121,667 \div (0.93$ motor efficiency $\times 0.94$ inverter efficiency$) = 139,175$ kWh.

Constant-speed kWh = $(73.8$ hp $\times 8,760$ h/yr $\times 0.746$ kW/hp$) \div 0.94$ motor efficiency $= 513,064$.

Saving in kWh/yr = $513,064 - 139,175 = 373,889$.

Reduction in power cost = $373,889$ kWh/yr $\times \$0.06$/kWh $= \$22,433$/yr.

Payback calculation

An economic evaluation of the application of a.c. adjustable-speed drive for the foregoing example is:

Equipment cost:

Add a.c. inverter with bypass	$19,000
Delete a.c. motor starter	−2,500
Net additional cost	$16,500

A quick, simple measure of economic attractiveness is gross payback period (*GPP*). *GPP* = additional investment/annual saving. In this case, *GPP* = $16,500 $\div \$22,433$/yr $= 0.736$ yr.

To realize the saving possible from reduced-speed operation, many recently built cooling towers have been equipped with two-speed motors for fan drives. In such a case, the kWh required per the table remain the same (at 121,667), but the constant-speed kWh is reduced some-

what. The two-speed drive operates at times at full speed (full hp) and other times at half speed (one-eighth hp). This results in an annual kWh consumption (based on the data in the table) of:

Full-speed (full hp) operating hours = $22 + 378 + 2,380 = 4,617$ h.

Half-speed (one-eighth hp) operating hours = $1,973 + 1,706 + 464 = 4,143$ h.

Two-speed kWh/yr = $[(73.8$ hp $\times 4,617$ h/yr $+ 73.8$ hp $\times \frac{1}{8} \times 4,143$ h/yr$) 0.746$ kW/hp$] \div 0.94$ motor efficiency $= 300,744$ kWh/yr.

The kWh saved/yr = $300,744 - 121,667 = 179,077$ kWh/yr.

The power cost reduction = $179,077$ kWh/yr $\times \$0.06$/kWh $= \$10,745$/yr.

Added-equipment costs:	
A.C. inverter with bypass	$19,000
Single-speed a.c. motor	3,700
Total additions,	$22,700
Deleted-equipment costs:	
Two-speed motor starter	$4,500
Two-speed a.c. motor	6,500
One set of motor power cables	800
Total deletions,	$11,800
Net additional cost	**$10,900**

Therefore, *GPP* = $10,900 $\div \$10,745$/yr $= 1.01$ yr. Reduced water evaporation and less fan-drive maintenance will provide additional savings in both examples.

Other features

Automatic control of fan speed is achieved by conventional process instrumentation. Measurement of the return- or cold-water temperature provides the control signal for speed reference to the fan's inverter power unit (Fig. 2). Manual operation is also possible to cover control malfunction and other abnormal operations.

Reverse operation (to prevent ice buildup on fan blades) is accomplished electronically within the inverter unit, and can be started manually, or automatically at a set low temperature. When extreme reliability is required, redundant controls (via the bypass contactor in Fig. 2) allow constant-speed operation. The transfer can be automatically or manually controlled.

Further savings in power-unit costs are possible from operating multiple motors from a single power supply (as also shown in Fig. 2). In this arrangement, all the motors run at the same speed.

Limiting motor starting torque avoids the "slamming" associated with full-voltage starting and stopping. Gradual startings and lower fan speeds extend equipment life and lessen maintenance costs. Lower blade-tip velocity also lessens fan noise and vibration. A 3:1 speed reduction is about equal to a noise reduction of 20–25 dBA.

The author

James D. Johnson, as Manager—Distribution and A.C. Drives Sales, General Electric Co.'s Speed Variator Products Operation (Erie, PA 16531), is responsible for market and product development for a.c. adjustable-speed inverter drives. Previously, he was a sales engineer and an applications engineer in the company's industrial sales division. Holder of a B.S. in mechanical engineering from the University of Cincinnati, he is a registered engineer in the State of Ohio.

Optimizing controls for chillers and heat pumps

Here is a control strategy that integrates
the various parts of these systems to arrive
at the most energy-efficient operation.

Béla G. Lipták, Consultant

☐ In devising a control strategy for a chiller or heat pump, the amount of energy used must be kept to a minimum, while meeting the users' load requirements. A control strategy will be devised to accomplish this. First, however, the thermodynamics of refrigeration processes will be reviewed, along with conventional chiller control systems and their optimization.

Just as conventional pumps can lift water from a low to a high elevation, so can refrigeration machines "pump" heat from a low to a high temperature. Fig. 1 illustrates this, where the heat pump removes Q_L amount of heat from the chilled water at the cost of investing W amount of work, and delivers Q_H amount of heat to the warmer cooling-tower water.

On the right side of Fig. 1 is the idealized temperature-entropy cycle for the refrigerator, consisting of two isothermal and two isentropic (adiabatic) processes:

■ 1–2 is an adiabatic process through an expansion valve—the high-pressure subcooled liquid refrigerant becomes a low-pressure liquid/vapor mixture.

■ 2–3 is an isothermal process through an evaporator—this mixture becomes a superheated low-pressure vapor.

■ 3–4 is an adiabatic process through a compressor—the pressure of the refrigerant vapor is increased.

■ 4–1 is an isothermal process through a condenser—the vapor is condensed at constant pressure.

The liquid leaving the condenser is usually subcooled, whereas the vapor leaving the evaporator is usually superheated (by a controlled amount). The isothermal processes of the cycle are also isobaric.

The efficiency of a refrigerator is defined as the ratio of the heat removed from the process, Q_L, to the work required, W:

$$\text{Efficiency} = \frac{Q_L}{W} = \frac{T_L}{T_H - T_L}$$

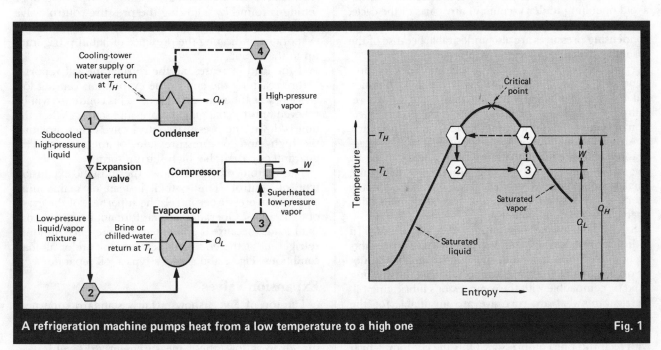

A refrigeration machine pumps heat from a low temperature to a high one   Fig. 1

Originally published October 17, 1983

Because the chiller efficiency is much more than 100% (the temperatures are in degrees absolute), the efficiency is usually called the coefficient of performance (COP).

If a chiller requires 1.0 kWh (3,412 Btu/h) to provide a ton of refrigeration (12,000 Btu/h), its COP is said to be 3.5. This means that each unit of energy introduced at the compressor will pump 3.5 units of heat energy to the cooling-tower water. Conventionally controlled chillers operate with COPs of 2.5 to 3.5. Optimization can double the COP by increasing T_L, decreasing T_H, and by other methods that will be described later.

The unit most frequently used in describing refrigeration loads is the ton. Several tons are referred to in the literature:

Standard ton	200 Btu/min
British ton	237.6 Btu/min
European ton (Frigorie)	50 Btu/min

Refrigerants

The refrigerant is the fluid that carries the heat from a low to a high temperature. The table lists the more frequently used refrigerants.

The data assume that the evaporator operates at 5°F, and the temperature of the cooling-water supply to the condenser will keep it at 86°F. (Other temperature levels would have illustrated the relative characteristics of the various refrigerants equally well.) It is generally good to avoid operating under vacuum in any parts of the cycle, because of sealing problems, but at the same time, high condensing pressures are also undesirable because of the resulting structural strength requirements.

From this point of view, the refrigerants between propane and methyl chloride have favorable characteristics. An exception is when very low temperatures are required. For such service, ethane can be the proper selection in spite of the resulting high system design-pressure.

Another consideration is the latent heat of the refrigerant. The higher it is, the more heat can be carried by the same amount of working fluid, and therefore, smaller equipment is needed. This has caused many users in the past to compromise with the undesirable characteristics of ammonia.

Safety is one of the most important considerations. In industrial installations, the desirability that refrigerants be non-toxic, non-irritating, and non-flammable cannot be overemphasized. It is also important that the working fluid be compatible with the compressor's lubricating oil. Refrigerants that are corrosive are unsuitable for the obvious reasons of higher first cost and maintenance.

Most working fluids listed in the table are compatible with reciprocating compressors. Only the last four, which

have high volume-to-mass ratios and low compressor discharge pressures, can be used with rotary or centrifugal machines. Considering all the advantages and drawbacks of the many refrigerants, Refrigerant-12 is suited for the largest number of applications.

Circulated fluids

In the majority of industrial installations, the evaporator is not used directly to cool the process. Rather, the evaporator cools a circulated fluid that is piped to cool the process.

For temperatures below where water can be used as a coolant, brine frequently is used. Weak brines may freeze, and strong brines, if they are not true solutions, may plug the evaporator tubes. For operation around 0°F, sodium brines are recommended; for services down to −45°F, calcium brines are best.

Care must be exercised in handling brines, because they are corrosive if not at the pH of 7 or if oxygen is present. In addition, brines will initiate galvanic corrosion between dissimilar metals.

Small industrial refrigerators

The top of Fig. 2 shows the direct-expansion type of control. Here a pressure-reducing valve maintains a constant evaporator pressure. The pressure setting is a function of load, and therefore these controls are recommended for constant-load installations only. The proper setting is found by adjusting the pressure-control valve until the frost stops just at the compressor end of the evaporator, indicating the presence of liquid refrigerant up to that point.

If the load increases, all the refrigerant will vaporize before reaching the end of the evaporator, causing low efficiency as the unit is "starved." This condition will be relieved only by changing the pressure setting. When the unit is down, the pressure-control valve closes, isolating the high- and low-pressure sides of the system. This guarantees a desirable high startup torque.

The bottom of Fig. 2 shows thermostatic-expansion type of control. This system, instead of maintaining evaporator pressure, controls the superheat of the evaporated vapors. Operation is therefore not limited to constant loads, because it guarantees the presence of liquid refrigerant at the end of the evaporator under all load conditions. Fig. 2 also shows a typical oil separator.

Expansion valves

The top of Fig. 3 shows a fairly standard superheat control valve. It detects the pressure into and the temperature out of the evaporator. If the evaporator pressure-drop is low, these measurements (the saturation

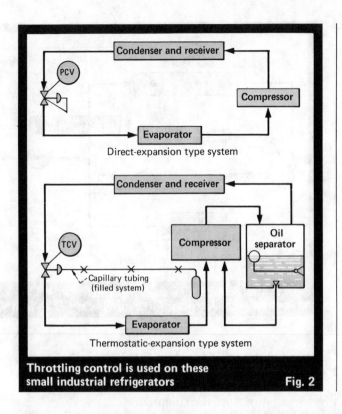

Throttling control is used on these small industrial refrigerators   **Fig. 2**

pressure and the temperature of the refrigerant) are an indication of superheat. The desired superheat is regulated by setting the spring in the valve operator, which together with the saturation pressure in the evaporator opposes the opening of the valve. The superheat-feeler-bulb pressure balances these forces when the unit is in equilibrium, operating at the desired superheat, usually 9°F.

An increase in process load results in an increase in the evaporator outlet temperature. An increase in this temperature raises the feeler-bulb pressure, which in turn further opens the superheat control valve. This greater flow from the condenser to the evaporator raises the saturation pressure and temperature, and the increased saturation pressure balances against the increased feeler-bulb pressure at a new (greater) valve opening at a new equilibrium. To adjust to an increased load condition, the evaporator pressure has been increased, but the amount of superheat (set by the valve spring) is kept constant.

Fig. 3 also shows the operation of the cooling-water regulating valve. This valve keeps the condenser pressure constant and, at the same time, conserves cooling water. At low condenser pressure, as when the compressor is down, the water valve closes. It starts to open when the compressor is restarted and the compressor dis-

Nomenclature

CHWP	Chilled-water pump	SC	Speed controller
CTWP	Cooling-tower water pump	SIC	Speed-indicating controller
FAH, FAL	Flow alarm, H = high, L = low	SP	Setpoint
FC	Fail closed	SV	Solenoid valve
FO	Fail open	T_c	Temperature of refrigerant in condenser inlet
FSH, FSL, FSHL	Flow switch, H = high, L = low, HL = high-low	T_e	Temperature of refrigerant in evaporator inlet
HLL	High-low limit switch	T_H	Temperature of cooling water at condenser exit, absolute
HWP	Hot-water pump		
LCV	Level-control valve	T_L	Temperature of chilled water at evaporator exit, absolute
M_{1-5}	Motor that drives equipment or "place" where energy is consumed, 1 = cooling-tower fans, 2 = cooling-water pumps, 3 = compressor, 4 = chilled-water pumps, 5 = hot-water pumps	T_{chwr}	Temperature of chilled-water return
		T_{chws}	Temperature of chilled-water supply
		T_{ctwr}	Temperature of cooling-tower water return
P-1	Pump	T_{ctws}	Temperature of cooling-tower water supply
P, P_{1-3}	Pressure, can be with or without subscript	T_{hws}	Temperature of hot-water return
PAH	Pressure alarm, H = high	T_{wb}	Temperature of wet bulb
PB	Pushbutton	TAH, TAL	Temperature alarm, H = high, L = low
PCV	Pressure-control valve		
PDIC	Pressure-differential indicating controller	TCV	Temperature control valve
		TIC	Temperature-indicating controller
PI	Pressure indicator	TSH, TSL	Temperature switch, H = high, L = low
PSH, PSL	Pressure switch, H = high, L = low		
Q_H	Amount of heat delivered to cooling-tower water	TT	Temperature transmitter
		TY	Temperature relay
Q_L	Amount of heat removed from chilled water	VPC	Valve position controller
		W	Work
RD	Rupture disk	XLS	Limit switch
S	Solenoid	XSCV	Superheat control valve

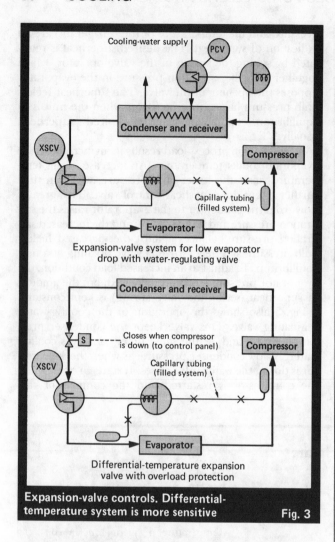

Expansion-valve system for low evaporator
drop with water-regulating valve

Closes when compressor
is down (to control panel)

Capillary tubing
(filled system)

Differential-temperature expansion
valve with overload protection

Expansion-valve controls. Differential-temperature system is more sensitive Fig. 3

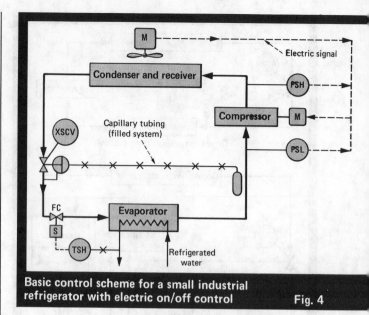

Basic control scheme for a small industrial refrigerator with electric on/off control Fig. 4

charge pressure reaches the setting of the valve. The water-valve opening follows the load, opening further at higher loads to maintain constant condenser pressure.

At very low temperatures, a small change in refrigerant vapor pressure is accompanied by a fairly large change in temperature. For example, at $-100°F$, a 5°F temperature change corresponds to a 0.3-psi variation of saturation pressure. Therefore, the use of a thermal bulb to detect indirectly the saturation pressure in the evaporator results in a more sensitive measurement. A differential-temperature expansion valve, taking advantage of this phenomenon, is on the bottom of Fig. 3.

On/off control of small industrial units

Fig. 4 shows the controls for a simple, small refrigerator. This system includes a conventional superheat control valve, a low-pressure-drop evaporator, a reciprocating on/off compressor and an air-cooled condenser.

This package maintains a refrigerated water supply to the plant within set limits.

The high-temperature switch, TSH, is the main control device. Whenever the temperature of the refrigerated water drops below a preset value (say, 38°F), the refrigeration unit is turned off, and when it rises to some other level (say, 42°F), it is restarted.

This on/off cycling control is accomplished by the temperature switch's closing the solenoid valve when the

water temperature is low enough. The closing of this valve causes the compressor suction-pressure to drop until it reaches the setpoint of the low-pressure switch, which in turn stops the compressor.

While the unit is running, the expansion valve maintains the refrigerant superheat constant, and the safety interlocks protect the equipment. These interlocks perform such functions as turning off the compressor if either the fan motor stops or the compressor discharge pressure becomes too high.

The setup in Fig. 4 can operate only at full compressor capacity; otherwise not at all. Such operation is referred to as two-stage unloading. When the cooling capacity of the unit is to be varied, instead of just turning it on and off, multi-step unloading of a reciprocating compressor can be used. In a three-step system the operating loads are 100, 50, and 0%; with five-step unloading they are 100, 75, 50, 25, and 0%.

Multistage refrigeration units

With industrial compressors, it is impractical to obtain a compression ratio outside the range of 3:1 to 8:1. This, of course, places a limitation on the minimum temperature that can be achieved by a single-stage refrigeration unit.

For example, if Refrigerant-12 were used, to maintain the evaporator at $-80°F$ and the condenser at 86°F (temperatures compatible with standard supplies of cooling water), the required compression ratio would be:

$$\text{Compression ratio} = \frac{\text{Refrigerant pressure at 86°F}}{\text{Refrigerant pressure } -80°F}$$

$$= \frac{108}{2.9} = 37$$

Such a compression ratio is obviously impractical, and therefore a multistage system is required.

Large industrial refrigerators

The refrigeration unit shown in Fig. 5, although far from a standard system (it is better instrumented than some commercial units), does contain some features typical of conventional industrial units, 500 tons and larger. Some of these features include: the capability for continuous load adjustment vs. stepwise unloading; the use of an economizer expansion-valve system; and employment of a hot-gas bypass to increase rangeability.

The unit in Fig. 5 provides refrigerated water at 40°F through the circulating header system of an industrial plant. The flowrate is fairly constant, and therefore process load-changes are reflected via the temperature of the returning refrigerated water. Under normal load conditions, this temperature is 51°F. As the process load decreases, the return-water temperature drops correspondingly. With the reduced load on the evaporator, TIC-1 gradually closes the suction vane or pre-rotation vane of the compressor. By throttling the suction vane, a 10:1 turndown ratio can be achieved. If the load drops below this ratio, the hot-gas bypass has to be activated.

The hot-gas bypass is automatically controlled by TIC-2. It keeps the constant-speed compressor out of surge by opening the bypass valve when the load drops to levels that approach surge conditions.

If the chilled-water flowrate is constant, the difference between chilled-water supply and return temperatures indicates the load. If a full load corresponds to a 15°F difference and the chilled-water-supply temperature is controlled by TIC-1 at 40°F, then the return water temperature detected by TIC-2 indicates the load.

If surge occurs at 10% load, this corresponds to a return water temperature of 41.5°F. To stay safely away from surge, TIC-2 is set at 42°F, corresponding to about 13% load. When the temperature hits 42°F, this valve starts to open, and its opening can be proportional to the load detected. This means that the valve is fully closed at 42°F, fully open at 40°F, and throttled in-between. Thus, it is theoretically possible to achieve a very high turndown ratio; in fact, the machine can be temporarily run on close to zero process load. Such operation can be visualized as a heat pump, transferring heat energy from the refrigerant to the cooling water since the process load is zero. Concomitantly, some of the refrigerant vapors are condensed, resulting in an overall lowering of operating pressures in the system.

Although the hot-gas bypass allows the chiller to operate at low loads without going into surge, it does so at an increased energy cost. As will be shown later, optimized control systems eliminate this waste through the use of variable-speed compressors.

The economizer shown in Fig. 5 can increase efficiency by 5 to 10%. This is achieved by reducing space requirements, saving compressor power consumption, reducing condenser and evaporator surfaces, etc. The economizer in Fig. 5 consists of a two-stage expansion valve with condensate collection chambers. When the load is above 10%, the hot-gas bypass system is inactive. Condensate collects in the upper chamber of the economizer, and drains under float level-control, driven by the condenser pressure.

The pressure in the lower chamber is the same as in the second stage of the compressor, and it, too, is drained into the evaporator under float level-control, driven by the pressure of the compressor's second stage. Economy is achieved due to the vaporization in the lower chamber. This precools the liquid that enters the evaporator and desuperheats the vapors that are sent to the compressor's second stage.

For commercial processes, there is a variety of refrigerants to meet various conditions

Refrigerant	Feature	Applicable compressor (R = reciprocating, RO = rotary, C = centrifugal)	Boiling point, °F, at atmospheric pressure	Evaporator pressure, psia, if operating temperature is 5°F	Condenser pressure, psia, if operating temperature is 86°F	Latent heat, Btu/lbm, at 18°F	Toxic (T), flammable (F), irritating (I)	Mixes and/or compatible with the lubricating oil	Chemically inert and noncorrosive	Remarks
Ethane	C_2H_6	R	−127	236	675	148	T, F	No	Yes	For low-temperature service.
Carbon dioxide	CO_2	R	−108	334	1,039	116	No	Yes	Yes	Low-efficiency refrigerant.
Propane	C_3H_8	R	− 48	42	155	132	T, F	No	Yes	
Refrigerant-22	$CHClF_2$	R	− 41	43	175	92	No	(*)	Yes	For low-temperature service.
Ammonia	NH_3	R	− 28	34	169	555	T, F	No	(†)	High-efficiency refrigerant.
Refrigerant-12	CCl_2F_2	R	− 22	26	108	67	No	Yes	Yes	Most recommended.
Methyl chloride	CH_3Cl	R	− 11	21	95	178	(‡)	Yes	(§)	Expansion valve may freeze if water is present.
Sulfur dioxide	SO_2	R	+ 14	12	66	166	T, I	No	(§)	Common to these refrigerants:
Refrigerant-21	$CHCl_2F$	RO	+ 48	5	31	108	No	Yes	Yes	a. Evaporator under vacuum.
Ethyl chloride	C_2H_5Cl	RO	+ 54	5	27	175	F, I	No	(‖)	b. Low compressor discharge-pressure.
Refrigerant-11	CCl_3F	C	+ 75	3	18	83	No	Yes	Yes	
Dichloromethane	CH_2Cl_2	C	+105	1	10	155	No	Yes	Yes	c. High volume-to-mass ratio across compressor.

(*) Oil floats on it at low temperatures.
(†) Corrosive to copper-bearing alloys.
(‡) Anesthetic.

(§) Corrosive in the presence of water.
(‖) Attacks rubber compounds.

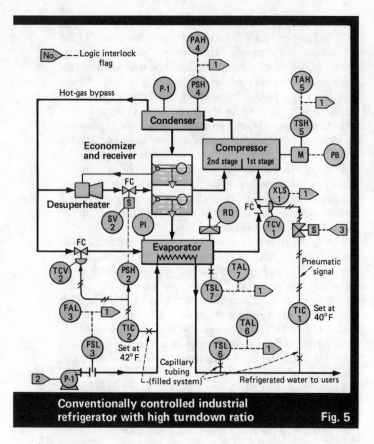

Conventionally controlled industrial refrigerator with high turndown ratio   Fig. 5

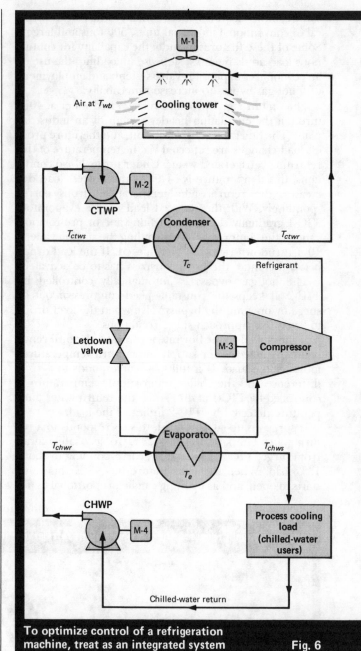

To optimize control of a refrigeration machine, treat as an integrated system   Fig. 6

When the load is below 10%, the hot-gas bypass is in operation, and the solenoid valve SV-2, actuated by the high-pressure switch PSH-2, opens. Some hot gas goes through the evaporator and is cooled by the liquid refrigerant, and some hot gas flows through the open solenoid valve. This second portion is desuperheated by the injection of liquid refrigerant upstream of the solenoid, a procedure that protects against overheating of the compressor motor.

Operating safety of the system is guaranteed by the following interlocks:

■ The first interlock system prevents the compressor motor from being started up if at least one of the following conditions exists, and also stops the compressor if any except the first condition occurs while it is running:

1. Suction vane is open, as detected by limit switch XLS-1.
2. Refrigerated-water temperature is dangerously low, approaching freezing, as sensed by TSL-6.
3. Refrigerated-water flowrate is low, as measured by FSL-3.
4. Evaporator temperature has dropped near the freezing point, as detected by TSL-7.
5. Condenser discharge pressure, and therefore pressure in the condenser, is high, as indicated by PSH-4.
6. Temperature of motor bearing and/or winding is high, as detected by TSH-5.
7. Lubricating-oil pressure is low (not shown).

■ The second interlock system guarantees that the following pieces of equipment are started or are already running upon starting of the compressor:

1. Refrigerated-water pump (P-1).
2. Lubricating-oil pump (not shown).

3. Water to lubricating-oil cooler, if such exists (not shown).

■ The third interlock usually assures that the suction vane is completely closed when the compressor is stopped.

Optimization of refrigeration machines

Fig. 6 shows that the cooling load from the process is carried by the chilled water to the evaporator, where it is transferred to the refrigerant, which takes it to the condenser. There it is passed on to the cooling-tower water, so that it might finally be rejected to the ambient air.

This heat-pump operation involves four heat-transfer substances (chilled water, refrigerant, cooling-tower water, air) and also four heat-exchanger devices (process heat exchanger, evaporator, condenser, cooling tower). The total system operating cost is the sum of the cost of

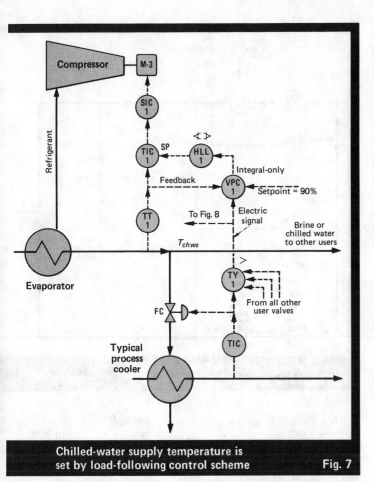

Chilled-water supply temperature is set by load-following control scheme   Fig. 7

circulating the four heat-transfer substances (M-1, M-2, M-3, M-4). In the traditional (unoptimized) control systems (e.g., Fig. 5), each of these four systems is operated independently, in an uncoordinated manner. In addition, conventional control systems are not designed to follow the load, but are operated at constant high levels of energy, with some energy wasted.

Load-following optimization eliminates this waste, while operating the above four systems as a coordinated single process. The goal is to continuously minimize the cost of operation. The supply and return temperatures of chilled and cooling-tower waters are controlled (but not necessarily constant), while the flowrates of chilled water, refrigerant, cooling-tower water and air are manipulated in response to the fixed variables.

Allowing the water temperatures to float in response to load and ambient temperature-variations eliminates the waste associated with keeping them at arbitrary fixed values, and the chiller's operating cost is drastically reduced, sometimes cut in half. The following describes the specific optimization loops and strategies used. First, we will cover the four control loops that maintain the four water temperatures (T_{chws}, T_{chwr}, T_{ctws}, T_{ctwr}) at their optimum values. Then, other strategies will be discussed.

T_{chws} optimization

The yearly operating cost of a chiller is reduced by 1.5 to 2% for each 1°F reduction in the temperature difference across it (T_c-T_e in Fig. 6). To minimize this difference, T_c must be minimized and T_e maximized. There-

fore, the optimum value of T_{chws} is the maximum temperature that will still satisfy all the loads. Fig. 7 illustrates the method of continuously finding and maintaining this value.

Nonetheless, an energy-efficient refrigeration system cannot be created by instrumentation alone, because sizing and selecting equipment also is important. For example, the evaporator's heat-transfer area should be maximized, so that T_e is as close as possible to the average chilled-water temperature in the evaporator. Similarly, at the compressor, the refrigerant flowrate should be matched to the load by motor-speed control, rather than by throttling.

TCV-1 and TCV-2 shown in Fig. 5 represent sources of energy waste (due to friction loss in these valves), while Fig. 7 shows the energy-efficient technique of motor-speed control. Variable-speed operation can be achieved by using variable-speed drives on electric motors or through steam-turbine drives. If several constant-speed motors are used, then all compressors should be driven to their maximum load (TCV-1 and TCV-2 in Fig. 5 fully open) except one, which is used to match the required load by throttling.

Fig. 7 shows the proper technique of maximizing the chilled-water supply's temperature in a load-following, floating manner. The optimization control loop guarantees that all users of refrigeration will always be satisfied, while the chilled-water temperature is maximized. This is done by selecting (with TY-1) the most-open chilled-water valve and comparing its transmitter signal with the 90% setpoint of the valve position controller, VPC-1. If even the most-open valve is less than 90% open, the setpoint of TIC-1 is increased; however, if the valve opening exceeds 90%, the TIC-1 setpoint is decreased.

Thereby, a condition is maintained such that all users are able to obtain more cooling if needed (by the further opening of their supply valves), while the header temperature is continuously maximized. The VPC-1 setpoint of 90% can be adjusted. Lowering it gives a wider safety margin, which might be required if some of the cooling processes served are critical. Increasing the setpoint maximizes energy conservation at the expense of the safety margin.

An additional side benefit of this load-following optimization strategy is that valve cycling is reduced and pumping costs are lowered, because all chilled-water valves in the plant are opened up as T_{chws} is maximized. The reduction in pumping costs is a direct result of all chilled-water valves' opening up and thereby lowering their pressure drops, while valve cycling is eliminated by moving the valve opening away from the unstable region near the closed position.

For the control system in Fig. 7 to be stable, it is necessary to use an integral-only controller for VPC-1, with an integral time that is tenfold that of the integral setting of TIC-1. Such a control mode is needed to keep the optimization loop stable when the valve-opening signal selected by TY-1 is either cycling or noisy. The high/low limit settings on the setpoint signal, HLL-1, to TIC-1 guarantee that VPC-1 will not drive the chilled-water temperature to unsafe or undesirable levels. Because these limits can block the VPC-1 output from affecting T_{chws}, it is necessary to protect against reset windup in

VPC-1. This is done through the external feedback signal shown in Fig. 7.

T_{chwr} optimization

The combined cost of operating the chilled-water pumps and the chiller itself is a function of the temperature drop across the evaporator, $T_{chwr}-T_{chws}$. Because an increase in this ΔT decreases compressor operating costs while also decreasing pumping costs, optimization strategy aims at maximizing this ΔT.

The temperature drop will be maximum when the chilled-water flowrate to the chilled-water users is minimum. (This assumes that the heat load is constant for the moment.) Even if the most-open chilled-water valve is not yet fully open, there is a choice of increasing the chilled-water supply temperature (set by TIC-1 in Fig. 7) or increasing the temperature rise across the users by lowering the ΔP (set by PDIC-1 in Fig. 8), or both. Increasing the chilled-water supply temperature reduces the yearly compressor operating cost by about 1.5% for each °F of temperature increase, while lowering the ΔP reduces the yearly pump operating cost by about $0.50/gpm for each psi. Both cost figures are widely used averages.

The setpoints of the two valve-position controllers (VPC-1 in Fig. 7 and 8) will determine if these adjustments are to occur in sequence or simultaneously. If both setpoints are the same, simultaneous action results, while if one adjustment is more economical than the other, then the setpoint of the corresponding VPC will be set lower.

This results in sequencing, which means that the more cost-effective correction will be fully exploited before the less effective one is employed. In Fig. 7 and 8, it was assumed that increasing T_{chws} is the more cost-effective step and therefore the VPC in Fig. 7 has a setting of 90%, while the VPC in Fig. 8 has one of 95%.

This controller is the cascade master of PDIC-1, which guarantees that the pressure difference between the chilled-water supply and return is always high enough to move water through the users, but never so high as to exceed their pressure ratings. The high and low limits are set on HLL-1, and VPC-1 is free to float this setpoint within these limits to minimize operating cost. (Cascade control is conecting several controllers in series, with the output of the master controller(s) becoming the setpoint(s) of the slave(s).)

In order to protect against reset windup when the output of VPC-1 reaches one of these limits, external feedback is provided from the PDIC-1 output signal. The VPC (in Fig. 8) is an integral-only controller with a setting of ten times that of TIC-1 in Fig. 7. This keeps the loop stable, even if the selected most-open-valve signal is noisy.

When the chilled-water pump station consists of several pumps and only one of them is variable-speed, pump increments are started up when PSH-1 signals that the pump-speed controller setpoint is at its maximum. When the load is dropping, excess pump increments are stopped on the basis of flow, as detected by FSL-2. To eliminate cycling, the excess pump increment is turned off only when the actual total flow corresponds to less than 90% of the capacity of the *remaining* pumps.

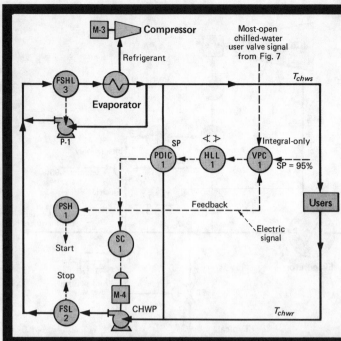

Floating the chilled-water flowrate to optimize the evaporator temperature differential   **Fig. 8**

This load-following optimization loop will float the total chilled-water flowrate to achieve maximum overall economy. To maintain efficient heat transfer and good turbulence within the evaporator, a small local circulating pump (P-1) is provided at the evaporator. This pump is started and stopped by FSHL-3, guaranteeing that the water velocity in the evaporator tubes will never drop below the adjustable limit of, say, 4 ft/s.

T_{ctws} optimization

Minimizing the temperature of the cooling-tower water is one of the most effective contributors to chiller optimization. Conventional control systems have been run with constant cooling-tower temperatures of 75°F or higher. Constant-utility operation is the worst enemy of efficiency and, therefore, optimization. Each 10°F reduction in the cooling-tower water temperature reduces the yearly operating cost of the compressor by about 15%.

Stated another way: If a compressor is operating at 50°F condenser water instead of 85°F, it will meet the same load but consume half the power. Operation at condenser water temperatures of less than 50°F is practical during the winter months, and savings exceeding 50% have been reported.

As shown in Fig. 9, an optimization control-loop is required to continuously maintain the cooling-tower water supply at an economical minimum temperature. This minimum temperature is a function of the wet-bulb temperature of the air. The cooling tower cannot generate a water temperature as low as the ambient wet-bulb one, but it can approach it. (The temperature difference between T_{ctws} and T_{wb} is called the approach.)

Fig. 9 illustrates that as the approach increases, the cost of operating the cooling-tower fans drops, while the costs of pumping and of compressor operation increase. Therefore, the total-operating-cost curve has a mini-

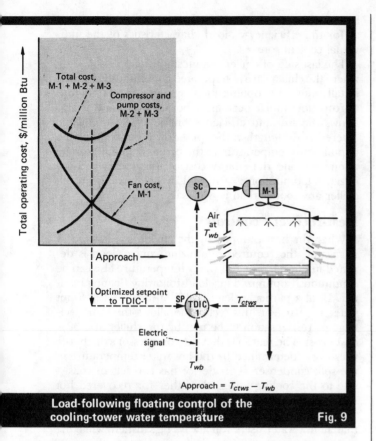

Load-following floating control of the cooling-tower water temperature   Fig. 9

mum, identifying the optimum approach that allows operation at an overall minimum cost. This ΔT automatically becomes the setpoint of TDIC-1. This optimum approach increases if the load on the cooling tower increases or if the wet-bulb temperature decreases.

If the cooling-tower fans have variable-speed drives, then the optimum approach is maintained by throttling. If the tower fans are two-speed or single-speed units, then the output of TDIC-1 incrementally starts and stops the fans to maintain the optimum approach.

Where a large number of cooling-tower cells comprise the total system, the water flowrates to the various cells should be automatically balanced as a function of the operation of the associated fan. In other words, flows to all cells whose fans are at high speed should be kept at identical high rates, while cells with low-speed fans should receive water at equal low flows. Cells with their fans off should be supplied with water at a minimum flowrate.

T_{ctwr} optimization

Fig. 10 shows that the combined cost of operating the cooling-tower pumps and the chiller compressor is a function of the temperature rise across the condenser. Because an increase in this rise increases compression costs while decreasing pumping costs, the combined cost curve has a minimum point. The ΔT corresponding to this minimum automatically becomes the setpoint of TDIC-1 in the optimized control loop in Fig. 10.

This controller is the cascade master of PDIC-1, which guarantees that the pressure difference between the supply and return cooling-tower water flowrates is enough to provide flow through the users, but never enough to

cause damage. The high and low limits are set on HLL-1. TDIC-1 freely floats this setpoint within these limits to keep the operating cost at a minimum.

To protect against reset windup (when the output of TDIC-1 reaches one of these limits), an external feedback is provided from the PDIC-1 output signal.

When the cooling-tower-water pump station consists of several pumps, and only one is variable-speed, pump increments are started when PSH-1 signals that the pump-speed controller setpoint is at its maximum. When the load is dropping, the excess pump increments are stopped on the basis of *flow*, as detected by FSL-2. To eliminate cycling, the excess pump increment is turned off only when the actual total flow corresponds to less than 90% of the capacity of the *remaining* pumps.

This load-following optimization loop floats the total cooling-tower water flowrate to achieve maximum overall economy. To maintain efficient heat transfer and good turbulence in the condenser, a small local circulating pump P-1, is provided at the condenser. This pump is started and stopped by FSLH-3, guaranteeing that the water velocity in the condenser tubes will never drop below the adjustable limit of, say, 4 ft/s. (This is the same strategy used for the evaporator in Fig. 8.)

Other chiller optimization methods

The controls described in Fig. 7–10 automatically minimize the operating cost of the refrigeration machine as a single integrated system. Additional steps toward reducing operating costs include:

If storage tanks are available, it is economical to produce the daily brine or chilled-water needs of the plant at night, when it is least expensive to do so. At night, ambient temperatures are lower, and electricity is less expensive in some areas.

Since, in a typical plant, coolant can be provided from many sources, another approach is to restructure the system in response to changes in loads, ambient conditions, and utility costs. For example, at some operating and ambient conditions, the cooling-tower water may be cold enough to meet the demands of the process directly. Alternatively, if the cooling-tower outflow is below the temperature required by the process, the chillers can be operated in a thermosiphon mode. Refrigerant circulation is then driven by the temperature differential rather than by a compressor.

When demand is low, it may be possible to save money by operating the chillers part-time at peak efficiency, rather than steadily at partial loading. Efficiency tends to be low at partial loads, because of losses caused by friction drop across suction dampers, pre-rotation vanes, and steam governors. Cycling is practical if the storage capacity of the distribution headers is large enough to avoid frequent stops and starts. For intermittent operation, data such as the heat to be removed and the characteristics of the available chillers are needed to determine the most economical operating strategy.

When chiller cycling is used, the thermal capacity of the chilled-water distribution system absorbs the load while the chiller is off. For example, a pipe distribution network with a volume of 100,000 gal represents a thermal capacity of about 1 million Btu for each degree of temperature rise. If the chilled-water temperature may

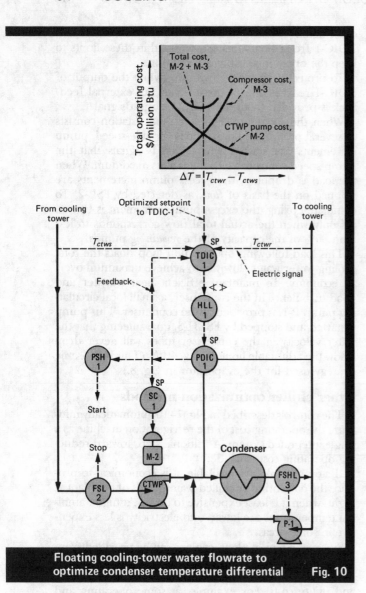

Floating cooling-tower water flowrate to optimize condenser temperature differential   Fig. 10

float 5° (say, from 40 to 45°F) before the chiller is restarted, this represents the equivalent of about 400 tons of thermal capacity. If the load happens to be 200 tons/h, the chiller can be turned off for 2 h at a time. If the load is 1,000 tons, the chiller will be off for only 24 minutes. The above illustrates the natural load-following time-proportioning nature of this scheme.

For longer rest periods

If the chiller needs a longer period of rest than the thermal capacity of the distribution system can provide, one has the option of:

1. Adding tankage to increase the water volume.
2. Starting up a second chiller, or distributing the load between chillers of different sizes by keeping some in continuous operation while cycling others. By continuously measuring the actual efficiency ($/ton) of each chiller, all loads can be met by the most efficient combination of machines for the load.

In plants with multiple refrigerant sources, the cost per ton of cooling can be calculated from direct measurements and used to establish the most efficient combination of units to meet present or anticipated loads. As with

boilers, this accounts for differences among units, as well as for the efficiency-vs.-load characteristics of the individual coolant sources.

The last step of chiller optimization is applicable only where the chillers are not operated continuously. In such installations, the optimization system "knows" before startup how much heat must be removed, and it also knows the size and efficiency of the available chillers. Therefore the length of the "pull-down" period (the time to pull down temperature to the process setpoint) can be minimized and the energy cost of this operation optimized. Optimization is achieved by not starting up the chiller any sooner than is necessary.

Heat-recovery controls

When the heat pumped by the chiller is recovered as hot water, the required optimizing control loop is depicted in Fig. 11. The hot-water temperature also can be continuously optimized in a load-following floating manner. If, at a particular load, 100°F instead of 120°F hot water is sufficient, this technique allows the same tonnage of refrigeration to be met by the chiller at a 30% lower cost. The reason is that the compressor's discharge pressure is determined by the hot-water temperature in the split condenser. This device has two sets of coils—one to the cooling tower, the other for recovered hot water. It can be run either way.

The optimization control loop in Fig. 11 guarantees that all hot-water users will always be satisfied, while the water temperature is minimized. TY-1 selects the most-open hot-water valve, and VPC-1 compares its transmitter signal to a 90% setpoint. If even the most-open valve is less than 90% open, the setpoint of TIC-1 is decreased; while if it exceeds 90%, the setpoint is increased. Thereby a condition is maintained such that all users are able to obtain more heat (by further opening their supply valves) if needed, while the header temperature is continuously optimized.

Fig. 11 also shows that an increasing demand for heat will cause the TIC-1 output signal to rise as its measurement drops below its setpoint. An increase in heat load will cause a decrease in the heat spill to the cooling tower, as the control valve TCV-1A is closed between 3 and 9 psi (typical range for pneumatic split-range valves). At a 9 psi output-signal, all the available cooling load is being recovered and TCV-1A is fully closed. If the heat load continues to rise (the TIC-1 output signal rises over 9 psi) this results in the partial opening of the "pay heat" valve, TCV-1B. Pay heat means that it has to be paid for, i.e., bought from an outside source. In this mode of operation, steam heat is used to supplement the freely available recovered heat.

A local circulating pump P-1 is started whenever the flow velocity is low, to prevent deposit formation in the tubes. P-1 is a small (10 to 15 hp) pump, operating *only* when the flow is low. The main cooling-tower pump (usually larger than 100 hp) is stopped when TCV-1A is closed.

Selection of optimum operating mode

The cost-effectiveness of heat recovery is a function of the outdoor temperature, the unit cost of energy from the alternative heat source, and the percentage of the

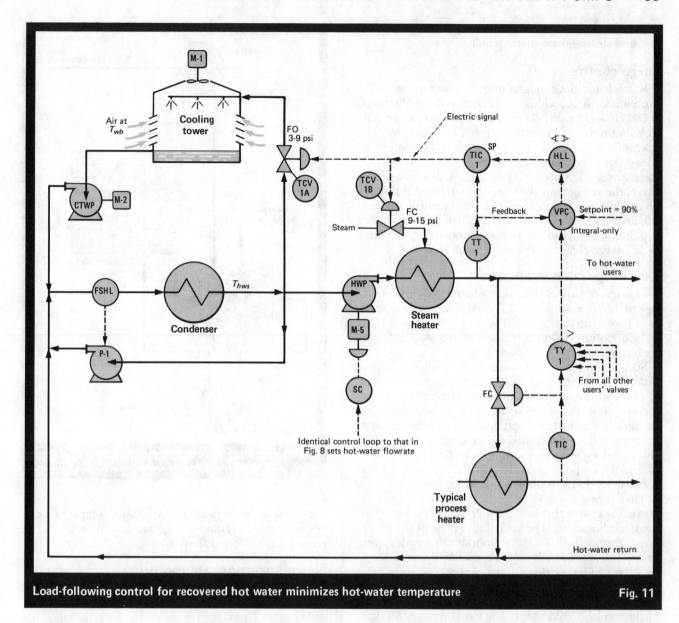

Load-following control for recovered hot water minimizes hot-water temperature   Fig. 11

cooling load that can be usefully employed as recovered heat. Calculations would show that if steam is available at $7/million Btu, and only half of the cooling load is needed as hot water, it is more economical to operate the chiller on cooling-tower water and use steam as the heat source when the outside air is below 65°F.

Conversely, when the outdoor temperature is above 75°F, the penalty for operating the split condenser at hot-water temperatures is no longer excessive; therefore, the plant should automatically switch back to recovered-heat operation. This cost-benefit analysis is a simple and continuously used part of the overall optimization scheme.

In such locations as the southern U.S. states, where there is no alternative heat source, such as a boiler or an outside steam line, another problem can arise because all the heating needs of the plant must be met by recovered heat from the heat pump. It is possible that on cold winter days there might not be enough recovered heat to meet this load. Whenever the heat load exceeds the cooling load and there is no other heat source available, an artificial cooling load must be added to the heat pump.

This artificial heat source can, in some cases, be the cooling-tower water itself. A direct heat exchange between cooling-tower and chilled water streams is of advantage not only in this case, but also when there is no heat load—but when there is a small cooling load during the winter. At such times, the chiller can be stopped, and the small cooling load can be met by direct cooling from the cooling-tower water that is at a winter temperature.

Retrofitting chiller controls

In designing a plant, it is easy to specify variable-speed pumps, provide the thermal capacity required for chiller cycling, or locate chillers and cooling towers at the same elevation and near to each other so that the cooling-tower-water pumping costs will be minimized. In existing plants, the inherent design limitations must be accepted and the system optimized without any equipment changes. This is quite practical, and can still produce savings exceeding 50%, but precautions are needed.

In optimizing existing chillers, the constraints are

surge, low evaporator temperature, economizer flooding, and steam-governor rangeability.

Surge control

At low loads, not enough refrigerant is circulated. This can initiate surge, with its associated violent vibrations.

Old chillers usually do not have automatic surge controls, but have only vibration sensors for shutdown. For operation at low loads, an antisurge control loop must be added. However, surge protection is always provided at the expense of efficiency. To bring the machine out of surge, the refrigerant flow must be increased if there is no real load on the machine. The only way to do this is to add artificial and wasteful loads (i.e., hot-gas or hot-water bypasses). Therefore, it is more economical to either cycle a large chiller or operate a small one.

Low evaporator temperature

Low temperatures can occur when an old chiller designed for operation at 75°F condenser water is optimized, and run on 45 or 50°F water (in the winter). This is the opposite of surge, because it occurs when refrigerant vaporizes at an excessively high rate.

Such a situation occurs because the chiller is able to pump twice the tonnage it was designed for, due to the low compressor discharge pressure. Here, the evaporator heat-transfer area becomes the limiting factor, and the only way to increase heat flow is to increase the temperature differential across the evaporator tubes. This shows up as a gradual lowering of refrigerant temperature in the evaporator until it reaches 32°F, and the machine shuts down to protect against ice formation.

There are two ways to prevent this. The first is to increase the evaporator heat-transfer area (a major equipment modification). The second is to prevent the refrigerant temperature in the evaporator from dropping below 33°F, by not allowing the cooling-tower water to cool the condenser down to its own temperature. The latter solution requires adding a temperature-control loop that prevents the chiller from taking full advantage of the available cold water. The loop throttles the flowrate, thereby causing a temperature rise.

Economizer flooding

On existing chillers, the economizer control valves, LCV-1 and LCV-2 in Fig. 12, are often sized on the basis that the refrigerant vapor pressure in the condenser, P_3, is constant and corresponds to a condenser water temperature of 75 or 85°F. When such units are operated with 45 or 50°F water, P_3 is reduced, and the pressure differentials across LCV-1 and LCV-2 are also less.

If this occurs—as it easily can—when the refrigerant circulation rate is high, the control valves will be unable to provide the necessary flowrate, and flooding of the economizer will occur (the flow is higher and the ΔP is lower than that used to size these valves.) The solution is to install larger valves, preferably external ones (located outside of the economizer) with two-mode control to eliminate offset. (Plain proportional controllers cannot maintain their setpoints as loads change, while the addition of the integral mode eliminates this offset.)

This is important in machines that were not originally designed for optimized, low-temperature condenser-

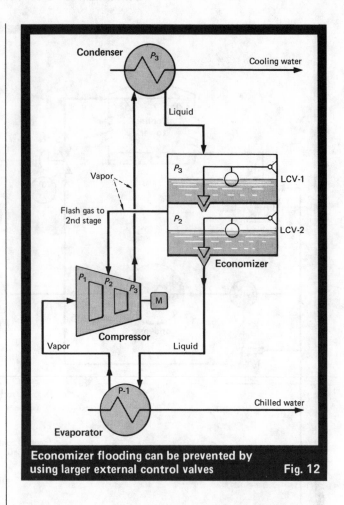

Economizer flooding can be prevented by using larger external control valves   Fig. 12

water operation, because otherwise the compressor can be damaged by liquid-refrigerant overflow from the economizer's flooded flash evaporator chamber.

Steam-governor rangeability

To optimize a steam-turbine-driven compressor, its rotational velocity must be modulated over a reasonably wide range. This is not possible with older machines, because they have quick-opening steam valves. A slight increase in lift from the fully closed position results in a substantial steam flow, and therefore a substantial rotational velocity. To try to throttle this steam flow makes the valve unstable and noisy. Valve characteristics can be changed from quick-opening. Wide rangeability can be obtained by welding two rings with V-notches to the seats of the existing steam-governor valves.

The author

Béla G. Lipták is president of Lipták Associates, 84 Old North Stamford Rd., Stamford, CT 06905. Tel: (203) 357-7614. He works as an instrumentation consultant on all phases of process-control projects. He received his degrees from The Technical University of Budapest, Stevens Institute of Technology, and The City College of New York. He has published about 75 technical articles and 11 books, the most recent of which is the revised edition of the "Instrument Engineers' Handbook on Process Measurement." He is a fellow of the Instrument Soc. of America and is a P.E. in several states.

Estimate air-cooler size

This program for the HP-41CV will help determine surface area, physical size and other parameters of air coolers.

Nadeem M. Shaikh, S. H. Landes, Inc.

☐ Often, at an early stage of a project, one must evaluate using air coolers, to see whether they can meet the cooling requirements of a process—and if they are feasible, to develop the cooler's capital investment cost, operating cost and the space requirements.

Several graphical techniques have been published for estimating air-cooler size. However, these methods do not provide any information about tubeside pressure drop, and analysis of horsepower versus surface and space requirements can be very laborious. The program presented in this article greatly reduces the calculation time and improves accuracy. Although written for the HP-41CV, it can be adapted for other programmable calculators. The main features of the program are:

■ It calculates the required extended surface, bay size, number and diameter of fans, driver horsepower, number of tubes, and tubeside velocity and pressure drop.

■ It makes calculations based upon the most commonly used 1-in. O.D., 14-BWG, ⅝ × 10, 2¼ in. △-pitch tube layout. The number of rows and corresponding extended area per square foot of bundle face area can be selected from Table III. This feature is quite helpful in analyzing the space requirement vs. horsepower. This is of particular interest in evaluating the air coolers for offshore facilities. The greater the number of rows, the higher the horsepower required and the smaller the space.

■ It uses a curve-fit type of equation to correlate the air-side film coefficient and air static-pressure-drop across the bundle with the air mass-velocity for the selected tube layout [1]. This eliminates certain input data requirements such as fin thickness, spacing, height, conductivity, etc., for the estimation of air film coefficient.

■ It calculates the unit size within the allowable pressure drop.

■ It includes a provision for inputting the desired outlet-air temperature. This helps in analyzing the economics of increase in surface versus decrease in horsepower requirement. Higher outlet-air temperature would mean a lower air flowrate and, therefore, less horsepower required to drive the fans. At the same time, the *LMTD* and overall heat-transfer coefficient will decrease, thereby requiring more surface.

■ The program also makes it possible to input the tubeside film-coefficient, which will override the calculations for film resistance and pressure drop. This feature helps in estimating the size of condensers. For example, in the case of light-hydrocarbons condensing service, a tubeside film coefficient can be input as 450 Btu/ (h)(ft²)(°F) or, in the case of steam condensers, a film coefficient of 1,000 Btu/(h)(ft²)(°F) can be used. Calculations for two-phase pressure drop and heat-transfer rates are quite lengthy and involved, and no attempt has been made to cover them in this program.

■ It assumes the specific heat of air as 0.25 Btu/ (lb)(°F), for preliminary estimation. The program calculates the air density ratio based upon the average air-temperature across the bundle.

■ And the program enables one to study the effect of fouling on horsepower and surface requirements.

The basic philosophy of the calculation procedure is for one to select a particular unit configuration and then test it to see if it will work. The program input requirements are listed in Table IV.

As a starting point, an overall heat-transfer coefficient is assumed. The values of U_x given in Tables I and II can help in quick convergence. The air temperature rise is then estimated by the following equation [1]:

$$\Delta t_a = \left(\frac{U_x + 1}{10} \right)\left(\frac{T_1 + T_2}{2} - t_1 \right)$$

The outlet-air temperature thus arrived at is used to calculate the *LMTD*. A correlation factor of 1 is used, assuming that the unit will have three or more passes. The extended surface, A_x, is then calculated by the equation:

$$A_x = \frac{Q}{(U_x)(\Delta T_m)}$$

The bundle face area is determined by the appropriate APSF (area per square foot) factor given in Table III.

Approximate overall heat-transfer coefficients for air coolers that are cooling liquids   **Table I**

Material	Heat-transfer coefficient, U_x, Btu/(h)(ft²)(°F)
Water	6.1
Brine, 75% water	4.7
50% ethylene glycol-water	5.1
Ammonia	5.14
Hydrocarbon liquids, viscosity at average temperature, cP	
0.2	4.7
0.5	4.2
1.0	3.5
2.5	2.6
4.0	1.6
6.0	1.2
10.0	0.6

Originally published December 12, 1983

Approximate overall heat-transfer coefficients for air coolers that are cooling vapors   Table II

Material	Heat-transfer coefficient, Btu/(h)(ft²)(°F)				
	50 psig	100 psig	300 psig	500 psig	1,000 psig
Light hydrocarbons	1.6	1.9	2.6	3.0	4.2
Medium hydrocarbons and organic solvents	1.9	2.3	3.3	3.5	3.6
Light organics	1.0	1.6	2.3	2.6	2.8
Air	1.0	1.4	2.1	2.3	2.4
Ammonia	1.0	1.6	2.3	2.6	2.8
Steam	1.0	1.4	2.3	2.8	3.0
Hydrogen	2.3	3.27	4.4	4.7	4.9

Data on air coolers for use with program   Table III

Fintube data for 1-in.-OD, 5/8-in. fin ht., 10 fins/in., 2 1/4-in. pitch layout

Extended area per foot		5.58 ft²
Area of ratio of fintube to 1-in.-O.D. bare tube		21.4
Extended area per ft² of bundle face area (APSF)		
	3 rows	89.1 ft²
	4 rows	118.8 ft²
	5 rows	148.5 ft²
	6 rows	178.2 ft²

Input variables for the program   Table IV

Variable	Units
Duty	Btu/h
Overall heat-transfer coefficient (guess)	Btu/(h)(ft²)(°F)
Inlet temperature of process stream	deg F
Outlet temperature of process stream	deg F
Stream flowrate	lb/h
Density of fluid	lb/ft³
Viscosity of fluid	centipoise
Specific heat of fluid	Btu/(lb)(°F)
Thermal conductivity of fluid	Btu/(h)(ft)(°F)
Maximum allowable pressure drop	psi
Ambient air temperature	deg F
Extended area per ft² of bundle face area (APSF)	ft²
Number of rows	dimensionless
Number of tubeside passes	dimensionless
Fouling factor	(h)(ft²)(°F)/Btu
Outlet air temperature (optional)	deg F
Tubeside film coefficient (optional)	Btu/(h)(ft²)(°F)

Storage configuration of HP-41CV calculator Table V

Size 050

Storage register	Contents	Storage register	Contents
01	U_x (guess)	25	N_t
02	T_1	26	W_t
03	T_2	27	N_p
04	t_1	28	G_t
05	t_2	29	P
09	T_m	30	U
10	A_x	31	Cp
11	APSF	32	N_{Re}
12	F_a	33	J
13	L	34	Sp. ht.
14	W	35	K
15	W_a	36	N_{Pr}
16	G_a	37	H_t
17	F_D	38	H_a
18	F_p	39	U_x (calculated)
19	N	40	H_t (given)
20	D_r	41	PDROP calculated
21	P_a	42	f
22	ACFM/fan	43	ΔP max. allowable
23	Fan h.p.	44	Fouling
24	t_2 (given)	45	No. of fans.

The width and length of the unit are based upon several operating and layout considerations.

The design assumes a minimum of two fans covering at least 40% of the face area. This provides somewhat of a safety factor against fan or driver failure and also allows a method of control (by fan staging).

A minimum of 40% fan coverage will ensure a good air distribution across the bundle. The maximum fan diameter is limited to 14 ft, and the tube length to 40 ft.

Once the unit dimensions are determined, the number of tubes is calculated as:

$$N_t = A_x / (APSF)(L)$$

The tubeside heat-transfer coefficient is calculated by the Colburn-factor heat-transfer equation for the turbulent-flow regime. The Colburn factor is given by [2]:

$$J = 0.023/N_{Re}^{0.2}$$

where: $N_{Re} > 10^4$, $0.77 < N_{Pr} < 160$, and $L/D > 60$. The convective heat-transfer coefficient is, then, given by the following equation:

$$H_t = J(C_p \rho V / N_{Pr}^{2/3})$$

The tubeside pressure drop is estimated by Darcy's pressure-drop equation [3], using Fanning's friction factor For the turbulent flow regime, Churchill's modification [4] of the Colebrook equation is:

$$\frac{1}{\sqrt{f}} = 2.457 \ln\left(\frac{1}{(7/N_{Re})^{0.9} + (0.27\epsilon/D)}\right)$$

which gives Fanning's friction factor as $2f$.

The entrance and exit losses on the tubeside are assumed to be 1½ velocity heads. The pressure drop per return is assumed to be 3 velocity heads.

The airside heat-transfer coefficient is determined by a curve-fit type of equation, giving the air film coefficient as a function of air mass-velocity for the selected tube layout.

The overall heat-transfer coefficient is then determined by the equation:

$$\frac{1}{U_x} = \left(\frac{1}{H_t}\right)\left(\frac{A_x}{A_i}\right) + r_{dt}\left(\frac{A_x}{A_i}\right) + r_m + \frac{1}{H_a}$$

For the selected tubes, $A_x/A_i = 21.4$, and the resistance of the metal is small as compared to the other resistances. Therefore:

$$\frac{1}{U_x} = \left(\frac{1}{H_t}\right)21.4 + (r_{dt})21.4 + \frac{1}{H_a}$$

This calculated value of U_x is compared with the assumed value and, if the absolute difference is greater than 1% of the assumed U_x, the calculations are repeated with a new value of U_x, which is an arithmetical average of the assumed and calculated values.

Program listing for HP-41CV for estimation of air-cooler parameters (SIZE storage registers to 050)   **Table VI**

Input data and initialize
```
01◆LBL "AC"
02 CLRG
03 "DUTY?"
04 PROMPT
05 STO 08
06 "UX?"
07 PROMPT
08 STO 01
09 "T1?"
10 PROMPT
11 STO 02
12 "T2?"
13 PROMPT
14 STO 03
15 "FLRT?"
16 PROMPT
17 STO 26
18 "RHO?"
19 PROMPT
20 STO 29
21 "VIS?"
22 PROMPT
23 STO 31
24 "SPHT?"
25 PROMPT
26 STO 34
27 "TCON?"
28 PROMPT
29 STO 35
30 "PDROP?"
31 PROMPT
32 STO 41
33 "TAIR?"
34 PROMPT
35 STO 04
36 "APSF?"
37 PROMPT
38 STO 11
39 "ROWS?"
40 PROMPT
41 STO 19
42 "PASS?"
43 PROMPT
44 STO 27
45 "FOULF?"
46 PROMPT
47 STO 44
48 0
49 "T2AIR?"
50 PROMPT
51 STO 24
52 0
53 "HT?"
54 PROMPT
55 STO 40
56 2
57 STO 45
```

Check if T_2 air is given
```
58◆LBL 03
59 RCL 24
60 X>0?
61 GTO 01
```

Calculate outlet temp of air
```
62 RCL 02
63 RCL 03
64 +
65 2
66 /
67 RCL 04
68 -
69 RCL 01
70 1
71 -
72 10
73 /
74 *
75 RCL 04
76 +
77◆LBL 01
78 STO 05
```

Calculate LMTD
```
79 RCL 02
80 RCL 05
81 -
82 RCL 03
83 RCL 04
84 -
85 X>Y?
86 X<>Y
87 STO 06
88 X<>Y
89 RCL 07
90 RCL 06
91 -
92 RCL 07
93 RCL 06
94 /
95 LN
96 /
97 STO 09
```

Calculate extended surface required
```
98 RCL 08
99 X<>Y
100 /
101 RCL 01
102 /
103◆LBL 09
104 STO 10
```

Calculate unit's length
```
105 RCL 11
106 /
107 STO 12
108 2
109 *
110 SQRT
111 40
112 X<>Y
113 X<=Y?
114 GTO 07
115 40
116◆LBL 07
117 STO 13
```

Calculate width
```
118 RCL 12
119 X<>Y
120 /
121 STO 14
```

Calculate no. of tubes
```
122 RCL 10
123 RCL 13
124 /
125 5.58
126 /
127 STO 25
```

Check if H_T is given
```
128 RCL 40
129 X>0?
130 GTO 04
```

Calculate tubeside mass-velocity
```
131 RCL 25
132 1/X
133 .0733
134 *
135 RCL 26
136 *
137 RCL 27
138 *
139 STO 28
```

Calculate velocity
```
140 RCL 29
141 /
142 STO 30
```

Calculate Reynolds number
```
143 RCL 29
144 *
145 .07
146 *
147 RCL 31
148 1488.6
149 /
150 /
151 STO 32
```

Calculate Fanning's friction factor
```
152 1/X
153 7
154 *
155 .9
156 Y↑X
157 .00193
158 +
159 1/X
160 LN
161 2.457
162 *
163 1/X
164 X↑2
165 2
166 *
167 STO 42
```

Calculate pressure drop
```
168 RCL 13
169 RCL 27
170 *
171 *
172 .07
173 /
174 4
175 *
176 RCL 27
177 3
178 *
179 1.5
180 +
181 +
182 2
183 /
184 32.2
185 /
186 144
187 /
188 RCL 29
189 *
190 RCL 30
191 X↑2
192 *
193 STO 43
194 1.05
195 /
196 RCL 41
197 X<>Y
198 X<=Y?
199 GTO 08
200 RCL 41
201 RCL 43
202 /
203 SQRT
204 1/X
205 RCL 10
206 *
207 GTO 09
```

Calculate Colburn factor, J
```
208◆LBL 08
209 RCL 32
210 .2
211 Y↑X
212 1/X
213 .023
214 *
215 STO 33
```

Calculate Prandtl number
```
216 RCL 34
217 RCL 31
218 1488.6
219 /
220 *
221 RCL 35
222 /
223 3600
224 *
225 STO 36
```

Calculate tubeside film-coefficient
```
226 .666
227 Y↑X
228 1/X
229 RCL 34
230 *
231 RCL 29
232 *
233 RCL 30
234 *
235 3600
236 *
237 RCL 33
238 *
239◆LBL 04
240 STO 37
```

Calculate air-mass velocity
```
241 RCL 08
242 .25
243 /
244 RCL 05
245 RCL 04
246 -
247 /
248 STO 15
```

Calculate air film coefficient
```
249 RCL 12
250 /
251 STO 16
252 .556
253 Y↑X
254 .107
255 *
256 STO 38
```

Calculate overall heat-transfer coefficient
```
257 1/X
258 RCL 37
259 1/X
260 24.6
261 *
262 +
263 RCL 44
264 24.6
265 *
266 +
267 1/X
268 STO 39
269 RCL 01
270 -
271 ABS
272 RCL 01
273 .01
274 *
275 X<>Y
276 X<=Y?
277 GTO 02
278 RCL 01
279 RCL 39
280 +
281 2
282 /
283 STO 01
284 GTO 03
285◆LBL 02
```

Calculate number and diameter of fans
```
286 RCL 12
287 .4
288 *
289 RCL 45
290 /
291 .785
292 /
293 SQRT
294 .5
295 +
296 INT
297 14
298 X<>Y
299 X<=Y?
300 GTO 05
301 2
302 ST+ 45
303 GTO 02
304◆LBL 05
305 STO 17
```

Calculate fan horsepower
```
306 RCL 16
307 1.802
308 Y↑X
309 7.084 E-8
310 *
311 STO 18
312 RCL 04
313 RCL 05
314 +
315 2
316 /
317 460
318 +
319 1/X
320 39.73
321 *
322 .0764
323 *
324 STO 20
325 1/X
326 RCL 18
327 *
328 RCL 19
329 *
330 STO 21
331 .222
332 RCL 15
333 *
334 RCL 20
335 /
336 RCL 45
337 /
338 STO 22
339 4000
340 /
341 .785
342 /
343 RCL 17
344 X↑2
345 /
346 X↑2
347 RCL 20
348 *
349 RCL 21
350 ÷
351 RCL 22
352 *
353 6370
354 /
355 .7
356 /
357 STO 23
```

Output
```
358 BEEP
359 FIX 2
360 "UX="
361 ARCL 01
362 PROMPT
363 "AX="
364 ARCL 10
365 PROMPT
366 "L="
367 ARCL 13
368 PROMPT
369 "W="
370 ARCL 14
371 PROMPT
372 "FANS="
373 ARCL 45
374 PROMPT
375 "FD="
376 ARCL 17
377 PROMPT
378 "FHP/P="
379 ARCL 23
380 PROMPT
381 "HT="
382 ARCL 25
383 PROMPT
384 "VEL="
385 ARCL 30
386 PROMPT
387 "DP="
388 ARCL 43
389 PROMPT
390 "T2AIR="
391 ARCL 05
392 PROMPT
393 END
```

The fan horsepower is estimated by using the air static-pressure drop and actual cubic feet per minute (ACFM) of air delivered by the equation:

$$\text{Fan horsepower} = \frac{(ACFM/\text{fan})(P_f)}{6{,}370 \times 0.70}$$

where

$$P_f = \Delta P_a + \left[\frac{ACFM}{4{,}000\left(\frac{\pi D^2}{4}\right)} \right]^2 (D_r)$$

and

$$\Delta P_Q = \frac{(F_P)N}{D_r}$$

Nomenclature

A_i	Inside surface of tube, ft^2
A_x	Outside extended surface of tube, ft^2
(APF)	External area per ft of fintube, ft^2
$(APSF)$	Area per square foot of bundle face area, ft^2
C_p	Specific heat at average temperature, Btu/(lb)(°F)
D	Inside dia. of tube, ft
D_r	Air density ratio
f	Fanning's friction factor
F_p	Pressure-drop factor
G_a	Airside mass velocity, lb$_{mass}$/(ft^2)(s)
G_t	Tubeside mass velocity, lb$_{mass}$/(ft^2)(s)
H_a	Airside film coefficient, Btu/(h)(ft^2)(°F)
H_t	Tubeside film coefficient, Btu/(h)(ft^2)(°F)
J	Colburn factor
k	Thermal conductivity, Btu/(h)(ft^2)(°F/ft)
L	Unit length, ft
$LMTD$	Log mean temperature difference, °F
N_p	Number of tube passes
N_{Pr}	Prandtl number
N_{Re}	Reynolds number
N_t	Number of rows of tubes in the direction of flow
ΔP_a	Air static pressure drop
P_f	Fan total pressure, in. H$_2$O
P_Q	Air static-pressure drop, in. H$_2$O
Q	Heat transferred, Btu/h
r_{dt}	Fouling factor
r_m	Metal resistance
T_1	Process fluid inlet temperature, °F
T_2	Process fluid outlet temperature, °F
Δt_a	Air-temperature rise, °F
t_1	Ambient air temperature, °F
t_2	Outlet temperature of air, °F
ΔT_m	Log mean temperature difference
U_x	Heat-transfer coefficient, Btu/(h)(ft^2)(°F)
V	Fluid velocity in tubes, ft/s

Greek letters

ϵ	Surface roughness, ft
μ	Fluid viscosity, cP
ρ	Fluid density, lb/ft

The air static-pressure-drop factor (F_p) is calculated by using another curve-fitting type of equation, relating air-face mass-velocity to the pressure-drop factor for the selected tube configuration.

The program listing shown in Table VI should be keyed into the calculator and stored on a magnetic card.

Example

Determine the size of an air cooler required to cool a 48 API hydrocarbon liquid from 250°F to 150°F. The physical properties at the average temperature, T = 200°F are: C_p = 0.55 Btu/(lb)(°F), μ = 0.51 cP, and k = 0.0766 Btu/(h)(ft^2)(°F/ft).

Heat load = 15 × 10^6 Btu/h; flowrate = 273,000 lb/h.

Fouling factor = 0.001; allowable pressure drop = 5 psi.

Ambient air temperature = 100°F.

Solution

Load the program into the calculator. Begin execution, and input the data as shown below:

Keystrokes	Display
[XEQ] [ALPHA] AC [ALPHA]	Duty?
15 EE X 6 [R/S]	Ux?
4.2 [R/S]	T1?
250 [R/S]	T2?
150 [R/S]	FLRT?
273000 [R/S]	RHO?
45.7 [R/S]	VIS?
.51 [R/S]	SPHT?
.55 [R/S]	TCON?
.076 [R/S]	PDROP?
5 [R/S]	TAIR?
100 [R/S]	APSF?
118.8 [R/S]	ROWS?
4 [R/S]	PASS?
3 [R/S]	FOULF?
.001 [R/S]	T2 AIR? (The program
[R/S]	will calculate
[R/S]	HT? t_2 Air and H_t if no input is given.)

Output

BEEP	Ux = 4.2
[R/S]	Ax = 50,070.27
[R/S]	L = 29.03
[R/S]	W = 14.52
[R/S]	FANS = 2.00
[R/S]	FD = 10.00
[R/S]	FHP/F = 22.23
[R/S]	NT = 309.06
[R/S]	VEL = 4.25
[R/S]	DP = 4.94
[R/S]	T2 Air = 152.0

References

1. Gas Processors Suppliers Assn. [GPSA], "Engineering Data Book," Sect. 9, Air Cooled Exchangers, Gas Processors Suppliers Assn., Tulsa, Okla., Rev. 1979.
2. Pierce, Bill L., Heat Transfer Colburn-Factor Equation Spans All Fluid-Flow Regimes, *Chem. Eng.*, Dec. 17, 1979, p. 113.
3. McCabe, Warren L., and Smith, Julian C., "Unit Operations of Chemical Engineering," McGraw-Hill, New York, 1967.
4. Churchill, Stuart W., Friction-Factor Equation Spans All Fluid-Flow Regimes, *Chem. Eng.*, Nov. 7, 1977, p. 91.
5. Brown, Robert, A Procedure for Preliminary Estimates, *Chem. Eng.*, Mar. 27, 1978, p. 108.
6. Kern, D. Q., and Kraus, A. D., "Extended Surface Heat Transfer," McGraw-Hill, New York, 1972.

The author

Nadeem M. Shaikh is a senior process engineer with S. H. Landes, Inc., 7272 Pinemont, Houston, TX 77040, phone (713) 460-7600, where he specializes in process design for the oil and gas industry and conducts courses in design and optimization of oil and gas processing facilities. He obtained a B.S. in physics and mathematics from the Forman Christian College, a B.S.Ch.E. from Punjab University, Lahore, and an M.S.Ch.E. from Tulsa University. He is a member of AIChE and of its fuels and petrochemical division.

Calculator program aids quench-tower design

An essential step in designing evaporative cooling equipment
is finding the adiabatic saturation temperature of the gas.
This program streamlines the computations.

William H. Mink, *Battelle Columbus Laboratories*

☐ Adiabatic humidification is commonly used in the chemical process industries as an economical means of cooling hot gases. In the design of such evaporative cooling systems as quench towers, the engineer must frequently determine the adiabatic saturation temperature of the gas from its temperature, pressure and composition.

If the gas is air at atmospheric pressure, psychrometric charts or tables make this task straightforward and relatively simple. Charts are even available to accommodate slight changes in composition and pressure.

However, if the gas composition varies significantly from that of air (as in the case of flue gas), or if the pressure varies significantly from atmospheric, a rather tedious iterative calculation is required. The calculation is derived as follows:

The saturation humidity, W_s, at the adiabatic-saturation temperature, t_s, is given by [1]:

$$W_s = W_t + \frac{h_g + h_w W_t}{r_s}(t - t_s) \tag{1}$$

where: W_t = humidity at temperature t, lb water vapor/lb dry air; t = air temperature, °F; t_s = saturation temperature, °F; r_s = latent heat of vaporization of water at the adiabatic-saturation temperature, Btu/lb; h_w = specific heat of water vapor, Btu/lb/°F; and h_g = specific heat of the dry gas, Btu/lb/°F.

The latent heat of vaporization can be expressed by [2]:

$$r_s = 91.86(705.56 - t_s)^{0.38} \tag{2}$$

The saturation humidity is also given by [1]:

$$W_s = \frac{18.016\, p_s}{M\,(P - p_s)} \tag{3}$$

which may be expressed as:

$$W_s = \frac{p_s}{(P - p_s)[1.6096(1 - f_{CO_2}) + 2.4428\, f_{CO_2}]} \tag{4}$$

where: p_s = vapor pressure of water at t_s, psia; P = total pressure, psia; M = molecular weight of dry gas; and f = mole fraction.

The vapor pressure of water can be calculated from [3]:

$$\log p_s = 15.092 - \frac{5079.6}{T} - 1.6908 \log T$$
$$- 3.193 \times 10^{-3}T + 1.234 \times 10^{-6}T^2 \tag{5}$$

where T = temperature, °R.

The specific heat of the gas can be determined using equations of the form [4]:

$$h = a + bt + ct^2 \tag{6}$$

For an N_2/O_2 mixture similar to air, $a = 0.239$, $b = 1.288 \times 10^{-5}$, and $c = 1.4 \times 10^{-9}$. For CO_2, $a = 0.2045$, $b = 6.135 \times 10^{-5}$, and $c = 5.8 \times 10^{-9}$. For H_2O, $a = 0.4633$, $b = 2.581 \times 10^{-5}$, and $c = 2.3 \times 10^{-8}$.

The mean specific heat is given by:

$$h\,(mean) = \frac{h_2 t_2 - h_1 t_1}{t_2 - t_1}. \tag{7}$$

The mean specific heat of the gas mixture can be determined using mole fractions:

$$h_{mix}\,(mean) = (1 - f_{CO_2} - f_{H_2O}\, h_{N_2,O_2} + f_{CO_2} h_{CO_2} + f_{H_2O} h_{H_2O}) \tag{8}$$

The adiabatic saturation temperature can now be calculated by the following iterative process:

1. Assume t_s.
2. Calculate r_s, p_s, and h, using Eq. (2), (4), (6), (7) and (8).
3. Calculate W_s, using Eq. (1).
4. Calculate W_s, using Eq. (3).
5. If W_s by Eq. (1) is not equal to W_s by Eq. (3), assume a new value of t_s and repeat the above.

This iterative process can be done quite easily using a programmable hand calculator. Such a program has been prepared for the Texas Instruments, Inc. TI-59

Originally published December 3, 1979

Program listing for adiabatic humidification calculation

Table I

Location	Code	Symbol	Location	Code	Symbol	Location	Code	Symbol	Location	Code	Symbol	Location	Code	Symbol
000	76	LBL	060	43	RCL	120	44	SUM	180	77	GE	240	30	TAN
001	11	A	061	43	43	121	10	10	181	48	EXC	241	24	CE
002	42	STO	062	69	OP	122	76	LBL	182	76	LBL	242	43	RCL
003	01	01	063	04	04	123	88	DMS	183	69	OP	243	36	36
004	91	R/S	064	43	RCL	124	01	1	184	98	ADV	244	69	OP
005	76	LBL	065	03	03	125	00	0	185	43	RCL	245	04	04
006	12	B	066	69	OP	126	22	INV	186	45	45	246	43	RCL
007	42	STO	067	06	06	127	44	SUM	187	69	OP	247	10	10
008	02	02	068	43	RCL	128	10	10	188	02	02	248	66	PAU
009	91	R/S	069	44	44	129	71	SBR	189	43	RCL	249	69	OP
010	76	LBL	070	69	OP	130	30	TAN	190	46	46	250	06	06
011	13	C	071	04	04	131	77	GE	191	69	OP	251	43	RCL
012	42	STO	072	43	RCL	132	88	DMS	192	03	03	252	02	02
013	03	03	073	04	04	133	71	SBR	193	43	RCL	253	75	–
014	91	R/S	074	69	OP	134	39	COS	194	47	47	254	43	RCL
015	76	LBL	075	06	06	135	77	GE	195	69	OP	255	10	10
016	14	D	076	98	ADV	136	88	DMS	196	04	04	256	95	=
017	42	STO	077	43	RCL	137	01	1	197	69	OP	257	42	STO
018	04	04	078	04	04	138	00	0	198	05	05	258	33	33
019	91	R/S	079	55	÷	139	44	SUM	199	43	RCL	259	43	RCL
020	76	LBL	080	01	1	140	10	10	200	10	10	260	02	02
021	15	E	081	00	0	141	76	LBL	201	99	PRT	261	33	X²
022	43	RCL	082	00	0	142	89	π	202	98	ADV	262	75	–
023	37	37	083	95	=	143	01	1	203	43	RCL	263	43	RCL
024	69	OP	084	42	STO	144	22	INV	204	52	52	264	10	10
025	01	01	085	07	07	145	44	SUM	205	69	OP	265	33	X²
026	43	RCL	086	43	RCL	146	10	10	206	03	03	266	95	=
027	38	38	087	01	01	147	71	SBR	207	43	RCL	267	42	STO
028	69	OP	088	85	+	148	30	TAN	208	53	53	268	34	34
029	02	02	089	01	1	149	77	GE	209	69	OP	269	43	RCL
030	43	RCL	090	04	4	150	89	π	210	04	04	270	02	02
031	39	39	091	93	.	151	71	SBR	211	69	OP	271	45	YX
032	69	OP	092	07	7	152	39	COS	212	05	05	272	03	3
033	03	03	093	95	=	153	77	GE	213	43	RCL	273	75	–
034	43	RCL	094	42	STO	154	89	π	214	48	48	274	43	RCL
035	40	40	095	06	06	155	01	1	215	69	OP	275	10	10
036	69	OP	096	08	8	156	44	SUM	216	01	01	276	45	YX
037	04	04	097	00	0	157	10	10	217	43	RCL	277	03	3
038	69	OP	098	00	0	158	76	LBL	218	49	49	278	95	=
039	05	05	099	42	STO	159	77	GE	219	69	OP	279	42	STO
040	98	ADV	100	10	10	160	93	.	220	02	02	280	35	35
041	69	OP	101	76	LBL	161	01	1	221	43	RCL	281	43	RCL
042	00	00	102	87	IFF	162	22	INV	222	50	50	282	10	10
043	29	CP	103	01	1	163	44	SUM	223	69	OP	283	85	+
044	43	RCL	104	00	0	164	10	10	224	03	03	284	04	4
045	41	41	105	00	0	165	71	SBR	225	43	RCL	285	06	6
046	69	OP	106	22	INV	166	30	TAN	226	51	51	286	00	0
047	04	04	107	44	SUM	167	77	GE	227	69	OP	287	95	=
048	43	RCL	108	10	10	168	77	GE	228	04	04	288	42	STO
049	01	01	109	71	SBR	169	71	SBR	229	69	OP	289	09	09
050	69	OP	110	30	TAN	170	39	COS	230	05	05	290	43	RCL
051	06	06	111	77	GE	171	77	GE	231	43	RCL	291	15	15
052	43	RCL	112	87	IFF	172	77	GE	232	11	11	292	75	–
053	42	42	113	71	SBR	173	24	CE	233	99	PRT	293	43	RCL
054	69	OP	114	39	COS	174	43	RCL	234	98	ADV	294	16	16
055	04	04	115	77	GE	175	10	10	235	98	ADV	295	55	÷
056	43	RCL	116	87	IFF	176	75	–	236	98	ADV	296	43	RCL
057	02	02	117	01	1	177	43	RCL	237	98	ADV	297	09	09
058	69	OP	118	00	0	178	02	02	238	91	R/S	298	75	–
059	06	06	119	00	0	179	95	=	239	76	LBL	299	43	RCL

Program listing for adiabatic humidification calculation (continued)   Table I

Location	Code	Symbol	Location	Code	Symbol	Location	Code	Symbol
300	.17	17	360	11	11	420	65	×
301	65	×	361	53	(	421	53	(
302	43	RCL	362	43	RCL	422	01	1
303	09	09	363	30	30	423	75	-
304	28	LOG	364	75	-	424	43	RCL
305	75	-	365	43	RCL	425	07	07
306	43	RCL	366	10	10	426	54	)
307	18	18	367	54	)	427	85	+
308	65	×	368	45	Y^x	428	53	(
309	43	RCL	369	93	.	429	43	RCL
310	09	09	370	03	3	430	26	26
311	85	+	371	08	8	431	65	×
312	43	RCL	372	65	×	432	43	RCL
313	19	19	373	43	RCL	433	33	33
314	65	×	374	29	29	434	85	+
315	43	RCL	375	95	=	435	43	RCL
316	09	09	376	42	STO	436	27	27
317	33	X^2	377	13	13	437	65	×
318	95	=	378	53	(	438	43	RCL
319	22	INV	379	43	RCL	439	34	34
320	28	LOG	380	23	23	440	85	+
321	42	STO	381	65	×	441	43	RCL
322	12	12	382	43	RCL	442	28	28
323	75	-	383	33	33	443	65	×
324	43	RCL	384	85	+	444	43	RCL
325	06	06	385	43	RCL	445	35	35
326	95	=	386	24	24	446	54	)
327	92	RTN	387	65	×	447	65	×
328	76	LBL	388	43	RCL	448	43	RCL
329	39	COS	389	34	34	449	03	03
330	43	RCL	390	75	-	450	95	=
331	12	12	391	43	RCL	451	55	÷
332	55	÷	392	25	25	452	43	RCL
333	53	(	393	65	×	453	13	13
334	43	RCL	394	43	RCL	454	85	+
335	06	06	395	35	35	455	43	RCL
336	75	-	396	54	)	456	03	03
337	43	RCL	397	65	×	457	95	=
338	12	12	398	43	RCL	458	42	STO
339	54	)	399	07	07	459	55	55
340	55	÷	400	85	+	460	75	-
341	53	(	401	53	(	461	43	RCL
342	43	RCL	402	43	RCL	462	11	11
343	31	31	403	20	20	463	95	=
344	65	×	404	65	×	464	94	+/-
345	53	(	405	43	RCL	465	92	RTN
346	01	1	406	33	33	466	76	LBL
347	75	-	407	85	+	467	48	EXC
348	43	RCL	408	43	RCL	468	43	RCL
349	07	07	409	21	21	469	02	02
350	54	)	410	65	×	470	42	STO
351	85	+	411	43	RCL	471	10	10
352	43	RCL	412	34	34	472	71	SBR
353	32	32	413	85	+	473	30	TAN
354	65	×	414	43	RCL	474	71	SBR
355	43	RCL	415	22	22	475	39	COS
356	07	07	416	65	×	476	61	GTO
357	54	)	417	43	RCL	477	69	OP
358	95	=	418	35	35	478	00	0
359	42	STO	419	54	)	479	00	0

Constants and printed messages must be stored in the data registers   Table II

Register contents	Register number	Register contents	Register number
0.	00	1.60785	31
0.	01	2.44283	32
0.	02	0.	33
0.	03	0.	34
0.	04	0.	35
0.	05	37173637.	36
0.	06	1316241314.	37
0.	07	1337241500.	38
0.	08	3613374135.	39
0.	09	1337243231.	40
0.	10	33362422.	41
0.	11	37006521.	42
0.	12	23000000.	43
0.	13	61153203.	44
0.	14	36133740.	45
15.092	15	37173033.	46
5079.6	16	4057006521.	47
1.6908	17	27140043.	48
0.003193	18	1337173563.	49
0.000001234	19	2714001635.	50
0.239	20	4500221336.	51
0.000012875	21	23413024.	52
.0000000014	22	1624374500.	53
0.204499	23	0.	54
0.00006135	24	0.	55
.000000058	25	0.	56
0.463255	26	0.	57
0.00002581	27	0.	58
0.000000023	28	0.	59
91.85747	29		
705.56	30		

Notes:
Registers 15-19 contain constants for Eq. (5). Registers 20-22, constants for Eq. (6), air. Registers 23-25, constants for Eq. (6), CO_2. Registers 26-28, constants for Eq. (6), H_2O. Registers 29-30, constants for Eq. (2). Registers 31-32, constants for Eq. (4). Registers 36-53, alphanumeric code.

model calculator and its accessory PC-100A printer.

The program is shown in Table I. Certain constants for the calculations and for the alphanumeric operations must be stored in the data registers. These are shown in Table II. Once the calculator has been programmed and the constants stored, both the program and data-register contents may be recorded on magnetic cards for future use.

Table III gives users' instructions for finding adiabatic saturation temperature and humidity.

After initiating the calculation, the printer prints out the entered data. Next, it prints the trial saturation temperature starting from 700°F, and continues printing the trial saturation temperatures until the correct value is reached. This temperature is then printed together with the saturated humidity.

Results of a sample calculation are shown in Table IV. In this example, the adiabatic saturation temperature and humidity are calculated for air (0% CO_2) at atmospheric pressure (0 psig). Psychrometric tables [1] give values of 82.0 and 0.02387, respectively, for saturation temperature and saturated humidity at these conditions.

Users' instructions for computing adiabatic saturation humidity and temperature   Table III

Step	Procedure	Enter	Press	Display/print
1	Enter program (use standard partitioning)			
2	Enter data	System pressure, psig	A	psig
		Entering gas temperature, °F	B	T,°F
		Entering humidity, lb water/lb dry gas	C	H
		CO_2 content, mole percent (based on moisture-free gas)	D	% CO_2
3	Calculate		E	Saturation temperature,°F
			R/S*	Saturated humidity, lb water/lb dry gas

*Required only if program is modified for use without printer

Small variations between the results calculated by this program and those available in tables are likely, since specific heats, latent heats, and partial pressures are calculated from polynomial formulas rather than from tabular data. However, the error in the saturation temperature is less than 0.5°F over a wide range of input data and is, for the most part, less than 0.2°F. The error in saturated humidity is less than $1\frac{1}{2}$%.

If the TI-59 is to be used without the printer, it will be necessary to make one change in the program:
1. Enter the program as listed in Table I.
2. Make a program addition by pressing the following keys:

GTO
2
0
1
LRN
*INS
R/S
LRN

With this change, the calculator will display each trial saturation temperature briefly, and halt with the final saturation temperature shown on the display. To show the saturated humidity, press R/S.

For HP-67/97 users

The operation of the Hp version of the program is identical to that of the TI version. The output is also the same except that the HP calculators do not have alphabetic output. Table V lists the HP version of the program, and Table VI shows the data storage required before running the program.

Results obtained from a sample calculation   Table IV

```
        ADIABATIC SATURATION

        0.      PSIG
      160.      T 'F
     0.005747   H
        0.      %CO2         (continued)

      700.      TEST         86.      TEST
      600.      TEST         85.      TEST
      500.      TEST         84.      TEST
      400.      TEST         83.      TEST
      300.      TEST         82.      TEST
      200.      TEST         82.9     TEST
      100.      TEST         82.8     TEST
        0.      TEST         82.7     TEST
       90.      TEST         82.6     TEST
       80.      TEST         82.5     TEST
       89.      TEST         82.4     TEST
       88.      TEST         82.3     TEST
       87.      TEST         82.2     TEST

         SAT. TEMP., 'F
            82.2

          SAT. HUMIDITY
     LB WATER/LB DRY GAS
       .0237024164
```

Listing of HP version of the program   Table V

Step	Key	Code		Step	Key	Code
001	*LBLA	21 11		030	0	00
002	PRTX	-14		031	0	00
003	ST+1	35-55 01		032	ST+0	35-55 00
004	R/S	51		033	*LBL4	21 04
005	*LBLB	21 12		034	1	01
006	PRTX	-14		035	0	00
007	STO2	35 02		036	ST-0	35-45 00
008	R/S	51		037	GSB2	23 02
009	*LBLC	21 13		038	X>0?	16-44
010	PRTX	-14		039	GTO4	22 04
011	STO3	35 03		040	GSB3	23 03
012	R/S	51		041	X>0?	16-44
013	*LBLD	21 14		042	GTO4	22 04
014	PRTX	-14		043	1	01
015	STx4	35-35 04		044	0	00
016	SPC	16-11		045	ST+0	35-55 00
017	R/S	51		046	*LBL5	21 05
018	*LBLE	21 15		047	1	01
019	1	01		048	ST-0	35-45 00
020	0	00		049	GSB2	23 02
021	0	00		050	X>0?	16-44
022	ST-0	35-45 00		051	GTO5	22 05
023	GSB2	23 02		052	GSB3	23 03
024	X>0?	16-44		053	X>0?	16-44
025	GTOE	22 15		054	GTO5	22 05
026	GSB3	23 03		055	1	01
027	X>0?	16-44		056	ST+0	35-55 00
028	GTOE	22 15		057	*LBL6	21 06
029	1	01		058		-62

Listing of HP version of the program (continued)   Table V

Step	Key	Code	Step	Key	Code	Step	Key	Code	Step	Key	Code	Step	Key	Code
059	1	01	093	X²	53	127	×	-35	161	+	-55	195	-	-45
060	ST-0	35-45 00	094	-	-45	128	+	-55	162	RCL5	36 05	196	.	-62
061	GSB2	23 02	095	STOB	35 12	129	10^x	16 33	163	RCLC	36 13	197	3	03
062	X>0?	16-44	096	RCL2	36 02	130	STOE	35 15	164	×	-35	198	8	08
063	GTO6	22 06	097	3	03	131	P⇄S	16-51	165	+	-55	199	Y^x	31
064	GSB3	23 03	098	Y^x	31	132	RCL1	36 01	166	P⇄S	16-51	200	RCL6	36 06
065	X>0?	16-44	099	RCL0	36 00	133	-	-45	167	i	01	201	×	-35
066	GTO6	22 06	100	3	03	134	RTN	24	168	RCL4	36 04	202	÷	-24
067	RCL0	36 00	101	Y^x	31	135	*LBL3	21 03	169	-	-45	203	RCL3	36 03
068	RCL2	36 02	102	-	-45	136	P⇄S	16-51	170	×	-35	204	+	-55
069	-	-45	103	STOC	35 13	137	RCL7	36 07	171	+	-55	205	RCLE	36 15
070	X<0?	16-45	104	RCL0	36 00	138	RCLA	36 11	172	P⇄S	16-51	206	RCL1	36 01
071	GTO9	22 09	105	4	04	139	×	-35	173	RCL9	36 09	207	RCLE	36 15
072	RCL2	36 02	106	6	06	140	RCL8	36 08	174	RCLA	36 11	208	-	-45
073	STO0	35 00	107	0	00	141	RCLB	36 12	175	×	-35	209	÷	-24
074	GSB2	23 02	108	+	-55	142	×	-35	176	P⇄S	16-51	210	1	01
075	GSB3	23 03	109	STOD	35 14	143	+	-55	177	RCL5	36 05	211	RCL4	36 04
076	*LBL9	21 09	110	P⇄S	16-51	144	RCLI	36 46	178	RCLB	36 12	212	-	-45
077	RCL0	36 00	111	RCL0	36 00	145	RCLC	36 13	179	×	-35	213	RCL8	36 08
078	SPC	16-11	112	RCL1	36 01	146	×	-35	180	+	-55	214	×	-35
079	PRTX	-14	113	RCLD	36 14	147	+	-55	181	2	02	215	÷	-24
080	RCLA	36 11	114	÷	-24	148	P⇄S	16-51	182	.	-62	216	RCL9	36 09
081	PRTX	-14	115	-	-45	149	RCL4	36 04	183	3	03	217	RCL4	36 04
082	R/S	51	116	RCL2	36 02	150	×	-35	184	EEX	-23	218	×	-35
083	*LBL2	21 02	117	RCLD	36 14	151	P⇄S	16-51	185	CHS	-22	219	+	-55
084	RCL0	36 00	118	LOG	16 32	152	.	-62	186	9	09	220	STOA	35 11
085	PRTX	-14	119	×	-35	153	2	02	187	RCLC	36 13	221	-	-45
086	RCL2	36 02	120	-	-45	154	3	03	188	×	-35	222	CHS	-22
087	RCL0	36 00	121	RCL4	36 04	155	9	09	189	+	-55	223	RTN	24
088	-	-45	122	RCLD	36 14	156	RCLA	36 11	190	RCL3	36 03	224	R/S	51
089	STOA	35 11	123	×	-35	157	×	-35	191	×	-35			
090	RCL2	36 02	124	RCL3	36 03	158	RCL6	36 06	192	+	-55			
091	X²	53	125	-	-45	159	RCLB	36 12	193	RCL7	36 07			
092	RCL0	36 00	126	RCLD	36 14	160	×	-35	194	RCL0	36 00			

Data storage locations before running the program   Table VI

Storage area	Value	Storage area	Value
Primary 0	800.0	Secondary 0	15.092
1	14.7	1	5079.6
4	0.01	2	1.6908
5	0.00002581	3	0.003193
6	91.85747	4	0.000001234
7	705.56	5	0.0000000014
8	1.60785	6	0.000012875
9	2.44283	7	0.204499
		8	0.00006135
		9	0.463255
			0.0000000058

References

1. Zimmerman, O. T., and Lavine, I., "Psychrometric Tables and Charts," Mack Printing Co., Easton, Pa., 1964.
2. Thakore, S. B., Miller, J. W., Jr., and Yaws, C. L., Heats of Vaporization, *Chem. Eng.*, Aug. 16, 1976, p. 85.
3. Patel, P. M., Schorr, G. R., Shah, P. N., and Yaws, C. L., Vapor Pressure, *Chem. Eng.*, Nov. 22, 1976, p. 159.
4. Hougen, O. A., and Watson, K. M., "Industrial Chemical Calculations," John Wiley and Sons, 1936.

Coolers for cryogenic grinding

Cryopulverizing is used to recycle and recover materials such as steel and rubber. Here, the author discusses how the cooling is done, and how coolers are sized.

Michael W. Biddulph, *University of Nottingham*

Air Products and Chemicals, Inc.

☐ A technique that is being increasingly used in recovering and recycling materials is cryogenic grinding, also known as cryopulverizing.

In this process, the material to be ground up has its temperature lowered so that it changes from a ductile to a brittle solid—i.e., its temperature is reduced to below the glass-transition temperature. When this has been achieved, grinding the material to a small particle size can be done more efficiently, due to reduced power consumption needed by the grinder. Another advantage of this method is that it yields a purer product in cases where one material must be separated from a mixture, such as recovering steel from automobiles. In addition, cryogenic grinding provides the possibility of efficient separation in cases where some materials in a mixture become brittle, while others do not [1-3].

The process has been used already for the treatment of a number of materials, including metal in scrap autos [1,4], nonferrous metals [5], automobile tires [6,7], plastics [3,6,7], copper wire [1,2], gums and waxes [8], color concentrates [8] and foodstuffs [8]. The method may also be useful in processing pharmaceutical products that have low softening points or are heat-sensitive [8].

The rubber crumb produced by the cryogenic grinding of scrap tires is very pure, since the steel and fabric are separated easily, and this crumb has found a number of uses, including road surfaces, sports surfaces, and as a filler for polymers [12].

Originally published February 11, 1980

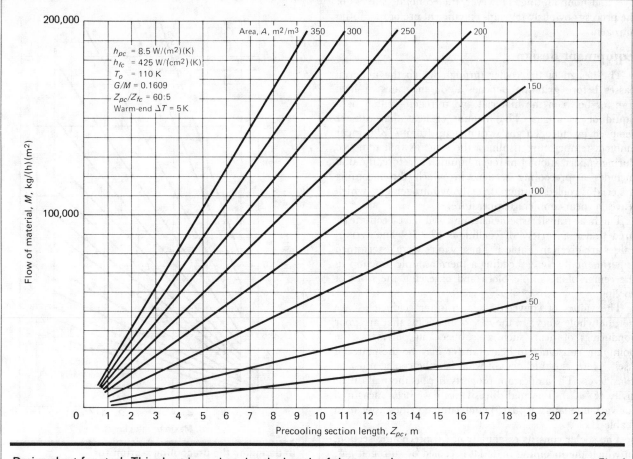

Design chart for steel. This chart is used to size the length of the precooler   Fig. 1

The advantage of this technology as applied to polymers probably rests on improved separation of mixed or closely combined materials. Some polymers (e.g., polypropylene, polyethylene) are embrittled by liquid nitrogen, while others (e.g., nylon, tetrafluoroethylene) remain ductile. Such a difference can be used in effectively separating these materials.

The stripping of polymeric insulation from copper wire has been done using cryopulverizing [1,2]. This is an example of a process in which one material (polymer) embrittles, while the other (copper) does not. Passage of the cooled wire through crushing rolls breaks the insulation, and sieving or elutriation separates the materials.

Pioneering work in cryopulverizing has been carried out in Belgium, where the Inchscrap process treats baled scrap automobiles by passing them on a conveyor through a long cooling tunnel [1]. Nitrogen consumption is 0.24 kg/kg of steel processed, with a throughput of 30 metric tons/h [4].

In the process, baled scrap is precooled with nitrogen gas, then immersed in liquid nitrogen. The material is shredded and dedusted, and high-alloy and carbon-steel components are separated magnetically. The nonferrous components in the scrap are not fragmented and are removed in a second installation. A pure, low-carbon steel product is screened into different-sized fractions.

The viability of any scrap treatment technique normally depends upon the economics of the process, which rest heavily on the value of the recovered material. The cheapest method of cooling materials to a low temperature is by the use of liquid nitrogen. This substance is readily available from air-separation plants, relatively inexpen-

Nomenclature

A Specific surface area, m^2/m^3

C_p Specific heat, $J/(kg)(K)$

G Flow of nitrogen, $kg/(h)(m^2)$

h_{pc} Individual convective heat transfer coefficient in precooling section, $W/(m^2)(K)$

h_{fc} Individual convective heat transfer coefficient in final-chilling section, $W/(m^2)(K)$

k Thermal conductivity, $W/(m)(K)$

M Flow of material, $kg/(h)(m^2)$

T_o Outlet temperature of material, K

Z_{pc} Length of precooling section, m

Z_{fc} Length of final chilling section, m

ρ Density of material, kg/m^3

ρ_b Bulk density of material, kg/m^3

sive, and nonpolluting. However, the economic success of the process obviously depends on efficient use of the liquid nitrogen.

Equipment design

The design of the cooler through which the material passes before grinding depends upon the rates of heat transfer between the material and nitrogen—either as a liquid or as a gas. The amount of heat that can be removed by nitrogen gas warming up from 77K to near ambient temperature is almost the same as that removed during vaporization of nitrogen liquid. Therefore, the unit includes a precooling section, in which the incoming material is cooled countercurrently by nitrogen gas, which leaves at near-ambient temperature.

The heat-transfer coefficients involved have been measured under a variety of conditions [3,9,10] and typical values are given in Table I. These values were determined experimentally by embedding a thermocouple junction in the center of a sample block and observing the rates of cooling.

The values of convective-heat-transfer coefficients are not the whole story. In the case of materials that are poor conductors of heat, such as rubber and polymers, the dominant resistance to heat transfer may be the material itself. Here, the thickness of what is to be cooled must be considered. The important property here is thermal diffusivity ($k/\rho C_p$). Thermal diffusivities have been measured for various materials [3,10], and sample values appear in Table II.

The cooler consists essentially of (a) precooling section, in which the incoming material is cooled by nitrogen gas, and (b) a final chilling section, where liquid nitrogen is splashed onto the material to bring it to the required final temperature. The boiling liquid in the final chilling section provides the precooling gas.

The cooler often is tunnel-shaped, material traveling through it on a conveyor belt. The gas may or may not be stirred by using fans of the type commonly found in food freezers. Alternatively, the cooler is of the inclined-rotary type, similar to conventional rotary dryers. In the latter style, the precooling section is shorter, due to the improvement in heat-transfer coefficients that results from better gas-solid contact.

Cooling-tunnel calculations

Design charts have been developed [9] for cryogenic coolers, and a typical example is shown in Fig. 1. This chart gives the required precooling section length as a function of the throughput of steel, with the specific surface area as a parameter. The conditions are near-ideal, with the nitrogen gas leaving at only 5K below ambient temperature. The nitrogen consumption predicted here is 0.1609 kg/kg steel. The final chilling section is about $1/60$ the length of the precooling section.

Cryogenic grinding of steel gives a product over 99% pure, and suitable for use in high-quality steel production. The regular-shaped fragments can be handled by continuous feeding systems. Bulk density is about three times that of conventionally fragmented scrap, reducing transportation costs. Copper content after separation is less than half the level usually found in conventionally fragmented scrap.

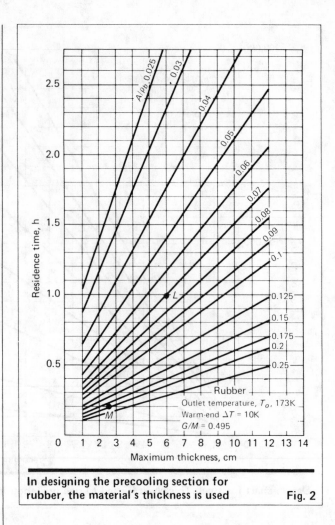

In designing the precooling section for rubber, the material's thickness is used   **Fig. 2**

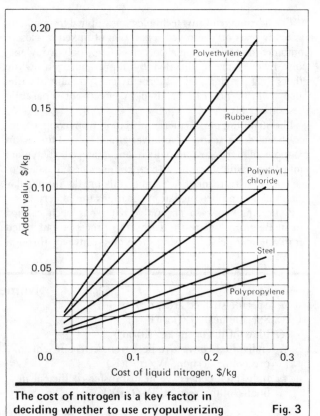

The cost of nitrogen is a key factor in deciding whether to use cryopulverizing   **Fig. 3**

Heat-transfer coefficients for cooling by nitrogen were measured experimentally		Table I
Conditions	Material	Heat transfer coefficient, $W/(m^2)(K)$
Cooling in liquid nitrogen	Steel	85–100
Cooling in a stream of nitrogen gas	Steel	5–10
Cooling in a stream of nitrogen gas	Rubber; polymers	20
Cooling by splashing liquid nitrogen	Steel	250–500

For nonconductive materials, thermal diffusivities are important in designing cooler		Table II
Material	Thermal diffusivity, m^2/h	
Teflon	0.001	(100K)
Teflon	0.0005	(273K)
Rubber	0.00075	(200K)
Rubber	0.00055	(300K)
Polypropylene	0.0008	(140K)
Polypropylene	0.00035	(300K)
Polyethylene	0.0022	(140K)
Polyethylene	0.0006	(300K)
Polyvinyl chloride	0.0007	(140K)
Polyvinyl chloride	0.0004	(300K)

In the case of cryogenic cooling of rubber (for example, automobile scrap tires), the thermal diffusivity of the rubber is much lower than that of steel, so, as noted before, the thickness of the pieces to be cooled must be taken into account. The same general form of cooler is used, and design charts in terms of required residence time have been produced [10]. A typical example is shown in Fig. 2. The dependence of residence time on the maximum thickness of material is demonstrated, and the parameter here is the ratio of the specific area, A, to the bulk density, ρ_b, of the material. On the graph, points M and L represent typical values for medium-sized automobile tires and large truck tires.

We may use these design methods to evaluate the cooling-tunnel requirements for a tire treatment plant. Consider a plant designed to process 15 medium-sized automobile tires/min, which amounts to 5.6 million tires/yr [7]. For efficient use of the liquid nitrogen coolant, it is desirable to have a 10K temperature difference between entering rubber and exiting nitrogen gas at the warm end of the cooler. The required residence time in the precooling section from Fig. 3 is calculated at about 12 min, and the residence time in the final chilling section is about 3.6 min [10]. If the tires were not precut before cooling, the length of the required tunnel would be about 46 m; the consumption of liquid nitrogen would be near-ideal at 0.495 kg N_2/kg rubber; the liquid-nitrogen injection point would be about three-fourths of the way along the tunnel from the rubber inlet.

If we wished to reduce the length of the cooling tunnel, which naturally would require more nitrogen, we could design for a 50K warm-end temperature difference. This would result in a cooling tunnel 37 m long. The nitrogen consumption would have risen to 0.55 kg N_2/kg rubber. For an even shorter tunnel, the plant could be operated with a 100K warm-end difference, requiring a tunnel of 28m and calling for a nitrogen usage of 0.64 kg N_2/kg rubber. Design charts for the above conditions have been developed [10]. The liquid nitrogen usage for a well-designed plant would be about 100 metric tons/d.

Process economics

The economic viability of the process depends on the added value for a particular material. This added value is the difference between the price that can be obtained for the product and the cost of the scrap processed. In each instance, there will be a minimum added value needed for viability. The more the actual added value exceeds the minimum added value, the more profitable the process.

An economic analysis based on this principle was carried out in particular with reference to scrap automobile tires [11]. Details of this lengthy analysis will not be given here; however, it was shown that the cost of the nitrogen is by far the most important determinant in setting minimum added value. This emphasizes the need for good cooler design. The analysis was presented in the form of charts based on nitrogen cost. An example of such a chart is shown in Fig. 3, which includes materials in addition to rubber.

References

1. Bilbrey, J. H., Jr., "Use of Cryogenics in Scrap Processing," Proc. Fourth Mineral Waste Utilization Symposium, Chicago, May 1974.
2. Bilbrey, J. H. Jr., and Valdez, E. G., Use of Cryogenics in Scrap Processing, Adv. in Cryogenic Engineering, Vol. 20, 1975, p. 411.
3. Biddulph, M. W., Cryogenic Embrittlement of Some Polymers, Conservation and Recycling, Vol. 1, No. 3/4 1977, p. 281.
4. Anon., Freezing Scrap Cuts Materials Separation Costs, Processing, p. 13, Aug. 1975.
5. Valdez, E. G., Dean, K. C., and Wilson, W. J., "Use of Cryogens to Reclaim Non-Ferrous Scrap Metals," United States Bureau of Mines, Pub. No. RI 7716, 1973.
6. Braton, N. R., and Koutsky, J. A. Cryogenic Recycling, Proc. Fourth Mineral Waste Utilization Symposium, Chicago, May 1974.
7. Mishra, I. B., Koutsky, J. A., and Braton, N. R., Cryogenic Recycling of Solid Wastes, Polymer News, Vol. 2, No. 9/10, 1975, p. 32.
8. Wary, J., and Davies, R. B., Cryopulverising, Chemtech, March 1976, p. 200.
9. Biddulph, M. W., Cryogenic Embrittlement of Steel, Conservation and Recycling, Vol. 1, No. 2, 1977, p. 221.
10. Biddulph, M. W., Cryogenic Embrittlement of Rubber, Conservation and Recycling, Vol. 1, No. 2, 1977, p. 169.
11. Allen, D. H., and Biddulph, M. W., The Economic Evaluation of Cryopulverising, Conservation and Recycling, Vol. 2, No. 3/4, 1978, p. 255.
12. Freeguard, G. F., University of Nottingham, private communication.
13. Fredericks, S. L., "An Economic Evaluation of the Use of Cryogenics in Rubber Tire Reclaiming." Conf. Proc. Environ. Aspects of Chem. Use in Rubber Process Operations, Akron, Ohio, March 12-14, 1975, pp. 442-452.

The author

Michael W. Biddulph is a professor in the department of chemical engineering, University of Nottingham, University Park, Nottingham NG7 2RD, U.K. His research has been mainly in cryogenics, especially in recovering waste materials. He has been employed by British Oxygen Co. as a development engineer in low-temperature air separation, and has worked for Chevron Research Co. in the U.S., writing design programs. He holds bachelor's and Ph.D. degrees in chemical engineering from the University of Birmingham, and is a member of AIChE.

Designing a

Cooling water is expensive to circulate. Reducing its flow—i.e., hiking exchanger outlet temperatures—can cut tower, pump and piping investment as much as one-third and operating cost almost in half.

Ralph A. Crozier, Jr., E. I. du Pont de Nemours & Co.

☐ Heat-exchanger-network optimization has been accomplished in large integrated plants, such as petroleum refineries. In many of the chemical process industries, however, a plant contains several individual processes, and network optimization, except on a limited basis, is not feasible.

So far, no one has developed similar procedures for designing and optimizing a cooling-water once-through-exchanger system.* This article attempts to fill this void by presenting a design basis that will produce a "near optimum" system.

A cooling-water system consists of four major components: heat exchangers, cooling towers, circulation piping and pumps. To optimize such a system, one must define the system interactions and apply these relationships to the simultaneous design of the aforementioned

*The term "once-through cooling water" usually distinguishes river water from cooling-tower water. In this article, "cooling water once-through-exchanger system" means that the water passes through only one exchanger before it returns to the tower.

When a process is large, single-line and integrated, reducing cooling cost by circulating water through several exchangers in series before returning it to the tower is very attractive, especially if the coolant is used for refrigeration or turbine condensers. Such design, however, has very limited application in the typical multi-process, multi-step chemical plant, because parts of it may be shut down while other parts are still operated.

equipment. This article develops criteria that for most applications allow one to ignore system interactions, and still design a "near optimum" system.

Cooling-water systems have long been designed by "rules of thumb" that call for fixing the coolant temperature-rise across all heat exchangers (usually 20°F) and setting the coolant inlet temperature to the heat exchanger at the site's wet-bulb temperature plus 8°F. These rules produce a workable cooling system; but, by taking the same coolant rise across all exchangers, regardless of the individual process outlet-temperatures, this cannot result in an optimized design.

The design method presented in this article replaces the "rules of thumb" with criteria that are easy to apply and that take into account the effect that the individual exchanger process outlet-temperatures have on cooling-system economics.

Economic analyses of actual processes have shown that cooling-system investment can be reduced by one third, and cooling-system operating cost by one half, if the proposed design criteria are used instead of the "rules of thumb." It has been found that the controlling economic factor for a cooling system is the quantity of

Originally published April 21, 1980

"near optimum" cooling-water system

water being circulated. Reducing the flow (raising the coolant outlet temperature of heat exchangers) significantly reduces cooling-tower, pump and piping investment, and operating cost, and only moderately increases the heat-exchanger investment. The overriding conclusion to be drawn is that cooling water is very expensive, and its conservation can result in significant savings.

Three system interactions

What are the system interactions and their relationships? When can they be ignored, and when must they be considered?

From Fig. 1, it can be seen that there are three coolant interactions: (1) between the outlet temperature of the heat exchanger and the inlet temperature of the cooling tower, (2) between the inlet temperature of the exchanger and the outlet temperature of the cooling tower, and (3) between total coolant flowrate and the size of connecting piping and circulating pumps.

First Interaction—In the Fig. 1 system, cooling water enters the shell side of the exchanger at a fixed temperature of T_1 and exits at a variable temperature of T_2; the process fluid passes through the tube side at fixed inlet and outlet temperatures of t_1 and t_2, respectively. The cooling tower dissipates heat to the ambient air through humidification (evaporation) and sensible heating.

Cooling-tower investment and operating cost are reduced by returning the water to the tower as hot as possible, because the hotter the water, the larger the temperature difference (thermal driving force) between it and the ambient air; hence, less heat-transfer area (tower packing) is required. Additionally, as the effluent air temperature is raised, so is the moisture content of the exiting air; hence, less air flow is necessary.

Upping the temperature of the water to the tower also cuts down the quantity of water circulated through the tower. From Eq. (1), it can be seen that for a fixed heat load and tower outlet temperature (T_{out}), the quantity of cooling water required to satisfy the heat balance is inversely proportional to the tower inlet temperature (T_{in}):

$$W_T = Q_T/[C(T_{in} - T_{out})] \qquad (1)$$

Boosting the tower inlet temperature means increasing the coolant temperature rise across the individual exchangers; hence, to reduce cooling-tower investment and operating cost, one must widen the coolant temperature rise across the individual exchangers.

A significant discovery has been that in most instances exchanger area is relatively insensitive to incremental changes in coolant outlet temperature. To understand this, one must recognize that a well designed exchanger must have adequate heat-transfer area, but it must also have sufficient fluid-flow area to satisfy the shell-and-tube pressure-drop restrictions.

Because coolant flowrate is inversely proportional to coolant temperature rise, a small coolant temperature rise produces a large increase in coolant flow. This means that the exchanger diameter must be increased, if the pressure-drop (maximum velocity) restriction is to be satisfied.

If the heat-transfer equation $Q = UA(LMTD)$ were solved for the small coolant temperature-rise geometry, it would show that the area needed for heat transfer is less than required by the pressure-drop limitation. In other words, for small coolant temperature rises, the exchanger's heat-transfer area would be excessive if determined by fluid-flow considerations.

If the coolant outlet temperature is sequentially increased and the required exchanger area is calculated for the different coolant temperature rises, a plot similar to that in Fig. 2 would be generated. Notice that the required exchanger area initially lessens, reaches a maximum, and then gains as the coolant temperature rise is increased.

The required exchanger area, as depicted in Fig. 2, passes through three distinct design regions. In the first (80°F to 115°F), the required area is controlled by fluid-flow phenomena. In the second (115°F to 158°F), the fluid-flow and heat transfer considerations are the same order of magnitude; that is, neither requirement is controlling. In the third (above 158°F), the heat-transfer requirement becomes controlling (reduced *LMTD*). Therefore, to minimize exchanger area, the design should be in the second (transition) region.

Fig. 2 was developed on the basis of full tube bundle. At low coolant temperatures, the better design would be to have the tube bundle partially filled, so as to maintain a constant tubeside velocity. In the transition region, however, the shell velocity was kept at a constant rate by reducing the baffle spacing as the coolant flow decreased.

Second Interaction—The tower outlet temperature (approach plus wet bulb) is the variable with which the cooling-tower designer has the least latitude, yet it is the most significant determinant of cooling-tower investment and operating cost. The tower outlet temperature is the "Achilles heel" of this study and the reason why only a "near optimum," rather than a truly optimized, cooling system is attained.

Specifying the coolant outlet temperature is difficult

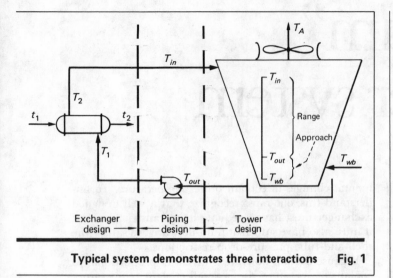

Typical system demonstrates three interactions   Fig. 1

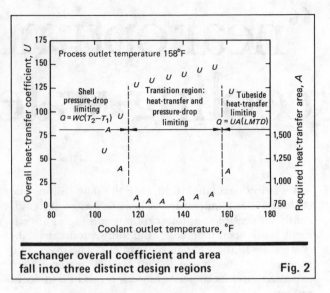

**Exchanger overall coefficient and area
fall into three distinct design regions   Fig. 2**

because it is bound on the lower side by the wet-bulb temperature and on the upper side by the minimum process outlet temperature. To minimize cooling-tower investment and operating cost, one needs to maximize the approach to wet-bulb temperature. However, to minimize heat-exchanger investment, one needs to maximize the approach to the minimum process outlet temperature. Therefore, the tower outlet temperature has to be a compromise between these two boundaries.

The importance of the cooling-tower and heat-exchanger designers' discussing the temperature limitations imposed on each before any calculations are started cannot be overemphasized. Usually, the lowest cooling temperature is required for a refrigeration system or a steam turbine condenser. Considerable savings can be realized if alternative cooling methods are provided for these services, rather than penalizing the entire cooling system for one, or both, of them.

Many cooling towers are designed by a rule of thumb that sets the tower outlet temperature equal to the site wet-bulb temperature plus 8°F. This can result in significant cooling-tower investment and operating cost penalties if process requirements do not demand cooling water at a temperature that low. As previously stated, cooling-tower investment and operating cost are minimized by maximizing the approach to wet-bulb temperature.

Third Interaction—Via Eq. (1), it can be shown that the system flowrate decreases as the tower inlet temperature is hiked. Therefore, as the coolant rise across the individual exchangers is increased, the supply and return piping, as well as the circulating pumps, can be reduced in size.

In summary, the investment and operating cost of the tower, pumps and piping will be reduced when the cooling-tower inlet temperature is increased. The exchanger investment may or may not be reduced as the coolant temperature rise is increased, depending on the relationship of the coolant outlet temperature and the exchanger design region (Fig. 2).

If the exchanger's optimum coolant outlet temperature is in the transition region, its area will be minimized, and system interactions can be ignored. If the

exchanger's coolant outlet temperature is not in the transition region, however, system interactions must be considered.

It has been empirically determined that if the exchanger *LMTD* is greater than 30°F when the coolant outlet temperature is set at the process outlet temperature or some maximum temperature (scaling limitation), the exchanger is in the transition region. If the *LMTD* is less than 30°F, the exchanger is in the heat-transfer-limiting region and the exchanger area will be very sensitive to coolant temperature changes. Therefore, if the *LMTD* is less than 30°F, system interactions must be considered; and a compromise between higher exchanger investment and lower cooling-tower investment must be determined.

Constraints on outlet temperature

Two temperature constraints limit the exchanger coolant outlet temperature. First, the wood fill of the conventionally designed cooling tower limits the maximum water temperature to 150°F. Second, the temperature from individual exchangers is set by process outlet temperatures or by tube fouling or material of construction.

Waterside deposits of suspended solids, mineral salts, biological growths, and corrosion products interfere with heat transfer. In the design calculations for sizing heat exchangers, the magnitude of the interference is the fouling factor. Attempts to quantitatively relate the factor to specific fouling conditions have been only partially successful; therefore, the numerical value assigned to different degrees of fouling by specific deposits has been based largely on experience.

Because there is no theoretical basis for selecting fouling factors at different coolant temperatures, the same factor was used at all temperatures in this study. To analyze the effect of this assumption on the conclusions, a cooling system for which the fouling factor was arbitrarily doubled was evaluated. It was found that the cooling tower, pump and piping investment and operating cost savings far outweighed the higher exchanger investment and operating cost. Therefore, the conclusions reached in the increased-fouling-factor evaluation were the same as those obtained by assuming a constant fouling factor. For simplification, only the latter evaluation is presented here.

Basis of the economic evaluations

The investment and operating cost for the cooling-system components were calculated as follows: heat exchangers, by a method similar to that described by Woods, Anderson and Norman [1]; cooling-tower pumps, by a proprietary correlation developed for electrically driven, vertical turbine water pumps operating at 880, 1,200 and 1,800 rpm; piping, by a proprietary correlation based on material and man-hours per linear foot (1- to 12-in. dia., "Cast Iron Pressure Pipe—Cement Lined," and 14- to 96-in. dia., "Concrete Pressure Pipe, American Water Works Assn., Spec. C-301 Prestressed"); and cooling towers, by two booklets of The Marley Co. [2,3].

Economic evaluations are based on cost plus return on investment. Using cost plus return, one can combine the investment and operating cost factors into a single variable.

Cost plus return is the total cost of owning and operating a cooling system for one year. Total investment (allocated, permanent, working capital, etc.) is included as a cost—that is, as an interest penalty. Operating cost (depreciation, maintenance, electrical, etc.) is added to the interest penalty. This total is defined as cost plus return:

$$C_T = C + \{xI/[(1 - a)(1 - i)]\} \qquad (2)$$

In Eq. (2), C = annual operating cost; x = required net return on investment; I = total investment; a = state and local taxes; and i = federal income tax.

Hereafter, cost means cost plus 10% net return on investment. The lowest-cost alternative proposal is the economic choice. All investment and operating costs represent the January 1976 *Engineering News Record Construction Cost Index* of 215 (1967 = 100).

Component design criteria

Heat-exchanger criteria are divided into two categories: *LMTD* greater and less than 30°F. Cooling-tower criteria are similarly divided: inlet temperature greater and less than 110°F.

Exchanger *LMTD* greater than 30°F

Fig. 3a shows system cost as a function of the coolant temperature rise for an exchanger whose process outlet temperature is 195°F. Two concepts in Fig. 3a are worth discussing:

1. Because this exchanger's *LMTD* is large, its area is relatively insensitive to changes in the coolant temperature rise. This phenomenon can be explained by a simple example.

It was shown previously that the area and overall heat-transfer coefficient of an exchanger in the transition region are fairly constant; hence, the required area is only a function of the *LMTD* (Fig. 2).

Assume two extremes, a large and a small *LMTD*. If the *LMTD* is large, narrowing it from 150°F to 140°F augments the exchanger area by 7%. On the other

hand, if the *LMTD* is small, lessening it from 50°F to 40°F boosts the exchanger area by 20%. Hence, as the *LMTD* is reduced, the exchanger area becomes more sensitive to the coolant outlet temperature.

2. Recall that as the coolant outlet temperature is raised, the coolant flow is reduced. Assume coolant investment to be $50/gpm (Fig. 3a). Hiking the coolant temperature rise from 20°F to 60°F reduces the coolant flow by 660 gpm, with a resultant lowering of the coolant investment by $33,000. Therefore, a significant investment saving can be realized for exchangers in this category if the coolant outlet temperature is raised to 150°F.

As can be seen in Fig. 3a, the system cost could be further reduced by increasing the coolant outlet temperature above 150°F. However, there are practical reasons for not going this high, although this limit is arbitrary and should be evaluated for a particular site.

The evaluation should consider: for the heat exchanger, such effects as (1) the use of exotic tube materials of construction, and (2) increased fouling stemming from the lower coolant velocity; for the cooling tower, such effects as (1) increased blowdown to remove mineral salts, and (2) costlier water treatment because of the higher temperatures.

Fig. 3b shows system cost as a function of coolant temperature rise for an exchanger whose process outlet temperature is 115°F. Again, the coolant cost drops as the coolant temperature rise increases. In this case, however, the effect of the *LMTD* on exchanger area is becoming evident. At a coolant outlet temperature of 125°F, multiple exchangers in series are required because of a temperature cross (coolant outlet temperature exceeds the process outlet temperature). Although the system cost for two exchangers in series is less, the savings are marginal, and one exchanger is preferred, with the coolant outlet and process outlet temperatures set equal to one another.

An economic evaluation of Fig. 3b at three different coolant outlet temperatures provides the first insight into the effect of coolant flowrate on system economics. It also establishes the basis for the practice of setting the process and coolant outlet temperatures equal to each other.

In Table I, the three cases represent three different coolant outlet temperatures: (1) 15°F coolant temperature rise, (2) equal to process outlet temperature, and (3) 5°F higher than the process outlet temperature.

Using Case 3 as the basis, the incremental investment penalties for Cases 1 and 2 are $33,700 and $6,100, respectively. The incremental cost penalties for Cases 1 and 2 are $15,000/yr and $3,000/yr, respectively.

From this analysis, one can see that by considering the system cost, a temperature cross is economically justified. But before deliberately designing a heat exchanger for a cross, one must realize that an error in the predicted outlet temperature of either stream will result in a thermodynamically unstable exchanger.

In summary, the design criteria for heat exchangers that have an *LMTD* greater than 30°F (calculated at the coolant outlet temperature specified by the criteria) are:

Single-pass geometries—The coolant outlet temperature

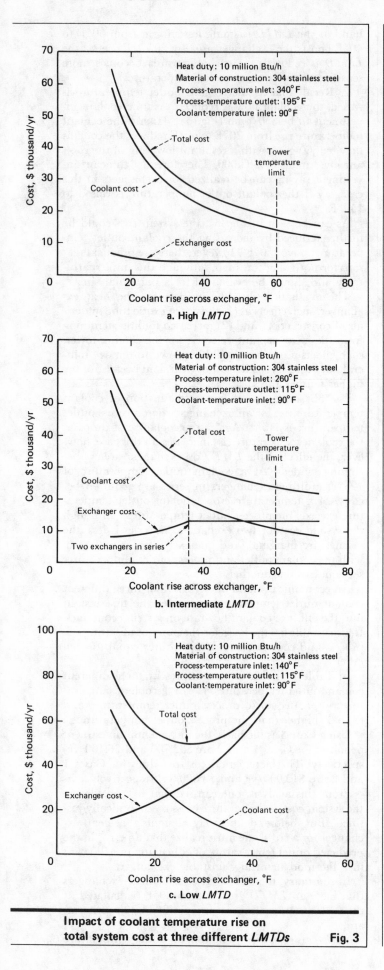

a. High *LMTD*

b. Intermediate *LMTD*

c. Low *LMTD*

Impact of coolant temperature rise on
total system cost at three different *LMTDs*   Fig. 3

should be set at the process inlet temperature minus
10°F, or at 150°F if the process inlet temperature is
greater than 160°F.

Multiple-pass geometries—The coolant outlet temperature should be set at the process outlet temperature, or
at 150°F if the process outlet temperature is greater
than 150°F.

Exchanger *LMTD* less than 30°F

Fig. 3c shows cost as a function of the coolant temperature rise for a process outlet temperature of 115°F
and an *LMTD* of less than 30°F (calculated at a coolant
outlet temperature of 115°F). Once again the coolant
cost decreases as the coolant temperature rise increases;
however, in this example, the higher cost of the heat
exchanger as the coolant outlet temperature is raised
has a significant effect on system economics.

This example differs from the previous two in yet
another respect. For the first time, the optimum coolant
outlet temperature depends on coolant economics. In
the previous examples, coolant costs so predominated
that an incremental increase or decrease in the coolant
cost did not change the design criteria. In this case,
however, if the coolant cost increases incrementally, one
wants to reduce the quantity of coolant; therefore, the
optimum coolant outlet temperature increases (or if the
coolant cost decreases, the optimum coolant outlet temperature becomes lower).

In Fig. 3c, the total cost curve has a minimum that is
a function of both the heat exchanger and coolant economics. Therefore, for an exchanger with an *LMTD* less
than 30°F, the exchanger and coolant interaction precludes setting the coolant outlet temperature *a priori* by
using the design criteria. For an exchanger in this category, the optimum coolant outlet temperature must be
found by constructing a graph similar to Fig. 3c.

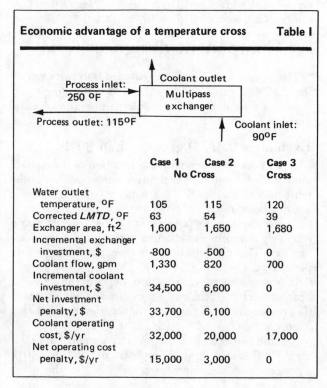

Economic advantage of a temperature cross			Table I
	Case 1 No Cross	Case 2	Case 3 Cross
Water outlet temperature, °F	105	115	120
Corrected *LMTD*, °F	63	54	39
Exchanger area, ft²	1,600	1,650	1,680
Incremental exchanger investment, $	-800	-500	0
Coolant flow, gpm	1,330	820	700
Incremental coolant investment, $	34,500	6,600	0
Net investment penalty, $	33,700	6,100	0
Coolant operating cost, $/yr	32,000	20,000	17,000
Net operating cost penalty, $/yr	15,000	3,000	0

Such a graph is generated by (1) using plant coolant cost data, or estimating the coolant economics via a Fig. 4a or 4b type graph; (2) determining the area requirements of the exchanger at different coolant temperature rises; (3) calculating the exchanger and coolant cost at the different temperature levels; (4) plotting the exchanger and coolant cost; and (5), at a given coolant temperature rise, summing the individual costs to obtain the system cost. The optimum coolant outlet temperature is that at which the total system cost is minimized. In Fig. 3c, this coolant outlet temperature happens to coincide with the design criteria temperature of 115°F.

The Step 1 graphical estimate of coolant economics is made as follows: Using Fig. 4a and assuming a tower having a 20°F range and a 91°F outlet temperature, one arrives at a total cost of $255,000/yr, or $31.25/h (8,000 h/yr). For the tower duty of 100 million Btu/h, one then calculates, using Eq. (2), a tower flowrate of 600,000 gal/h. Dividing $31.25/h by 600,000 gal/h, one obtains a coolant operating cost of 5.2¢/1,000 gal, which can be used to calculate the coolant economics at this temperature level.

Coolant piping and pumps

The connecting piping (cooling tower to heat exchanger) in the Fig. 5 cooling system is composed of large underground distribution headers and smaller takeoffs from central distribution points to individual heat exchangers. It was originally thought that these two piping configurations (one of straight runs, the other of many fittings and valves) would have different optimum design velocities. It was found, however, that the optimum velocities coincide.

From Fig. 6, it can be seen that the cost is insensitive to coolant velocity changes, and that the optimum velocity falls between 5 and 10 ft/s. (The cost-plus-return calculation of the Fig. 6 pipe and pump cost was based on the Fig. 5 cooling-system piping configuration.)

The insensitivity of system cost to the coolant velocity is the result of counterbalancing cost factors. Two extremes are possible in piping system design: (1) large pipe diameters (small friction losses) and small pumps with low horsepower requirements, or (2) smaller pipe diameters (large friction losses) and large pumps with high horsepower requirements.

In the first, the piping investment is high, and the pump investment and operating cost low. In the second, the reverse is true. When these two extremes are evaluated on a cost-plus-return basis, the total system costs are the same. This analysis will be true for all liquid piping systems that do not require exotic materials of construction.

The skill in designing coolant piping systems is in balancing flows through the various loops. To adequately analyze the flow distribution problem, one is forced to computer simulations of the piping systems (the cooling-water supply and return headers, and the piping to and from heat exchangers), choosing by trial and error the pipe diameter that matches the loop and system pressure drop. Such a piping configuration forms a series of parallel flow loops, in which, by definition, the pressure drops must be the same.

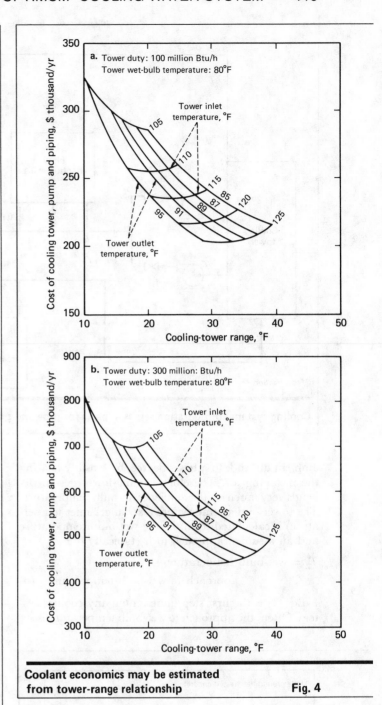

Coolant economics may be estimated from tower-range relationship   **Fig. 4**

The system pressure-drop requirement can be satisfied by increasing or decreasing the flowrate in the individual loops. If the piping in a particular loop is oversized, the flow through it will be excessive. To reduce the flow, a control valve must supply the pressure drop needed to achieve balanced pressure distribution. This means excessive wear and control-valve maintenance. Therefore, a properly designed cooling piping system will specify the correct pipe diameter for each loop that will balance the pressures without the need to resort to a control valve.

Cooling-tower design

As previously mentioned, a cooling tower dissipates process heat to the atmosphere by humidifying the

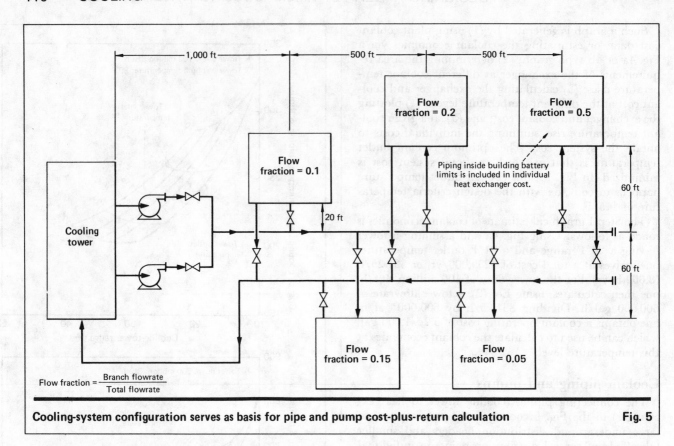

Cooling-system configuration serves as basis for pipe and pump cost-plus-return calculation   **Fig. 5**

ambient air and, to a lesser degree, by sensibly heating the incoming air. Therefore, cooling-tower design depends very much on the ambient wet-bulb temperature. The tower outlet temperature (exchanger inlet temperature) is calculated from the site wet-bulb temperature and an assumed approach to it (typically 8°F):

$$T_2 = \text{wet-bulb temperature} +$$

$$\text{approach to wet-bulb temperature} \quad (3)$$

Eq. (3) is the first step in designing any cooling system. Often, the approach to wet-bulb temperature is set arbitrarily, and this can result in considerable investment and operating cost penalties.

To analyze the magnitude of these cost differentials, consider the following example:

$$W_{gpm} = Q/[500\ C\ (T_{out} - T_{in})] \quad (4)$$

Here, $Q = 300 \times 10^6$ Btu/h; $C = 1$ Btu/(lb)(°F); $T_{in} = 105°F$ and $125°F$; and $T_{out} = 85°F$ (80°F wet bulb + 5°F approach) and 95°F (80°F wet bulb + 15°F approach).

Using Fig. 4b and 6, and subtracting, one can find the tower cost (Table II). There are several things in this example that should be particularly noted.

Although the tower heat duty in all four cases is the same and each represents a valid design, the cooling-system costs vary significantly. One can see that, for a fixed tower outlet temperature, boosting the tower inlet temperature materially reduces cooling-tower cost. At a fixed tower inlet temperature, widening the approach to wet-bulb temperature may or may not lower cooling system cost.

At a 125°F tower inlet temperature, varying the tower outlet temperature has a minor effect on the quantity of water circulated. At 105°F, however, the effect is quite significant.

Therefore, even though cooling-tower cost can be reduced by raising the tower outlet temperature, for low tower inlet temperatures, the increase in piping cost can exceed the reduction in cooling-tower cost. Above a tower inlet temperature of 110°F, the interaction between the tower inlet and outlet temperatures and the quantity of water circulated has minimal effect on system cost.

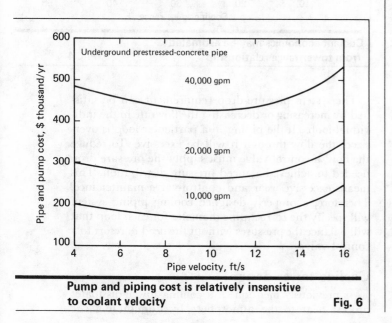

Pump and piping cost is relatively insensitive to coolant velocity   **Fig. 6**

Tower outlet temperatures are usually chosen arbitrarily because of lack of communication between the cooling-tower and heat-exchanger designers. Before heat exchangers can be sized, the coolant inlet temperature must be known. The cooling-tower designer will usually calculate the tower outlet temperature based on either refrigeration or turbine condenser requirements because these exchangers take the lowest-temperature coolant. The tower designer then gives this coolant outlet temperature (exchanger inlet temperature) to the exchanger designer, who ordinarily does not question the low coolant temperature (although it may not be needed) because it makes the job easier.

A better procedure would be for the tower and exchanger designers to simultaneously develop the optimum approach to wet-bulb temperature by: (1) assuming a reasonable approach to the wet-bulb temperature based on both the minimum refrigeration and process outlet temperatures, (2) following the heat-exchanger guidelines to establish the coolant outlet temperatures, (3) calculating the tower inlet temperature based on these outlet temperatures, (4) using Fig. 4a or 4b to ascertain whether the assumed approach is the optimum, and (5) considering putting the low-temperature-coolant users on a separate cooling-tower cell, if the approach assumed is not optimum.

The advantages of this procedure are that it is quick, does not require heat-exchanger or cooling-tower sizing, and makes both the exchanger and tower designers aware of the system limitations.

Fig. 4a and 4b are for cooling-tower heat duties of 100 million and 300 million Btu/h for a site wet-bulb temperature of 80°F. The tower range (tower inlet minus outlet temperature) is plotted against total tower cost. Note that for a fixed tower outlet temperature total cost decreases as the range widens. Increasing the range requires higher coolant temperature rises across the individual heat exchangers. Also notice that for a tower inlet temperature above 110°F, the different approach to wet-bulb (or tower outlet) temperature for a particular tower inlet-temperature has little effect on the total cost.

In summary, if, after the application of the heat-exchanger guidelines, the tower inlet temperature is less than 110°F, the cooling-tower and heat-exchanger interactions are important, and the approach to wet-bulb temperature must be found by trial and error. On the other hand, if the tower inlet temperature is greater than 110°F, system interactions can be ignored, and a first approximation to the tower outlet temperature should be the site wet-bulb temperature plus 10°F.

Applying the guidelines

It has been shown that cooling-system optimization requires reducing the coolant flowrate through the system. How the application of the proposed design guidelines to a cooling system can result in considerable investment and operating cost savings will now be illustrated (Table III).

To simplify the analysis, a cooling system composed of five heat exchangers will be considered. Each ex-

Wider approach temperature cuts tower cost			Table II	
	Case 1	Case 2	Case 3	Case 4
Tower inlet temperature, °F	105	105	125	125
Tower outlet temperature, °F	85	95	85	95
Coolant flowrate, gpm	30,000	60,000	15,000	20,000
Cost of tower, pumps and piping (from Fig. 4B), $/yr	700,000	810,000	500,000	430,000
Cost of pumps and piping, (from Fig. 6), $/yr	340,000	560,000	220,000	270,000
Cost of tower, $/yr	360,000	250,000	280,000	160,000

changer has a heat duty of 18 million Btu/h, water as the shell-side coolant, and ethylene glycol as the process fluid. This cooling system is evaluated under two different circumstances: (1) at a fixed coolant temperature rise of 20°F, or a 10°F approach to the process outlet temperature if a 20°F coolant rise results in a temperature cross, and (2) by means of the design guidelines that have been presented.

First application example

The first exchanger in this application example has a process outlet temperature greater than 150°F. Therefore, the design guidelines call for a coolant outlet temperature of 150°F. A check of the LMTD at the specified temperature levels reveals that the LMTD is greater than 30°F; therefore, system interactions can be ignored and the exchanger coolant outlet temperature set at 150°F.

The next three exchangers have process outlet temperatures less than 150°F; therefore, the design guidelines require that the coolant outlet temperatures of these exchangers be set at their process outlet temperatures. A check of the LMTDs at the specified temperatures shows that they are greater than 30°F for these three exchangers. Again, system interactions can be ignored, and the coolant outlet temperatures of the second, third and fourth exchangers can be set at 113°F, 131°F and 113°F, respectively.

The process outlet temperature of the fifth exchanger is 104°F. If this exchanger's coolant outlet temperature is set at 104°F, a check at this temperature shows that the LMTD is less than 30°F. This means that, for the fifth exchanger, system interactions are important and must be considered. The cooling-tower inlet temperature, based on the first four exchangers, is calculated to be 121°F; therefore, as a first approximation, the tower outlet temperature (exchanger inlet temperature) should be set at the site wet-bulb temperature plus 10°F.

Finding the optimum outlet temperature of the coolant from the fifth exchanger requires generating a graph similar to Fig. 3c. The procedure is: (1) assume a coolant outlet temperature for the fifth exchanger; (2) calculate the tower inlet temperature for the five exchangers; (3) determine the coolant cost from Fig. 4a, using the calculated inlet temperature and an assumed outlet temperature of site wet-bulb temperature plus 10°F; (4) divide the total coolant cost by the tower heat

	Process temperature, °F		Rule-of-thumb design			Guidelines design				Incremental differences	
			Coolant		Exchanger investment, $		Coolant		Exchanger investment, $	Coolant flow, gpm	Exchanger investment, $
Exchanger	In	Out	Outlet, °F	Gpm		LMTD, °F	Outlet, °F	Gpm			
(1)	212	151	108	1,800	35,000	62	150	600	39,000	-1,200	4,000
(2)	212	113	103	2,400	45,000	52	113	1,600	52,000	-800	7,000
(3)	194	131	108	1,800	41,000	51	131	900	47,000	-900	6,000
(4)	158	113	103	2,400	50,000	33	113	1,600	63,000	-800	13,000
(5)	140	104	94	6,000	70,000	23	104	2,600	86,000	-3,400	16,000
				14,400	241,000			7,300	287,000	-7,100	46,000

Summary of cost saving gained from application of design guidelines   **Table III**

Cost comparisons	Rule-of-thumb design	Guidelines design
Tower inlet temperature, °F	100	115
Tower outlet temperature, °F	88	90
Coolant flowrate, gpm	14,400	7,300
Tower, pump and piping investment, $*	1,100,000	700,000
Water-treatment investment, $	230,000	220,000
Total coolant investment, $	1,330,000	920,000
Incremental coolant investment, $		-410,000
Incremental exchanger investment, $		46,000
Total investment saved, $		-364,000
Coolant operating cost, $/yr	210,000	110,000
Increased water-treatment cost, $/yr	0	10,000
Total coolant operating cost, $/yr	210,000	120,000
Total operating cost saved, $yr		-90,000

*These investment figures were obtained from plots similar to those in Fig. 4a and 4b, except for the ordinate being investment instead of cost. Investment vs. tower-range graphs for tower duties of 100, 200, 300 and 400 million Btu/h may be requested from the author.

Water treatment economics for higher coolant outlet temperature   **Table IV**

	Tower inlet temperature, °F	
	120	140
Circulation rate, gpm	10,000	6,000
Tower outlet temperature, °F	90	90
Tower inlet temperature, °F	120	140
Makeup rate, gpm	400	450
Blowdown, gpm	100	150
Concentration cycles	4	3
Investment, $/gpm	50	64
Suspended-solids removal, $/gpm	16	30

	Investment cost, $		Incremental investment, $
Tower investment*	500,000	384,000	-116,000
Wastewater treatment of blowdown	120,000	135,000	15,000
Suspended-solids removal	160,000	180,000	20,000
Net incremental investment, $			-81,000

	Operating cost, $/yr		Incremental cost, $/yr
Tower cost†	150,000	90,000	-60,000
Water for makeup at 10¢/1,000 gal	20,000	22,500	2,500
Chemicals	48,000	54,000	6,000
Net incremental operating cost, $/yr			-51,500

*Cooling-tower includes pumps and piping.
†Operating cost is taken as $0.03/1,000 gal of capacity.

duty and multiply this result by the heat duty of the fifth exchanger; (5) calculate the required heat-exchanger area, using the assumed cooling-tower temperatures; (6) plot the exchanger and coolant costs; (7) repeat Steps 1 through 6 at two different coolant outlet temperatures.

The coolant outlet temperature at which the system cost is lowest should be chosen. In this example, the optimum coolant outlet temperature is 122°F. For a 122°F coolant outlet temperature, an 18°F cross occurs. This requires multiple shells in series; therefore, the coolant outlet temperature was set at 104°F. For an exchanger in this category, one should consider going to a single-pass tube geometry or to a spiral exchanger, so that true countercurrent flow will be achieved and the temperature cross eliminated. By eliminating the temperature cross, the coolant outlet temperature of 122°F can be specified.

This example shows why the cooling-tower outlet temperature is the "Achilles heel" of this study. To be rigorous, the fifth heat exchanger should be resized, using the optimum coolant outlet temperature calculated but with the coolant inlet temperature reduced by 4°F. If exchanger-area requirements are greatly reduced by the lower coolant inlet temperature, the tower and exchanger interaction is significant, and the foregoing Steps 1 through 6 should be repeated at the lower coolant inlet temperature.

Recalculating this exchanger for three different coolant outlet temperatures is, however, time-consuming and may not be necessary. Because the four exchangers with the *LMTD* greater than 30°F will be unaffected by a slight reduction in coolant inlet temperature, what is really being sought is how significant is the cost of the fifth exchanger to the total system cost.

Through engineering judgment and application of the proposed guidelines, one can design a partially, nearly or wholly optimized cooling system, depending on how much effort one is willing to put out.

1. If one applies the guidelines only to exchangers that have an *LMTD* greater than 30°F, a partial optimization will be attained.

2. If one is willing to generate curves similar to Fig. 3c for exchangers with an *LMTD* less than 30°F, a nearly optimum design will be gained.

3. If one is willing to recalculate the exchangers with an *LMTD* of less than 30°F at different coolant inlet temperatures, then a wholly optimized system will be designed.

Inspection of Fig. 4a shows, however, that the last procedure is not necessary if the calculated tower inlet temperature is greater than 110°F, or if the exchangers that require the low-temperature coolant do not contribute significantly to the system heat duty. If the calculated tower inlet temperature is less than 110°F, the last procedure is necessary.

Returning to the example (Table III), if one used the fixed coolant temperature-rise case as the economic basis, it can be seen that the guideline design would reduce the total coolant flowrate by 49% (14,400 gpm vs. 7,300 gpm), the operating cost by 43% ($90,000/yr), and the total investment by 27% ($364,000). This example again demonstrates that a higher exchanger investment can reduce cooling-system investment and operating cost.

It is important to maintain a water velocity that minimizes the deposition of suspended solids. In the initial design of an exchanger, it is desirable to select the baffle spacing or number of tube passes to get a water velocity above 5 ft/s. This velocity should never be less than 2 ft/s. One should recognize that raising the coolant temperature for an existing exchanger results in a lower coolant velocity, which, if below 2 ft/s, intensifies the tendency for suspended solids to deposit. Besides increasing the fouling factor, the sludge deposits accelerate the pitting of tube surfaces.

Water-treatment costs

It is usually at about this point that skeptics ask, "What about the higher water-treatment cost?" The following answers this question:

Water-treatment economics—Water-treatment requirements do rise if the cooling-tower inlet temperature is increased. The actual cost of water treatment at any operating temperature is related to the quality of the makeup water and its particular treatment requirements. Obviously, waters having high concentrations of sludge or dissolved mineral salts will require more treatment. Usually, the magnitude of the increased treatment cost will be minimal for the same source of makeup water.

Mineral salts—The commonly present mineral salt,

calcium carbonate, has a limited and inverse solubility (less soluble at higher temperatures), and thus requires more treatment to prevent scale deposits as the water temperature is raised. Increased addition of sulfuric acid should resolve most problems. However, if the volume of blowdown to control the concentration of mineral salts resulting from increased evaporation in the cooling tower must be augmented, the added cost of wastewater treatment for the greater volume of blowdown could represent an appreciable water-treatment cost increase.

Suspended solids—The tendency of suspended solids (turbidity) in the cooling water to deposit on tube surfaces increases as the coolant temperature is raised. Such deposition can be prevented by the use of a sludge dispersant.

Biological growth—Organic growth speeds up as the coolant-water temperature increases. The resulting slime deposits that develop on tube surfaces can reduce heat transfer and tend to collect more of the suspended solids than might otherwise deposit from a cooling-temperature increase alone. This type of fouling can be controlled by addition of free chlorine or a nonoxidizing biocide.

Accounting for water-treatment costs

Table IV compares system cost between a cooling tower receiving water at temperatures of 120°F and 140°F. Using the 120°F temperature case as the economic basis, it can be shown that the tower investment is reduced by $116,000 by going to 140°F, but the water-treatment investment is boosted $35,000 ($15,000 for water-treatment facilities and $20,000 for augmented blowdown facilities).

Therefore, at the 140°F cooling-tower return-water temperature, the net investment is reduced by $81,000. Additionally, an operating cost savings of $60,000/yr can also be realized. However, this $60,000/yr savings is reduced by the increase in water-treatment chemicals and makeup-water cost of $8,500/yr. Therefore, the net-operating-cost reduction is $51,500/yr.

References

1. Woods, D. R., Anderson, J. J., and Norman, S. L., Evaluation of Capital Cost Data: Heat Exchangers, *Can. J. of Chem. Eng.*, Vol. 54, December 1976.
2. The Marley Co., "Managing Waste Heat with the Water Cooling Tower," 1970.
3. The Marley Co., "Cooling Tower Fundamentals and Application Principles," 1969.

The author

Ralph A. Crozier, Jr., is a project engineer with E. I. du Pont de Nemours & Co. (Engineering Dept., Louviers 5321, Wilmington, DE 19898). Responsible for process design of petrochemicals facilities, he is also involved in energy conservation and is an active member of the Du Pont Heat Exchanger Standards Committee.

Holder of B.S. and M.S. degrees in chemical engineering from the University of Delaware, where he is a part-time Ph.D. candidate in mechanical engineering, he is a registered engineer in the State of Delaware and has previously had several articles published.

Section III
Heating and Insulation

Choosing economic insulation thickness

Optimum insulation thickness depends on the designer's selection of factors for balancing economic and engineering criteria. Here are some guidelines being used by experts in the field.

Alan R. Koenig, Jim Walter Research Corp.

☐ What is the most economic thickness of insulation? Unfortunately, there is no single answer to this question, as an almost infinite number of physical, mechanical and economic variables are involved. However, with rapidly escalating energy costs, getting a good answer is becoming more important.

Equations for estimating heat loss

There are a number of methods for estimating the theoretical heat loss through insulation. A particularly useful one is the graphical solution, which allows the designer to gauge the relative effects of several variables.

The three modes of heat transfer include losses by conduction through the insulation, radiation losses to the air, and convection losses caused by the movement of the air. Because the heat lost through the insulation must equal that transferred to the surroundings:

$$Q \text{ Conduction} = Q \text{ Radiation} + Q \text{ Convection} \quad (1)$$

In the graphical solution, various values for the surface temperature of the insulation are assumed, and heat loss at the chosen temperatures is determined. The intersection of (a) the curves of the heat loss by conduction and (b) the heat losses by radiation plus convection represents the surface temperature and heat loss condition set that satisfies Eq. (1).

The heat loss by conduction is a function of the thermal conductivity and thickness of the insulation components.

For a flat surface:

$$Q_{cd} = (T_h - T_s)/[(l_1/k_1) + (l_2/k_2)] + \cdots \quad (2)$$

For pipe:

$$Q_{cd} = \frac{T_h - T_s}{[r_3\ln(r_2/r_1)/k_1] + [r_3\ln(r_3/r_2)/k_2]} + \cdots \quad (3)$$

When the insulation is homogeneous, the denominators in Eq. (2) and (3) reduce to a single term because the thermal conductivity is uniform.

The heat lost to the air by radiation is given by:

$$Q_r = 0.1713\varepsilon\left[\left(\frac{T_s + 459.6}{100}\right)^4 - \left(\frac{T_a + 459.6}{100}\right)^4\right] \quad (4)$$

The heat loss by convection is a function of the temperature difference between the insulation surface and the air, and the speed at which the air passes over the surface [1]:

$$Q_{cv} = C\left(\frac{1}{d}\right)^{0.2}\left(\frac{1}{T_{av}}\right)^{0.181}(\Delta T)^{1.266}(1 + 1.277v)^{1/2} \quad (5)$$

Originally published September 8, 1980

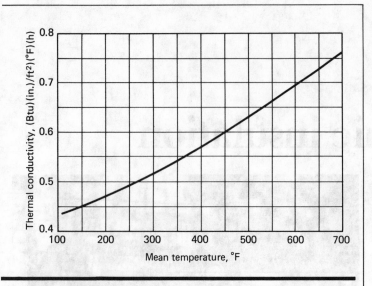

Perlitic insulation curve can be considered linear   Fig. 1

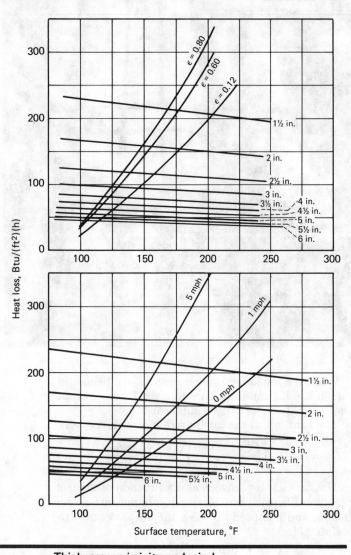

Thickness, emissivity and wind factors affect insulation heat loss   Fig. 2

Here, C is a constant that depends on the shape of the surface. Values of this constant are: 1.016 for horizontal cylinders, 1.235 for long vertical cylinders, 1.394 for vertical plates, 1.79 for horizontal plates facing up, and 0.89 for horizontal plates facing down.

Somewhat simpler but less exact is Langmuir's equation:

$$Q_{cv} = 0.296(T_s - T_a)^{5/4}(1 + 1.277v)^{1/2} \qquad (6)$$

Example of graphical method

The graphical solution is best explained by an example. The design data are: iron pipe size—8.00 in.; operating temperature—800°F; average ambient air temperature—80°F; wind velocity—1 mph; insulation—Celotemp 1500; jacketings—aluminum, white paint, stainless steel; and emissivities—for bare pipe, 0.80; aluminum jacketing, 0.12; white painted canvas, 0.60; stainless steel, 0.80.

The thermal conductivity curve for a perlitic high-temperature insulation is given in Fig. 1. Values for thermal conductivity vs. mean temperature follow a second-order equation; however, over a relatively small (100°F) temperature range, the curve can be considered linear. Therefore, two points can define the line when the heat loss through a particular thickness of this insulation is being determined. Inside and outside diameters of pipe insulation are given in ASTM C-585, "Inner and Outer Diameters of Rigid Thermal Insulation for Nominal Sizes of Pipe and Tubing" [2].

For an insulation O.D. of 15 in. and I.D. of 8.7 in., and an insulation thermal conductivity of 0.58 (Btu)(in.)/(ft^2)(°F)(h) (from Fig. 1, at a mean temperature of 450°F), the heat loss by conduction through 3-in.-thick perlitic pipe insulation covering an 8-in.-dia. pipe at a surface temperature of 100°F would be 99 Btu/(ft^2)(h), calculated via Eq. (3).

Similarly, heat losses by conduction for other insulation thicknesses are plotted in Fig. 2.

The next step is to calculate the heat losses stemming from radiation and convection. A separate curve is required for each jacketing material. Generally, three points can adequately define such a curve.

If $T_s = 100°F$, $\varepsilon = 0.12$, and $T_a = 80°F$, then $Q_r = 3$ Btu/(ft^2)(h), via Eq. (4).

Via Eq. (6), with $v = 1$ mph, $Q_{cv} = 19$ Btu/(ft^2)(h). Therefore, $Q_r + Q_{cv} = 22$ Btu/(ft^2)(h).

Heat losses for surface temperatures of 150°F and 200°F, and for emissivities of 0.60 and 0.80, are also calculated, and the results plotted in Fig. 2. The intersections between the conductive and the radiant-plus-convective loss curves represent the theoretical heat losses for the system at various surface temperatures.

The velocity of air passing over insulation will significantly affect heat loss. Also plotted in Fig. 2 are heat losses calculated via Eq. (6) at wind velocities of 0, 1 and 5 mph, with insulation having a surface emissivity of $\varepsilon = 0.12$.

Other causes of heat loss

System designers agree that a design margin has to be added to theoretically calculated values of heat loss and

Nomenclature

C	Constant that depends on shape of the surface used in calculating heat transfer by convection	r_2	Radius of an intermediate layer of pipe insulation system composed of more than one type of insulation, in.
d	Diameter of outermost layer of pipe insulation system, in.	r_3	Radius of outermost layer of pipe insulation system, in.
i_1	Annual projected fuel inflation rate, decimal	S	Incremental insulation investment, \$/ft
i_2	Current interest rate, decimal	T_a	Surrounding air temperature, °F
k	Thermal conductivity of an insulation layer at a particular mean temperature, (Btu)(in.)/(ft^2)(°F)(h)	T_{av}	Average temperature of insulation surface and surrounding air, °F
l	Thickness of an insulation layer, in.	T_h	Hot-face temperature of insulation, °F
n	Time period, yr	T_s	Cold-face temperature of insulation, °F
P	Positive cash flow, \$/ft	ΔT	Difference between surface temperature and surrounding air temperature, °F
Q_{cd}	Heat transfer by conduction, Btu/(ft^2)(h)	v	Surrounding air velocity, mph
Q_r	Heat transfer by radiation, Btu/(ft^2)(h)	ε	Surface emissivity of outermost layer of insulation system
Q_{cv}	Heat transfer by convection, Btu/(ft^2)(h)		
r_1	Radius of innermost layer of pipe insulation system, in.		

surface temperatures to account for workmanship, thermal expansion, intrusions into the insulation and possible loss of performance with time. The magnitude of this factor will depend on personal judgment based on experience, and has been known to range from 25 to 300%.

Small openings in insulation can allow significant heat loss. Such openings occur at joints between insulation half-sections, around valves and flanges, at gouges and breaks in insulation, and are also caused by differences in thermal expansion between pipe and insulation.

In the example system, a gap as small as $\frac{1}{16}$ in. between adjacent sections of insulation can cause a heat loss of up to 37 Btu/(ft)(h). And heat losses from the seam between half-sections of insulation with a closure gap of $\frac{1}{16}$ in. could be as large as 99 Btu/(ft)(h).

When heated, metal pipes expand more than insulation does. Also, when first heated, all high-temperature insulation tends to shrink. This opposing expansion and contraction can cause fissure cracks. In the sample problem, these cracks theoretically could allow as much as 200–300 Btu/(ft)(h) from an insulated carbon steel pipe. Generally, the magnitude of this loss will be much smaller because the cracks will rarely extend completely through the insulation.

Moisture in insulation will also hike heat losses, and water is present in almost all thermal insulation. It is difficult to accurately estimate the heat loss that can be attributed to moisture. The loss will depend not only on the insulation's water content but also on its porosity, the temperature profile through it, and the type and condition of the jacketing. One rule of thumb states: A 1% increase in moisture content will boost thermal conductivity by 5%.

Economics of insulation thickness

In some cases, technical and safety considerations will primarily determine the insulation thickness. Most often, however, economic criteria will prevail. Insula-tion investment is judged in most companies by economic criteria via a fixed accounting method, which can significantly affect the choice of insulation thickness. Some of the more common methods include simple payback, minimum annual cost, and maximum present value of cash flow.

Simple payback—In its simplest form, this method compares the savings from reduced heat loss to the cost of the insulation system—with the return determined in months or years. The common criterion for the payback period is from one to three years.

Table I presents heat-loss data calculated for the sample pipe-insulation system, having aluminum jacketing—with a wind speed of 1 mph. Process heat is assumed to cost \$3.06 per million Btu, based on oil at \$20/bbl (heating value taken as 140,000 Btu/gal), and 90% conversion efficiency.

Heat losses were obtained from Fig. 2. These values were increased by dividing them by a design factor of 0.75, to account for the additional heat losses discussed. Finally, the heat losses per hour were converted to an annual basis, assuming an operating time of 6,000 h/yr.

Estimated installed costs of the insulation systems, in $\frac{1}{2}$-in. increments, are given in Table II. The 1.15 piping complexity factor represents average installation.

The additional cost for incremental insulation thickness is divided by incremental cost savings to obtain the Table I payback periods. If the payback criterion selected were two years, an insulation thickness of 3 in. would be chosen.

With fuel costs increasing rapidly, some estimators choose to increase the current cost of fuel in the simple payback calculation by an escalation factor. This is commonly done by assuming a linear annual-percentage increase in fuel costs, and using this percentage in the calculation of the escalator factor:

$$\text{Fuel-cost escalator} = \frac{(1 + i_1)^n - 1}{i_1 n} \qquad (7)$$

Simple payback at an assumed heat cost of $3.06/million Btu   Table I

Insulation thickness, in.	Annual cost heat loss, $/(ft) (yr)	Annual heat savings, $/(ft) (yr)	Cost of insu-lation, $/ft	Incre-mental cost, $/ft	Pay-back, yr
0	193.51	–	–	–	–
1½	15.12	178.39	7.08	7.08	0.04
2	12.18	2.94	8.98	1.91	0.94
2½	10.10	2.08	10.94	1.96	0.94
3	8.84	1.26	12.32	1.38	1.10
3½	7.31	1.53	16.36	4.04	2.64
4	6.82	0.49	18.81	2.45	5.00
4½	6.30	0.52	21.52	2.71	5.21
5	5.91	0.39	23.79	2.27	5.82
5½	5.66	0.25	26.05	2.26	9.04
6	5.36	0.30	28.90	2.85	9.50

Estimated total installed costs of the sample insulation   Table II

Insulation thickness, in.	Costs, $/ft				
	Insulation	Jacketing*	Installation	Total	With complexity factor† of 1.15
1½	2.95	0.61	2.60	6.16	7.08
2	4.15	0.66	3.00	7.81	8.98
2½	5.38	0.73	3.40	9.51	10.94
3	6.13	0.78	3.80	10.71	12.32
3½	8.39	0.84	5.00	14.23	16.36
4	9.87	0.89	5.60	16.36	18.81
4½	11.57	0.94	6.20	18.71	21.52
5	12.89	1.00	6.80	20.69	23.79
5½	14.21	1.04	7.40	22.65	26.05
6	16.03	1.10	8.00	25.13	28.90

*Based on $0.20/ft²

†Piping complexity factor accounts for extra costs associated with covering elbows, valves, etc. In practice, this factor can range from 1.0 to 1.6, depending on the complexity of installation and size of the line.

Assuming an annual fuel-cost rise of 10%/yr for 10 years, the average cost of the sample fuel would be $4.88/million Btu. On this basis, a two-year payback would call for an insulation thickness of 3.5 in.

Minimum annual cost—In this accounting method, all costs are annualized, then discounted at an assigned interest rate. The insulation chosen is that which yields the minimum annual cost:

$$\text{System cost} = \text{insulation cost} + \text{energy cost} \quad (8)$$

Insulation cost commonly includes installed cost and cost of maintenance (usually a percentage of initial insulation cost). Costs normally associated with energy include the initial cost of the equipment for producing the desired form of energy, the maintenance cost of this equipment, and direct costs for the fuel, with escalations for inflation. Depreciation on equipment and insulation costs is usually also taken into account.

To annualize the cost of the insulation, its initial cost is multiplied by a capital recovery factor:

$$\text{CRF} = \frac{i_2(1 + i_2)^n}{(1 + i_2)^n - 1} \quad (9)$$

The cost of the heating equipment is obtained by multiplying the heat loss per linear foot, the cost of the equipment to produce the heat, and the capital recovery factor. The maintenance cost of the heating equipment is determined by multiplying the heat loss per linear foot times the cost of the equipment to produce the heat times the yearly maintenance cost percentage of the equipment. The cost of heat = annual heat loss × first-year cost of heat × 1 − (tax rate/100) × Discount Factor (DF).

$$\text{DF} = (\text{CRF}) \sum_{n-1}^{n} \frac{(1 + i_1)^{n-1}}{(1 + i_2)^n} \quad (10)$$

If the fuel inflation rate equals the interest rate, Eq. (10) reduces to:

$$\text{DF} = (\text{CRF})n/(1 + i_2) \quad (11)$$

The tax saving is computed by multiplying, by the tax rate, the investment costs for the insulation and the heat-generating equipment, and dividing the product by the expected life of the equipment.

The total annual cost for the installation is the sum of the original cost of the insulation and its maintenance cost, the cost of the heating equipment and its maintenance cost, and the cost of the heat—less the tax saving.

Discounted payback—This is the time required for a capital investment to generate discounted, after-tax cash flows equal to the initial investment. The investment payback in years is calculated with Eq. (12):

$$n = \frac{\ln P/[(P - Si_2)}{\ln(1 + i_2)} \quad (12)$$

Using this method, the economic thickness of the insulation for the sample problem would generally be 3½ in.

References

1. Heilman, R. H., "Surface Heat Transmission," Trans. Am. Soc. Mech. Engrs., Vol. 51, 1929, p. 257.
2. "Annual Book of ASTM Standards, Part 18," American Soc. for Testing and Materials, Philadelphia, 1979.

The author

Alan R. Koenig is head of a research section at Jim Walter Research Corp. (10301 Ninth St. North, St. Petersburg, FL 33702). He has been with the Celotex Industrial Products Div. for 13 years. Previously, he had been with General Electric Corp. and Gulf Oil Corp. Active in the ASTM C-16 (thermal insulation) committee, he is a member of Keramos, honorary ceramic engineering society, and Sigma Xi, honorary engineering society. He holds B.S. and M.S. degrees in ceramic engineering from Georgia Institute of Technology.

Preventing burns from insulated pipes

Personnel safety depends not only on pipe insulation thickness, but on the insulation jacket. The difference between a shiny and a dull jacket-surface can mean the difference between a serious burn and mild discomfort.

M. McChesney and *P. McChesney*, *Fuel Save Associates*

☐ Recently the writers were asked to advise on the insulation of steam headers—the main steam line into which boilers are connected—to ensure that the insulation was not only economic but also safe. Because of the very high steam-temperatures involved, together with the new tough laws on health and safety at work, it was imperative that there should be no risk of skin burns to those who came into contact with the insulated headers.

Steam headers are usually large-diameter pipes (up to 30-in. dia.) and in this particular case we were asked to consider both 10- and 20-in. pipes. Their insulation is essential not only to conserve heat but also to avoid high boiler base-loads, which increase the fuel bill out of all proportion to their size.

Insulation guidelines

There are general guidelines for header insulation laid down by various organizations. In the U.S., those given by NIMA (National Insulation Manufacturers Assn.) are typical [1]. Specifically, large-diameter hot pipes should be insulated with curved preformed sections, which should have an outer covering to ensure, among other reasons, that the insulation is neither damaged nor crushed and that it is fully protected from the weather.

This outer finish depends upon where the pipes are. For example, if they are indoors, the finish can be lightweight canvas; but if outdoors, the finish must be completely waterproof. This is obvious, and yet what is not often realized is that once wet, insulation may never become dry. Even a steam header at 1,000°F may have wet insulation because some of the insulation will be below 212°F; the heat transfer through the insulation will only completely dry it as far as the 212° isotherm, leaving the outermost insulation still damp. Insulation containing water is only about a tenth as effective as the same insulation when completely dry.

Although both weather-barrier mastic and aluminum (or perhaps even stainless steel) jacketing are used for outdoor weather protection, the metal jacketing is to be preferred since there is a chance it will allow water vapor to escape outwards.

A common header insulant is calcium silicate: it has high compressive and flexural strength, it meets the specifications for prevention of stress-corrosion cracking of austenitic stainless steels and, being a mixture of lime and silica reinforced with fibers, it can readily be moulded into preformed sections. Weather protection of calcium silicate is essential since not only can it hold about 70% of its own weight of water without losing rigidity, but it has the astonishing capability of being able to hold about 350% of its own weight of water without actually dripping (showing that it is wet)!

Before attempting to calculate the economic thickness of insulation—and this proved to be a vexing problem for reasons we shall give in a later article—the writers considered the following problems: 1. What is the

Originally published July 27, 1981

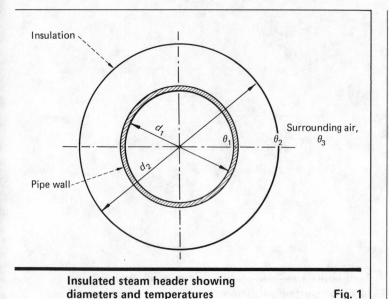

Insulated steam header showing diameters and temperatures   **Fig. 1**

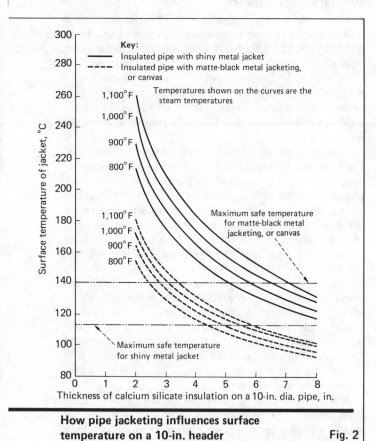

How pipe jacketing influences surface temperature on a 10-in. header   **Fig. 2**

maximum allowable "safe" temperature that the insulation surface can have such that if it is touched, either accidentally or on purpose, there is no possibility of skin burns occurring? 2. How does this surface temperature depend upon which type of insulation outer cover is used?

Effect of insulation outer-cover

It comes as a surprise to some people that the temperature of the insulation surface depends not only upon the pipe temperature and the thickness and thermal properties of the insulant, but also upon its surface condition—specifically its surface emissivity. A roughened matte surface will have a higher emissivity than a shiny metallic surface, meaning that the matte surface will have a *higher* radiation heat loss but a *lower* surface temperature than the metallic surface. There is therefore a conflict here because if we cover the insulation with canvas then we obtain a "low" surface temperature (good for personnel protection) but a "high" heat loss (bad for fuel conservation), whereas metal-jacketing the insulation will save fuel but perhaps cause skin burns to anyone unfortunate enough to touch the jacketing.

The effect of different surface emissivities is not small—we shall show that the insulation surface temperature can differ by almost 100°F, depending solely on the surface finish. Not only is this difference in temperature important from the insulation economics viewpoint, but it is crucial from the safety point of view since the rate at which skin tissue is damaged by heat depends exponentially on the temperature to which the skin is raised when in contact with a hot surface.

The heat-balance equation

Fig. 1 shows an insulated header carrying steam at a constant temperature. Even with insulation present the pipe will lose heat and, in the steady state (when the surface temperature of the insulation does not change), the heat-balance equation per foot length of pipe can be written:

| Heat transfer from hot pipe through insulation by radial conduction | = | Heat loss from insulation surface by convection to the surrounding air | + | Heat loss from insulation surface by radiation to the surrounding air |

Expressed mathematically:

$$\frac{2\pi k(\theta_1 - \theta_2)}{\ln(d_2/d_1)} = \pi d_2 h_c (\theta_2 - \theta_3) +$$

$$\pi d_2 \varepsilon \sigma [(\theta_2 + 460)^4 - (\theta_3 + 460)^4] \quad (1)$$

Steady-state heat-balance equations often appear in technical articles. In most cases it is all too common to see the radiation-loss term on the right-hand side either ignored altogether or else incorporated into the convection heat-loss to give a surface coefficient that is sometimes treated as a constant and sometimes (in the hope of increasing accuracy) taken as a variable. While this is a perfectly satisfactory procedure in many industrial situations, it can (as in the present case) lead to errors in the overall heat-loss calculation (which is likely not too important). However, it can also lead to errors in safety considerations, which are of paramount importance.

Numerical studies of laminar natural convection based upon the nondimensional heat-transfer equation

$$\left(\begin{array}{c}\text{Nusselt}\\\text{number}\end{array}\right) = \text{Constant} \cdot \left(\begin{array}{c}\text{Grashof}\\\text{number}\end{array} \times \begin{array}{c}\text{Prandtl}\\\text{number}\end{array}\right)^n$$

show that an excellent, convenient and accurate approximation for the convective heat-transfer coefficient for a horizontal hot circular pipe is

$$h_c = 0.270(\theta_2 - \theta_3)^{0.25} d_2^{-0.25}$$

In these equations the values of k and σ are taken as constants, while the values of d_1, d_2, θ_1, θ_3, ε are to be specified. This leaves the equation to be solved for θ_2, which is the all-important insulation surface-temperature.

The heat-balance equation cannot be solved algebraically since it is a quartic whose lower-order terms have non-integral powers, but it can be solved on a computer using the Newton-Raphson iterative technique. The results are shown in Fig. 2 and 3.

The substantial difference between the jacket surface-temperatures for the two chosen values of surface emissivity is very striking indeed. The value $\varepsilon = 0.09$ is a representative figure for shiny aluminum jacketing, while the value $\varepsilon = 0.9$ is typical of canvas either covered with thick dust or, better, painted with a matte oil-bound paint. For example, this difference in surface emissivity alone can reduce the jacket surface-temperature from 286°F to 192°F for a pipe temperature of 1,100°F on a 20-in.-dia. header having 2-in.-thick insulation.

This drop in surface temperature due to differences in surface emissivity decreases as the insulation thickness increases or the pipe temperature decreases, but it is still greater than 30°F even for a pipe temperature as "low" as 800°F with an insulation thickness as great as 6 in. Although a 30°F temperature difference is unimportant in terms of the economics of insulation, it is very important indeed from the safety viewpoint since it can make all the difference between the possibility of skin burn and no injury at all.

The "safe" temperature

What is the value of an acceptable surface temperature to ensure that no skin burn will occur if that surface is touched? On reading through the literature, one finds that there is no single answer to the question. Malloy [2] remarks that "metallic surfaces at temperatures in excess of 145°F can cause skin burns but a specific figure is not available as this would depend upon the person concerned, the roughness of the metallic surface and the thermal properties of the insulation material." The British Standard, BS 4086 [3], states that for hot *domestic* (not industrial) equipment, the following are the maximum surface temperatures if skin burn is to be avoided:

■ Metallic surface that will be gripped, 130°F.
■ Metallic surfaces that will be deliberately touched but not gripped, 140°F.
■ Metallic surfaces that may be accidentally touched but not gripped, 220°F.
■ For vitreous or plastic materials the permitted maximum temperatures are higher.

The British Standard Code of Practice, CP 3005 [4], recommends that in industrial situations, the maximum "safe" temperatures are:

■ Metallic surfaces that can be reached from the floor, 135°F
■ Non-metallic surfaces that can be reached from the floor, 150°F

This range of "safe" temperatures is not helpful to the plant engineer concerned with insulation economics and safety because it is too large. If he accepts the

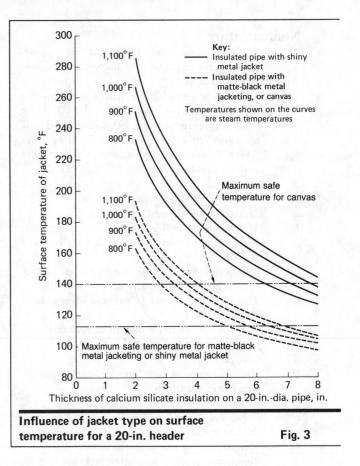

Influence of jacket type on surface temperature for a 20-in. header   Fig. 3

higher temperatures, he may practice good economics but cause skin burn to personnel.

Physiological studies of skin burn

Physiologists have studied pain and injury due to hot and cold surfaces for many years and have explained some—but by no means all—of the very complicated processes that occur within injured skin tissue. They make an important and clear distinction between pain and injury. In the context of skin burn, it is believed that the 21 just noticeable differences in pain intensity are in fact independent of the time taken for the skin to respond to a heat stimulus. However, the essential ingredient for injury, as opposed to pain, after exposure to heat (not necessarily actual physical contact with a hot surface) is a raising of the skin temperature to a threshold level of 113°F.

This temperature level, and of course higher temperatures, must be sustained for "a sufficient time"—not only the actual contact time with the hot insulation surface but also the remainder of the time that the skin remains above 113°F after physical contact has been broken.

The fact that the skin can continue to be damaged while cooling down is a very important consideration. In everyday life when we either blow upon a burn or hold it under a cold-water tap, we are instinctively trying to accelerate the cooling process.

Physiologists have found that equal heat loads (Btu/ft²) do not produce the same skin damage. Damage depends upon the *rate* at which the heat is transferred to the skin—the higher this rate, the greater the injury.

Nomenclature

A The pre-exponential factor in the tissue damage-rate equation (evaluated from experimental data as $10^{11.479}$)

b_{INSUL} Thermal-penetration coefficient of the insulation jacketing: 67.1 Btu ft^{-2} h$^{-1/2}$ °F^{-1} for shiny aluminum or matte-black-painted aluminum and 0.264 Btu ft^{-2} h$^{-1/2}$ °F^{-1} for canvas

b_{SKIN} Thermal-penetration coefficient of skin at normal temperature 3.69 Btu ft^{-2} h$^{-1/2}$ °F^{-1}

B b_{SKIN}/b_{INSUL}

d_1 Outer dia. of steam header pipe, ft

d_2 Outer dia. of insulation jacketing, ft

h Heat-transfer coefficient of barrier between skin and hot insulation surface jacketing, Btu h^{-1} ft^{-2} °F^{-1}

h_C Convective heat-transfer coefficient of insulation jacketing for still air, Btu h^{-1} ft^{-2} °F^{-1}

k Thermal conductivity of calcium silicate, 0.05 Btu h^{-1} ft^{-1} °F^{-1}

k_{SKIN} Thermal conductivity of skin at normal body temperature, 0.192 Btu h^{-1} ft^{-1} °F^{-1}

K Constant

Q Heat flux into the skin from hot insulation surface jacketing, Btu h^{-1} ft^{-2}

t Time, h

x Depth below skin surface, ft

ε Surface emissivity of the insulation jacketing

θ_1 Temperature of the steam in header, °F

θ_2 Temperature of the insulation jacketing, °F

θ_3 Temperature of still air (70°F)

θ_4 Instantaneous skin surface temperature, °F

θ_5 Normal surface temperature of the skin (95°F)

$\theta(x,t)$ Temperature of the skin as a function of depth x and time t

σ Stefan's radiation constant, 0.171 × 10^{-8} Btu h^{-1} ft^{-2} °R^{-4}

ϕ Nondimensional variable (exp $\phi \simeq (1 + \phi)$

Ω Degree of skin injury; $\Omega = 1$ is condition for complete trans-epidermal necrosis

X $b_{SKIN}x/2k_{SKIN}\sqrt{t}$

erf   Error function

exp   Exponential function

Mathematical modelling of skin heating

Let us try to model a situation where some exposed skin comes into contact with the hot surface of the pipe insulation—as idealized to the thermal contact between two semi-infinite bodies with a nonsteady conduction heat-transfer between them. This is a heat-transfer problem that has been solved by Hsu [5]; the result is:

$$\theta(x,t) = \theta_4[1 - \text{erf}(X)] + \theta_5 \cdot \text{erf}(X) \qquad (2)$$

where

$$\theta_4 = \frac{\theta_2 + B\theta_5}{1 + \cdot B} \qquad (3)$$

On elimination of θ_4, this gives:

$$\theta(x,t) = \frac{\theta_2 + B\theta_5}{1 + B} - \left(\frac{\theta_2 - \theta_5}{1 + B}\right)\text{erf}(X) \qquad (4)$$

The heat flux into the skin is:

$$Q = \frac{b_{SKIN}}{\sqrt{\pi t}} \cdot (\theta_4 - \theta_5) \qquad (5)$$

$$= \frac{1}{\sqrt{\pi}\sqrt{t}} \cdot \left(\frac{1}{\frac{1}{b_{SKIN}} + \frac{1}{b_{INSUL}}}\right) \cdot (\theta_2 - \theta_5) \qquad (6)$$

The meaning of Eq. (2) and (3) is that, at the moment of contact, the interface (i.e., the skin's surface) rises *instantaneously* to the temperature θ_4, and thereafter there is a heat flow from the hot insulation surface into the skin. This flow, according to Eq. (5) and (6), is instantaneously infinite but thereafter decreases to a finite value.

Let us use these equations to determine the value of the instantaneous skin temperature θ_4 for the two insulation surface coverings already discussed. For an aluminum jacket covering, B has a value of about 0.055, so that

$$(\theta_4)_{METAL} = \frac{\theta_2 + 5 \cdot 2}{1.055} \simeq \theta_2 \qquad (7)$$

This shows that the instantaneous skin temperature is always close to the insulation surface temperature, which represents a considerable hazard indeed. However, for a canvas jacket, B has a value of about 14, so that

$$(\theta_4)_{CANVAS} = \frac{\theta_2 + 1330}{15} \simeq 89 + \frac{\theta_2}{15} \qquad (8)$$

meaning that the instantaneous skin temperature is always slightly—but not significantly—greater than 95°F, provided that we are concerned with hot and not cold surfaces; this is excellent from the safety viewpoint.

In addition, Eq. (6) shows that

$$\frac{Q_{METAL}}{Q_{CANVAS}} = 14 \qquad (9)$$

which means that the heat-transfer rate, for equal time intervals and equal values of $(\theta_2 - \theta_5)$, from the metal to the skin is 14 times that from the canvas to the skin.

These results, once again, confirm everyday experience, which tells us that it is safer to touch (or even grip) a hot canvas surface than an aluminum surface at the same temperature. Such confirmation gives us confidence in our mathematical model. Unfortunately, however, if we compare the actual numerical values predicted by the above equations with the values obtained in experiment by the physiologists, then there are considerable discrepancies.

In terms of body reaction, Table I shows what happens when the skin temperature is raised from a level where the normal sensation of warmth alters to a sensation of burning and pain, followed by a rising level of pain and increased injury, to severe and then intolerable pain, and finally numbness—with irreversible destruction of the skin tissue and loss of all sensation. Experiments reported by Stoll [6] consider the problem of a *variable*-thermal-property material (which is what skin is) being subjected to various heat fluxes so that its temperature rises slowly (measured in seconds) to a high value, but certainly not instantaneously as predicted by Eq. 2, 3 and 4. This slow rate of skin temperature rise observed in the real-life situation could be the result of one or more of a number of effects, such as a heat-flow contact resistance (a barrier to heat flow) arising because of the presence of sweat or dirt on either or both the skin and the insulation surface.

Modelling a skin heat-flow barrier

A first attempt at a mathematical model is the thermal conduction interaction between two semi-infinite materials with a barrier of constant heat-transfer coefficient between them. Such a problem has been solved by Schneider [7], who gives the solution

$$\frac{\theta(x,t) - \theta_5}{\theta_4 - \theta_5} = 1 - \mathrm{erf}(X) -$$

$$\left[\exp\left\{\frac{hx}{k_{SKIN}} + \frac{h^2 t}{b_{SKIN}}\right\}\right] \cdot \left[1 - \mathrm{erf}\left(X + \frac{h\sqrt{t}}{b_{SKIN}}\right)\right] \quad (10)$$

When there is no barrier present, $h \to \infty$ and Eq. (10) becomes identical to Eq. (2). If we concern ourselves with short times of exposure (measured in seconds rather than hours) and zero tissue depth, i.e. the surface of the skin, then using the mathematical approximations

$$\mathrm{erf}(X) \simeq X \quad \text{and} \quad \exp \phi \simeq 1 + \phi$$

we obtain Eq. (10) in the form

$$\frac{\theta(0,t) - \theta_5}{\theta_4 - \theta_5} = \frac{h\sqrt{t}}{b_{SKIN}} - \left(\frac{h\sqrt{t}}{b_{SKIN}}\right)^2 + \left(\frac{h\sqrt{t}}{b_{SKIN}}\right)^3 \quad (11)$$

For a thin film of moisture, taking values of h lying between 1 and 10 Btu ft^{-2}h^{-1}°F^{-1} and time intervals between 1 and 4 seconds, we find that the right-hand side of Eq. (11) has values lying between 4.5×10^{-3} and 9×10^{-2}, so that the skin temperature is close to its original value of 95°F. This means that the presence of even a thin barrier to heat flow can dramatically alter the value of the instantaneous skin temperature, reducing it from the high and dangerous values predicted, for example, by Eq. (7). It is arguable that considerations such as these, together with the fact that the value of b_{SKIN} is known to double at high skin temperatures due to vaso-dilation (increased blood flow producing increased thermal conductivity of the skin), cause the discrepancy between the predictions of Eq. 4 and 6 and the values actually observed, in the laboratory, by the physiologists.

The physiologists have empirically related skin-damage rate to the skin-temperature-versus-time behaviour in the form:

$$\text{Tissue damage rate} = \frac{d\Omega}{dt} = A \exp\left(-\frac{K}{\theta}\right) \quad (12)$$

where the total damage is

$$\Omega = \int_{HEATING} \frac{d\Omega}{dt} \cdot dt + \int_{COOLING} \frac{d\Omega}{dt} \cdot dt \quad (13)$$

such that $\Omega = 1$ is the condition for complete transepidermal necrosis. The integration limits in both cases are those appropriate to the threshold temperature for

Sensation felt	Value of skin temperature, °F
No sensation	82– 93
Warm to hot	93–113
Threshold of pain	113
Increasing pain	113–126
Severe pain	126–133
Intolerable pain	133–140
Numbness	Above 140

How level of pain varies with skin temperature   Table I

injury (113°F) and the peak skin temperature actually reached.

Numerical data given by Stoll have been curve-fitted by the writers to give the equation

$$\frac{d\Omega}{dt} = 10^{11.479} \exp\left(-\frac{3618 \cdot 3}{\theta}\right) \quad (14)$$

which means that very small differences in skin temperature can make large differences in the rate at which tissue damage occurs. For example, an increase in skin temperature from 140°F to 145°F more than doubles the rate of skin damage. Clearly, all that is required to bridge the heat-transfer analysis given in our mathematical models to the physiological data is to evaluate Eq. (13) using Eq. (14) which relates the skin temperature to time, and then use $\Omega = 1$ as the skin-burn criterion. Unfortunately, it is this last step that, as yet, cannot be taken because there is no heat-transfer theory that reliably predicts how the skin temperature varies with time (as opposed to measuring it with live human beings in the medical laboratory and thereafter performing numerical integration, as has been done by the physiologists).

Proposals for "safe" temperatures

In spite of this failure to bring together heat-transfer theory and medical observation, we can draw upon some of the conclusions of the above analysis to frame tentative criteria for "safe" temperatures. Because of the uncertainty of the mathematical modelling, coupled with the most important consideration of ensuring safety to personnel, *these criteria must err very much on the side of caution* because of the extreme sensitivity of the damage rate to skin temperature. It was for just these reasons that the writers returned to the more exact heat-balance Eq. 1 for pipe heat-loss rather than that normally used in typical insulation-thickness studies.

Since Eq. (7) shows that the instantaneous skin temperature in the idealized heat-transfer analysis is always very close to the metal-jacketed-insulation surface temperature, and since the threshold of pain and injury is 113°F, we propose that for an aluminum jacket the surface temperature should never be higher than 113°F. However, in the case of the canvas jacketing, we recognize the lower instantaneous skin temperature and heat-transfer rate predicted by the idealized theory of Eq. (8) and (9) and therefore we can allow a higher insulation surface temperature.

Steam temperature, °F	Minimum insulation thickness, in.	
	10-in. pipe	20-in. pipe
800	2½	3
900	3	3½
1,000	3½	4
1,100	3½	4

Amount of calcium silicate insulation to ensure its surface temperature does not exceed 140°F   Table II

There is of course the possibility of an inherent danger in this slower heat-transfer rate because the skin temperature, although rising more slowly, can still rise to a value higher than initially appreciated and sensed, and skin damage will occur during the contact period combined with the noncontact period during which the skin is cooling down from its peak temperature back to 113°F. Since physiologists tell us that skin is destroyed "instantaneously" once its temperature reaches 162°F, we propose that the maximum temperature of the canvas jacket covering the insulation surface should be 140°F.

Effect of proposals on insulation thickness

One immediate and important consequence of these proposals comes from referring to Fig. 2 and 3. This is that even *6 in. of calcium silicate* over either the 10- or the 20-in. header is totally insufficient to reduce the temperature of the shiny aluminum jacket to the "safe" value we propose (113°F). It is also clear from the graphs that, depending on the value of the pipe temperature θ_1, it will be necessary to increase the thickness of the calcium silicate insulation up to what are likely to be unacceptable values (as much as 12 in.), or, more sensibly, to increase significantly the surface emissivity of the metal jacket.

There will inevitably be an increase in the surface emissivity of any initially shiny aluminum jacketing as it oxidizes by natural "weathering," but this is still not sufficient since the surface emissivity of heavily oxidized aluminum is only 0.31. It will therefore be necessary to paint the jacketing with a very thick layer of high-temperature-resistant matte ("flat") oil-bound paint to raise the surface emissivity to 0.9 or higher.

It might seem surprising that a thick coat of paint alone can cause such a great change in heat output and surface temperature. However, this is a well-known effect with central heating panels; tests have shown that painting these panels with ordinary oil-bound paints and enamels has little or no effect, irrespective of the color chosen, since all have high emissivities when dry. However, a coat of metallic paint on a central heating panel will reduce the overall heat output by up to 25% because of the reduction of 50% or more in the radiant heat output. In addition, these tests show that the heat output is substantially restored by two thick coats of clear varnish! Since central heating panels are really convectors rather than radiators and operate at temperatures far lower than the steam headers considered here,

we have no difficulty in appreciating the great effect that the surface emissivity has on the surface temperature and heat output because of the Stefan-Boltzmann heat-loss equation, which contains the fourth power of the radiating-surface temperature.

Referring to Fig. 2 and 3 and the proposed criteria given for the "safe" temperature of aluminum jacketing with a thick matte finish (or, alternatively, a canvas jacket), we can draw up Table II which shows the minimum thicknesses of calcium silicate insulation that must be provided to ensure no hazard to personnel.

It must be remembered, of course, that this is only part of the story, because consideration must be given to reducing the heat loss from the steam headers to an acceptable value. The writers therefore turned their attention to calculating the economic thickness of insulation—subject to the proposed "safe" temperatures proposed here—using the ECON and ETI procedures, and after considerable computer analysis were obliged to abandon this orthodox methodology and adopt a totally different approach to sizing header insulation. The reasons for this form the contents of the article on p. 139.

References

1. Perry, R. H., and Chilton, C. H., "Chemical Engineers' Handbook," 5th ed., McGraw-Hill, New York, 1973.
2. Malloy, J. F., "Thermal Insulation," Van Nostrand-Reinhold, New York, 1969.
3. British Standards Institute, BS.4086: Recommendations for maximum surface temperature of heated domestic equipment, 1966.
4. British Standards Institute, Code of Practice 3005: Thermal insulation of pipework and equipment, 1969.
5. Hsu, S. T., "Engineering Heat Transfer," Van Nostrand, New York, 1963.
6. Stoll, A. M., "Advances in Heat Transfer," Vol. 4, "Heat Transfer in Biotechnology," Academic Press, 1967.
7. Schneider, P. J., "Conduction Heat Transfer," Addison-Wesley, Reading, Mass., 1974.

The authors

Malcolm McChesney is founder and Senior Associate of Fuel Save Associates, a group of energy consultants. He is also Senior Lecturer in Energy Studies and Thermodynamics in the Dept. of Mechanical Engineering, University of Liverpool, P.O. Box 147, Liverpool, L69 3BX, U.K. He has written over 20 research papers, three books, and articles for the Encyclopaedia Britannica and *Scientific American.*

Peter McChesney is a college student and Associate (computing) in Fuel Save Associates. He is the son of Malcolm McChesney. Since the age of 12 he has published articles on analog and digital circuitry, ranging from analog to digital converters through musical synthesizers to microprocessor-based hardware. He was a U.K. North-West prize winner in the 1979 Young Engineer for Britain Competition, for designing and constructing a multiplexed digital system for monitoring failure in a chain of steam traps.

Calculating heat loss or gain by an insulated pipe

This program for the HP-97 can be used to quickly calculate the insulation surface temperature and heat loss or gain by a pipe, while all other factors are varied.

Frank S. Schroder, 3M Co.

☐ Determining the most economical insulation system for hot or cold pipes has long been difficult to pin down without the aid of a computer because of the interdependence of the three modes of heat transfer. Charts and tables are available, but permit the evaluation of a limited range of only a few variables.

The accompanying program (figure) is developed for use with the Hewlett-Packard HP-67 or HP-97 programmable calculator, and is intended as a tool to help the engineer evaluate the parameters of: fluid temperature; ambient air temperature; thermal conductivity of the insulating material; pipe diameter; insulation thick-

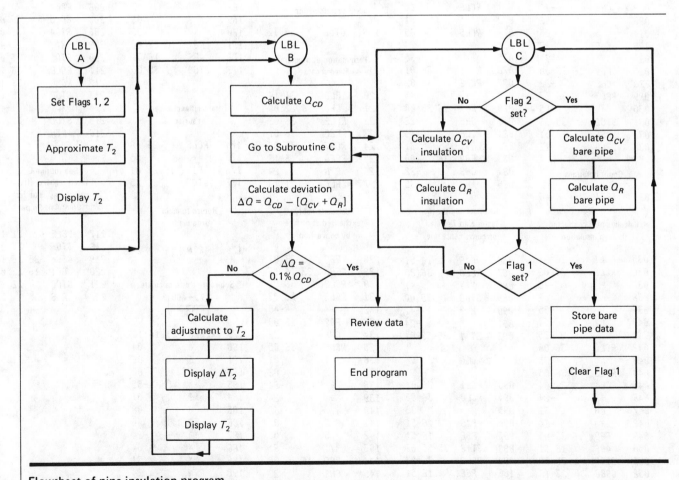

Flowsheet of pipe insulation program

Originally published January 25, 1982

Program listing   Table I

Step	Key	Code	Step	Key	Code	Step	Key	Code	Step	Key	Code	Step	Key	Code
Initialize registers and flags			*Calculate ΔQ (error); if less than 0.1% Q_{cd}, terminate program*			*Limit adjustment to T_2 to less than $(T_2 - T_3)$*			146	.	-62	*Sub-routine b ($\pi D_2/12$)*		
001	*LBLA	21 11				101	RCL3	36 03	147	2	02			
002	0	00	053	RCL7	36 07	102	-	-45	148	7	07	191	*LBLb	21 15 13
003	STOI	35 46	054	X≠Y	-41	103	F0?	16 23 00	149	x	-35	192	RCL5	36 05
004	SF1	16 21 01	055	-	-45	104	CHS	-22	150	GSBb	23 16 12	193	Pi	16-24
005	SF2	16 21 02	056	STOA	35 11	105	X>0?	16-44	151	RCL2	36 02	194	x	-35
006	.	-62	057	PSE	16 51	106	GTOB	22 12	152	RCL3	36 03	195	1	01
007	7	07	058	RCL7	36 07	107	1	01	153	-	-45	196	2	02
008	STOE	35 15	059	.	-62	108	F0?	16 23 00	154	x	-35	197	÷	-24
009	F≠S	16-51	060	1	01	109	CHS	-22	155	F0?	16 23 00	198	x	-35
010	1	01	061	%	55	110	RCL3	36 03	156	ABS	16 31	199	RTN	24
011	STOI	35 01	062	ABS	16 31	111	+	-55	157	STO8	35 08			
012	EEX	-23	063	RCLA	36 11	112	STO2	35 02						
013	3	03	064	ABS	16 31	113	PSE	16 51	*Calculate heat exchanged at surface by radiation*			*Sub-routine c interchanges T_1 and T_2, D_1 and D_2, e_p and e_i for purpose of calculating heat exchanged by uninsulated pipe, on first iteration*		
014	STO0	35 00	065	X≤Y?	16-35									
015	F≠S	16-51	066	GTOD	22 14	*If overcorrection has occurred, reduce correction exponent E*			158	RCL2	36 02			
									159	GSBa	23 16 11			
First estimate of T_2			*Calculate adjustment in T_2*			114	RCLE	36 15	160	RCL3	36 03			
016	RCL1	36 01				115	.	-62	161	GSBa	23 16 11	200	*LBLc	21 16 13
017	RCL3	36 03	067	RCLA	36 11	116	1	01	162	-	-45	201	RCL1	36 01
018	-	-45	068	RCL7	36 07	117	-	-45	163	RCLC	36 13	202	RCL2	36 02
019	X<0?	16-45	069	÷	-24	118	STOE	35 15	164	x	-35	203	STO1	35 01
020	SF0	16 21 00	070	RCL4	36 04				165	GSBb	23 16 12	204	X≠Y	-41
021	F0?	16 23 00	071	RCLE	36 15	*Start recalculation*			166	.	-62	205	STO2	35 02
022	CHS	-22	072	Y^x	31				167	1	01	206	RCL4	36 04
023	.	-62	073	RCL5	36 05	119	GTOB	22 12	168	7	07	207	RCL5	36 05
024	5	05	074	÷	-24				169	3	03	208	STO4	35 04
025	5	05	075	x	-35	*Termination of program Review data*			170	x	-35	209	X≠Y	-41
026	Y^x	31	076	RCL1	36 01				171	F0?	16 23 00	210	STO5	35 05
027	F0?	16 23 00	077	RCL2	36 02	120	*LBLD	21 14	172	ABS	16 31	211	RCLB	36 12
028	CHS	-22	078	-	-45	121	CF0	16 22 00	173	STO9	35 09	212	RCLC	36 13
029	RCL3	36 03	079	F0?	16 23 00	122	RCL2	36 02				213	STOB	35 12
030	+	-55	080	CHS	-22	123	PSE	16 51	*Total heat exchanged at surface*			214	X≠Y	-41
031	STO2	35 02	081	RCLE	36 15	124	RCL7	36 07				215	STOC	35 13
032	PSE	16 51	082	Y^x	31	125	PREG	16-13	174	RCL8	36 08	216	RTN	24
033	*LBLB	21 13	083	F0?	16 23 00	126	RTN	24	175	+	-55			
			084	CHS	-22	127	*LBLC	21 13	176	F1?	16 23 01			
Count iterations			085	x	-35				177	GTOE	22 15	*Sub-routine E resets registers after calculating value for uninsulated pipe*		
						Calculate heat exchanged at surface by convection								
034	ISZI	16 26 46	*Calculate adjustment factor F to T_2 based on convergence rate*						*Return to main program B*			217	*LBLE	21 15
						128	F2?	16 23 02				218	STO6	35 06
Calculate heat conducted through insulation			086	P≠S	16-51	129	GSBc	23 16 13	178	RTN	24	219	GSBc	23 16 13
			087	RCL0	36 00	130	RCL2	36 02				220	CF1	16 22 01
035	RCL1	36 01	088	X≠Y	-41	131	RCL3	36 03	*Sub-routine a to calculate $\left(\frac{T_x+460}{100}\right)^4$*			221	GTOC	22 13
036	RCL2	36 02	089	STO0	35 00	132	-	-45				222	R/S	51
037	-	-45	090	X≠Y	-41	133	F0?	16 23 00						
038	2	02	091	÷	-24	134	CHS	-22	179	*LBLa	21 16 11			
039	x	-35	092	ST+1	35-55 01	135	RCL5	36 05	180	4	04			
040	Pi	16-24				136	1	01	181	6	06			
041	x	-35	*Calculate new T_2*			137	2	02	182	0	00			
042	RCL0	36 00				138	÷	-24	183	+	-55			
043	x	-35	093	RCL1	36 01	139	÷	-24	184	1	01			
044	RCL5	36 05	094	RCL0	36 00	140	.	-62	185	0	00			
045	RCL4	36 04	095	x	-35	141	2	02	186	0	00			
046	÷	-24	096	P≠S	16-51	142	5	05	187	÷	-24			
047	LN	32	097	PSE	16 51	143	Y^x	31	188	4	04			
048	÷	-24	098	ST+2	35-55 02	144	F0?	16 23 00	189	Y^x	31			
049	F0?	16 23 00	099	RCL2	36 02	145	CHS	-22	190	RTN	24			
050	ABS	16 31	100	PSE	16 51									
051	STO7	35 07												
052	GSBC	23 13												

Typical values of thermal conductivity of insulations, Btu/(h) (ft²) (°F/ft)						Table II
	Temperature, °F					
	0	100	200	300	400	500
Fiberglass	0.010	0.011	0.020	0.023	0.031	0.043
Fiberglass - 850[1]	0.036	0.017	0.022	0.028	0.037	—
Magnesia (85%)	—	0.039	0.041	0.043	0.046	0.049
Phenolic foam[2,3]	—	—	—	0.018	—	—
Mineral wool[3]	—	—	—	0.034	—	—
Urethane foam	—	0.023	0.032	—	—	—

[1] CertainTeed

[2] Accotherm

[3] Limits not specified

Typical values of emissivity			Table III
Material	Temperature range, °F	Emissivity (e)	
Black body	—	1.000	
Aluminum—highly polished	440-1,070	0.039-0.057	
—oxidized	390-1,110	0.110-0.190	
Copper—polished	242	0.023	
—heat-treated at 1,110 °F	390-1,110	0.570	
Cast iron—heat-treated at 1,100 °F	390-1,110	0.640-0.780	
Steel, oxidized at 1,100 °F	390-1,110	0.790	
Nickel—polished	74	0.045	
Paints—black lacquer	76	0.875	
—white lacquer	100-200	0.800-0.950	
—oil paints (16 colors)	212	0.920-0.960	
—10%Al, 22% lacquer	212	0.520	
—26%Al, 27% lacquer	212	0.300	
Paper	66	0.924	

ness; and emissivity of the insulation's surface. In addition, the loss or gain of a bare, uninsulated pipe is calculated for use as a reference.

The program

This program (Table I) is based on the premise that the heat conducted through the insulation, Q_{cd}, must balance the heat exchanged at the outer surface of the insulation by convection, Q_{cv}, and radiation, Q_r. For simplicity, it is assumed that the temperature of the inside of the insulation is equal to the temperature of the material inside the pipe.

The external surface temperature of the insulation is the unknown dependent variable. An estimate of this temperature is made, and the heat transferred by each of the three modes is calculated. The magnitude of the

difference between Q_{cd} and $Q_{cv} + Q_r$ is determined, and is used to adjust the estimated surface temperature. This process is repeated until the figure for the heat transferred through the insulation by conduction agrees within 0.1% with the figure for the heat exchanged by convection and radiation outside the insulation.

This usually takes from three to five iterations for hot pipes, and a little more for cold ones. One of these iterations is used to calculate the heat flux of the bare, uninsulated pipe. Theoretically, requiring agreement to less than 0.1% is unwarranted, as the input data are seldom known with that degree of accuracy, and constants in the equations indicate a somewhat lower degree of precision. It is justified only in that it ensures that the calculator routine is not the limiting factor in the calculation.

User instructions for pipe insulation program				Table IV
Step	Instruction	Input Data	Keys	Output Data
1.	Clear program		f CL PRGM	
2.	Clear all storage registers		f CL REG	
3.	Enter program, either by key or by card			
4.	Store input data in primary registers:			
	Thermal conductivity, Btu/(h) (ft²) (°F) (ft)	K_m	STO 0	
	Temperature inside pipe, °F	T_1	STO 1	
	Temperature of air, °F	T_3	STO 3	
	Diameter of pipe, in.	D_1	STO 4	
	Diameter of insulation, in.	D_2	STO 5	
	Emissivity of pipe (0.79 for steel)	e_p	STO B	
	Emissivity of insulation	e_i	STO C	
5.	Run program		A	
6.	Intermediate displays—While program is running, the following information will be displayed:			
	a) Estimated surface temp. of insulation, °F			T_2
	b) ΔQ, disagreement between Q_{cd} and $Q_{cv} + Q_r$, Btu/ft			ΔQ
	c) Correction to surface temp., °F			ΔT_2
	a, b, and c will be repeated with each iteration until $\Delta Q \leqslant 0.1\% \, Q_{cd}$			
7.	At the conclusion of the program, all primary storage registers are reviewed (or printed with HP-97) to display all input and calculated data.			
8.	The calculator comes to rest with Q_{cd}, heat of conduction, in Btu/(h) (ft) in the display.			

Nomenclature

K_m Mean thermal conductivity of insulation, Btu/(h)(ft^2)(°F)(ft) (see Table II)

T_1 Temperature of material inside pipe, °F

T_2 Temperature of outside surface of insulation (to be calculated by program), °F

T_3 Temperature of ambient air, °F

D_1 Outside diameter of pipe, in.

D_2 Outside diameter of insulation, in.

e Emissivity (see Table III)

e_p Emissivity of bare pipe

e_i Emissivity of outside surface of insulation

Q_{cd} Heat conducted through insulation, per hour, per foot of length, Btu/(h)(ft)

Q_{cv} Heat exchanged at surface of insulation by convection, per hour, per foot of length, Btu/(h)(ft)

Q_r Heat exchanged at surface of insulation by radiation, per hour, per foot of length, Btu/(h)(ft)

E Exponent used in re-estimating T_2

F Correction factor used in re-estimating T_2

n Number of iterations (including one to calculate values for uninsulated pipe) required to balance Q_{cd} against $Q_{cv} + Q_r$, within 0.1%.

List of storage registers   Table V

Register	Value
0	K_m
1	T_1
2	T_2
3	T_3
4	D_1
5	D_2
6	$(Q_{cv} + Q_r)$ uninsulated
7	Q_{cd}
8	Q_{cv}
9	Q_r
10	ΔT
11	F
12-19	Unused
A	ΔQ
B	e_{pipe}
C	$e_{insulation}$
D	Unused
E	E (exponent)
I	n

At the conclusion of the program, all primary storage registers may be displayed or printed out, showing the input and calculated data.

This program has been used successfully to determine the effectiveness of various insulation systems under a variety of conditions, including both high and low temperatures. Results agree with published data.

If conditions are found for which the program does not readily converge, check successive intermediate displays of T_2, ΔQ and ΔT_2, to check the convergence. It is also possible to print out or display n, the number of iterations, to see if this is excessive.

The equations used in this program are:

$$Q_{cd} = \frac{2\pi K_m (T_1 - T_2)}{\ln(D_2/D_1)}$$

$$Q_{cv} = 0.27 \left(\frac{T_2 - T_3}{D_2/12}\right)^{0.25} \left(\frac{\pi D_2}{12}\right)(T_2 - T_3)$$

$$Q_r = 0.173e(\pi D_2/12) \times \left[\left(\frac{T_2 + 460}{100}\right)^4 - \left(\frac{T_3 + 460}{100}\right)^4\right]$$

The definitions of terms and units are shown above. Typical values for conductivity and emissivity are listed in Tables II and III.

Steam line example

A 3-in. (3.500-in. O.D.) steel-pipe steam line carries 150 psig steam at 366°F, and is insulated with 1 in. of fiberglass insulation having a conductivity of 0.028 Btu/(h)(ft^2)(°F)(ft), and a cloth cover, painted with flat paint, having an emissivity of 0.94. The steel pipe has an emissivity of 0.79. The temperature of the surrounding air is 80°F.

Following the procedures as outlined in Table IV, it is determined that the heat loss will be 97.63 Btu/(h)(ft), and the surface temperature will be 115.2°F.

If the surface is painted with 26% aluminum paint with an emissivity of 0.30, instead of the flat paint, the heat loss will be 91.55 Btu/(h)(ft) and the surface temperature will be 130.8°F. Compare this to a heat loss of 872.4 Btu/(h)(ft) for the bare, uninsulated pipe. These calculations require four and five iterations respectively, and take less than 2 min each. Table V lists registers.

It is left for the user to determine the relative benefits of each pipe insulation system, which will depend on the prevailing economics in the local area.

For TI-58/59 users

The TI version of the program appears in Table VI. User instructions, along with data and results for the first example, are found in Table VII. Running the TI version is similar to running the HP version, but different registers are used.

In entering the data, be careful to use the diameter of the insulation (D_2), not its thickness. In the first example, the insulation is 1 in. thick, and the pipe O.D. is 3.5 in. Therefore, D_2 is 3.5 + 1 + 1 = 5.5 in. Do not use 1 in. or 4.5 in.—these certainly are incorrect!

Program listing for TI version   Table VI

Step	Code	Key	Step	Code	Key	Step	Code	Key
000	76	LBL	009	43	RCL	018	59	59
001	11	A	010	05	05	019	01	1
002	00	0	011	65	×	020	42	STO
003	42	STO	012	89	π	021	11	11
004	25	25	013	55	÷	022	01	1
005	86	STF	014	01	1	023	52	EE
006	01	01	015	02	2	024	09	9
007	86	STF	016	95	=	025	42	STO
008	02	02	017	42	STO	026	10	10

Step	Code	Key	Step	Code	Key	Step	Code	Key	Step	Code	Key	Step	Code	Key
027	43	RCL	091	52	EE	155	43	RCL	219	43	RCL	283	24	24
028	01	01	092	99	PRT	156	01	01	220	11	11	284	61	GTO
029	32	X:T	093	76	LBL	157	75	-	221	65	×	285	12	B
030	43	RCL	094	12	B	158	43	RCL	222	43	RCL	286	76	LBL
031	03	03	095	02	2	159	02	02	223	10	10	287	32	X:T
032	67	EQ	096	65	×	160	95	=	224	95	=	288	22	INV
033	95	=	097	53	(	161	22	INV	225	42	STO	289	86	STF
034	77	GE	098	43	RCL	162	87	IFF	226	53	53	290	00	00
035	22	INV	099	01	01	163	00	00	227	44	SUM	291	22	INV
036	76	LBL	100	75	-	164	33	X²	228	02	02	292	52	EE
037	95	=	101	43	RCL	165	94	+/-	229	43	RCL	293	98	ADV
038	22	INV	102	02	02	166	76	LBL	230	02	02	294	43	RCL
039	87	IFF	103	54	)	167	33	X²	231	99	PRT	295	02	02
040	00	00	104	65	×	168	45	Yˣ	232	43	RCL	296	99	PRT
041	85	+	105	89	π	169	43	RCL	233	02	02	297	43	RCL
042	76	LBL	106	65	×	170	24	24	234	75	-	298	07	07
043	22	INV	107	43	RCL	171	95	=	235	43	RCL	299	99	PRT
044	86	STF	108	00	00	172	42	STO	236	03	03	300	91	R/S
045	00	00	109	55	÷	173	56	56	237	95	=	301	76	LBL
046	43	RCL	110	53	(	174	22	INV	238	42	STO	302	13	C
047	03	03	111	53	(	175	87	IFF	239	52	52	303	22	INV
048	75	-	112	43	RCL	176	00	00	240	22	INV	304	87	IFF
049	53	(	113	05	05	177	34	ΓX	241	87	IFF	305	02	02
050	53	(	114	55	÷	178	94	+/-	242	00	00	306	44	SUM
051	43	RCL	115	43	RCL	179	42	STO	243	35	1/X	307	71	SBR
052	03	03	116	04	04	180	56	56	244	94	+/-	308	45	Yˣ
053	75	-	117	54	)	181	76	LBL	245	42	STO	309	76	LBL
054	43	RCL	118	23	LNX	182	34	ΓX	246	52	52	310	44	SUM
055	01	01	119	54	)	183	43	RCL	247	76	LBL	311	43	RCL
056	54	)	120	95	=	184	20	20	248	35	1/X	312	02	02
057	45	Yˣ	121	42	STO	185	55	÷	249	00	0	313	75	-
058	93	.	122	07	07	186	43	RCL	250	32	X:T	314	43	RCL
059	05	5	123	22	INV	187	07	07	251	43	RCL	315	03	03
060	05	5	124	87	IFF	188	65	×	252	52	52	316	95	=
061	54	)	125	00	00	189	53	(	253	67	EQ	317	22	INV
062	95	=	126	25	CLR	190	43	RCL	254	42	STO	318	87	IFF
063	42	STO	127	50	I×I	191	04	04	255	77	GE	319	00	00
064	02	02	128	42	STO	192	45	Yˣ	256	12	B	320	52	EE
065	61	GTO	129	07	07	193	43	RCL	257	76	LBL	321	94	+/-
066	23	LNX	130	76	LBL	194	24	24	258	42	STO	322	76	LBL
067	76	LBL	131	25	CLR	195	54	)	259	01	1	323	52	EE
068	85	+	132	71	SBR	196	55	÷	260	85	+	324	65	×
069	43	RCL	133	13	C	197	43	RCL	261	43	RCL	325	01	1
070	03	03	134	43	RCL	198	05	05	262	03	03	326	02	2
071	85	+	135	07	07	199	65	×	263	95	=	327	55	÷
072	53	(	136	75	-	200	43	RCL	264	42	STO	328	43	RCL
073	53	(	137	43	RCL	201	56	56	265	02	02	329	05	05
074	43	RCL	138	58	58	202	95	=	266	87	IFF	330	95	=
075	01	01	139	95	=	203	42	STO	267	00	00	331	45	Yˣ
076	75	-	140	42	STO	204	55	55	268	43	RCL	332	93	.
077	43	RCL	141	20	20	205	43	RCL	269	43	RCL	333	02	2
078	03	03	142	50	I×I	206	10	10	270	03	03	334	05	5
079	54	)	143	32	X:T	207	42	STO	271	75	-	335	95	=
080	45	Yˣ	144	43	RCL	208	54	54	272	01	1	336	22	INV
081	93	.	145	07	07	209	43	RCL	273	95	=	337	87	IFF
082	05	5	146	65	×	210	55	55	274	42	STO	338	00	00
083	05	5	147	93	.	211	42	STO	275	02	02	339	53	(
084	54	)	148	00	0	212	10	10	276	76	LBL	340	94	+/-
085	95	=	149	00	0	213	55	÷	277	43	RCL	341	76	LBL
086	42	STO	150	01	1	214	43	RCL	278	99	PRT	342	53	(
087	02	02	151	95	=	215	54	54	279	93	.	343	65	×
088	76	LBL	152	50	I×I	216	95	=	280	01	1	344	93	.
089	23	LNX	153	77	GE	217	44	SUM	281	22	INV	345	02	2
090	22	INV	154	32	X:T	218	11	11	282	44	SUM	346	07	7

(Continued) Table VI

Step	Code	Key	Step	Code	Key	Step	Code	Key	Step	Code	Key	Step	Code	Key
347	65	×	373	61	GTO	399	42	STO	425	04	4	451	43	RCL
348	43	RCL	374	42	STO	400	43	43	426	06	6	452	04	04
349	59	59	375	46	46	401	22	INV	427	00	0	453	48	EXC
350	65	×	376	43	RCL	402	87	IFF	428	54	)	454	05	05
351	53	(	377	03	03	403	00	00	429	55	÷	455	42	STO
352	43	RCL	378	42	STO	404	15	E	430	01	1	456	04	04
353	02	02	379	47	47	405	50	I×I	431	00	0	457	43	RCL
354	75	−	380	71	SBR	406	76	LBL	432	00	0	458	21	21
355	43	RCL	381	61	GTO	407	15	E	433	54	)	459	48	EXC
356	03	03	382	94	+/−	408	85	+	434	45	Y×	460	22	22
357	54	)	383	85	+	409	43	RCL	435	04	4	461	42	STO
358	95	=	384	43	RCL	410	08	08	436	95	=	462	21	21
359	22	INV	385	46	46	411	95	=	437	42	STO	463	92	RTN
360	87	IFF	386	95	=	412	42	STO	438	45	45	464	76	LBL
361	00	00	387	65	×	413	58	58	439	92	RTN	465	10	E'
362	71	SBR	388	43	RCL	414	87	IFF	440	76	LBL	466	43	RCL
363	50	I×I	389	22	22	415	01	01	441	45	Y×	467	43	43
364	76	LBL	390	65	×	416	10	E'	442	22	INV	468	42	STO
365	71	SBR	391	43	RCL	417	92	RTN	443	86	STF	469	06	06
366	42	STO	392	59	59	418	76	LBL	444	02	02	470	71	SBR
367	08	08	393	65	×	419	61	GTO	445	43	RCL	471	45	Y×
368	43	RCL	394	93	.	420	53	(	446	01	01	472	22	INV
369	02	02	395	01	1	421	53	(	447	48	EXC	473	86	STF
370	42	STO	396	07	7	422	43	RCL	448	02	02	474	01	01
371	47	47	397	03	3	423	47	47	449	42	STO	475	61	GTO
372	71	SBR	398	95	=	424	85	+	450	01	01	476	13	C

User instructions and example for TI version

Table VII

Step	Instruction	Input data	Example	Key	Output
1.	Enter program by key or card				
2.	Store input data:				
	Thermal conductivity, Btu/(h)(ft^2)(°F)(ft)	K_m	0.028	STO 00	
	Temperature inside pipe, °F	T_1	366	STO 01	
	Temperature of air, °F	T_3	80	STO 03	
	Diameter of pipe, in.	D_1	3.5	STO 04	
	Diameter of insulation, in.	D_2	5.5	STO 05	
	Emissivity of pipe	e_p	0.79	STO 21	
	Emissivity of insulation	e_i	0.94	STO 22	
	Calculation factor	—	0.7	STO 24	
3.	Run program			A	
4.	Intermediate printed output is surface-temperature estimates				T_2
5.	Final printout is:				
	Final surface temperature, °F	T_2			115.2
	Heat loss, Btu/(h)(ft)	Q			97.63

The output tape for the above example is:

```
102.4388870
111.7797835
115.0801308
115.1713215

115.1713315
97.63171797
```

References

1. Brown, G. G., et al., "Unit Operations," John Wiley & Sons, Inc., New York, 1950, pp. 427, 444, 459, 460 & 584.
2. CertainTeed Corp. Bulletin, "850° Snap-on Fiberglass Pipe Insulation," Mar. 1978.
3. Armstrong Cork Co., "Accotherm Pipe Insulation," Bulletin 15P, Nov. 1977.

The author

Frank S. Schroder is an Engineering Specialist in the Central Research Engineering Department, 3M Center, PO Box 33221, St. Paul, MN 55133; tel: 612-733-1657. He received a B.S. degree in chemical engineering from Michigan Technological University, Houghton, Mich. and previously worked for Union Carbide, Silicones Div.

Insulation without economics

To calculate the economic thickness of insulation, you must predict such items as future interest rates and fuel costs. It may be more reasonable to calculate an "acceptable heat-loss" thickness.

M. McChesney *and* P. McChesney, *Fuel Save Associates*

☐ Thermal insulation should be the simplest, most generally accepted, cost-effective method of saving energy immediately available to a plant owner. However this clearly is not the case to many—perhaps too many—owners, because to them it is just another capital investment; as such its purchase must ensure a return on the investment.

Their argument is simple and direct: if the plant owner can make a 17% profit on his salable products whereas insulation returns only, say, 12%, then surely it makes no economic sense to buy the insulation—since the more profitable thing to do is to expand plant capacity to produce more salable products.

Having reached this conclusion, the plant owner then can salve his conscience by reducing energy wastage solely by "good housekeeping" involving minimal or even no capital expenditure. After all, there is no shortage of advice and guidance available on how to save 10% of plant energy bills without spending *any* money!

Economic thickness of insulation (ETI)

Because of this "economic" attitude toward insulation, it became necessary to evaluate how much was actually "economic," and this gave rise to the concept of the "economic thickness of insulation." This concept has been extensively discussed on both sides of the Atlantic, being analyzed in the technical literature over the years. There are now lengthy books and manuals available that show the (sometimes bemused) plant owner how to calculate this economic thickness by using tables, graphs or computers. All the plant engineer needs to do. is to input data into the tables, etc., and out comes the "economic thickness."

In the U.S.A., the history of the tables highlights their problems. In 1949 a committee was established by Union Carbide Corp. and West Virginia University to reduce the economic-thickness calculation to a simple procedure in which, following the McMillan analysis [4], the heat factors were separated from the cost factors. A manual was produced by a forerunner of TIMA (Thermal Insulation Manufacturers Assn.), with nomographs and charts; TIMA itself apparently produced its own version around 1960.

Since that time the data and their presentation underwent modification and, around 1973, programs called ECON-I (tables, charts, worked examples),

Originally published May 3, 1982

ECON-II (marketing manual) and R-ECON (retrofitting) became available. However it was felt by some that these programs had averaged too many variables to nominal values and, in addition, the minimum insulation depreciation (amortization) period was too long. This last objection is important since it showed that plant owners were *not* prepared to regard insulation as a long-term investment—an attitude that is still common today.

Insulation, if properly maintained, can have a very long lifetime (30 years or more) but the pipework around which it is placed may be part of a plant that rapidly becomes obsolescent. Since in many cases insulation is not recovered, but scrapped, it makes no sense to amortize it over a longer period of time.

Accordingly, around 1976 a refinement of these tables was produced by York Research Corp. for the Federal Energy Administration and called ETI (Economic Thickness for Industrial Insulation). It consisted of 10 sections covering fuel costs (which had become a major consideration since the 1974 OPEC price rise), insulation costs, condensation control and retrofitting. All the information was given in tabular, graphical or nomographical form and also in a mathematical appendix.

However, this apparently was not suitable for all needs, and from 1976 to 1980 numerous suggestions were made to make the ETI program more useful. A recent revision has been made incorporating these suggestions, as a joint venture between TIMA, NICA (National Insulation Contractors Assn.) and Louisiana Technical University, the resulting manuals being called ETIH (Economical Thickness of Insulation for Hot Surfaces) and ETIC (Cold Surfaces).

Difficulties with calculating ETI

Over the years, the senior writer has used these tables and watched the brave attempts to present the calculational procedure in a digestible form without sacrificing too much accuracy. Undeniably the use of discounted-cash-flow analysis does complicate the calculation, and the incorporation of future fuel-price increases adds further difficulty. Possibly because of the complexity of these economic factors (but this is only a guess), in the U.K. the Department of Energy produced a slim book of graphs showing the heat loss from unit lengths of pipes of various diameters carrying various thicknesses of different types of insulation and giving a calculational procedure for obtaining the economic thickness of insulation.

However, the book makes no allowance whatever for either discounted cash flows or future fuel-price increases. These graphs have the merit of simplicity, but the writers have found them inaccurate because they are plotted on log-log scales.

In this article we use several of the commonly adopted methods for calculating the economic thickness, and show that a real dilemma arises in trying to decide what is the actual thickness that is *economic*. In

fact, we wonder whether there is such a quantity at all!

Put another way, we have a sneaking suspicion that the "economic" thickness of insulation is more or less what any plant engineer wants it to be. The reason is twofold: there are inevitably uncertainties attached to the heat-input data, but these can be controlled to a considerably greater extent than the uncertainties (or, put more euphemistically, the greater range of choice) in the economic-data inputs. Thereafter we propose a simple criterion for determining the *sensible* thickness of insulation (not economic thickness, since economics is taken right out of the problem altogether, at least in explicit form), that puts the decision-making back where it really belongs—in the hands of the plant engineer and not of the company accountant!

A full analysis of insulation economics requires that at least 20 input-data variables be assigned; these can be grouped under four headings for the case of pipe insulation.

Insulation factors

1. Cost of installed insulation, of thickness t, per linear foot $C_{I(t)}$.
2. Thermal conductivity of insulation, k_I.
3. Thermal resistance of insulation surface, R_S.
4. Pipe diameter (nominal), d_1.
5. Ambient temperature, θ_3.
6. Ambient wind speed.
7. Pipe temperature, θ_1.
8. Amortization period of insulation, n.
9. Pipe-complexity factor.
10. Maintenance and insurance costs.

For simplicity we shall assume a pipe-complexity factor of unity, ignore maintenance and insurance costs, and assume zero wind speed.

Fuel factors

11. Type of fuel and cost, C_F.
12. Expected annual price rise of fuel expressed as a decimal, f.

Heat-producing-plant factors

13. Efficiency of conversion of fuel to heat, E.
14. Number of hours of operation per year, N.
15. Capital investment in heat-producing plant.
16. Amortization period of heat-producing plant.

For simplicity we shall completely ignore the economic aspects of the heat-producing plant. Normally the existence of insulation can reduce the size of the needed heat-producing plant, and this represents an incremental positive cash flow. We make the simplifying assumption that the decision to insulate does not affect the plant capacity—as would be the case for insulation retrofit.

Economic factors

17. Cost of money, i.e., return on investment in insulation required, i.
18. Tax rate.
19. Cost of money to finance heat-producing plant.

Nomenclature

A_I	External surface area of insulation jacketing, ft^2
A_P	External surface area of bare pipe, ft^2
C_F	Cost of fuel, 2.83×10^{-6}, $ per Btu
$C_{H(t)}$	Total cost of heat loss per linear foot of pipe covered with insulation of thickness t, for an entire operational year, $ ft^{-1} yr^{-1} for money valued at this moment of time $$= \frac{NC_F}{E}\frac{Q_{(t)}}{L}$$
$C_{I(t)}$	Cost of installed insulation per linear foot of pipe, of thickness t_I inches, $ ft^{-1} for money valued at this moment of time
d_1	Nominal diameter of pipe, ft
d_2	Nominal outer diameter of insulation jacketing, ft
E	Efficiency of conversion of fuel to heat expressed as a decimal, 0.83
f	Annual increase in the cost of fuel expressed as a decimal, 0.1
F	Heat-loss rate per ft^2 of insulated pipe divided by heat-loss rate per ft^2 of same pipe bare of insulation
h_C	Convective heat-transfer coefficient of insulation surface jacketing for still air, Btu h^{-1} ft^{-2} °F^{-1}
h_I	Total surface heat-transfer coefficient of insulation surface jacketing for still air, Btu h^{-1} ft^{-2} °F^{-1}
h_P	Total surface heat-transfer coefficient of bare pipe for still air, Btu h^{-1} ft^{-2} °F^{-1}
h_R	Radiative heat-transfer coefficient of insulation surface jacketing for still air, Btu h^{-1} ft^{-2} °F^{-1}
i	Cost of money; return on investment required expressed as a decimal
k_I	Thermal conductivity of the pipe insulation, Btu ft h^{-1} ft^{-2} °F^{-1}
L	Length of pipe, ft
n	Amortization period of the insulation (number of years over which the insulation economics is to be evaluated), years
N	Number of hours of operation of the fuel-to-heat conversion equipment per year, 8,760 h
$Q_{(t)}$	Rate of heat loss from pipe covered with insulation of thickness t_I inches, Btu h^{-1}
r_1	Nominal radius of the pipe, ft
r_2	Nominal radius of insulation plus jacketing, ft
R_I	Thermal resistance of the insulation and its jacketing, h ft^2 °F (Btu)$^{-1}$
R_P	Thermal resistance of the bare pipe surface, h ft^2 °F (Btu)$^{-1}$
R_S	Thermal resistance of the insulation jacket surface, h ft^2 °F (Btu)$^{-1}$
t	Thickness of insulation and jacketing, in.
[USPWF]	Uniform-series-present-worth factor, $$\frac{(1+i)^n - 1}{i(1+i)^n}$$
ε	Surface emissivity of the insulation surface jacketing or of the bare pipe, 0.9
θ_1	Temperature of steam in pipe: also temperature of pipe wall, °F
θ_2	Temperature of insulation surface jacketing, °F
θ_3	Temperature of still air (70°F)
σ	Stefan's radiation constant, 0.171×10^{-8} Btu h^{-1} ft^{-2} °R^{-4}

20. Economic model used for determining the economic thickness of insulation.

It must be stressed that the simplifications that we make in no way detract from the overall conclusions drawn. What they do is avoid obscuring the important issues in a morass of arithmetic—they alter number values but not decisions.

The heat-loss equations

Referring to Fig. 1, consider a horizontal pipe of nominal diameter d_1, covered with insulation of thickness t, and of thermal conductivity k_I, carrying dry saturated steam at temperature θ_1. Even with insulation present, the pipe will lose heat, and in the steady state, when the surface temperature of the insulation does not change, the rate at which heat is lost per linear foot of the pipe is:

$$\frac{Q_{(t)}}{L} = \frac{\theta_1 - \theta_3}{\frac{1}{2\pi k_I}\ln(d_2/d_1) + \frac{1}{\pi d_2 h_I}} \quad (1)$$

Strictly, this equation is a simplification because:
■ It assumes that the convective heat-transfer coefficient from steam to the inner wall of the pipe is infinitely large compared with h_I, the sum of the convective and radiative heat-transfer coefficients from the insulation surface to the ambient air. This is an excellent approximation because any heat loss from the steam will cause it to condense, and the condensation heat-transfer coefficient (for film-wise condensation) lies in the range 1,000–2,000 Btu hr^{-1} ft^{-2} °F^{-1}, which is about one thousand times greater than h_I.

■ It ignores the thermal resistance of the pipe wall, which is usually an excellent approximation in most heat-transfer cases other than thick-walled vats.

In Eq. (1) the value of h_I is given by:

$$h_I = h_C + h_R$$

where:

$$h_C = 0.270(\theta_2 - \theta_3)^{0.25} d_2^{-0.25} \quad (2)$$
$$h_R = \varepsilon\sigma[(\theta_2 + 460) + (\theta_3 + 460)] \times [(\theta_2 + 460)^2 + (\theta_3 + 460)^2] \quad (3)$$

Clearly, to evaluate Eq. (2) and (3) we need to know the value of the insulation surface temperature θ_2. The calculation and importance of θ_2 have been described

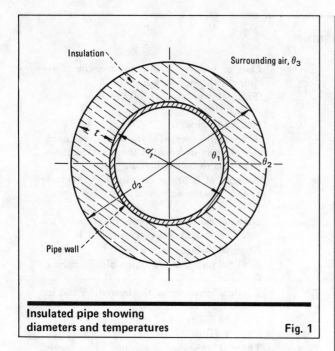

Insulation

Surrounding air, θ_3

t

d_1

θ_1

θ_2

d_2

Pipe wall

Insulated pipe showing diameters and temperatures   **Fig. 1**

Thickness of calcium silicate insulation, t, in.	Surface temperature of insulation θ_2, °F	Heat loss rate per linear foot $Q_{(t)}/L$ Btu h^{-1}ft^{-1}	Dollar value of heat lost per linear foot per year of operation $C_{H(t)}$ ($) ft^{-1}year^{-1}	Installed cost of insulation per linear foot $C_{I(t)}$ ($)ft^{-1}
0	500	3,708	111	–
½	210	756	22.7	2.0
1	160	479	14.5	4.1
1½	136	358	10.7	6.2
2	122	292	8.74	8.4
2½	113	249	7.47	10.5
3	107	220	6.58	12.6
3½	102	198	5.93	14.7
4	99	181	5.42	16.8
4½	95	167	5.02	18.9
5	92	156	4.69	21.0
5½	90	147	4.42	23.1
6	89	140	4.18	25.2

500° F, 8-in.-dia. horizontal pipe with calcium silicate insulation   Table I

by McChesney and McChesney, p. 127, [1]. Using the methods described in that article, the writers have solved Eq. (1), (2), (3) for the specific case of a straight horizontal pipe of nominal 8-in. dia., carrying dry saturated steam at 500°F, located in still air, lagged with various thicknesses of calcium silicate insulation. In addition, the boiler is assumed to be 83% efficient, operating for a full 8,760 hours per year, using fuel costing $2.84 per million Btu.

Results are shown in Table I, in which the installed cost of insulation is an acceptable simplification. Installed cost is much discussed in the technical literature, and different methods suggested for determining it. The writers have examined manufacturers' quoted costs for an 8-in. pipe with rigid calcium silicate and noted that the cost per linear foot of the actual insulating material is essentially linear with increasing thickness. However, when the cost of layering, jacketing and installation is added, this linearity disappears and a "kinked" straight line is obtained. The values used in Table I are simply those from the "best" straight line through the actual installed costs. Once again, this simplification only affects the numbers calculated from the economic models but not the conclusions drawn from them.

In the article already referred to, [1], we have shown that it is desirable purely from the safety-to-personnel point of view to ensure that the insulation surface should be either a canvas jacket (if indoors) or a metal jacket with a very thick layer of high-temperature-resistant matte ("flat") oil-bound paint; either of these will give the high radiation surface emissivity used in this article (0.90). In addition, the insulation surface-temperature should not exceed 140°F for canvas or 113°F for metal, to prevent skin-burn if the surface is touched. This means that the safety thickness of insulation—and this is the minimum that *must* be fitted to our 8-in. pipe—is 1½ in. (canvas cover) or 2½ in. (metal jacket).

In some of the economic models we use discounted cash flows—specifically the uniform-series present-worth factor:

$$[\text{USPWF}] = \frac{(1 + i)^n - 1}{i(1 + i)^n} \qquad (4)$$

This factor is used to discount—to the present time—a uniform series of expenses or receipts occurring annually for n years into the future. It is often referred to as the Net Present Value of future cash flows, and is sufficiently well known not to require further explanation.

What is less well known is how to incorporate fuel price increases into discounted cash flows, i.e., to allow for both the discount factor for money and the increase in the price of fuel over the amortization period. We have already defined and evaluated in Table I the quantity $C_{H(t)}$, which is the dollar value of the heat lost per linear foot per year of operation. This annual cost increases as fuel costs increase, by a multiplication factor $(1 + f)$ for the first year, but it must be discounted by an interest factor $(1 + i)$. For succeeding years we write Present Value of the total dollar value of the heat lost per linear foot for the n years as:

Present value

$$= C_{H(t)} \left[\left(\frac{1 + f}{1 + i}\right) + \left(\frac{1 + f}{1 + i}\right)^2 + \cdots \left(\frac{1 + f}{1 + i}\right)^n \right]$$

$$= C_{H(t)} \left[\frac{\left(\frac{1 + f}{1 + i}\right)^{n+1} - \left(\frac{1 + f}{1 + i}\right)}{\left(\frac{1 + f}{1 + i}\right) - 1} \right] \qquad (5)$$

since the series is a simple geometrical progression of n terms. We have assumed that the value of f is constant, so that the quantity in the square brackets in Eq. (5) is a modified [USPWF] uniform series present-worth factor that allows for a uniform increase in the price of fuel over the lifetime of the insulation.

In addition there are two quite distinct methods of

Typical calculations for Economic Model 1   Table II

Thickness of calcium silicate insulation, t, in.	Value of $n\,C_{H(t)} + C_{I(t)}$	
	$n = 2$ years	$n = 5$ years
½	47.29	115.23
1	32.79	75.83
1½	27.67	59.87
2	25.78	52.00
2½	25.33*	47.74
3	25.66	45.40
3½	26.45	44.23
4	27.54	43.80*
4½	28.83	43.59
5	30.28	44.39
5½	31.83	45.07
6	33.47	46.01

*The minimum value in each column is the ETI.

ETI as a function of amortization period   Table III

Model 1		Model 2	
Insulation amortization period, n years	Economic insulation thickness, t, inches	Insulation thickness fitted, t, inches	Amortization period, n years
2	2½	2	1.30
3	3	2½	1.80
4	3½	3	2.31
5	4	3½	2.84
6	4½	4	3.38
7	5	4½	3.93
		5	4.62
		5½	5.03
		6	5.59

Typical calculations for Economic Models 3 and 4   Table IV

	Model 3 Value of		Model 4 Value of	
	$[\text{USPWF}]\,C_{H(t)} - C_{I(t)}$		$[\text{USPWF}] \times [C_{H(t_a)} - C_{H(t_b)}] - [C_{I(t_b)} - C_{I(t_a)}]$	
Insulation thickness inches, t,	$n = 2$ years, $i = 8\%$	$n = 5$ years, $i = 12\%$	$n = 2$ years, $i = 8\%$	$n = 5$ years, $i = 12\%$
½	38.4	79.6	12.7	27.8
1	21.5	47.6	4.34	10.9
1½	12.9	32.5	1.45	5.09
2	7.26	23.2	0.169*	2.49
2½	2.92*	14.4	−0.519	1.10
3	−0.765	11.2		0.262*
3½		2.84*		−0.278
4		−0.714		
4½	All negative		All negative	All negative
5		All negative		
5½				
6				

*The ETI is given by the smallest positive number in each column.

looking at increasing insulation thickness, these being the nonincremental method and the incremental method, the differences between which should be obvious from the worked examples for the various economic models used.

Various economic models

Model 1 for economic thickness—In this method a table is drawn up showing the total cost of the heat lost for the entire lifetime of the insulation, added to the cost of the insulation (both per linear foot), for increasing thicknesses of insulation and the minimum value found by inspection.

Algebraically this is expressed as finding the minimum in the quantity:

$$n\,C_{H(t)} + C_{I(t)}$$

Table II shows typical calculations, while Table III, Columns 1 and 2 summarize the minimum values obtained as a function of n, the chosen amortization period of the insulation.

Model 2—If it is felt that an incremental method would be more accurate, i.e., that each additional thickness (increment) of insulation must pay for itself in terms of fuel-cost savings, then the criterion becomes simply:

$$n[C_{H(t_a)} - C_{H(t_b)}] = C_{I(t_b)} - C_{I(t_a)}$$

where t_a and t_b are any two thicknesses of insulation commercially available (not necessarily consecutive). Table III, in Columns 3 and 4, shows values of the amortization period obtained from this equation for given insulation thicknesses.

Model 3—Neither of the above methods allows for the time value of money (discounting the cash flows) nor the inevitable increase in the price of fuel. If the time value of money only is considered, then in a nonincremental analysis, although each thickness of insulation will result in a fixed amount of heat lost per linear foot, the dollar value of this heat lost will vary over the amortization period of the insulation because of the varying cost of money. The criterion in this nonincremental case now becomes:

$$[\text{USPWF}]\,C_{H(t)} - C_{I(t)} > 0$$

In evaluating this inequality we must decide not only on the length of the amortization period but also on the value of money. We consider the two cases for which i has the values 8% and 12%.

Table IV, in Columns 2 and 3, shows typical calculations while Table V, Columns 2 and 3, summarizes the economic thicknesses resulting from this method of analysis.

Model 4—The equivalent incremental method of analysis is obtained from the inequality:

$$[C_{H(t_a)} - C_{H(t_b)}][\text{USPWF}] - [C_{I(t_b)} - C_{I(t_a)}] > 0$$

Table IV, Columns 4 and 5, shows typical calculations while Table V, Columns 4 and 5, summarizes the economic thicknesses resulting from this method of economic analysis.

The final step is to include the factor given in Eq. (5) for the increase in the cost of fuel.

Model 5—In the nonincremental case, the economic thickness of insulation is given by the inequality:

$$\frac{\left(\dfrac{1+f}{1+i}\right)^{n+1} - \left(\dfrac{1+f}{1+i}\right)}{\left(\dfrac{1+f}{1+i}\right) - 1}\, C_{H(t)} - C_{I(t)} > 0$$

Table VI, in Columns 2 and 3, shows typical calculations while Table V, Columns 6 and 7, shows economic thicknesses resulting from this method of analysis. It is clear that in addition to having to specify the amortization period of the insulation and the value of money, we also need to specify the annual increase in the cost of fuel—which, if anything, is even more imponderable than the value of money. Values of f have changed in a fluctuating manner since 1974, and there is much in the economic literature about these fluctuations. We shall assume a value of 10%—with little justification for doing so!

Model 6—The final method is to make an equivalent incremental analysis allowing for the fuel price increase, and this is given by the inequality:

$$\left[\frac{\left(\dfrac{1+f}{1+i}\right)^{n+1} - \left(\dfrac{1+f}{1+i}\right)}{\left(\dfrac{1+f}{1+i}\right) - 1}\right][C_{H(t_a)} - C_{H(t_b)}] -$$

$$[C_{I(t_b)} - C_{I(t_a)}] > 0$$

Table VI, Columns 4 and 5, shows typical calculations, and Table V, Columns 8 and 9, shows economic thicknesses resulting from this method of analysis.

ETI calculations and models

If we examine Tables III and V, we see how dependent is the economic thickness on the choice of the economic factors n, f and i. For example, as n varies from 2 to 7 years, then several of the models show that the economic thickness varies from $2\frac{1}{2}$ to $4\frac{1}{2}$ in. The difference in cost between those two thicknesses is $8.40 per foot. Also considering an amortization period of 5 years, then the economic thicknesses are:

$3\frac{1}{2}$ in. from Model 4 ($i = 8\%$, $f = 0$)
4 in. from Model 1
$4\frac{1}{2}$ in. from Model 3 ($i = 8\%$, $f = 0$)
$4\frac{1}{2}$ in. from Model 6 ($i = 8\%$, $f = 10\%$)
$5\frac{1}{2}$ in. from Model 5 ($i = 8\%$, $f = 10\%$)

The difference in cost between these two extremes is again $8.40 per foot and these cost differences will represent very substantial sums of money for several hundred feet of insulation.

In addition it must be remembered that the values of i and f used here are no more than "reasonable" values chosen from a range of "equally reasonable" values, all of which would give different thicknesses of insulation. The amortization period has been deliberately restricted to a maximum of seven years and it might be thought that in view of the durability of insulation and its cheap maintenance (usually taken as an annual cost of about 1% of the original capital investment) a longer period should be chosen. As far as this article is con-

A comparison of ETI as a function of amortization period given by Economic Models 3, 4, 5, and 6   Table V

Insulation amortization period n years	Economic thickness of insulation, t, in.							
	Model 3 Nonincremental		Model 4 Incremental		Model 5 Nonincremental		Model 6 Incremental	
	$i = 8\%$, $f = 0$	$i = 12\%$, $f = 0$	$i = 8\%$, $f = 0$	$i = 12\%$, $f = 0$	$i = 8\%$, $f = 10\%$	$i = 12\%$, $f = 10\%$	$i = 8\%$, $f = 10\%$	$i = 12\%$, $f = 10\%$
2	$2\frac{1}{2}$	$2\frac{1}{2}$	$2\frac{1}{2}$	$2\frac{1}{2}$	3	3	$2\frac{1}{2}$	$2\frac{1}{2}$
3	$3\frac{1}{2}$	3	3	3	4	$3\frac{1}{2}$	3	3
4	4	$3\frac{1}{2}$	3	$3\frac{1}{2}$	$4\frac{1}{2}$	$4\frac{1}{2}$	4	$3\frac{1}{2}$
5	$4\frac{1}{2}$	4	$3\frac{1}{2}$	$3\frac{1}{2}$	$5\frac{1}{2}$	5	$4\frac{1}{2}$	4
6	$4\frac{1}{2}$	4	4	4	$5\frac{1}{2}$	5	$4\frac{1}{2}$	4
7	5	$4\frac{1}{2}$	$4\frac{1}{2}$	4	>6	$5\frac{1}{2}$	5	$4\frac{1}{2}$

Typical calculations for Economic Models 5 and 6   Table VI

Insulation thickness, t, in.	Model 5 Value of $B[C_{H(t)}] - C_{I(t)}$ †		Model 6 Value of $B[C_{H(t_a)} - C_{H(t_b)}] - [C_{I(t_b)} - C_{I(t_a)}]$ †	
	$n = 2$ years, $i = 8\%$	$n = 5$ years, $i = 12\%$	$n = 2$ years, $i = 8\%$	$n = 5$ years, $i = 12\%$
$\frac{1}{2}$	44.6	105	15.0	37.2
1	25.4	63.9	5.33	15.0
$1\frac{1}{2}$	15.8	44.6	2.00 ·	7.35
2	9.57	33.0	0.516*	3.93
$2\frac{1}{2}$	4.85	24.9	−0.277	2.10
3	0.929*	18.6		1.01
$3\frac{1}{2}$	−2.52	13.4		0.295*
4	↑	8.88	↑	−0.193
$4\frac{1}{2}$	All negative	4.87	All negativr	↑
5		1.22*		All negative
$5\frac{1}{2}$	↓	−2.18		↓
6		−5.38		

*The ETI is given by the smallest positive number in each column.

†Where: $B = \left[\dfrac{\left(\dfrac{1+f}{1+i}\right)^{n+1} - \left(\dfrac{1+f}{1+i}\right)}{\left(\dfrac{1+f}{1+i}\right) - 1}\right]$

cerned, the reason for short amortization periods is merely to ensure that the insulation thicknesses would lie within the usual single-layer and likely double-layer costing.

However, there is the more fundamental reason already referred to at the beginning of this article—many plant owners treat insulation as just another investment to be judged by the same criteria as all other plant investments. In addition they argue that quite often energy costs are less than 10% of total manufacturing costs and if there is any money to spare it should go to reducing the remaining 90% of costs. But, there are fallacies in this point of view.

Dollar savings resulting from insulation are totally predictable and without risk; they are guaranteed, whereas investment in more plant or new products is certainly not so predictable.

Energy savings through insulation can directly release cash flows to be spent where they can do most good, which is certainly not in heating the local atmosphere and drains! Every unit of heat saved is a direct addition to income and thus available for investment.

Although energy costs may be only 10% or less of total manufacturing costs, they represent a much higher proportion of controllable costs (actual production costs), unlike costs of buildings, capital and labor.

If this reasoning is accepted, then plant owners should cease to treat insulation as just another investment judged on a simple Payback Period or Return on Investment (ROI) basis, since neither of these gives credit for dollar savings that will continue to accrue long after the insulation has paid for itself. The only way they can do this is to use the discount methods shown here, provided—and this is an important proviso—they know their economic factors and prefer a specific economic model. After all, a reasonable question to ask is "which economic model is right?"

In fact, there is no single answer to this question, since all the models are "right"—being no more than different methods of mathematically modeling the economic influence of money and time. Economists do argue about the accuracy of mathematical models and there seems little consensus among them; this is one reason why we turned back to purely engineering considerations in proposing alternative criteria for assessment of insulation.

The purpose of this article is not to undermine confidence in the computer programs that exist for evaluating the economic thickness, because these programs are a great help if the economic factors are known (a point we return to at the end of this article). However it must be admitted that this is not always the case.

It has been the personal experience of the senior writer that in some cases the plant owners or their chief engineers have been undecided as to the exact value of n and i to use and therefore have tried a range of "reasonable" values. They inevitably ended up with a "reasonable" but confusing range of "economic" thicknesses and were undecided how to determine which one to use—as exemplified in Table V.

In addition, there are other uncertainties, such as the rising cost of insulant material and increasing labor costs, both of which will result in higher installed costs. Also in many new-plant (as opposed to retrofit) situations there will be a variety of pipe sizes, lengths, bends, flanges and valves, all of which must be allowed for. If the fluid temperatures are very high, then costly multi-layered insulation is required.

Finally, if an outside insulation contractor is employed, he will likely give a lump-sum cost estimate for the entire insulation work, leaving the plant engineer with little hope of checking its reasonableness by using computer methods.

In fact, the senior writer has had the benefit of frank talks with insulation contractors who, when "put in the hot seat," admit that their costing procedures do not go by the book, because they do not have the staff or time to use the computer programs for every pipe length or flange. One remarked that the time and cost of using the programs "would be commercial suicide"!

Insulation without economics

How, then, can a decision be made on the appropriate insulation thickness to use if the economic factors are not known accurately? One very direct way of resolving this dilemma is to remove all these economic factors from the problem. This is quite possible because it has long been known that the heat-loss factors and the economic factors appear separately in the computer programs; so that if the economic factors are thrown out, then the insulation thickness is decided by heat-loss considerations alone. The result then becomes not an "economic" thickness of insulation, but what could very well be termed an "acceptable heat-loss" thickness of insulation.

The acceptable heat-loss thickness

It should be recalled that although pipe insulation is sold per linear foot, the heat loss from an insulated (or bare) pipe is a *surface-area effect*, measured in Btu h^{-1} ft^{-2}.

It is easy to show that Eq. (1) when written on this surface-area basis becomes:

$$\frac{Q_{(t)}}{A_I} = \frac{\theta_1 - \theta_3}{\frac{d_2}{2k_I}\ln\left(\frac{d_2}{d_1}\right) + \frac{1}{h_I}} \qquad (6)$$

$$= \frac{\theta_1 - \theta_3}{R_I + R_S} \qquad (7)$$

while the heat loss per unit area from a bare pipe (zero insulation thickness) is:

$$\frac{Q_{(0)}}{A_P} = \frac{\theta_1 - \theta_3}{1/h_P} \qquad (8)$$

$$= \frac{\theta_1 - \theta_3}{R_P} \qquad (9)$$

Let us propose a criterion:

$$\frac{Q_{(t)}}{A_I} = F\left(\frac{Q_{(0)}}{A_P}\right) \qquad (10)$$

Eq. (10) is the algebraic way of stating that the actual heat loss per square foot from the insulated pipe is F percent (where F is expressed as a decimal) of the heat loss from the same pipe when bare of insulation. On this basis Eq. (10) becomes:

$$\frac{1}{R_I + R_S} = \frac{F}{R_P}$$

or:

$$R_I = \frac{R_P}{F} - R_S$$

or:

$$\frac{d_2}{2k_I}\ln\left(\frac{d_2}{d_1}\right) = \frac{R_P}{F} - R_S$$

Whence, replacing diameters by radii:

$$r_2 \ln\left(\frac{r_2}{r_1}\right) = \frac{k_I R_P}{F} - k_I R_S \qquad (11)$$

The writers have made a numerical study of the right-hand side of Eq. (11) and shown that (in all but the most unusual circumstances):

$$\frac{R_P}{F} \gg R_S$$

so that we have the accurate approximation equation

Thickness of calcium silicate insulation, t, in.	Surface temperature of insulation, θ_2, °F	Heat-loss rate per unit area, $Q_{(t)}/A$, Btu h^{-1}ft^{-2}	Heat-loss rate per unit area, compared with that from a bare pipe, %
0	500	1,770	100
½	210	321	18.1
1	160	183	10.3
1½	136	124	7.0
2	122	92.8	5.2
2½	113	73.2	4.1
3	107	59.9	3.4
3½	102	50.3	2.8
4	99	43.2	2.4
4½	95	37.6	2.1
5	92	33.2	1.9
5½	90	29.6	1.7
6	89	26.7	1.5

500°F, 8-in.-dia. horizontal pipe with calcium silicate insulation — Table VII

$$r_2 \ln \frac{r_2}{r_1} = \frac{k_I R_P}{F} \qquad (12)$$

Since the insulation thickness is given by

$$t = r_2 - r_1 \qquad (13)$$

then, provided that we can solve Eq. (12) for given values of k_I, F and R_P, we can determine this acceptable-heat-loss thickness of insulation.

It is often written in the technical literature that the solution of an equation of the form of Eq. (12) is "difficult," and accordingly extensive tables of presolved equations are given or graphical solutions drawn. Both of these packaged solutions are unnecessary, since the equation can be quickly and easily solved on the simplest of hand-held programmable calculators without the necessity of either interpolation or extrapolation.

Choosing values for F and k

Before discussing these simple programs, it may be of value to discuss appropriate values of F and k_I. If we return to our numerical study of the 8-in. steam main at 500°F, then we can readily determine the percentage heat loss per unit area from it, when it is covered with different thicknesses of insulation. Table VII shows these values, clearly illustrating the law of diminishing returns below about 3% heat-loss rate. It is suggested that F should never be greater than 5%, while 3% appears satisfactory in many situations—however, there is a free choice of F for whoever uses the program.

Choosing insulation materials

The value of k_I is determined by the choice of insulant and its average operating temperature. It is surprising how many plant engineers are still specifying the same pipe-insulation materials that they did 10 years ago. One consequence of this is that they may be paying more for their insulation than is necessary. In fact the writers, in order to draw attention to this point, have quite deliberately used the "wrong" (possibly more expensive) insulation in their calculations on the 8-in.-dia. pipe at 500°F, since calcium silicate is quite likely to be more expensive than rigid preformed glass fiber.

There is, in fact, a choice of insulants at 500°F, such as the mineral wools (slag, rock and glass), foamed glass, 85% magnesia, and diatomite. A discussion of suitable insulants for various temperature ranges has been given by Harrison and Pelanne [2] and also by Probert and Giani [3], and need not be repeated here. In many cases the choice seems to be between calcium silicate and glass fiber, the latter having several advantages over the former. Calcium silicate is widely available up to 3 in. and sometimes 4 in. and additionally up to 5 in. by special order, but apparently is not available up to 6 in.

Thick calcium silicate is heavy and in a retrofit situation on existing pipework that is close to a wall or ceiling or in a duct, the sheer bulk of the additional insulation may be too great. In addition, the existing pipe supports or hangers may not be strong enough to hold the extra weight.

Glass-fiber preformed rigid pipe section is available up to 6-in. thickness as a single layer; it has a low moisture absorption and its low chloride content makes it compatible with stainless steel. Additionally, it has better chemical resistance to both acids and alkalis than does calcium silicate. Another important advantage is that it has a lower thermal conductivity than calcium silicate, which means smaller insulation thicknesses and a more light-weight installation because of its lower density.

The writers were interested to read of a one-mile-long 14-in. steam main at 445°F recently installed in a U.S.A.F. base. It was insulated with $6\frac{1}{2}$ in. of preformed single-layer glass fiber having a factory-applied all-service jacket underneath and field-applied smooth aluminum jacket. It was claimed that if calcium silicate had been used, it would have had to be double-layered, which would have increased contract costs by at least 20%.

The actual value of k_I for a specified insulant to be used in the TI-57 program given in this article is best obtained from manufacturers' data sheets, because not only does the value of k_I depend upon the average temperature $(\theta_1 + \theta_3)/2$, but also upon the method of production, and only the manufacturer can supply this information. However for a *rough* "ball-park approximation" the following are suitable average values of the thermal conductivity, k_I.

Calcium silicate, k_{CaSi}:

$$k_{CaSi} = 3.33 \times 10^{-2} + 8.75 \times 10^{-6}\frac{\theta_1 + \theta_3}{2} +$$
$$2.38 \times 10^{-8}\frac{\theta_1 + \theta_3}{2}^2 \qquad (14)$$

Glass fiber, $k_{gl\,fib}$:

$$k_{gl\,fib} = 1.25 \times 10^{-2} + 3.95 \times 10^{-11} \times$$
$$\left(\frac{\theta_1 + \theta_3}{2} + 460\right)^3 \qquad (15)$$

For example at $\theta_1 = 500°F$ and $\theta_3 = 70°F$ the above formulas give:

Program 1: Bare pipe resistance — Table VIII

Location	Code	Instruction	Location	Code	Instruction
0	331	RCL 1	25	43	(
1	75	+	26	331	RCL 1
2	04	4	27	65	−
3	06	6	28	07	7
4	00	0	29	00	0
5	85	=	30	44	)
6	23	x^2	31	45	÷
7	75	+	32	332	RCL 2
8	05	5	33	44	)
9	03	3	34	35	y^x
10	00	0	35	04	4
11	23	x^2	36	25	1/x
12	85	=	37	55	X
13	55	X	38	83	°
14	43	(	39	02	2
15	331	RCL 1	40	07	7
16	75	+	41	85	=
17	09	9	42	25	1/x
18	09	9	43	55	X
19	00	0	44	333	RCL 3
20	44	)	45	45	÷
21	55	X	46	334	RCL 4
22	335	RCL 5	47	85	=
23	75	+	48	81	R/S
24	43	(	49	71	RST

Program 2: Insulation thickness — Table IX

Location	Code	Instruction	Location	Code	Instruction
0	333	RCL 3	25	84	+/−
1	321	STO 1	26	75	+
2	331	RCL 1	27	331	RCL 1
3	55	X	28	85	=
4	43	(	29	323	STO 3
5	331	RCL 1	30	65	−
6	45	÷	31	331	RCL 1
7	330	RCL 0	32	85	=
8	44	)	33	40	|x|
9	13	ln x	34	76	x⩾t
10	65	−	35	71	RST
11	332	RCL 2	36	333	RCL 3
12	85	=	37	321	STO 1
13	45	÷	38	65	−
14	43	(	39	330	RCL 0
15	43	(	40	85	=
16	331	RCL 1	41	55	X
17	45	÷	42	01	1
18	330	RCL 0	43	02	2
19	44	)	44	85	=
20	13	ln x	45	81	R/S
21	75	+	46	322	STO 2
22	01	1	47	482	Fix 2
23	44	)	48	71	RST
24	85	=			

$$k_{CaSi} = 3.33 \times 10^{-2} + 8.75 \times 10^{-6} \times$$
$$[(500 + 70)/2] + 2.38 \times 10^{-8} [(500 + 70)/2]^2$$
$$= 0.0377 \text{ Btu ft h}^{-1} \text{ ft}^{-2} \text{ }°\text{F}^{-1} \qquad (16)$$

and:
$$k_{gl\,fib} = 1.25 \times 10^{-2} + 3.95 \times 10^{-11} \times$$
$$\left[\left(\frac{500 + 70}{2}\right) + 460\right]^3$$
$$= 0.0288 \text{ Btu ft h}^{-1} \text{ ft}^{-2} \text{ }°\text{F}^{-1} \qquad (17)$$

It may be of value to note that in this article all values of k_I have units of Btu ft h^{-1} ft^{-2} °F^{-1}. However, many technical journals and manufacturers' data sheets give values of k_I in Btu in. h^{-1} ft^{-2} °F^{-1}. The relationship between those two units is very simple and involves a factor of 12; thus:

■ Multiply a k_I value in Btu in. h^{-1} ft^{-2} °F^{-1} by 1/12 to convert it to Btu ft h^{-1} ft^{-2} °F^{-1}

■ Multiply a k_I value in Btu ft h^{-1} ft^{-2} °F^{-1} by 12 to convert it to Btu in. h^{-1} ft^{-2} °F^{-1}

For example, the above value for glass fiber is

$$k_{gl\,fib} = 0.0288 \text{ Btu ft h}^{-1} \text{ ft}^{-2} \text{ }°\text{F}^{-1}$$
$$= 0.0288 \times 12$$
$$= 0.346 \text{ Btu in. h}^{-1} \text{ ft}^{-2} \text{ }°\text{F}^{-1}$$

Computer programs description

Eq. (12) and (13) have been programmed for the Texas Instruments TI-57 calculator. Because of the limited number of steps (50) in this calculator, it has been necessary to break down the calculation of the acceptable-heat-loss thickness of insulation into two programs. Owners of the more powerful TI-59 calculator or its

equivalent will be able to combine the two programs given here into one.

Program 1: Bare-pipe resistance

This program, shown in Table VIII, evaluates the right-hand side of Eq. (12), $k_I R_p/F$, by first calculating:

$$R_P^{-1} = 0.27\left(\frac{\theta_1 - 70}{d_1}\right)^{0.25} + 0.154 \times 10^{-8} \times$$
$$(\theta_1 + 990)[(\theta_1 + 460)^2 + (530)^2]$$

and thereafter the value of $k_I R_p/F$, where k_I is to be obtained from manufacturers' data sheets or, failing that, from Eq. (14) or (15), and F is decided by the user.

Before the program is run, the following calculator store locations must be initialized:

STO 1	θ_1
STO 2	d_1
STO 3	k_I
STO 4	F
STO 5	0.154×10^{-8}

The program is run from Location 0 and, after the data have been entered, it can be run by

■ Resetting the calculator by pressing the **RST** key.

■ Pressing the **R/S** key.

■ Waiting for a steady display showing the value of $k_I R_p/F$ that is to be used in Program 2.

*Program 1 will run directly on the TI59, but Program 2 will not. To run Program 2 on the TI59, add LBL A at the start of the program, substitute A for RST in Step 35, and insert the "accuracy required" figure in the t-register (rather than in STO 7). After the data are entered, press **A** to run—Editor.

Program 2: Insulation thickness

This program shown in Table IX solves Eq. (12) and (13) together and yields the insulation thickness t_I *in inches*. Before the program is run, the following store locations must be initialized:

STO 0	r_1
STO 1	1
STO 2	$k_I R_P/F$
STO 3	1
STO 7	Accuracy required (usually 0.001)

The program is run from Location 0 after the data have been entered (see Program 1); the steady display gives the insulation thickness in inches.

When the calculation is complete, in order to repeat the procedure for another value of $k_I R_P/F$ it is necessary to:
1. Enter the new value of $k_I R_P/F$ on the keyboard.
2. Press the **R/S** key.

As an example of the use of the two programs, let us return to the problem of the 8-in.-dia. horizontal pipe, in 70°F still air, carrying dry saturated steam at 500°F, i.e. $\theta_1 = 500°F$; $\theta_3 = 70°F$; $d_1 = \frac{2}{3}$ ft; and $r_1 = \frac{1}{3}$ ft.

For calcium silicate, from Eq. (16)
$$k_{CaSi} = 0.0377 \text{ Btu ft h}^{-1} \text{ ft}^{-2} \text{ }°F^{-1}$$
Example 1: $F = 5\% = 0.05$
  Program 1 gives $k_I R_P/F = 0.183$
  Program 2 gives $t = 1.83$ in.
Clearly 2 in. of calcium silicate insulation would be chosen.

If greater heat conservation is required, then we have:
Example 2: $F = 3\% = 0.03$
  Program 1 gives $k_I R_P/F = 0.305$
  Program 2 gives $t = 2.83$ in.
Clearly 3 in. of calcium silicate insulation would be chosen.

It is interesting to calculate glass-fiber insulation thickness for the same pipe: in this case $k_{gl\ fib} = 0.0288$ Btu ft h^{-1} ft^{-2} °F^{-1} from Eq. (17).
For $F = 5\%$
  Program 1 gives $k_I R_P/F = 0.140$
  Program 2 gives $t = 1.45$ in.
This means that $1\frac{1}{2}$ in. of glass fiber would give the same percentage heat loss as 2 in. of calcium silicate.
For $F = 3\%$
  Program 1 gives $k_I R_P/F = 0.233$
  Program 2 gives $t = 2.25$
This result means that $2\frac{1}{2}$ in. of glass fiber would give the same percentage heat loss as would 3 in. of calcium silicate.

These results not only show the simplicity of the proposed criterion for evaluating the insulation thickness but also confirm the thinner and lighter insulation resulting from the use of glass fiber as opposed to calcium silicate. However, it must not be forgotten that glass fiber needs particularly strong jacketing to avoid mechanical abuse since it is considerably less rigid than calcium silicate.

Another benefit of the proposed criterion is that the dollar cost of the heat loss from any thickness of insulation may readily be found from Program 1, since it is easy to show that:

Cost of heat loss per unit area of insulated pipe, dollars per square foot per hour $= \dfrac{k_I C_F (\theta_1 - \theta_3)}{(k_I R_P/F) E}$

Conclusion

In conclusion, the writers would like to reaffirm their confidence in the usefulness of the existing manuals and books that use discounted-cash-flow analysis of insulation economics for those cases where the parameters *n, i* and *f* are known. For example, company accountants may *dictate* what values of *n* and *i* are to be used, based upon their economics-oriented viewpoint of all engineering investments. Very often the government lays down the value of *f* that must be used if tax credits are to be claimed. In these cases there is no uncertainty, and the discounted-cash-flow tables can, and should, be used.

However, if the plant engineer is allowed some say in the decision-making, he may care to put aside the economic arguments of the accountants and use the very simple method described in this article—and feel that his intuition and experience about what is an acceptable percentage heat loss from a pipe, and its cost, are just as good a way of looking at the problem of specifying insulation thickness as any other.

References

1. McChesney, M., and McChesney, P., *Chem. Eng.*, Vol. 88, July 27, 1981, p. 58.
2. Harrison, M. R., and Pelanne, C. M., *Chem. Eng.*, Vol. 84, Dec. 19, 1977, p. 61.
3. Probert, S. D., and Giani, S., *Applied Energy*, Vol. 2, p. 83 (1976).
4. McMillan, J. B., Heat Transfer Through Insulation, paper presented to American Soc. of Mechanical Engineers, New York, N.Y., Dec. 6, 1926.

Effects of insulation on refractory structures

Adding insulation to reduce heat loss may lower a refractory's mechanical strength and result in failure. Here is how to prevent this from happening.

Gary J. Nagl, Air Resources, Inc.

☐ With increased emphasis on energy conservation in industrial processes, more and more refractory installations are being designed with backup insulation to reduce heat loss. Unfortunately, many such structures are engineered without considering the effect of insulation on the structural integrity of the refractory. Thus, many refractory installations fail, which ups maintenance costs and can offset any energy saving due to the use of insulation.

Softening

Most refractories are a combination of silica and alumina, and do not have a definite melting point. There are a few exceptions—such as silica, corundum and magnesite brick—that consist of essentially pure oxides. Typical refractories soften over a wide temperature range, in which both solid and liquid are present. Fortunately, for fireclay and high-alumina refractories the liquid remains extremely viscous at temperatures exceeding 2,000°F.

Under these conditions, the refractory will deform when subjected to compressive loading, the degree of deformation depending on the temperature and the amount of loading. When the brick temperature is increased to a point at which the viscosity of the liquid is very low, the brick will deform under its own weight.

Generally, refractory walls and arches rarely fail because of compressive stress. This is due to the temperature gradient that exists across the refractory. Under certain temperatures and with certain types of refractory, the hot face of the refractory may be pyroplastic (or soft) and unable to support a compressive load. However, due to the temperature gradient, the cooler portion of the brick may have sufficient strength to support the entire load.

An example of this is illustrated in Fig. 1. Here, a refractory wall consisting of 9 in. of superduty fireclay brick is subjected to a hot-face temperature of 2,600°F. Assuming that the refractory becomes pyroplastic above the temperature of 2,200°F, approximately 1¾ in. of the brick will be unable to support the load, while the remaining 7¼ in. should be sufficiently strong to carry the entire load.

Heat loss from the wall would be approximately 2,200

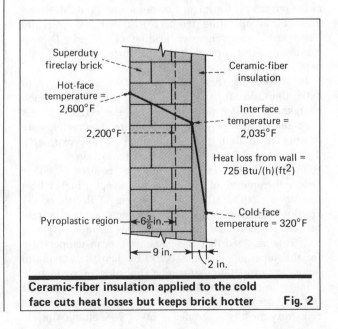

With the proper temperature gradient, the cooler part of the brick handles compression **Fig. 1**

Ceramic-fiber insulation applied to the cold face cuts heat losses but keeps brick hotter **Fig. 2**

Originally published October 18, 1982

Exceeding these temperatures can result in structural damage of the refractory   Table I

Refractory	Maximum mean operating temperature, °F
Superduty fireclay	1,950–2,200
60% Al_2O_3	2,225–2,300
70% Al_2O_3	2,400–2,500
80% Al_2O_3	2,425–2,525
90% Al_2O_3	2,600–2,800

In replacing superduty fireclay with higher-grade materials, the cost must be considered   Table II

Refractory	Relative cost
Superduty fireclay	1.00
60% Al_2O_3	1.82
70% Al_2O_3	2.45
80% Al_2O_3	3.09
90% Al_2O_3	7.59

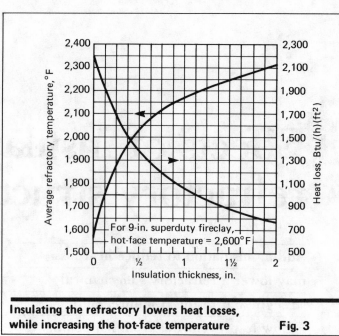

Insulating the refractory lowers heat losses, while increasing the hot-face temperature   Fig. 3

$Btu/(h)(ft^2)$. A considerable portion of this heat can be conserved by applying insulation on the cold face, as shown in Fig. 2. There, 2 in. of ceramic-fiber insulation was applied to the cold face of the refractory, which reduced the heat loss to approximately 770 $Btu/(hr)(ft^2)$. However, the interface temperature of the refractory has increased to approximately 2,010°F, resulting in over three-quarters of the brick being pyroplastic.

This installation will not have enough structural strength to give maximum service life. The result is that the money saved by conserving heat may be offset by the increased maintenance cost of replacing the refractory.

Mean temperature

Thus, attention must be given not only to the maximum service temperature of the refractory but also to the mean temperature, which is indicative of the structural strength of the refractory. Refractory manufacturers publish maximum mean operating temperatures that are generally based on a comprehensive loading of 25 psi. Typical maximum mean operating temperatures for commonly used refractories appear in Table I.

In considering these temperatures, it is necessary to make allowances for spalling, slagging and abrasion. Such phenomena reduce the thickness of the refractory and raise the mean operating temperature. Generally, safe practice is to limit this temperature to approximately 100°F below the recommended maximum value.

In Fig. 2, the designer would be required to either reduce the amount of insulation and accept a higher heat loss or to upgrade the refractory. Fig. 3 illustrates the relationship between insulation thickness, average refractory temperature and heat loss, for this example.

If it is assumed that the maximum mean temperature for the superduty brick is 2,200°F, then the maximum amount of insulation that can be applied is approximately 1 in., with a resulting heat loss of 1,050 $Btu/(h)(ft^2)$. If the designer desires to decrease the heat loss further, the refractory must be upgraded to 60 or 70% alumina brick,

which will cost over twice as much as superduty brick. The cost of the energy saved must offset the increase in material cost. The cost of various grades of refractories relative to superduty brick are shown in Table II.

Tie-backs and suspended walls

The effect of insulation on the operating temperatures of the refractory becomes more critical when tie-back or suspended-wall constructions are used. In both cases, the metallic shelves or clips that are used for structural support of the wall are embedded in the refractory. Thus, these devices are subjected to the operating temperatures of the refractory. As more insulation is applied to a suspended or tie-back refractory wall and the temperature of the brick increases, the maximum service temperature of the embedded metallic may be exceeded, resulting in structural failure of the wall.

For example, consider an overinsulated suspended wall. A 9-in. suspended superduty brick with 1 in. of insulation is exposed to a hot-face temperature of 2,600°F. The metallic support shelf extends 3 in. into the brick, and hence if it is not air-cooled, it will be subjected to the refractory temperatures. Approximately 2½ in. of the shelf will be overheated, resulting in certain failure of the wall. In the design and installation of suspended and tie-back walls, one must ensure that the metallic components are not overheated.

With realization that backup insulation on walls or arches will reduce the structural integrity of the refractory, the proper refractory and insulation thickness can be selected. This saves energy and gives long refractory life.

The author

Gary J. Nagl is vice-president, Energy Systems Div., of Air Resources, Inc., 600 N. First Bank Dr., Palatine, IL 60067. Tel: (312) 359-7810. With the firm since 1973, he is responsible for engineering design and construction, and has developed energy-recovery and particulate-incineration methods for petroleum-coke calcination. Previously, he worked for Universal Oil Products Co. as development engineer, process coordinator and engineering research and development coordinator. He received his B.Ch.E. degree from the University of Illinois and is a member of AIChE and the American Assn. of Cost Engineers, among other societies.

Winterizing process plants

Maintaining fluid temperatures without overheating is a problem in many CPI plants. Here are techniques to provide effective protection while keeping costs down. Included is an update on materials, and an explanation of some of the pitfalls that await the unwary.

E. Fisch, *The Halcon SD Group, Inc.*

☐ The need to winterize chemical-process-industries (CPI) plants to prevent freezing or undesirable condensation has long been recognized.

Over the years, the procedures for winterizing, as well as the required installation techniques, have been reduced almost to "cookbook" standards. Recently, however, certain changes have mandated a more thorough analysis of winterizing procedures. These include:

1. The rising cost of all forms of energy and, in particular, a great emphasis on conservation of steam, so as to avoid costly boiler additions.

2. Increasing field-labor costs, and the increasing availability of pretraced or preinsulated tubing components and custom-designed, preformed insulation and housings, particularly for various types of instruments.

3. Development of easy-to-use, reliable electrical tracing systems and components, as well as a change in the relative costs of electricity and steam in many areas.

4. Increased automation of plants, which makes less manpower available for inspection and adjustment of tracing systems.

5. A greater use of electronic instrumentation, which is more susceptible to damage from overheating arising from the use of tracing systems.

Winterization techniques

There are five basic winterizing techniques:

1. Designing and operating systems so as to avoid the need for protection against freezing. This includes design techniques—e.g., the inclusion of self-draining lines and warmup bypasses, valve placement, and operating procedures intended to drain stagnant lines or ensure continuity of flow.

2. Heat tracing using steam, electricity or circulating liquids as the heat source.

3. Installing internal heating coils.

4. Locating systems within heated buildings or nonfreezing environments.

5. Adding only insulation, to provide protection for lines with low flowrates.

System design and operation

Owing to the high costs of installing and operating heat tracing, it is worth making the additional effort to eliminate unnecessary tracing, and worth incurring the costs of modifications that make tracing unnecessary. Among the techniques:

Bypass lines — Bypass lines may be added around equipment to maintain circulation when block valves are closed for equipment maintenance. This is an almost universally applicable method for winterizing cooling-water systems, considering these factors:

- Cooling-water-circulation systems are, for all practical purposes, as reliable as steam systems. They normally employ multiple pumps and often include spare pumps with steam turbine drives, or alternative power feeders to ensure reliability.

- Cooling water can be circulated during maintenance operations, without flammability or toxicity hazards.

- There is generally sufficient freeboard in cooling-tower sumps to drain headers in an extreme emergency.

- Under certain circumstances, e.g., when using water for cooling pump seals and bearings, the addition of bypass lines may interfere with maintenance, and the use of tracing on these small lines may also raise the temperatures to undesirable levels. In these cases, block valves should be located at the headers or the subheaders to protect the water supply lines when the pumps must be removed for service. At such times, small amounts of water must be drained from the lines.

Recirculation lines — Recirculation lines may be provided to maintain backflow through the non-operating pumps. This is illustrated in Fig. 1. This technique, in principle, applies to all pumps handling fluids substantially above their freezing points. However, the practice should be limited to more critical systems such as condensate pumps, cooling water pumps and continuously running flush or seal pumps. In these services, pump stoppage generally becomes immediately apparent to the operators, and starting of the spare pump would have a high priority. Further, these systems are generally kept in operation, even during shutdowns.

A similar result may be achieved by drilling a hole in the pump-discharge check valves. This practice requires special care during installation and maintenance to ensure that the check valve (or its replacement) does indeed have the required hole.

Elimination of some tracing — Tracing may be eliminated on lines that are self-draining and that contain fluids not very close to their freezing points.

Diaphragm seals — Diaphragm seals may be used for instruments on lines that would not otherwise require steam tracing.

Dip pipes — Internal dip pipes, rather than external loop seals, may be used in vessels.

Block valve location — Block valves may be placed so as to eliminate dead legs or permit lines to be self-draining. This requires coordination with operational procedures to drain lines at shutdowns or when they are not in use. This is an operational judgment, balancing operating-labor availability versus cost savings. Lines that often fall into this category include:

■ Startup and rerun lines that would require the inclusion of double blocks and bleeds, if these are not already needed for other reasons.

■ Long product-transfer lines that normally stay in operation as long as the plant is running.

Use of traps — Traps may be used in steam lines to prevent accumulation of condensate even when block valves are closed (since eventually they will leak).

Common insulation of warm and cold lines — This, of course, can only be done if there is assurance that the warm line will always remain warm, as in the case of steam and condensate lines at hose stations. A typical arrangement is shown in Fig. 2.

Use of thermosyphon circulation — Tracing may be eliminated where advantage can be taken of thermosyphon circulation. This technique can be used for pump seal piping and for some level instruments.

Heat tracing

Unless another reliable method of winterization is used, heat tracing is required to provide the necessary protection against heat loss. Generally, heat tracing provides protection against winter temperatures, but it may be required all year round, or even indoors, if the temperatures to be maintained are high enough.

Although this article is largely devoted to heat tracing techniques, it should be recognized that the installation and operating costs of heat tracing are considerable, and

alternative techniques should be used wherever feasible to minimize these costs. J. T. Lonsdale and J. E. Mundy [1], have shown costs on the order of $15/ft for installation of the steam distribution and condensate collection systems, and $20/ft for the actual tracing, *plus* $4.23/(yr)(ft) for maintenance and operating costs (assuming steam at $5/1,000 lb, and an operating time of 8,760 h/yr). The saving resulting from eliminating even 100 ft of steam tracing comes to:

Installation cost	$2,000
Capital equivalent of 3-yr operating costs	1,300
Total equivalent capital	$3,300

If elimination of the tracing avoids the need for insulation, or if the tracing requires the use of heat transfer cement, the saving in installation cost would be 30 to 50% greater than noted above.

Internal coils and tracers

The design of internal coils for winterizing of tanks and vessels is, in principle, similar to the design of heat tracing systems. However, detailed design procedures are outside the scope of this article. This topic has been thoroughly covered by D. K. Kumana and S. P. Kothari [2].

Internal tracers are used to a limited degree. These have been employed for many years on transfer lines carrying heavy crudes, or similar materials. However, the potential for leaks into the process stream, as well as the impossibility of inspecting joints in the tracer, make this technique relatively impractical. Furthermore, the development of heat transfer cement has provided better means of maintaining high process temperatures with external tracing.

Heated buildings

The location of plants or equipment within heated buildings, or the running of piping inside heated ductwork, is occasionally a valid means of winterizing equipment and piping. This is often done for limited areas such as centrifuge rooms, analyzer rooms, water wells, etc., so as to avoid the costs of tracing, as well as to provide weather protection for operation or maintenance. In extreme cases, such as under subarctic conditions, entire plants may be enclosed.

In either case, when flammable or noxious materials are present, consideration must be given both to providing adequate ventilation (and the heating of makeup air), and to considering the need for, and cost of, upgrading the electrical hazard classification from Div. II (flammable vapors present only in the event of mishap) to Div. I (flammable vapors normally present).

The burial of lines and valves is also used, and is almost universally applied for winterizing fire-water and drinking-water lines. However, the burial of process lines is generally not suitable because of potential hazards in the event of leakage.

Insulation

Adding insulation, without steam tracing, is an economical way to provide against freezing. Generally, local experience will indicate if the approach is feasible. Conventional heat-loss calculations described later in this article can also be used. An additional check can be made by using published tables of time for freezing in a stagnant line, such as those of F. F. House [3]. In evaluating the use of insulation for winterizing, the expected period of low flow or flow interruption should be considered. This is especially important for small-diameter lines, which will freeze more rapidly than larger ones.

Design—systems review

The importance of a thorough review of the systems to ensure adequate winterization cannot be overemphasized. Failure to provide adequate systems, or failure to operate them properly, will result in freezing or undesirable condensation. These can damage piping, instruments and equipment, and interrupt plant operation.

Attention to winterization must be given at all stages of a project—from inception to engineering flowsheets (or piping and instrument diagrams), and then to detailed engineering design and installation. Finally, the systems must be adequately tested in conjunction with the overall plant-commissioning procedures before being placed in operation.

Stage I: Establishing basic criteria

A review of both process and utility systems is necessary to establish winterizing requirements and applicable temperature criteria.

First, the tracing criteria for the various systems are selected, including type of tracing (steam or electric), need for heat transfer cement or spacers, required steam pressures or electric-heat input rates, as well as maximum temperature limitations for both the items being traced and the tracing materials.

Then, an initial evaluation is made of the application of techniques other than tracing; this is further refined and finalized during the preparation of the piping and instrumentation diagrams.

The evaluation of temperature limitations of both the process materials and the tracers themselves should be made at this preliminary stage to ensure that proper selections of tracing systems are made—from both technical and economic viewpoints. Among the guidelines to be established:

Heat stability — The stability of the process material to the tracing medium temperatures during periods of low flow should be considered. This will indicate the maximum heating-medium temperature, as well as any need for spacers. It may even dictate that a low-pressure steam tracer with heat transfer cement should be used, even though a higher-pressure steam tracer without heat transfer cement would be less costly and would provide the necessary protection against freezing. When corrosive materials are being handled, the exposure of the piping itself to higher temperatures (and attendant higher corrosion rates) can also be significant.

Maximum temperatures — The maximum process temperatures to which the tracer will be exposed will dictate the selection of the tracer material.

Process temperatures should take into account the possibility of hot water flushing, or steaming out, of lines, that could expose the tracers to temperatures

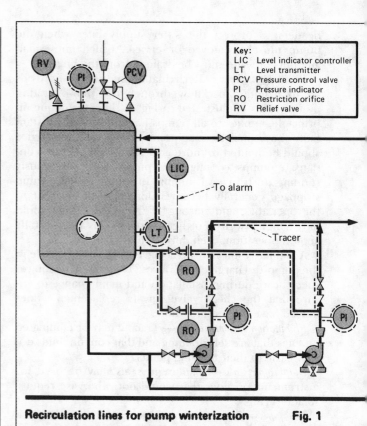

Recirculation lines for pump winterization   Fig. 1

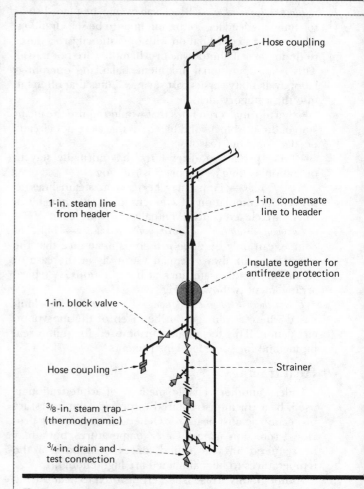

Typical use of common insulation for winterizing   Fig. 2

higher than the anticipated process operating (or other normally expected) temperatures.

The upper limit for using copper tubing is 400°F (equivalent to 250-psig steam). Above this temperature, steel tubing or pipe is required. Copper tubing is widely used when high-temperature materials are not required, because of the ease with which it is bent, and its resistance to scale formation in the system. However, the availability and cost of copper varies, so at various times steel tracers are used instead, particularly outside the U.S. This can be done either by the total substitution of steel tubing for copper tubing, or by using mostly steel or iron pipe (with copper tubing sections retained for valve wrapping and for other applications where appreciable bending is required).

When electric tracers are used, the process temperatures should be included in the requisitions issued to suppliers to ensure that the proper tracer type is selected. The widely used temperature-limiting type of tracer is generally available in several grades, with typical limitation ranges of 150 – 175°F, 200 – 250°F, and temperatures up to 500°F. Electric tracers of the resistance type, generally used for higher-temperature applications, also require that the designer consider maximum temperatures, so that the appropriate sheath and electrical insulation materials will be used.

Ambient conditions — In establishing the winterizing requirements, the selection of the minimum-temperature/maximum-wind-velocity combination also requires careful attention. The minimum winter temperature used for building heating design is *not* suitable as the basis for process winterization design. The selection of the building design temperatures often includes an economic balance of excessive equipment cost versus the relatively minor, infrequent and short-term inconvenience of an underheated building. The economic consequences of a single short-term freezeup of even a few process lines are serious enough to call for "once in a century" minimum temperatures as the basis of process-system winterization design.

In establishing the basic design criteria for the plant, adequate provision should be made for utilities requirements for heat tracing; these are often neglected. A quick guideline for estimating steam tracing requirements allows 10 lb/h of steam usage for each tracer, with one tracer required for every 60 – 70 ft of traced pipe. (This allows for tracing of instruments as well as piping, and includes distribution losses and normal trap leakage.) Power requirements for electric tracing of piping carrying water-like materials are typically 3 to 6 watts/ft of traced pipe for the normal operating load, with a peak load of 5 to 10 watts/ft.

With higher-freezing-point materials, the electric-power and steam consumption requirements could be considerably greater, since a large number of lines could require multiple tracers.

Stage II: Reviewing P&I diagrams

The flowsheets should show the type and extent of tracing, including instruments, lines that require makeup of heat losses during normal operation, and lines that require freeze protection during maintenance shutdowns. Such would include process lines, service or process-water lines, safety-shower supply lines, steam and condensate lines not otherwise protected, non-self-draining steam traps or liquid drainers, high-pourpoint fuel oil lines, etc.

The flowsheets should be carefully reviewed to identify and show tracing on the normally stagnant portions of lines where design alternatives or operational procedures cannot eliminate tracing. Note particularly:

■ Pumps that are spared, up to and including block valves.

■ Spare filters, or similar devices that are spared.

■ Control valve or meter bypasses, and nondrainable portions of control valve or meter stations.

■ Instrument leads, and instruments on liquid and gas lines that contain liquid that may freeze or condensable vapors. This includes instruments both on lines and on equipment in steam, steam condensate, cooling water, compressed air or hot vapor service.

■ Instruments on vessels or towers, including level transmitters and gage glasses, pressure and differential transmitters and gages (together with their leads).

■ Sample taps and sampling systems.

Note that for sampling systems and differential-pressure transmitters in gas or vapor service, the criterion for heat tracing is normally to prevent condensation, rather than to avoid freezing.

■ Pressure and vacuum relief valves and gage hatches.

■ Seal pots.

■ Nondraining portions of lines, such as loop seals (unless special operating provisions are made for draining at shutdown). Note that plans must be made for continuously refilling critical loop seals or seal pots, since tracing may vaporize the liquid in the seal.

The flowsheets should include all provisions for alternatives to tracing, where needed. Items that should be shown on flowsheets include:

■ Lines that should be selfdraining.

■ Minimum-flow bypasses.

■ Recirculation lines (or drilled check valves) at pumps.

■ Traps in steam lines not provided with header block-valves and drains.

■ Header block valves and adequate drains in steam lines, and in condensate lines that are not traced.

■ Limits of enclosures and underground line risers.

■ Double block and bleed valves.

■ Block valve locations, where necessary for winterization. These must be supplemented by appropriate drains—it must be assumed that block valves will leak.

■ Instrument location notes, where close coupling or free drainage is necessary for winterizing.

■ Seal diaphragms on instruments.

■ Lines to be insulated together.

■ Items to be connected to emergency power, such as critical electric tracers or enclosure heaters.

At this time, the preliminary evaluations of alternative winterizing techniques (made during the initial process design) should be finalized.

Stage III: Detailed design and review

Items that should particularly be checked during the reviews of the piping design drawings or model include:

■ Location of maintenance block valves and minimum-flow bypasses, to ensure that the piping downstream of the block valves can be fully drained and that the distances between the valves and the bypasses are not excessive.

■ Location of block valves to ensure that dead legs will be of minimum length (or be selfdraining), at risers from underground piping, at double blocks and bleeds, at header takeoffs, and at branches for spare pumps, filters and similar items.

■ Installation of sufficient drain valves and traps in steam and condensate lines, and in other lines requiring drainage for freeze protection.

Instrument drawings and installation details should also be reviewed to ensure that:

■ Diaphragm seals are used where specified.

■ Instruments, particularly in untraced lines, are specified to be traced where required.

■ Instruments that are to be close coupled, or those with selfdraining leads, are properly located.

■ Multitube bundles are specified (and requisitioned) that use steel tubing when tracing exceeds the upper limit for copper (400°F or 250-psig saturated steam).

■ Instruments are not exposed to excessive tracing temperature (generally limited to 200°F maximum).

The design specification for steam tracing should adequately define the sizing of the steam-supply and condensate-collection system, maximum tracer lengths, number of tracers required for adequate protection, and tracing details. Guidelines and specific recommendations for these and other steam-tracing criteria have been very adequately covered by F. F. House [4].

Steam-tracing routing drawings and schedules should be reviewed to ensure that:

■ A pump and its spare have independent tracers, so that maintenance will not interfere with operability. This also applies to spare filters, etc.

■ Bucket and other non-selfdraining traps are traced.

■ An adequate number of spare connections have been provided on the steam supply and condensate manifolds. A minimum of 10% spare, or at least three per manifold, is recommended.

■ Maximum allowable tracer lengths are not exceeded, particularly in long rack runs.

■ Steam supply and condensate manifolds are not excessively long or placed to interfere with operation or maintenance. Locations for manifolds should allow space for racks or other supports for steam and condensate leads and returns.

■ Where appropriate, tracers requiring operation year-round should be put on manifolds separate from those for tracers that can be shut off during warm weather. This will facilitate steam conservation.

■ Vertical steam subheaders are installed on towers to provide convenient sources of supply for instruments on the upper parts of towers.

■ Electric tracers are not connected to those lighting panels that are controlled by photocells or "master" lighting switches, and there is an adequate power supply at the local panels to meet tracing requirements.

■ Electric tracers or enclosure heaters on critical lines, e.g., fire-water supplies, are taken from emergency power supplies. (There may be additional systems that require emergency power.)

Preinsulated tubing running between traced lines and the manifolds for steam and condensate

■ Where circulation systems are used for tracing, there are adequate provisions for venting tracing supply and return lines.

The design of the steam tracing system should include a tracing schedule. This schedule should list, for each steam supply manifold or steam supply line:

■ The steam tag number.

■ The lines and instruments served by the tracer.

■ The condensate manifold (or trap) tag number.

■ The manifold steam pressure.

■ The manifold location drawing.

■ The steam tracer routing drawing or isometric.

This steam tracing schedule should be a project drawing kept in reproducible form so that it may be updated to reflect modifications made either during construction or during plant operations. The schedule can be part of the "home office engineering" or be made by the field engineer or construction supervisor.

For complex systems using electrical or circulating-fluid tracing, an analogous tracing schedule should be made. (For simple systems, it may be enough to show the tracing on flowsheets or wiring diagrams and layouts.)

Stage IV: During construction

The extent of tracing should be reviewed to verify that it covers, as designed, all items requiring winterization. Items to be particularly noted include:

■ Dead legs downstream of recirculation lines and upstream of block valves at branches, headers and control stations.

■ Control valve and meter station bypasses.

■ Pump seal systems, to ensure that where thermosyphon action is used to maintain circulation, there is sufficient head available.

■ Actual location of close-coupled or free-draining instruments. Occasionally, the necessary locations are not maintained because of visibility problems, or inconvenience in mounting.

The workmanship and the installation procedures should be reviewed when the work is started. It is especially important to inspect the installation early, to ensure that proper tools (such as tube benders) and materials (such as wire and bands) are being used, and that the installers are thoroughly skilled in the techniques necessary. Wherever feasible, dedicated crews should be assigned to work on steam tracing. Points that should be particularly checked:

■ Tubing bends, which should be smooth and free of kinks or crimps.

■ Straight runs, which should be free of dents and lie tightly against the piping.

■ Ties, which should be installed at the specified intervals.

■ Tracing on pumps, which should be left clear to permit convenient maintenance.

■ Electrical tracer cables of the non-self-temperature-limiting type, to ensure that they are not overlapped (to avoid hot spots, and possible burnout).

■ Underground risers and wall openings, to ensure that the tracers go below the frost line or into the heated building space.

■ Installation of loops and break unions at flanges, and wherever else necessary, to enable removal of instruments, valves, equipment, etc. The loops should be checked to be sure they are deep enough to clear the insulation after it is installed.

■ Spacers, which should be installed where required, at proper intervals to prevent tubing from touching the pipe being traced.

■ Changes from the tracing design drawings to suit the field conditions. These should be kept at a minimum, and adequately recorded for incorporation in the "as built" tracing schedules.

■ Temporary tagging on tracers prior to hookup to the steam and condensate leads, as well as final tagging at the manifolds.

■ A visual nonoperating continuity check of the tracers. This should be made by "walking the lines" prior to installing insulation; even partial insulation can interfere with this check.

■ Electrical tracing should be checked for resistance to ground, both before and after insulation is applied.

■ Permanent tagging of tracers, to ensure that the tags are in the proper location and securely fastened.

Where heat transfer cement is needed, one must take care to ensure that it is properly applied. Failure to do this will result in loss of effectiveness of the cement, with the subsequent development of cold spots and line blockages The manufacturer's instructions must be carefully followed, particularly with regard to:

■ The cleanliness of the surfaces to which the cement is to be applied. Wire brushing may be required to remove loose scale, rust, paint, grease, oil, etc.

■ The temperature of the material during application. Normal warehouse storage temperatures may be below

	Freezing point, °F	Allowable Temp., °F Min.	Allowable Temp., °F Max.	Max. process temp., °F
Utilities				
Process water	32	40	–	120
LP and MP condensate	32	40	–	350
HP condensate	32	40	–	460
Safety showers	32	40	100	100
Feeds				
10% Caustic	5	40	170	Ambient
Solvent	60	70	170	120
Process Streams				
Reactor effluent	320	340	400	375
Centrifuge feed	140	140	250	200
Product	252	300	350	350
Product column bottoms	375	420	600	450
Heavy ends	425	475	600	500
Process air–water vapor	32	40	250	350

Sample summary of tracing requirements as determined in

the minimum recommended for cement application, and preheating may be required. The storage period for the heat transfer cement should not exceed the manufacturer's recommendations.

■ The need for the priming of aluminum or galvanized surfaces, unless the cement grade has been specifically selected to be compatible with these materials.

■ The grade of cement being used; the cement must be suitable for the anticipated service temperature. To the maximum extent possible, the application of the cement should be immediately followed by the installation of the insulation and its weatherproof covering. If this is not done, temporary protection against the weather must be provided.

■ Curing, if required. The manufacturer's instructions for curing should be carefully followed, with necessary periods allowed for air drying, and suitable precautions taken against overheating and boiling. (Not all grades of heat transfer cement require curing.)

Stage V: Testing and startup

Mechanical testing and checkout should be conducted *prior* to placing the tracing system in operation.

Steam tracing testing should preferably be done using steam, supplied through the normal lines and manifolds. During the test, the following should be verified:

■ The continuity of each tracer from the steam supply manifold to the condensate manifold. (This could previously have been done by visual inspection.) This will also

a survey done for a particular process

| Heat transfer cement? | Required steam pressure, psig | | | | | | Steam-manifold pressure, psig | Remarks |
| | Winter operation | | | Spring and fall operation | | | | |
	1-1½ in.	2 in.	3 in.	1-1½ in.	2 in.	3 in.		
No	10	10	10	Not required			30	
No	10	10	10	Not required			30	
No	10	10	10	Not required			30	Use steel tubing
No	Electric			Required (electric)			Electric	Use electric tracing, leave on all year.
No	10	10	10	Not required			Electric or 30	Use electric tracing in headers and storage areas. For process-area branches use spacers with steam tracing.
No	10	10	10	10	10	10	30	Use spacers, 1½ in. min. insulation for 2 in. and 3 in.
Yes	145	145	140	135	140	130	150	1½ in. min. insulation for 3 in.; 2 tracers for 2 in. and 3 in.
Yes	10	10	10	10	10	10	30	
Yes	70	75	70	65	70	65	75	Use 2 tracers for 3 in.
Yes	100	375	300	375	360	350	400	Use steel tubing tracer. 1½ in. min. insulation all sizes. Two tracers for 2 in. and 3 in.
Yes	Use hot oil			Use hot oil			Hot oil	Use steel tubing with hot-oil tracing.
No	10	10	10	Not required			30	Trace instruments only.

verify that the tubing is free from flat spots or other damage.

■ The tracers are free from leaks at the tubing joints and the manifold connections.

■ The traps are operable and freely pass condensate without blowing steam.

Steam tracing that employs copper or stainless steel tubing does not need to be flushed out, since the tubing, as received from the factory, is normally clean and remains clean. But where steel tubing is used for tracers, it is advisable to blow down each tracer to remove rust and scale prior to placing the tracers in full operation.

Steam is the preferred medium for testing since leaks are more easily detected, and trap operation can be verified in conjunction with the testing. If steam is not available (or cannot be left continuously on in the event of freezing weather), air testing may be used. However, this requires additional labor, makes it more difficult to find leaks, and will not fully verify trap operation.

For tracing systems using circulating fluids, air can be used for continuity and leak testing of the system. Water may be used to facilitate detection of leaks, but one must allow for its presence (or removal) in order to prevent dilution or contamination of the circulating fluid.

For electric tracing systems, the tracers should be tested by turning on the electric power and seeing that:

■ The cables do heat up.
■ All pilot lights are operating.
■ Switches are properly labeled.

■ The thermostats are operable and adjusted to the proper settings. (Where the thermostats monitor process temperatures, the settings may have to be adjusted after the systems are put in operation and actual process temperature can be measured.)

After the mechanical testing has been completed and the plant is being made ready for preoperation and testing, the tracing systems should be commissioned and any final checkout and adjustment made.

The tracing systems should be fully operational prior to startup of the system being protected. Often, the construction work is scheduled to be completed just prior to the onset of cold weather, and the commissioning of the tracing systems is on the critical path for startup. However, even when this is not the case, commissioning of the tracing system should be given a high priority, to ensure that it will be ready when needed.

As shown in the tracing design procedures, given elsewhere in this article, the tracing generally makes up for heat loss only, and a reduction in the insulation efficiency may make the tracing ineffective. Any moisture that has been absorbed by the insulation will cause a substantial loss of its efficiency. Because of this, it is very desirable to turn on the tracing as early as possible, to vaporize any moisture trapped in the insulation. Indeed, the leak testing of the steam tracers (using steam), and the continuity testing of electrical tracers, can be a first step in this dryout.

The dryout can take up to several weeks, since the heat

Nomenclature

A,B,C	Constants for wind-factor equation
a	Thermal conductance, tracer to pipe, without heat transfer cement, Btu/(h)(°F)(ft of pipe)
b	Thermal conductance, tracer to pipe, with cement, Btu (h)(°F)(ft of pipe)
C_p	Specific heat of hot medium, Btu/(lb)(°F)
d_i	Inside diameter of insulation, in.
d_o	Outside diameter of insulation, in.
h_a	Film heat-transfer coefficient to air (corrected for wind), Btu/(h)(°F)(ft²)
h_c+h_f	Combined convection and radiation heat transfer coefficient, Btu/(h)(ft²)(°F/ft)
K_i	Thermal conductivity of insulation, Btu/(h)(ft²)(°F/ft)
kW	Kilowatts
L_L	Total pipeline length, ft
Q	Heat loss per ft of pipe, Btu/(h)(ft)
Q_t	Total heat lost from pipeline, Btu/h
T_a	Average temperature of pipe and tracer, °F
T_{air}	Air temperature, °F
$T_{med.avg.}$	Average temperature of hot medium, °F
T_{mi}	Inlet temperature of hot medium, °F
T_{mo}	Outlet temperature of hot medium, °F
T_p	Temperature in pipe, °F
T_s	Outside surface temperature of insulation, °F
T_{wc}	Number of tracers required with heat transfer cement
T_{woc}	Number of tracers required without heat transfer cement
W	Flowrate of hot medium, lb/h
W_f	Wind factor

input rate of the tracers is very low. Fortunately, this process is considerably speeded up by the circulation of hot fluids in the traced piping. Generally, this takes place routinely as part of the preoperational checkout of the process systems.

However, certain lines, notably those to and from storage areas, may not be amenable to circulation, and the dryout of the insulation may not be completed as part of the preoperational checkout. Such lines, particularly if they are to handle higher-freezing-point material, should be given particular attention. Among the methods for heating to ensure dryout for such lines are prolonged flushing with hot water, or blowing with hot air or inert gas. The tracer(s) may also be temporarily connected to higher-pressure steam. If there are any doubts, a plug of insulation should be removed at suspected cold spots and the pipe metal temperature verified (being certain that the tracer itself does not touch the thermometer).

When the process is made operational, a final check of steam traps, electrical circuits and pilot lights, and high-point vents in circulating systems should be made.

If hot oil is the circulating fluid, this check of high-point vents is especially important, since any residual

moisture in the tracers will be vaporized. As the circulation of process fluids in the system being traced is initiated, final adjustment should be made of thermostats and steam pressure (if variable). Where steel tubing is used, the strainers should be checked for accumulation of rust, several times a shift during the first few days of operation.

When the first cold snap occurs, a check should be made of the steam tracing system, since any tracers that have inadvertently not been turned on will fill up with condensate backing up from the manifolds and may very well freeze and burst. The process systems should also be given a thorough check to detect cold spots and plugging, and the necessary repairs made.

Winterization design criteria

It should be recognized that the basic function of winterizing systems is to prevent the fluid temperature of the system being winterized from falling below that value that would cause undesirable freezing or condensation. Where tracing systems are used, the primary function of the tracing is to add sufficient heat to the process fluid to make up for heat losses through the insulation. This is the theoretical basis for selection of the heat tracing design. The equations representing this (when a circulating-fluid heating medium is used) have been presented by W. W. Blackwell [5]:

$$Q=2\pi K_i(T_a-T_s)/\ln(d_o/d_i) \tag{1}$$
$$Q=h_a(\pi d_o/12)(T_s-T_{air}) \tag{2}$$
$$X=\ln(d_o/d_i)(h_a)(d_o/12)/2K_i \tag{3}$$
$$T_s=(T_a+XT_{air})/(X+1) \tag{4}$$
$$Q_t=QL_L \tag{5}$$
$$W=Q_t/C_p(T_{mi}-T_{mo}) \tag{6}$$
$$T_{woc}=Q/a(T_{med.avg.}-T_p) \tag{7}$$
$$T_{wc}=Q/b(T_{med.avg.}-T_p) \tag{8}$$
$$h_c+h_r=564/(d_o)^{0.19}[273-(T_s-T_{air})] \tag{9}$$
$$W_f=A+B(T_s-T_{air})+C(T_s-T_{air})^2 \tag{10}$$
$$h_a=(h_c+h_r)W_F \tag{11}$$

For the case of steam as the heating medium, the equations for the flowrate and number of tracers are:

$$W=Q_t/h_a \tag{12}$$
$$T_{wot}=Q/a(T_s-T_p) \tag{13}$$

Where electricity is the source of heat, the equations for selecting the heater cable are:

$$\text{watts/ft}=Q/0.3405 \tag{15}$$
$$\text{or, kilowatts/ft}=Q/3,405 \tag{16}$$

A convenient way of utilizing the above equations is graphically illustrated in Fig. 3a and b, which was done for ⅜-in. tubing in conjunction with the steam tracing requirements for a particular process. These give a clear picture of the heat losses as a function of process temperature, for winter and nonwinter conditions and for various pipe sizes and insulation thicknesses. They also show the heat input capability of single and multiple tracers, with and without heat transfer cement, as a function of heating medium temperature (or its equivalent as steam pressure). Typical results of the survey are shown in the table.

In reviewing the data in the table and applying the general principles governing the use of tracing, the following ensued:

■ The minimum insulation thicknesses given were compared with the insulation thicknesses shown in the project specification to ensure that these thicknesses were not, in any case, less than those established by steam tracing requirements.

■ Although the steam headers were originally established at 400 psig, 150 psig and 75 psig to meet the process heating levels, it was deemed desirable to add a low-pressure system to effectively utilize steam recoverable from condensate flash, and reduce exposure temperatures of the solvent piping so as to minimize corrosion. Although 10-psig steam would have been a suitable level for the low-pressure steam tracing, 30 psig was selected to avoid the need for "high backpressure" traps that would have increased the spare-parts stocking requirements.

■ Electric heating was selected for the entire safety-shower system, as well as for the caustic system in the storage area, and for headers.

■ Electric tracing was considered impractical for the caustic lines at the process equipment. Steam tracing with spacers was specified for these services, to reduce metal temperatures.

■ Provisions were made in the hot oil system for the hot oil tracing requirements.

■ Steel tracing tubing requirements were defined not only for the hot-oil and 400-psig tracers, but for wherever else the process temperature made this necessary.

■ Those services requiring heat transfer cement and multiple tracers were defined.

■ Two sets of 30-psig steam supply and condensate return manifolds were installed in each area. The utility lines were connected to one set and the process lines to the other, facilitating conservation of steam during non-winter periods.

As illustrated in the preceding example, the equations and their derived graphs, and tabular data, can be used to establish the following tracing criteria:

■ Tracing-medium temperatures, steam pressures or heat input ratings.

■ Need for heat transfer cement.

■ Need for multiple tracers for a given system as a function of size.

■ Changes in tracing operating modes, to meet seasonal temperature variations. This includes the installation of thermostats to control tracer operation, as well as segregation of tracer supplies for different process sytems.

■ Need for spacers to avoid excessive process temperatures.

■ Need for thermostats to limit process temperatures.

■ Maximum tracer exposure temperatures. This will indicate the need for steel tubing in fluid systems, and the type of sheathing for electric tracers. When self-limiting electric tracing is used, selection of the particular heating cable should also take into account the variation of heat output rate with temperature. A typical curve (which varies from manufacturer to manufacturer) is shown in Fig. 4. It should be noted that this curve is based on metallic pipe.

When plastic pipe is used, the operating temperature of the heating element will increase. This requires that the nominal (base temperature) output rating of the tracer be upgraded by a factor of 0.7 to allow for this.

■ Temperature changes, in the process fluids, that may occur under various operating conditions. This criterion is particularly useful, since it will verify changes occurring under normal conditions and indicate the feasibility of employing operational procedures to eliminate tracing or, conversely, the need for tracing so as to maintain process fluid temperature even if freeze protection is not required.

Generally, the use of tracing systems to add heat to a process (in lieu of a heat exchanger) is not economically practical, or otherwise feasible. However, rigorous analysis of the heat tracing system will clearly indicate the amount of heat that can be input into the system by external tracing. This can be of value in eliminating external heat exchangers under some conditions, and in ensuring that the heat added by tracing can be accommodated within the process.

Choosing a winterizing method

In considering plant winterization, steam tracing is usually first thought of. It is the oldest and most widely practiced method and, indeed, the term "steam tracing" has almost become synonymous with "heat tracing" and even "winterization." However, as shown in this article, several types of heat tracing—and, under many circumstances, other methods of winterization—are both more economical and technologically superior. Choosing the best method for a particular plant, or a particular area within a plant, requires careful analysis of process constraints and ambient conditions, as well as economics and system reliability.

Winterization that does not require heat tracing, but that depends on design or operating techniques, is almost invariably the most economical choice. However, this demands that the required details be incorporated in the design, and that the appropriate operating procedures be followed. If there is any uncertainty, heat tracing should be installed, since the consequences of freezing can be quite costly, and even hazardous.

Should heat tracing be required, selection of the heating method should consider both the first cost (the installed cost of the tracing system) and the long-term operating cost (the cost of energy consumed, plus maintenance costs). The economics will depend on the plant size, the temperature(s) required and the relative costs of steam and electricity.

The availability of "free" excess steam will greatly affect the economics, but it should be recognized that, in reality, there is no "free" steam.

In general, electrical tracing is favored for simple piping systems and lower temperatures (below 140°F), while steam is favored for higher temperatures (above 200°F), and for systems containing items that require frequent maintenance. Circulating-media systems are usually the most expensive choice and are used only when the process or ambient conditions make steam or

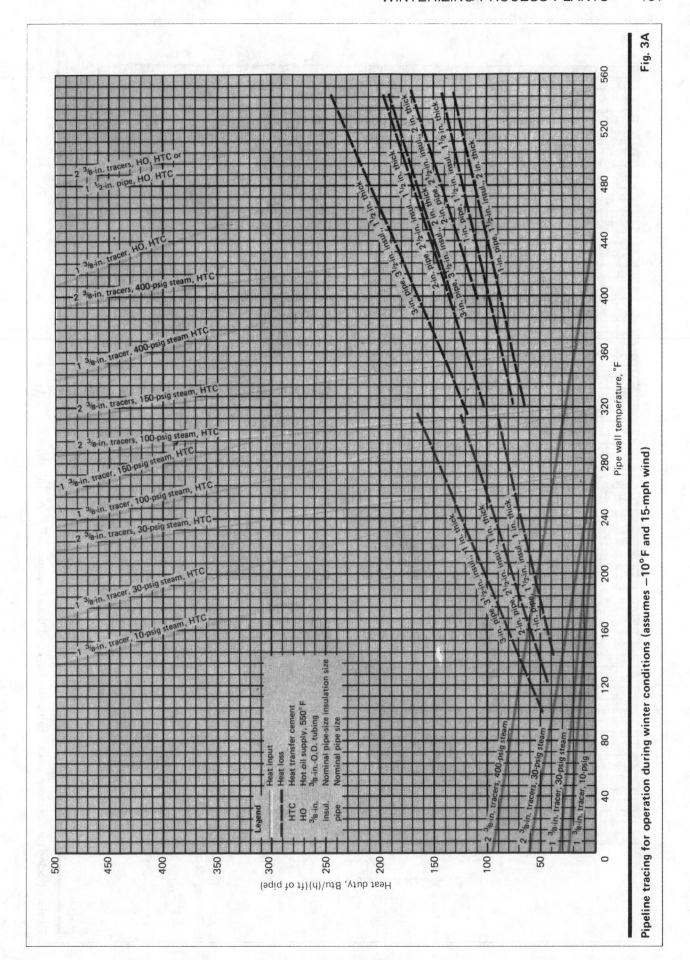

Legend

	Heat input
	Heat loss
HTC	Heat transfer cement
HO	Hot oil supply, 550°F
3/8-in.	3/8-in.-O.D. tubing
insul.	Nominal pipe-size insulation size
pipe	Nominal pipe size

Pipe wall temperature, °F

Heat duty, Btu/(h)(ft of pipe)

Pipeline tracing for operation during winter conditions (assumes −10°F and 15-mph wind)

Fig. 3A

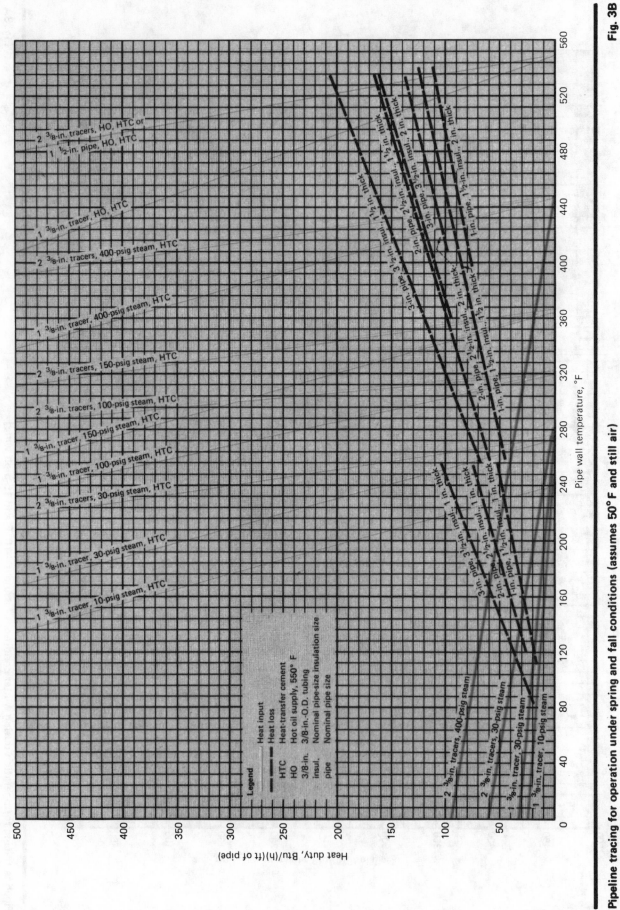

Pipeline tracing for operation under spring and fall conditions (assumes 50°F and still air)

Fig. 3B

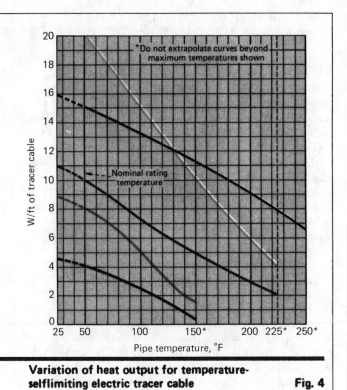

Variation of heat output for temperature-selflimiting electric tracer cable   **Fig. 4**

electric tracing unsuitable. A summary of the particular characteristics of each of these systems follows.

Steam tracing

Except in extremely cold environments, such as might be encountered in arctic locations, steam tracing is generally applicable. The minimum practical steam pressure for tracing is 25 psig, and this level is satisfactory when minimum temperatures do not fall below 5 to 10°F. Successful system operation has been observed with 35-psig steam at temperatures falling to below −30°F, even though 50-psig steam would normally be a recommended minimum pressure for these temperatures. The stability of steam tracing in very cold climates requires a high level of workmanship, and this may be a factor in choosing between steam tracing and tracing with a circulating medium or with electricity, the latter two of which are much less subject to failure due to extreme cold.

Tracing with steam is very reliable, and will function even if leaks develop or a trap blows steam (although either of these occurrences will waste steam and increase the operating cost). During plant maintenance, steam-tracing systems are the easiest and safest to bypass, and this is a basic reason for their widespread use. Furthermore, steam is most often the most reliable utility, and a steam system has the additional advantage of being relatively easy to check for failures.

On the negative side, steam tracing requires labor with suitable skills to ensure that it is installed satisfactorily. The steam supply and condensate return systems—usually consisting of headers, subheaders, manifolds, individual feeders, trays, strainers and valves—are cumbersome and may take up considerable space. Steam systems are high-maintenance items. Traps must be closely monitored in order to reduce steam wastage.

Trap failures often result in freezing, causing bursting of the tracer and, often, insulation damage. In cold climates, condensate return lines can freeze, which can result in a loss of tracers. In steam tracing, temperatures cannot be controlled precisely, and overheating is not easily avoided. Although spacers can be used to avoid direct contact between the tracer and traced pipe, this adds to the cost and is not totally effective in eliminating overheating. This makes steam tracing undesirable for heat-sensitive processes that require close temperature control. Sometimes, it is necessary to use jacketed pipe and equipment to ensure adequate protection while avoiding excessively high steam temperatures.

In theory, tracers can be turned on and off in response to seasonal temperature variations. However, in practice, the best that can be realistically achieved is turning off the tracers in the spring, then turning them on again in the fall, with considerable wastage of steam for much of the year. In situations where *some* of the tracers must be left on all year (for higher-freezing-point materials), often *all* are left on, with even more steam wastage.

As a result of the increasing cost of steam, thermostatically operated valves for steam tracers have become available. These automatically shut off the steam whenever the ambient temperature (or process temperature) rises above a selected value. Although these have not been widely used, they can reduce steam-tracing operating costs appreciably, and should gain wider use as steam costs continue to rise.

Steam is not practical for heat tracing where very high temperatures are required, because the saturation pressure of steam increases rapidly as the temperature increases. This makes the cost of a steam tracing system prohibitive for high-temperature tracing.

In handling hazardous vapors, care must be taken that the surface temperature of the heat tracer, hence the saturation temperature of the steam, does not exceed 80% of the ignition temperature of the gas involved. This can limit the pressure of the steam and may require multiple tracers, jacketed piping, or the use of electric temperature-selflimiting tracers, described later.

Circulating-media systems

In high-temperature applications where steam would be impractical, hot-oil systems are frequently used, but are economic only if hot oil is required for major process heating.

Systems using a circulating antifreeze are often used in arctic climates where there are freezing problems in steam condensate systems that makes these unsuitable.

Circulating-media systems are expensive to install and operate, and are also subject to failure because of pump stoppages or loss of circulating-medium inventory. In addition, inventories of oil or antifreeze must be maintained to permit continued operation in the event of leakage, and leakage can be hazardous (particularly with hot oil).

Because they are the most expensive, heat tracing systems using circulating media are specified only where special process, or ambient, conditions are involved. Besides providing protection at temperatures both above and below the range of applicability of steam tracing systems, circulating-media systems can easily

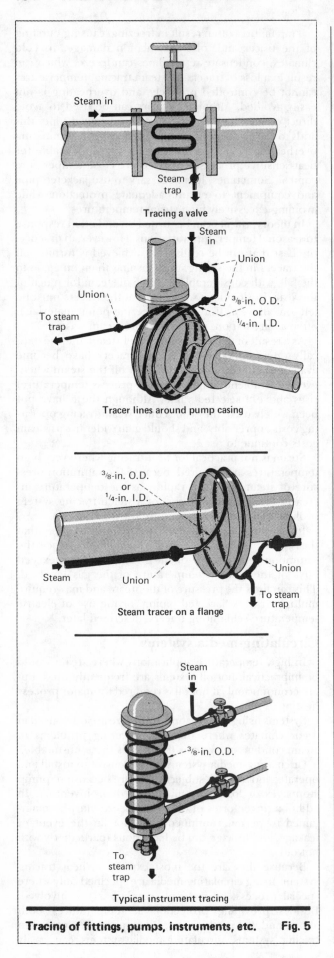

Tracing a valve

Tracer lines around pump casing

Steam tracer on a flange

Typical instrument tracing

Tracing of fittings, pumps, instruments, etc.   Fig. 5

avoid overheating, since the temperature of the flowing medium can be kept close to the required process temperature.

Electric tracing

Electric tracers are available in three styles, having different (though somewhat overlapping) ranges of applicability. With selection of the appropriate style, electric tracing is usable over a very wide range of temperatures; from those below that for which steam is suitable to those above that for which hot oil is applicable. The major advantage of electrical tracing is the ease with which it can be thermostatically controlled, either by ambient temperature thermostats or by thermostats directly in contact with the pipe surface. This use of thermostats can effect substantial savings in operating costs when not needed, as well as effectively minimizing (or even totally avoiding) overheating of the traced pipe.

Electric tracing systems can also be installed with pilot lights to provide a simple means of determining whether the system is functioning. Electric tracers are available not only in tape or cable form but also in blanket form and in special custom shapes. The tracers are made for general-purpose and all types of Div. II hazardous-electrical-area classifications. However, particular attention should be given to surface temperatures. In a recent application involving ethylene, even though the tracing system had the appropriate hazardous-area classification, surface temperatures were above 80% of the gas ignition temperature, and special provisions had to be made.

In outlying plant locations, steam is often not available, whereas electricity, at least for lighting, almost always is. Thus, use of electric tracing there has a very considerable cost advantage.

Although, in many plants, loss of power supply is more likely than loss of steam, electric tracers can survive the loss of power, and exposure to freezing temperatures, without damage.

Despite the many advantages of electric tracing, including—in many cases—lower costs, steam tracing is still the primary choice in CPI plants. Even though recent advances in insulation, jacketing and sheathing materials have made damage by chemical attack less likely, the potential hazards associated with the temporary dismantling of tracing for maintenance often make electric tracing the second choice.

Although steam tracers are easily provided with break unions to permit maintenance and removal of valves, equipment and instruments, as well as the installation of jumpers, there is no convenient equivalent of a break union available for electric tracers. Furthermore, the extra wraps and loops of cable needed to permit maintenance are subject to damage, and potentially hazardous.

In evaluating the use of electric heat tracing for winterization, the choice exists among the three basic styles, namely, series-resistance conductors, parallel-resistance cable, and temperature-selflimiting heating tape. The specific features of each of these types follow.

Series-resistance conductors

This style of electric tracer consists of a pair of conductors in a tape or cable, connected in series. Due to their

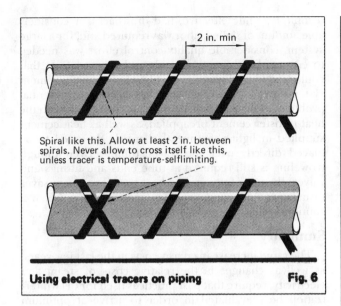

2 in. min

Spiral like this. Allow at least 2 in. between spirals. Never allow to cross itself like this, unless tracer is temperature-selflimiting.

Using electrical tracers on piping Fig. 6

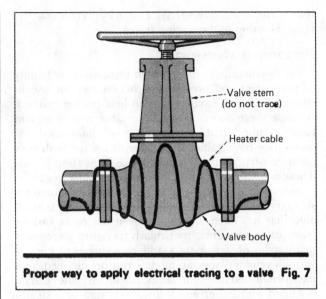

Valve stem (do not trace)

Heater cable

Valve body

Proper way to apply electrical tracing to a valve Fig. 7

nature, each section must be specifically designed to be a fixed length with a predetermined resistance, selected to yield a desired heat rate that is constant over its entire length. The tracing system requires fairly accurate pre-engineering, and factory assembly of custom designed sections. Field modification is generally impractical, and maintaining inventories of individual spare cables is often necessary. This type of tracer is subject to total failure in the event of conductor breakage (at even a single point) and is subject to burnout, particularly if overlapping is not avoided during installation. Thermostatic control is necessary to avoid process overheating and to reduce the frequency of burnout.

Despite their inherent disadvantages, series-resistance electric tracers are a suitable choice for high-temperature, high-input-rate tracing since their construction allows for the use of high-temperature insulating material, and conductors suitably sized to handle high currents.

Parallel-resistance cable

This style consists of two insulated parallel conductors wrapped with a fine-resistance-wire coil that is alternately fastened to each of the conductors at intervals of about 2 to 3 ft, depending on the desired heat-input rate. The conductors and resistance wire are then encased in a protective sheath.

Various sheath and insulating materials are available, as are variations in resistance wire size and contact intervals. These enable the making of heater tapes with maximum exposure temperatures ranging between 200 and 600°F, and heat input rates generally between 2 and 15 watts/ft. This type of tape is relatively flexible and can be used for wrapping valves and instruments but, like series-resistance cables, is subject to burnout at overlapping points.

Because of the parallel conductor construction, failure of one of the thin heating wires will cause loss of heat input over only a limited length. The parallel construction permits field cutting to the necessary length, and only limited preengineering is necessary. Since only parallel conductors have to carry the full current, the maximum length of tracer can often be 200 ft or more,

and the desired heat input is not appreciably affected by tracer length.

Since the heat input rate is constant, thermostatic control of surface temperatures is essential if overheating of the traced pipe is to be avoided.

Temperature-selflimiting heating tape

The most recently developed style of electric tracer cable is the selflimiting type. This style consists of two parallel conductors embedded in a semiconductive core, which has the unique property of rapidly increasing electrical resistance with increasing temperatures until the core material becomes nonconductive, stopping the generation of heat entirely. This "cutoff" temperature can be adjusted by changes in the formulation of the core material, and tapes are designed with maximum cutoff temperatures between 150 and 275°F.

This selflimiting feature has significant advantages. Local overheating can be avoided, and there is automatic conservation of power consumption as external temperatures and pipe flowrates vary, even if no thermostats are installed. This style of tape is specified with a nominal heat input rate at about 50°F. This nominal rate can be between about 3 and 30 watts/ft. Selflimiting heating tapes have maximum exposure limits, generally between 180 and 350°F, depending on the jacket material used. With appropriate selection of materials, this type of tape is also available with short-term limits as high as 500°F. These high-exposure temperature ratings are necessary when either steaming-out of lines, or other high-temperature short-term exposure, is anticipated.

The selflimiting construction of this style of tracer makes it essentially immune from burnout even if the tape is overlapped. Further, in the case of lower temperature ratings, this tape is quite thin and does not require any increase in diameter of the insulation, as would be necessary with steam tracers.

As with parallel-resistance heaters, tracer length has little effect on heat input rate. Maximum length can be 100 ft or more. However, since the electrical conductivity of the tape increases as the temperature drops, the power supply must allow for the current inrush during cold startups, and should have a capacity greater than the nominal rating of the tracing system.

Preformed accessories

For 15 years, there has been an increasing availability of preformed and preinsulated accessories for use in steam, electric or circulating-fluid heat tracing systems. Because these accessories reduce labor and time, and result in more reliable and more easily maintained systems, they have gained wide acceptance for both new plant construction and maintenance in existing plants. These accessories fall into the following main categories:

Instrument enclosures — In addition to conventional insulated meter boxes, premolded foam insulation is available that has been specifically designed to fit various types of instruments, particularly pressure gages and many styles of transmitters. Different shapes to suit the various vendors' specific models can be supplied, and have preformed holes for valves, drains, etc. They can be obtained with either an electrically-heated or a steam-heated element designed to provide protection against freezing without exposing the instruments to excessive temperatures.

Besides providing freeze protection, these enclosures protect against corrosion and splashes. Most important, they can easily be removed without damage, and are available for quick reinstallation at the conclusion of recalibration or other maintenance operations.

Pretraced tubing and tubing bundles — Pretraced, preinsulated and prejacketed tubing and tubing bundles are frequently the most cost-effective choice for tracing instrument impulse lines, analyzer taps, sample connections or other small-diameter lines, and for use as steam supply and condensate return leads.

Bundles are available with a single tube, or with multiple tubes of various sizes and materials. The multiple tubes are available with the tubes in contact, or separated with a layer of insulation to prevent overheating. In addition, preinsulated bundles with electrical tracing are also available, having the tracers of the temperature self-limiting type or constant-heat-input type, with or without thermostatic control.

The most commonly used bundle combinations are:
- ⅜-in. copper tube (for steam) with either one ½-in. 316 SS tube (for pressure meters or sample lines) or two ½-in. 316 SS tubes (for flow meters).
- ½-in. copper tube for steam supply and condensate runs between manifolds and users.

Bundles with two, three or four ⅜-in. copper tubes are also occasionally used for steam supply and condensate collection lines. Bundles with two ⅜-in. copper tubes and one ½-in. 316 SS tube are also useful for tracing instrument impulse lines, so that the same bundle can be used for steam tracing and condensate return.

Use of the preinsulated bundles greatly reduces the amount of labor and inconvenience in handling loose insulation materials, and the bundles are particularly advantageous in maintenance work, since the tracers are almost immediately available for use.

Preapplied heat transfer cement

When heat transfer cements were first introduced, they were formulated to be applied by troweling. This technique, while effective, had drawbacks. A considerable amount of skilled labor was required and, for a large system, considerable quality-control effort was needed to ensure the absence of voids and other defects that could lead to cold spots. Preapplied heat transfer cement systems were developed to overcome these potential problems. These include heating panels or coils with the heat transfer cement preapplied, as well as heat cement supplied in light sheet-metal channels designed to be placed directly over the tracer tubing. Even though troweling is still required at tube ends and fittings and valves, the use of preapplied heat transfer cement saves installation costs and provides greater assurance of a sound installation.

Summary

The rapid increase in energy costs in the last few years, as well as changes in the relative costs of steam and electricity, require that past practices used for CPI winterization be reevaluated in order to arrive at optimum designs. Because of higher energy costs, greater use should be made of winterization techniques not based on heat tracing. Where tracing is required, sound design requires careful evaluation of the relative merits of steam, electricity and other heating media, as well as the savings that can be made by using recent developments in heat tracing materials. These can lead to substantial cost savings, not only in new plant construction but in the maintenance of existing plants. Also, newly developed materials and accessories offer additional savings as well as increased effectiveness and reliability.

Acknowledgment

The author wishes to acknowledge the assistance of his colleagues at Scientific Design Co., not only in the preparation of this article but for their efforts over the years in evaluating new techniques and developing improved specifications and more effective procedures for plant winterization.

References

1. Lonsdale, J. T., and Mundy, J. E., Estimating Pipe Heat Tracing Costs, *Chem. Eng.*, Nov. 29, 1982, p. 89.
2. Kumana, D. K., and Kothari S. P., Predict Storage Tank Heat Transfer Precisely, *Chem. Eng.*, Mar. 22, 1982, p. 127.
3. House, F. F., Winterizing Chemical Plants, *Chem. Eng.*, Sept. 11, 1967, p. 177.
4. Ibid, p. 173.
5. Blackwell, W. W., Estimate Heat Tracing Requirements for Pipelines, *Chem. Eng.*, Sept. 6, 1982, p. 115.

The author

Eugene Fisch is a project manager at The Halcon SD Group's Scientific Design division, 2 Park Ave., New York, NY 10016, tel. (212) 689-3000, where he has been engaged for more thant 30 years in a variety of activities associated with the design and construction of chemical plants worldwide. He holds a B.S. degree in chemical engineering from City College of New York, and is a licensed professional engineer in New York.

Estimate heat-tracing requirements for pipelines

Because the equations for tracing contain two unknown quantities, this calculator program contains a rapid procedure for determining the film heat-transfer coefficient to air. The program then continues to establish the remaining design values for heat tracing.

W. Wayne Blackwell, Ford, Bacon & Davis, Texas Inc.

□ Pipelines containing liquids are often heat-traced to prevent the liquids from freezing or becoming too viscous to flow. Pipelines handling gases are sometimes heat-traced to prevent components or water vapor in the gases from condensing.

The program to be described will allow us to rapidly calculate the heat loss and tracing requirements for any given pipeline, using hot oil or other fluid medium as the heat source for the tracer. The program was written for the Texas Instruments TI-59 programmable calculator, to be used with the PC-100C printer.

This line-tracing program:
- Calculates surface temperature of insulated pipe.
- Calculates heat transferred per 100 ft of pipe.
- Calculates total heat transferred.
- Determines flowrate for hot media.
- Estimates number of heat tracers required without heat-transfer cement.
- Estimates number of heat tracers required with heat-transfer cement.

The program can be used without the printer because most results are stored in the TI-59 registers.

Equations for heat, flow and temperature

The program solves the following equations[†]:

$$Q = 2\pi K_i(T_a - T_s)/\ln(d_o/d_i) \qquad (1)$$

$$Q = h_a(\pi d_o/12)(T_s - T_{air}) \qquad (2)$$

$$X = \ln(d_o/d_i)(h_a)(d_o/12)/2K_i \qquad (3)$$

$$T_s = (T_a + XT_{air})/(X + 1) \qquad (4)$$

$$Q_t = QL_L \qquad (5)$$

$$W = Q_t/C_p(T_{mi} - T_{mo}) \qquad (6)$$

$$T_{woc} = Q/a(T_{med.avg.} - T_p) \qquad (7)$$

$$T_{wc} = Q/b(T_{med.avg.} - T_p) \qquad (8)$$

[†]Eq. (1) and (2) are from Ref. 1; Eq. (7) and (8), Ref. 2; and Eq. (9), Ref. 3.

$$h_c + h_r = 564/(d_o)^{0.19}[273 - (T_s - T_{air})] \qquad (9)$$

$$W_F = A + B(T_s - T_{air}) + C(T_s - T_{air})^2 \qquad (10)$$

$$h_a = (h_c + h_r)W_F \qquad (11)$$

Kern [1] and others have demonstrated that the heat transferred through an insulated pipe encounters four resistances: (1) film resistance on inside wall of pipe, (2) heat resistance through pipe wall, (3) heat resistance through insulation, and (4) air film resistance on outside of insulation. The first two resistances are normally very small, and have been neglected in this program.

For this program, Eq. (1) and (2) were equated, and the terms rearranged to form Eq. (4). Since Eq. (3) and (4) involve two unknowns (h_a and T_s), an initial value of h_a is assumed and T_s calculated. The program then calculates a new value of h_a from Eq. (9), and Eq. (4) is resolved for a new T_s. This procedure is repeated until the film heat-transfer coefficient changes less than 0.01 from the previous calculation.

After T_s has been determined, the program continues to calculate Q, W, and the number of tracers, with and without transfer cement, required to maintain pipeline temperatures. Heat losses are based on a 20-mph wind speed, but may be adjusted for zero wind speed, as will shortly be explained.

Using the program

Table I lists the detailed program-operating instructions. After entry of the program (Steps 000 to 361) into program memory, and entry of the required constants in Storage Registers 18, 19, 20, 23, 24 and 25 (as outlined in Table III), the program and contents of the storage registers are down-loaded onto magnetic cards. Once this information has been thus stored, the program is ready for use.

The user need only read in the magnetic cards, store pertinent data in Storage Registers 0 through 10, and press **A** to begin the calculations (Table II). Usually a first guess of about 4 for h_a speeds up convergence and

Originally published September 6, 1982

167

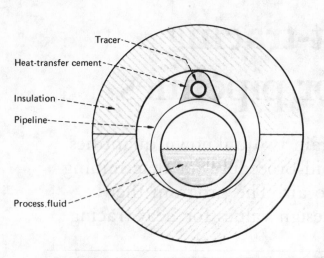

Single tracer

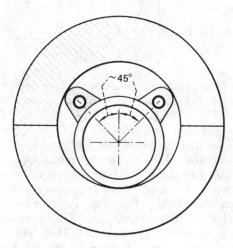

Two tracers

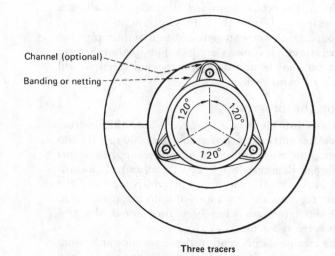

Three tracers

**Configuration for heat tracers
depends on number required**

Program for calculating total heat transferred, flowrate

Step	Code	Key	Step	Code	Key	Step	Code	Key
Start program			059	11	11	120	54	)
			060	55	÷	121	95	=
000	76	LBL	061	43	RCL	122	35	1/X
001	11	P	062	12	12	123	65	×
002	69	OP	063	95	=	124	05	5
003	00	00	064	23	LNX	125	06	6
004	43	RCL	065	65	×	126	04	4
005	18	18	066	43	RCL	127	95	=
006	69	OP	067	10	10	128	42	STO
007	01	01	068	65	×	129	22	22
008	43	RCL	069	43	RCL	130	43	RCL
009	19	19	070	11	11	131	14	14
010	69	OP	071	55	÷	132	75	-
011	02	02	072	01	1	133	43	RCL
012	43	RCL	073	02	2	134	04	04
013	20	20	074	55	÷	135	95	=
014	69	OP	075	02	2	136	42	STO
015	03	03	076	55	÷	137	27	27
016	69	OP	077	43	RCL	138	43	RCL
017	05	05	078	08	08	Calculate wind factor		
018	43	RCL	079	95	=			
019	02	02	080	42	STO	139	23	23
020	85	+	081	16	16	140	42	STO
021	43	RCL	082	65	×	141	26	26
022	03	03	083	43	RCL	142	43	RCL
023	95	=	084	04	04	143	27	27
024	55	÷	085	85	+	144	65	×
025	02	2	086	43	RCL	145	43	RCL
026	95	=	087	13	13	146	24	24
027	42	STO	088	95	=	147	95	=
028	21	21	089	55	÷	148	44	SUM
029	85	+	090	53	(	149	26	26
030	43	RCL	091	43	RCL	150	43	RCL
031	01	01	092	16	16	151	27	27
032	95	=	093	85	+	152	33	X²
033	55	÷	094	01	1	153	65	×
034	02	2	095	54	)	154	43	RCL
035	95	=	096	95	=	155	25	25
036	42	STO	097	42	STO	156	95	=
037	13	13	098	14	14	157	44	SUM
038	43	RCL	099	76	LBL	158	26	26
039	05	05	100	13	C	159	43	RCL
040	85	+				160	26	26
041	43	RCL	Calculate $h_c + h_r$			161	65	×
042	06	06				162	43	RCL
043	95	=	101	43	RCL	163	22	22
044	42	STO	102	11	11	164	95	=
045	12	12	103	45	YX			
046	85	+	104	93	.	Calculate h_a		
047	53	(	105	01	1			
048	43	RCL	106	09	9	165	42	STO
049	07	07	107	95	=	166	28	28
050	65	×	108	65	×	167	93	.
051	02	2	109	53	(	168	00	0
052	54	)	110	02	2	169	01	1
053	95	=	111	07	7	170	32	X:T
054	42	STO	112	03	3	171	43	RCL
055	11	11	113	75	-	172	10	10
056	76	LBL	114	53	(	173	75	-
057	12	B	115	43	RCL	174	43	RCL
			116	14	14	175	28	28
Calculate T_s			117	75	-	176	95	=
058	43	RCL	118	43	RCL	177	50	I×I
			119	04	04			

Tracer

Heat-transfer cement

Insulation

Pipeline

Process fluid

~45°

Channel (optional)

Banding or netting

120° 120° 120°

of heat-transfer medium, and number of heat tracers   Table I

Step	Code	Key	Step	Code	Key	Step	Code	Key	Step	Code	Key	Step	Code	Key	Step	Code	Key
178	77	GE	208	75	-	242	06	06	276	95	=	Print T_{woc}			336	04	4
179	15	E	209	43	RCL	243	55	÷	277	42	STO	306	69	OP	337	03	3
180	76	LBL	210	04	04	244	43	RCL	278	29	29	307	04	04	338	01	1
181	14	D	211	54	)	245	09	09	279	65	×	308	32	X:T	339	05	5
			212	95	=	246	55	÷				309	69	OP	340	69	OP
Calculate variables			213	42	STO	247	53	(	Constant a						341	04	04
182	69	OP	214	15	15	248	43	RCL	280	93	.	310	06	06	342	32	X:T
183	00	00	215	69	OP	249	02	02	281	03	3	311	43	RCL	343	69	OP
184	03	3	216	00	00	250	75	-	282	09	9	312	29	29	Print T_{wc}		
185	07	7	217	03	3	251	43	RCL	283	03	3	313	65	×	344	06	06
186	03	3	218	04	4	252	03	03	284	95	=	Constant b			345	98	ADV
187	06	6	219	69	OP	253	54	)	285	35	1/X	314	04	4	346	91	R/S
188	69	OP	220	04	04	254	95	=	286	65	×	315	93	.	347	76	LBL
189	04	04	221	43	RCL	255	42	STO	287	43	RCL	316	05	5	348	15	E
190	43	RCL	222	15	15	256	17	17	288	15	15	317	08	8	Readjust h_a		
191	14	14	223	69	OP	257	02	2	289	95	=	318	95	=	349	43	RCL
192	69	OP	224	06	06	258	07	7	290	85	+	319	35	1/X	350	28	28
193	06	06	225	01	1	259	01	1	291	93	.	320	65	×	351	42	STO
194	43	RCL	226	04	4	260	04	4	292	09	9	321	43	RCL	352	10	10
195	28	28	227	03	3	261	06	6	293	09	9	322	15	15	353	61	GTO
196	65	×	228	07	7	262	03	3	294	09	9	323	95	=	354	12	B
197	89	π	229	04	4	263	02	2	295	95	=	324	85	+	355	91	R/S
198	65	×	230	01	1	264	03	3	296	59	INT	325	93	.	356	76	LBL
199	43	RCL	231	02	2	265	69	OP	297	32	X:T	326	09	9	357	10	E'
200	11	11	232	03	3	266	04	04	298	03	3	327	09	9	Print data registers		
201	55	÷	233	69	OP	267	43	RCL	299	07	7	328	09	9	358	00	0
202	01	1	234	04	04	268	17	17	300	04	4	329	95	=	359	22	INV
203	02	2	235	43	RCL	269	69	OP	301	03	3	330	59	INT	360	90	LST
204	65	×	236	15	15	270	06	06	302	03	3	331	32	X:T	361	91	R/S
205	53	(	237	65	×	271	43	RCL	303	02	2	332	69	OP	End program		
206	43	RCL	238	43	RCL	272	21	21	304	01	1	333	00	00			
207	14	14	239	00	00	273	75	-	305	05	5	334	03	3			
			240	95	=	274	43	RCL				335	07	7			
			241	69	OP	275	01	01									

reduces run time. After a program run, intermediate results are maintained in unused Storage Registers 11 through 29. See Table III for all stored information.

In this program, the combined convection and radiation heat-transfer coefficients are corrected by a wind factor to calculate h_a. If designing for zero wind conditions, enter 1.0 in Storage Register 23, 0 in Storage Registers 24 and 25, and run the program as usual.

Thermal conductance values, a and b, used in this program are for $\frac{1}{2}$-in. tracer lines. The user may substi-

Nomenclature

A,B,C	Constants for wind-factor equation
a	Thermal conductance, tracer to pipe, without heat-transfer cement, Btu/(h)(°F)(ft of pipe)
b	Thermal conductance, tracer to pipe, with cement, Btu/(h)(°F)(ft of pipe)
C_p	Specific heat of hot medium, Btu/(lb)(°F)
d_i	Inside diameter of insulation, in.
d_o	Outside diameter of insulation, in.
h_a	Film heat-transfer coefficient to air (corrected for wind), Btu/(h)(°F)(ft²)
$h_c + h_r$	Combined convection and radiation heat-transfer coefficient, Btu/(h)(°F)(ft²)
K_i	Thermal conductivity of insulation, Btu/(h)(ft²)(°F/ft)
L_L	Total pipeline length, ft
Q	Heat lost per ft of pipe, Btu/(h)(ft)
Q_t	Total heat lost from pipeline, Btu/h
T_a	Average temperature of pipe and tracer, °F
T_{air}	Air temperature, °F
$T_{med.avg.}$	Average temperature of hot medium, °F
T_{mi}	Inlet temperature of hot medium, °F
T_{mo}	Outlet temperature of hot medium, °F
T_p	Temperature in pipe, °F
T_s	Outside surface temperature of insulation, °F
T_{wc}	Number of tracers required with heat-transfer cement
T_{woc}	Number of tracers required without heat-transfer cement
W	Flowrate of hot medium, lb/h
W_F	Wind factor

User instructions　Table II

Step	Procedure	Enter	Press	Display
1.	Read in both magnetic cards, Sides 1, 2 and 4		CLR	1, 2, 4
2.	Store data in registers R0 through R10	$R0 = L_L$		Data
		$R1 = T_p$		
		$R2 = T_{mi}$		
		$R3 = T_{mo}$		
		$R4 = T_{air}$		
		$R5 = $ Pipe O.D.		
		$R6 = $ Tracer allowance		
		$R7 = T_k$		
		$R8 = K_i$		
		$R9 = C_p$		
		$R10 = h_a$ (est.)		
3.	Press A to begin computations		A	T_{wc}
4.	Option: Press E' for printout of data registers		E'	0

Contents of data registers　Table III

0. Line length, ft
1. T_p, °F
2. T_{mi}, °F
3. T_{mo}, °F
4. T_{air}, °F
5. Pipe O.D., in.
6. Tracer allowance, in.
7. T_k, in.
8. K_i, Btu/(h)(ft²)(°F/ft)
9. C_p, Btu/(lb)(°F)
10. h_a (trial calculation, Btu/(h)(°F)(ft²)
11. d_o, in.
12. d_i, in.
13. $T_{avg.\ inside}$, °F
14. T_s, °F
15. Q, Btu/(h)(ft)
16. X
17. W, lb/h
18. 2724311700*
19. 3735131517*
20. 3500332230*
21. $T_{med.\ avg.}$, °F
22. $h_c + h_r$, Btu/(h)(°F)(ft²)
23. 2.814*
24. −0.0003885714*
25. −0.0000012857*
26. W_F
27. $T_s − T_{air}$, °F
28. h_a', Btu/(h)(°F)(ft²)
29. $T_{med.\ avg.} − T_p$, °F

*Constants that must be stored on magnetic card before program execution (first time only).

tute constants for other-sized tracers, as given in Table IV. Constant *a* occupies Program Steps 280 through 283, and *b* occupies Steps 314 through 317. Constants for other-sized tracer lines may be keyed into the same area of the program.

An example

Estimate the number of tracers required to maintain 100 ft of 6-in.-dia. process line at 500°F. Hot tracing medium is available at 625°F, and has a heat capacity of 0.53 Btu/(lb)(°F). The process line is covered with 2.5 in. of insulation whose thermal conductivity is 0.037 Btu/(h)(ft²)(°F/ft). Design this system for 0°F air tem-

Thermal conductance values for tracer lines　Table IV

Tube size, in.	Constant, a	b
3/8	0.295	3.44
1/2	0.393	4.58
5/8	0.490	5.73

See Eq. (7) and (8)

User-defined keys　Table V

A – Starts program

B – Calculates T_s (internal)

C – Calculates $h_c + h_r$ (internal)

D – Calculates Q, Q_t, W, and number of tracers (internal)

E – Readjusts h_a for new trial (internal)

E' – Prints data registers

perature and 20-mph winds. The tracing medium is to be returned at 550°F. Use ½-in. tracers.

Enter the problem variables into the calculator:

Variable	Register	Variable	Register
$L_L = 100$	(R0)	Tracer	
$T_p = 500$	(R1)	allowance* = 1.25	(R6)
$T_{mi} = 625$	(R2)	$T_k = 2.5$	(R7)
$T_{mo} = 550$	(R3)	$K_i = 0.037$	(R8)
$T_{air} = 0$	(R4)	$C_p = 0.53$	(R9)
Pipe O.D. = 6.065	(R5)	h_a (trial) = 4.0	(R10)

Press Key A to run the program. The results print as:

```
LINE TRACER PGM
 18.84804484        TS
234.8519058          Q
23485.19058        BTUH
590.8224045        LB/H
           7.      TWOC
           1.       TWC
```

The estimated number of tracers without the heat-transfer cement for this example is seven, while using a heat-transfer cement reduces the required number to one. Circulation rate of the tracing medium is 590.8 lb/h, and heat lost from 100 ft of pipeline is 23,485 Btu/h.

Users not having a printer may recall most of the calculated results from the data registers (see Table III). The value displayed after program execution is T_{wc}.

For HP-67/97 users

The HP version closely follows the TI program. Table VI offers the HP program listing, and Table VII provides user instructions for the HP version. Table VIII lists the contents of the HP data registers.

*Allow approximately 1¼ in. between the pipe and insulation to accommodate the ½-in. tracer line and heat-transfer cement. For three, or more, tracers, allow twice this value. Smaller tracers may require only ⅞ to 1 in. of space. Tracers are normally spaced equidistant around the pipe (see illustration), and are run parallel to the pipe. A final run with the calculator program may be made after the total number of tracers and spacing has been established.

Program listing for HP version   **Table VI**

Step	Key	Code	Step	Key	Code	Step	Key	Code	Step	Key	Code	Step	Key	Code
001	*LBLA	21 11	040	2	02	079	STOE	35 15	118	2	02	157	9	09
002	P≷S	16-51	041	÷	-24	080	RCL4	36 04	119	÷	-24	158	9	09
003	RCL2	36 02	042	2	02	081	P≷S	16-51	120	RCL4	36 04	159	⊤	-55
004	RCL3	36 03	043	÷	-24	082	RCL4	36 04	121	P≷S	16-51	160	INT	16 34
005	+	-55	044	P≷S	16-51	083	-	-45	122	RCL4	36 04	161	DSP0	-63 00
006	2	02	045	RCL8	36 08	084	STOB	35 12	123	-	-45	162	PRTX	-14
007	÷	-24	046	÷	-24	085	P≷S	16-51	124	x	-35	163	RCL2	36 12
008	P≷S	16-51	047	STOD	35 14	086	RCL7	36 07	125	P≷S	16-51	164	4	04
009	STO6	35 06	048	RCL4	36 04	087	STO0	35 00	126	STO5	35 05	165	.	-62
010	P≷S	16-51	049	x	-35	088	RCLB	36 12	127	PRTX	-14	166	5	05
011	RCL1	36 01	050	P≷S	16-51	089	RCL8	36 08	128	P≷S	16-51	167	8	08
012	+	-55	051	RCL3	36 03	090	x	-35	129	RCL0	36 00	168	x	-35
013	2	02	052	+	-55	091	ST+0	35-55 00	130	x	-35	169	1/X	52
014	÷	-24	053	RCLD	36 14	092	RCLE	36 15	131	PRTX	-14	170	RCL5	36 05
015	P≷S	16-51	054	1	01	093	X²	53	132	RCL9	36 09	171	x	-35
016	STO3	35 03	055	+	-55	094	RCL9	36 09	133	÷	-24	172	.	-62
017	P≷S	16-51	056	÷	-24	095	x	-35	134	RCL2	36 02	173	9	09
018	RCL5	36 05	057	STO4	35 04	096	ST+0	35-55 00	135	RCL3	36 03	174	9	09
019	RCL6	36 06	058	*LBLC	21 13	097	RCL0	36 00	136	-	-45	175	9	09
020	+	-55	059	RCL1	36 01	098	RCLE	36 15	137	÷	-24	176	+	-55
021	P≷S	16-51	060	.	-62	099	x	-35	138	PRTX	-14	177	INT	16 34
022	STO2	35 02	061	1	01	100	STOC	35 13	139	STO1	35 45	178	PRTX	-14
023	P≷S	16-51	062	9	09	101	RCLA	36 11	140	P≷S	16-51	179	DSP2	-63 02
024	RCL7	36 07	063	Yˣ	31	102	-	-45	141	RCL6	36 06	180	R/S	51
025	2	02	064	2	02	103	ABS	16 31	142	P≷S	16-51	181	*LBLE	21 15
026	x	-35	065	7	07	104	.	-62	143	RCL1	36 01	182	RCLC	36 13
027	+	-55	066	3	03	105	0	00	144	-	-45	183	STOA	35 11
028	P≷S	16-51	067	RCL4	36 04	106	1	01	145	STOB	35 12	184	GTOB	22 12
029	STO1	35 01	068	P≷S	16-51	107	X≤Y?	16-35	146	.	-62	185	R/S	51
030	*LBLB	21 12	069	RCL4	36 04	108	GTOE	22 15	147	3	03	186	*LBLe	21 16 15
031	RCL1	36 01	070	-	-45	109	*LBLD	21 14	148	9	09	187	DSP9	-63 09
032	RCL2	36 02	071	-	-45	110	RCL4	36 04	149	3	03	188	FREG	16-13
033	÷	-24	072	x	-35	111	PRTX	-14	150	x	-35	189	DSP2	-63 02
034	LN	32	073	1/X	52	112	RCLC	36 13	151	1/X	52	190	RTN	24
035	RCLA	36 11	074	5	05	113	PI	16-24	152	P≷S	16-51	191	R/S	51
036	x	-35	075	6	06	114	x	-35	153	RCL5	36 05			
037	RCL1	36 01	076	4	04	115	RCL1	36 01	154	x	-35			
038	x	-35	077	x	-35	116	x	-35	155	.	-62			
039	1	01	078	P≷S	16-51	117	1	01	156	9	09			

User instructions for HP version   **Table VII**

Store the following data:		
Pipeline length, ft	L_L	STO 0
Temperature in pipe, °F	T_p	STO 1
Inlet temperature, hot medium, °F	T_{mi}	STO 2
Outlet temperature, hot medium, °F	T_{mo}	STO 3
Air temperature, °F	T_{air}	STO 4
Pipe OD, in.		STO 5
Tracer allowance, in.		STO 6
Insulation thickness, in.	T_k	STO 7
Thermal conductivity of insulation, Btu/(h)(ft²)(°F/ft)	K_i	STO 8
Heat capacity of hot medium, Btu/(lb)(°F)	C_p	STO 9
Air film coefficient, estimate, Btu/(h)(°F)(ft²)	h_a	STO A
Exchange registers		P ⇌ S
Store constants:	2.814	STO 7
(See note below)	−3.885712E−4	STO 8
	−1.285E-6	STO 9

Run program with key **A**

Printed output is:

Surface temperature of insulation, °F	T_s
Heat loss per foot of pipe, Btu/h	Q
Total heat loss, Btu/h	Q_t
Flow rate of hot medium, lb/h	W
Number of tracers required:	
without transfer cement	T_{woc}
with transfer cement	T_{wc}

Note: When designing for zero wind conditions, enter the following constants in the place of those given above, 1.0, 0 and 0 in secondary registers 7, 8 and 9.

Contents of data registers—HP version   Table VIII

HP	TI		HP	TI	
0	26	W_F	D	16	X
1	11	d_o, in.	E	22	$h_c + h_r$, Btu/(h)(°F)(ft²)
2	12	d_i, in.	S0	0	line length, ft
3	13	$T_{avg\ inside}$, °F	S1	1	T_p, °F
4	14	T_s, °F	S2	2	T_{mi}, °F
5	15	Q, Btu/(h)(ft)	S3	3	T_{mo}, °F
6	21	$T_{med,\ avg}$, °F	S4	4	T_{air}, °F
7	23	2.814	S5	5	pipe O.D., in.
8	24	-0.0003885714	S6	6	tracer allowance, in.
9	25	-0.0000012857	S7	7	T_k, in.
A	10	h_a (trial calculation, Btu/(h)(°F)(ft²))	S8	8	K_i, Btu/(h)(ft²)(°F/ft)
B	29	$T_{med,\ avg} - T_p$, °F	S9	9	C_p, Btu/(lb)(°F)
C	28	h'_a, Btu/(h)(°F)(ft²)			

References

1. Kern, D. Q., "Process Heat Transfer," pp. 18–20, McGraw-Hill, New York, 1950.
2. Kohli, I. P., *Chem. Eng.*, Mar. 26, 1979, p. 163.
3. Kuong, J. F., *Chem. Eng.*, July 25, 1960, p. 146.

Estimating pipe heat-tracing costs

For the engineer faced with recommending tracing, here are data and a method for estimating installed and operating costs of steam and common electric systems.

Joseph T. Lonsdale, Raychem Corp., and Jerry E. Mundy, Procon International, Inc.

□ Until a decade ago, pipes in process plants were traced almost exclusively with steam. Since then, higher energy and labor costs, and improvements in electric heat tracers, have led to increased application of electric heat tracing.*

Steam tracing's chief advantage is its high reliability. Leaks and failed traps may waste a lot of energy, but these problems do not prevent the tracing from maintaining a line at or above the desired temperature. Steam tracing can heat a line quickly, particularly when thermal contact between the pipe and tracer is enhanced by heat-transfer cement.

Its major disadvantages are high costs, and the inability to closely control pipe temperature. Installation is labor-intensive, and often represents 70–80% of total system cost. Operating cost is high because of the difficulty of controlling temperature, despite advances in temperature-sensitive, self-actuated valves. Energy lost from failed steam traps, and the maintenance required to replace such traps, add to operating cost.

Types of electrical tracing

The three most common types of electrical tracing are mineral-insulated cable, zone heaters and self-regulating heaters. The first has been available for over 40 years, the others for about a decade.

Mineral-insulated cable consists of an outer sheath (of copper, stainless steel, Inconel, or some other metal), insulation of magnesium oxide (hence, the name), and one or more heating conductors (Fig. 1a).

Its major advantage is its exceptionally wide temperature range (up to 1,500°F, with special alloys). If a line is to be maintained above 500°F, there are few practical alternatives, this being beyond the range of existing zone and self-regulating heaters. Such temperatures could be maintained with steam tracing, but would require high-pressure steam (above 600 psi for 500°F, which is not practical). Another advantage is that it resists impact well.

Mineral-insulated cable's chief disadvantage is that it

*Only the more common types of electric tracing—mineral-insulated cable, zone heaters and self-regulating heaters—are discussed. Less frequently used types of tracing, such as skin-effect current tracing, circulating-liquid systems and impedance heating, are not considered.

Originally published November 29, 1982

Typical construction of three common types of electric tracing for process piping   Fig. 1

cannot be cut to length or spliced in the field. This makes design, installation and maintenance more difficult with it than with other types of electric tracers. Also, because it is a series heater, an entire circuit is lost if any section fails.

The zone heater consists of two insulated bus wires wrapped with a thin (38–40 American Wire Gage) Nichrome heating wire and covered with polymer insulation. The heating wire is connected to alternate bus wires at nodes every 12 to 48 in. The distance between nodes constitutes a heating zone. A metallic braid or outer jacket, or both, is usually optional (Fig. 1b).

The main advantage of this tracer is that it can be cut to length in the field. However, one must be aware of node location when cutting, because there will be no heating from the cut to the nearest node. Another advantage is that it operates on standard voltages.

Because it is jacketed with polymer, the zone heater is more flexible than the mineral-insulated cable. However, it usually is not recommended for service above

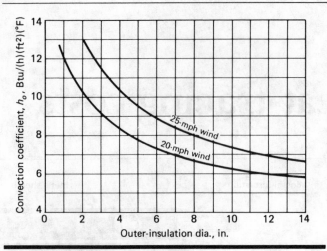

Forced-convection coefficient for flow normal to cylindrical surface   **Fig. 2**

400°F. Also, it is more likely to fail from impact or abuse, because of its thin heating element, than other types of electric tracers.

The self-regulating heater consists of two current-carrying bus wires connected by a conductive polymer over which a polymeric insulator is extruded (Fig. 1c). A metallic braid or fluoropolymeric outer jacket, or both, is optional.

This heater operates on standard voltages, and may be cut to length in the field without concern about nodes (because it has none). However, its main feature is its self-regulation. The resistance of the conductive polymer core between the bus wires increases as the temperature of the core rises. This causes the power output to decrease with rising pipe temperature, and vice versa.

Power output changes independently at each point on the heater. If the temperature of a section of pipe drops off, only the tracing there will respond, by decreasing its resistance and increasing its heat output. This helps prevent low-temperature failures.

Still another important feature of this heater is that it cannot destroy itself with its own heat output, as can a constant-wattage heater. When the latter is trapped inside the insulation or crossed over itself, it will continue

to generate a constant amount of heat and may reach a temperature that destroys it. In such a situation, the self-regulating heater will automatically cut back its heat output.

The self-regulating heater's major disadvantage is its limited temperature range: its maximum heat-maintenance temperature being 300°F, and its maximum intermittent exposure temperature being 400°F.

Evaluating system requirements

The first step in designing a heat-tracing system is to calculate the heat loss from the insulated pipe. This can be done via Eq. (1):

$$Q = \frac{(T_p - T_a)}{(1/\pi D_2 h_o) + (\ln(D_2/D_1)/2\pi K)} \tag{1}$$

Because it is necessary to design for maximum heat loss, the ambient temperature, T_a, in Eq. (1) should be the minimum expected, and the heat-transfer coefficient, h_o, should be chosen for high wind. The heat-transfer coefficient is usually calculated from an empirical relationship between the Nusselt number and the Reynolds number [2]. Fig. 2, which was calculated from this relationship, simplifies finding h_o.

The tracing selected must, of course, be able to replace the maximum heat loss. For example, the heat loss from Line 5 (Table I) is, via Eq. (1), 30 Btu/h/ft, or 8.8 W/ft. A constant-wattage heater must provide this output to supply the required heat. A self-regulating heater capable of at least 8.8 W/ft at 120°F would be adequate, because the power output of this type of heater varies as a function of pipe temperature.

The power output from a steam tracer is calculated from Eq. (2) [1]:

$$Q = C_t(T_s - T_p) \tag{2}$$

Values for C_t are furnished by Kohli [1].

For ½-in. copper tubing without heat-transfer cement, $C_t = 0.393$ Btu/F/ft. Using 50-psi (298°F) steam would result in a heat output of 70 Btu/ft/h on a 120°F pipe, via Eq. (2). This could be decreased to 53 Btu/ft/h with ⅜-in. tracing, $C_t = 0.295$ Btu/(h)(°F)(ft), and reduced still further through the use of lower-pressure steam; however, both of these changes would decrease the allowable circuit length and increase installed costs.

Example list of lines to be traced, with data					Table I
Line no.	Pipe dia., in.	Insulation thickness, in.	Pipe length, ft	Set temperature, °F	Maximum exposure temperature, °F
1.	4	2	120	50	100
2.	4	2	100	50	100
3.	3	1.5	50	140	150
4.	4	2	170	80	90
5.	3	1.5	75	120	200
6.	3	1.5	125	120	200
7.	2	2	100	500	700

Nomenclature

C_t	Heat-transfer coefficient from steam tracer to traced pipe, Btu/(h)(ft)(°F)
D_1	Inside dia. of insulation, ft
D_2	Outside dia. of insulation, ft
h_o	Heat-transfer coefficient from outside of insulation to air, Btu/(h)(ft²)(°F)
K	Insulation thermal conductivity, Btu /(h)(ft) (°F)
Q_p	Heat flow from pipe, Btu/ft/h
Q_t	Heat flow from tracer, Btu/ft/h
T_a	Temperature of air, °F
T_p	Temperature of pipe, °F
T_s	Temperature of steam, °F

Unit costs and method for estimating the installed cost of steam tracing   Table II

The example system consists of 6,000 ft of process piping in an area of 100 ft by 200 ft and of three levels. Steam is available on the lowest level. The average circuit length of steam tracing is 125 ft. The maximum allowable circuit length is 150 ft.

Description and cost	Tracing served, ft	Cost per foot of tracing, $
a. Steam-distribution lines: 300 ft, bottom level 600 ft, 2nd level 600 ft, 3rd level 1,500 ft × $25/ft=$37,500	6,000 (entire system)	6.25
b. Insulation of steam-distribution lines: 1,500 ft × $10/ft=$15,000	6,000 (entire system)	2.50
c. Steam-supply lines: 20 ft × $16/ft=$320	One circuit, 125 ft	2.56
d. Insulation of steam-supply lines: 20 ft × $8/ft=$160	One circuit, 125 ft	1.28
e. Tracers (1/2-in. copper tubing with all fittings): $10/ft	1 ft	10.00
f. Steam-trap assemblies, including 1 trap, 2 valves and 6 ft of piping: $300 per assembly	One circuit, 125 ft	2.40
g. Condensate-collection system: 1,500 ft × $20/ft=$30,000	6,000 (entire system)	5.00
h. Insulation of condensate-collection system: 1,500 ft × $10/ft=$15,000	6,000 (entire system)	2.50
i. Condensate-return lines: 20 ft × $16/ft=$320	One circuit, 125 ft	2.56
j. Insulation of condensate-return line: 20 ft × $8/ft=$160	One circuit, 125 ft	1.28
Total cost per foot of tracing:		$36.33

One steam tracer can provide much more heat than necessary. This is an advantage in that one tracer can handle a wide range of pipe sizes and set temperatures.

Estimating installed costs

The total installed cost of steam tracing is usually 50–150% greater than that for equivalent electric tracing, principally because of the labor required to install the steam-supply, tracing and condensate-return lines.

The following technique will produce a preliminary-type estimate (±25%) of the costs of a steam or electric tracing system.

The lines to be traced, along with the data necessary to calculate the heat loss from each, should be listed (as in Table I).

Estimates of the average circuit length of both the steam and electric systems are necessary for analyzing installed costs. This length is a function of the piping layout and the allowable maximum circuit length of the tracer. For steam tracing, this maximum is usually 150 ft for ½-in. copper tubing, and 60 ft for the ⅜-in. size. For electric tracers, the maximum typically varies from 200–800 ft, depending on the type of tracer, power output required and operating voltage.

The average circuit length of long, straight piping will approach the maximum allowable circuit length of the tracer selected. However, with increasing piping complexity, other considerations (such as the length and position of each line) will limit the actual tracer circuit-lengths.

The best way to estimate the average circuit length is

to choose a number of lines from the line list and actually design the tracer circuits for these lines.* The lines chosen should approximate the entire system in length and complexity.

Steam and condensate lines

Steam-tracing installed costs strongly depend on the lengths of the steam-supply and condensate-return systems. Accurate estimates of these lengths are the keys to a reliable installed-cost estimate. For estimating purposes, the piping connecting the mains and the tracing should be handled in two parts:

1. *The piping required to bring the steam and condensate mains to within 20–40 ft of the lines to be traced.* The lengths of these steam-distribution and condensate-collection lines can be estimated by sketching them on the plan view. If the process area consists of several levels (there are three in the example), the piping that runs vertically, and that on all the levels, must be considered. Items a and g in Table II represent the costs of these lines. The insulation of the lines is also a relevant cost (items b and h in Table II). The required diameters of the steam-distribution lines can be estimated from Table III. Condensate-collection lines are generally somewhat smaller than steam-distribution lines.

2. *The steam-supply lines, which take the steam from the distribution lines to the tracing, and the condensate-return lines, which convey the condensate from the steam traps to the collection*

*A discussion of designing such circuits is beyond the scope of this article; assistance can be obtained from steam-trap and electric-tracing manufacturers. For steam-tracing design, see Kohli [1] and Bertram, C. G., Desai, V. J., and Interess, E., Designing Steam Tracing, *Chem. Eng.*, Apr. 3, 1972, p. 74.

Sizing steam-distribution lines	Table III
No. of 1/2-in. tracers	**Line size, in.**
1—2	3/4
3—5	1
6—15	1 1/2
16—30	2
31—60	2 1/2

lines. The average length of these lines is estimated by designing them to and from a number of tracing circuits. (The tracing circuits designed previously to estimate the average circuit length may be used for this purpose.) The costs of these lines for the example are noted in items c and i in Table II. Their insulation (items d and j) must also be considered.

After the piping lengths between the mains and the tracing have been estimated, the total average cost of tracing per foot of traced pipe can be determined from standard data in Table II. Of course, installed costs will vary with labor rate and productivity, supervisory efficiency, material cost and piping complexity.

Installed cost of electric tracing

Estimating the installed cost of an electric system is easier than estimating that of a steam system, because fewer parts are involved. The cost of the tracer itself constitutes only a small part of the total cost of an electrical system (Table IV). The average distance from the electric-distribution panel to the tracing circuit (i.e., the average power-feeder cable length) is critical to installed cost, particularly in hazardous areas, where the wiring must be run in conduit at a cost as high as $10/ft; elsewhere, the power cable can be run in trays at considerably less cost. This distance also is best estimated from a plan view.

If power and space are already available at a panel of an existing motor-control center, the cost of supplying power to the panel will be negligible. If, however, power

Installed cost of example electric tracing system		Table IV

Example system consists of 6,000 ft of piping in an area 100 ft by 200 ft, three levels. Power is available 50 ft from area. Average circuit length is 250 ft, and maximum allowable length is 400 ft. Area is Class 1, Div. 2 (hazardous).

Description and cost	Tracing served, ft	Cost/ft of tracing, $
a. Power-supply panel: $8,000	6,000 (entire system)	$1.33
b. Panel, 48 circuits: $5,000	6,000 (entire system)	$0.83
c. Power supply to circuits: 300 ft X $10/ft=$3,000	One circuit, 250 ft	$12.00
d. Thermostat (line-sensing): $250	One circuit, 250 ft	$1.00
e. Electric tracer: $4.50/ft	1 ft	$4.50
Total cost per foot		$19.66

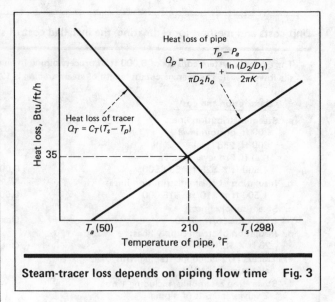

$$Q_p = \frac{T_p - P_a}{\frac{1}{\pi D_2 h_o} + \frac{\ln(D_2/D_1)}{2\pi K}}$$

Heat loss of pipe

Heat loss of tracer
$Q_T = C_T(T_s - T_p)$

Steam-tracer loss depends on piping flow time   Fig. 3

has to be brought from a distance and transformed down to the proper voltage, supplying power can be the most expensive part of an electric system.

Installation time for electric tracing can vary from 4 h/100 ft to more than 10 h/100 ft, depending on piping complexity and type of tracer. Self-regulating heaters can be installed somewhat faster than zone heaters, because the installer need not be concerned with nodes when making connections, or with the tracer destroying itself if it is either caught inside the insulation or crossed over itself. Mineral-insulated cable takes the longest to install, because it is more difficult to bend, cannot be crossed over itself, and each prefabricated length must be fitted precisely on each circuit (usually by spiraling the last 25%, because a safety factor is included in the length of each circuit).

The components of an electrical system are easily estimated. How the parts can be combined to obtain an average cost per foot of tracing is shown in Table IV.

Estimating operating costs

Operating costs—energy and maintenance—can be large enough to influence the selection of a system. The energy-cost tradeoff lies between a less expensive but less efficient form of energy (steam) and a more expensive but more efficient form (electricity).

Total operating cost is most affected by the desired traced-line temperatures, energy cost and the percentage of time that fluid flows in the traced piping. When fluid is not flowing, the energy consumption of a steam tracer can be calculated by the simultaneous solution of Eq. (1) and (2). However, this is more easily determined graphically, as in Fig. 3.

At an ambient temperature of 50°F, the equilibrium temperature is 210°F, and as Fig. 3 indicates, the tracer heat loss is 35 Btu/ft/h. If there is no flow 50% of the time, the heat loss will be 17.5 Btu/ft/h. When fluid is flowing at 130°F, the tracer heat loss can be calculated directly via Eq. (2): $Q_t = [0.393 \text{ Btu}/(\text{ft})(\text{h})(°F)] (298 - 130) = 66 \text{ Btu/h/ft}$. As there is flow 50% of the time, the energy consumption will be 33 Btu/ft/h.

When there is no flow in the traced line, the steam

tracing transfers more energy than is necessary and keeps the line at a higher than desired temperature. When there is flow, the lower temperature of the line draws significantly more energy from the tracer. The energy lost from the steam-distribution and -supply lines can be calculated by means of Eq. (1), using the steam temperature as the pipe temperature.

The energy use of operating traps can vary from 700 to 2,000 Btu/h, depending on the steam pressure and type of trap. Traps for steam tracing usually operate at the low end of this scale; for the example, therefore: 1,000 Btu/h for one trap per 125 ft of tracing = 8 Btu/h/ft.

A failed trap can waste 25,000 Btu/h. The percentage of time that a trap is likely to be passing steam is a critical variable. In a good maintenance program, traps will be checked at least every three months and replaced as necessary. At a typical failure rate of 3%/mo, failures will average about $4\frac{1}{2}$%, with the energy loss = 0.045 (25,000 Btu/h/125 ft) = 9.0 Btu/ft/h.

If steam leaks are reported and promptly fixed, the waste from them can be kept to a minimum. Annual maintenance cost for the example steam tracing system is determined as follows:

1. Check traps four times per year (10 min each, a trap at every 125 ft of tracing): (40/60) ($25/h/125 ft) = $0.13/ft/yr.
2. Replace trap (inline replaceable traps are assumed) every third year: $40/(125 × 3) = $0.11/ft/yr.
3. Repair leaks and blow strainers (one leak every two years): $40/(125 × 2) = $0.16/ft/yr.
4. Total maintenance cost per foot of tracing (adding items 1, 2 and 3) = $0.40 yr.

For the example steam tracing system, the calculated energy transfer, in Btu/h per foot of traced pipe, is as follows: traced line, no flow—17.5; traced line, with flow—33.0; steam-distribution lines—8.8; steam-supply lines—5.0; steam traps, operating—8.0; and steam traps, failed—9.0; for a total of 81.3 Btu/ft/h. The annual cost of 50-psi steam (latent heat of vaporization = 912 Btu/lb) per foot of traced pipe is calculated as follows: [(81.3 Btu/ft/h)/(912 Btu/lb)] ($5/1,000 lb steam) (8,800 h/yr) = $3.92/ft/yr, which is about 11% of the total installed cost.

Therefore, total operating cost = $4.32/ft/yr.

Note the dependency between maintenance and energy costs. If, for example, trap inspections were cut to one per year, the maintenance cost would be reduced by $0.10/ft of tracing but the energy cost would increase by $1.30/ft. A trap inspection program is usually cost-effective.

Operating cost of electric tracing

The energy used by an electric tracing system can be calculated directly from Eq. (1). The pipe temperature will be the set temperature, because the thermostat will cycle the heater to maintain this temperature.* The thermostat should be set slightly below the typical flow-

*Line-sensing thermostats are not always the preferred method of temperature control, particularly for freeze protection. Often, an ambient-sensing thermostat is used to turn off an entire panel when the ambient temperature exceeds 40°F. This can offer a considerable installed-cost saving over using a line-sensing thermostat for each circuit. Self-regulating heaters, which decrease their power drawing as piping becomes warmer, can keep extra energy consumption to a minimum.

ing-fluid temperature, so that the heater will be off when fluid is flowing.

Via Eq. (1), the energy use for the example is 15.2 Btu/ft/h, or 4.44 W/ft. Because fluid is flowing half the time, the actual use is 2.22 W/ft. Therefore, the energy operating cost per foot of tracing is: (2.22 W/ft)($0.04/1,000 W/h)(8,800 h/yr) = $0.78/yr/ft.

Maintenance cost per foot of tracing (checking thermostat and meggering circuit once a year) is: 0.5 h ($25/h/200-ft circuit) = $0.06/yr/ft.

Therefore, total operating cost = $0.84/ft/yr. The $0.84/ft/yr for electric tracing vs. the $4.32/ft/yr for steam tracing is typical for the example pipe size and maintenance temperature.

To compare total operating costs, an analysis must be made for each temperature range. The operating-cost advantage of electric tracing will be less at higher set temperatures. Pipes less than 6 in. dia. are usually kept heated at much less cost with electric tracing. For holding pipes 12 in. and up above 200°F, steam tracing may be cheaper to operate, depending on energy costs and the percentage of time that fluid is flowing.

Sometimes, the availability of excess low-pressure steam is given as justification for using steam tracing. This is not always valid. Consider Table II and IV: the steam system costs $16.67/ft more to install than the electric system. Assuming no cost for steam, and maintenance cost reduced by 50% (because of little concern about wasting steam), the operating cost advantage of the steam system becomes $0.64/ft/yr (i.e., $0.84 − $0.20). This represents a pretax return of only 3.8% on the extra money spent to install the steam system.

References

1. Kohli, I. P., Steam tracing of pipelines, *Chem. Eng.*, Mar. 26, 1979, p. 156.
2. Eckert, E. R. G., "Heat and Mass Transfer," McGraw-Hill, New York, 1963.

The authors

Joseph T. Lonsdale is Manager of Technical Services for the Chemelex Div. of Raychem Corp. (300 Constitution Dr., Menlo Park, CA 94025), responsible for new-product and new-application testing. (Raychem pioneered conductive-polymer self-regulating technology.) Also an adjunct professor at Golden Gate University, he has had 10 years experience in the chemical industry and has taught thermodynamics and unit operations at Northeastern University. He received a B.S. from Northeastern and an M.S. from Purdue University (both in chemical engineering), and an M.B.A. from Harvard University.

Jerry E. Mundy is Vice President, Regional Manager of Business Development for Procon International, Inc. (50 UOP Plaza, Des Plaines, IL 60016), an engineering and construction company serving the chemical process industries. He has had 17 years of experience in engineering and business development with both operating and construction firms. He holds a B.S. in mechanical and industrial engineering from the University of Missouri and has recently completed an executive M.B.A. program at the University of Houston.

Section IV
Condensers

Fog formation in low-temperature condensers

A sulfuric acid mist that might corrode through process piping is an example of a dangerous fog. The following information on predicting and combating fog in low-temperature condensers could prevent a lot of trouble.

Lidia LoPinto, Consultant

☐ When low-temperature condensers are used to recover small concentrations of vapor in a stream of noncondensables, fogging can occur. A fog is a suspension of fine droplets, ranging from less than 1 micron up to 10 microns in diameter. A fog this fine is difficult to capture with conventional equipment such as mist eliminators. If the fog is an aerosol having a corrosive or reactive condensate, it might eventually settle and create corrosion problems downstream.

Fogging is not often reported as a problem because it is seldom looked for. Its presence becomes known only when yield losses, or pollution or corrosion problems become apparent.

A fog can be formed in a condenser when the ratio of noncondensable to condensable vapor is high and the temperature differential is high. One example is in the recovery of trace solvents from effluent gases, using refrigerated condensers for air-pollution-control purposes. Under these conditions, the heat-transfer rate is much greater than the mass-transfer rate, so that the vapor undergoes "quick chilling" and condenses midstream before reaching the condenser wall. A fog will probably form, depending on whether or not nuclei of condensation are available.

When does fog become visible?

Supersaturation is measured by the ratio of the actual vapor pressure of the condensable vapor to the equilibrium vapor pressure at the given temperature:

$$S = \frac{P}{P_{\infty(T)}} \qquad (1)$$

where S = supersaturation ratio; P = vapor pressure, mm Hg; and $P_{\infty(T)}$ = equilibrium vapor pressure at temperature T.

Before fog can become visible by the light-scattering test (Tyndall effect), the mixture will reach a certain level of supersaturation. The level at which a visible fog occurs is called the critical supersaturation ratio, S_{cr}. This value is characteristic of the particular vapor and

Typical values of critical supersaturation ratio

	T, K	S_{cr}
Methyl alcohol	270.0	3.20 ± 0.10
Ethyl alcohol	273.2	2.30 ± 0.05
n-Propyl alcohol	270.4	3.05 ± 0.05
Nitromethane	252.2	6.05 ± 0.75
Water	263.7	4.85 ± 0.08

Source: Ref 3, p. 18.

noncondensable mixture, and varies with temperature. Some typical values, measured by cloud-chamber technique, are shown in Table I.

These values correspond to fairly clean gases. If dusts or other nuclei are present, S_{cr} approaches 1.

S_{cr} can be estimated using Fenhel's formula in [1]:

$$S_{cr} = \text{Exp}\left[84\frac{M}{\rho_L}\left(\frac{\sigma}{T}\right)^{1.5}\right] \qquad (2)$$

where σ = surface tension, dynes/cm
ρ_L = liquid density, lb/ft^3
T = temperature, degrees Rankine
M = molecular weight of condensable

S_{cr} can be measured by experimental means. Note that it decreases with temperature.

For practical purposes, it is sufficient to assume that fog will be formed when $S \geq 1$ [2]. If free gaseous ions exist in the gas, and the droplets become charged, fogging may occur even when $S < 1$. This is discussed in Ref. 3.

Estimating supersaturation

If we calculate the maximum value of S that will occur in a system, we can see whether this is large enough to cause fogging.

A formula for the calculation of S for condensation in

Originally published May 17, 1982

a tube is derived by Amelin [3] from differential heat- and mass-transfer formulas:

$$S = \left(\frac{T-T_2}{T_1-T_2}\right)^{K\delta}\left(\frac{P_1-P_2}{P_{\infty(T)}}\right) + \frac{P_2}{P_{\infty(T)}} \quad (3)$$

where T = temperature at any point in the condenser tube (use consistent units, K or °R)

T_1 = initial gas temperature

T_2 = temperature of the condensation surface

P_1 = vapor pressure of condensable at initial temperature T_1 (use consistent absolute units)

P_2 = vapor pressure of condensable at wall temperature T_2

$P_{\infty(T)}$ = equilibrium vapor pressure at T

$$K = 1 + \left(\frac{P-P_\eta}{P-P_2}\right)\left(\frac{F_n}{F_w}\right) \quad (4)$$

where, as defined in Ref. 3, p. 107:

P = vapor pressure at T

P_η = saturated vapor pressure at the temperature of the nuclei (corrected for droplet curvature). Assume it is equal to $P_{\infty(T)}$ for initial estimate.

F_n = surface area of nuclei/unit pipe length

F_w = surface area of pipe/unit pipe length

$$\delta = \left(\frac{C\rho K_D}{K_T}\right)^m \quad (5)$$

(from Ref. 3, p. 106)

where C = heat capacity of gas, Btu/lb°F

ρ = density of gas, lb/ft^3

K_D = diffusivity of vapor in condensable, ft/h (see box)

K_T = thermal conductivity, Btu/(min) (ft^2) (°F/ft)

and m = a dimensionless index, 0.6 for turbulent flow, 0.7 for laminar flow

Function $S = f(T)$ has a maximum. By calculating S at the maximum, we can see whether fogging occurs, without calculating the whole supersaturation curve, because if fogging occurs at all, it will occur at S_{max}.

To determine S_{max}, $P_{\infty(T)}$ is replaced by the expression:

$$P_{\infty(T)} = e^{C-E/T} \quad (6)$$

(from Ref. 3, p. 2)

where $E = Q/1.987$

and Q = molar heat of vaporization, Btu/lb mole

$S = f(T)$ is differentiated and set equal to 0. P_2 is eliminated, as it is generally negligible compared with P_1. This yields an expression for T_{max} (the temperature at S_{max}):

$$T_{max} = \frac{E \pm \sqrt{E^2 - 4\delta ET_2K}}{2K\delta} \quad (7)$$

if we take the positive root (see Ref. 3, p. 102).

T_{max} is then substituted into the supersaturation expression $S = f(T)$ to obtain S_{max}. If $S_{max} > 1$, then fogging is probably occurring. The fogging will not be visible unless $S_{max} > S_{cr}$, by definition.

Cures for fogging

■ **Reduce ΔT and increase A**—If the temperature differential across the tubes is reduced, and the surface area is increased, the heat flow remains similar, but the driving force for heat transfer is decreased, and the ratio of mass transfer to heat transfer is increased. This "gradual cooling" approach is the most conventional cure cited in various references.

It can be expensive, however. These exchangers will often be handling very corrosive substances and will require exotic materials of construction. In one such design I handled, the condenser tubes were specified to be Hastelloy. When enough area was added to prevent fogging, the price of the condenser nearly doubled.

■ **Use a mist eliminator**—There are some high-performance mist eliminators on the market that can remove particles in the "fogging" range. However, these units require large areas and high pressure drops. Again, costs can soar if exotic materials of construction are specified. If this solution is sought, I recommend that the equipment should be thoroughly tested on a bench scale.

■ **Seed the gas stream with condensation nuclei**—To produce drops that can be captured with a conventional mist eliminator or settler. This is a less-conventional approach (Ref. 3, p. 107), but it is probably less costly than other methods. Testing is recommended to ensure that the scheme will work for the particular chemicals and prevailing conditions.

■ **Filter the gases**—Almost all industrial gases will carry mists and dusts that will provide nuclei for fog condensation. The greater the concentration of these dusts, the lower the value of S_{cr}. Qualitative experimental observations were done by Amelin to confirm this. Fog may in some cases be inhibited by filtration of the gases prior to condensation. A clue to the existence of a dust problem is when a fog is reported but the supersaturation expression yields values of less than the calculated value of S_{cr}.

■ **Other methods**—Some somewhat-less-practical solutions are discussed in the literature. One presents a method of inhibiting fog formation by stringing a heated wire across the center of the exchanger tube. An electrical current will provide the heat.

Gas diffusivity

K_D data are generally scarce. Perry's Handbook, Section 14 [1], contains good methods for estimating gas diffusivity. One way to measure the ratio of K_D to K_T is by taking the wet- and dry-bulb temperatures and employing the formula:

$$\lambda(H_s - H) = \frac{hc}{K_G}(T - T_W)$$

where λ = latent heat of vaporization, Btu/lb

H_s = equilibrium concentration of vapor at temperature T, lb/lb of dry gas

H = actual concentration of vapor, lb/lb dry gas

T = dry-bulb temperature, °F

T_W = wet-bulb temperature, °F

hc = film heat-transfer coefficient, Btu/(lb) (°F) (ft^2)

K_G = mass transfer coefficient, lb/(h)(ft^2)(atm)

and $\dfrac{hc}{K_G} = C\left(\dfrac{K_T}{C\rho K_D}\right)^{2/3}$

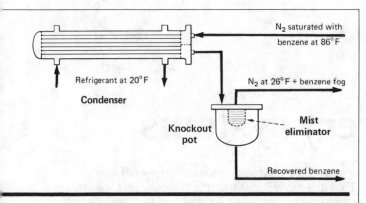

Fog reduces efficiency of benzene recovery   Fig. 1

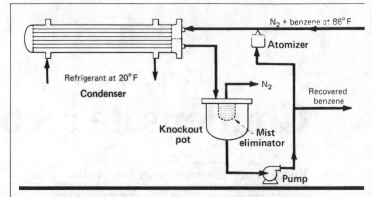

Fogging is reduced by seeding with benzene mist   Fig. 2

Benzene fog in a nitrogen stream

A benzene-recovery condenser is designed to knock down 83% of the benzene from a nitrogen stream (see Fig. 1). Nitrogen saturated with benzene enters the tube bundle at 86°F and exits at 26°F. The refrigerant temperature is 20°F. Check to see whether fogging is a problem, and propose a solution. The pressure is 1 atm, the tubes are 1 in., and the gas velocity through the tubes is 30 ft/s.

A vapor-pressure expression for benzene is correlated from data in Perry [1]:

$$\log P = 6.89272 - 1203.53/(219.8 + t)$$

where $t = $ °C, $P = $ mm Hg (absolute)
$T_1 = 86$°F, $P_{\infty(T_1)} = 119$ mm Hg $= P_1$
$T_2 = 20$°F, $P_{\infty(T_2)} = 17$ mm Hg $= P_2$
$T_{av} = 56$°F, $P_{\infty(T_{av})} = 53$ mm Hg

Average molecular weight, $M = 32$

Gas density from ideal gas law $= 0.085$ lb/ft$^3 = \rho$

Thermal conductivity of $N_2 = 0.009$ Btu/(h)(ft^2)(°F/ft) $= K_T$

Diffusivity of benzene in $N_2 = 0.0104$ cm^2/s or 0.0402 ft^2/h $= K_D$

(above value from Gilliland equation, in Perry, Section 14.0 [1].

Eq. 5 yields:

$$\delta = \frac{C \rho K_D^{0.6}}{K_T} = 0.243, \text{ (turbulent flow)}$$

Eq. 7 yields:

$$T_{max} = \frac{E \pm \sqrt{E^2 - 4\delta E T_2 K}}{2K\delta}$$

where $K = 1.0$ (no nuclei being introduced)
$E = Q/1.987$
Q for benzene $= 14,430$ Btu/lb mole
$= 7,262$ °R^{-1}
$T_{max} = 488$°R or 28°F

$$S_{max} = \left(\frac{T_{max} - T_2}{T_1 - T_2}\right)^\delta \frac{(P_1 - P_2)}{P_{\infty T_{max}}} + \frac{P_2}{P_{\infty T_{max}}}$$

$$S_{max} = \left(\frac{28 - 26}{86 - 26}\right)^{0.243} \left(\frac{119 - 17}{23}\right) + \frac{17}{23}$$

$$S_{max} = 2.67$$

Therefore, fogging is occurring, since $S_{max} > 1.0$.

A simple and effective solution is obtained by recycling some of the recovered benzene through an atomizer to provide condensation nuclei so that large, recoverable droplets are formed (see Fig. 2).

To find the value of K, select a value of $S_{max} \leq 1$ and substitute in Eq. 4:

$$K = 1 + \left(\frac{2.67(23) - 23}{2.67(23) - 17}\right)\left(\frac{F_n}{F_w}\right) = 2.45$$

P_η should be determined experimentally from Eq. 4, but, in this case, we can approximate $P_\eta = P_{\infty T_{max}}$:

Hence $F_n/F_w = 1.67$.

For a 1-in. tube, $F_w = 0.26$ ft^2/ft; therefore, $F_n = 0.43$ ft^2/ft. The total area of nuclei/s to be provided $= 0.43$ ft^2/ft $\times 30$ ft/s $= 12.9$ ft^2/s. Assuming spherical droplets 100μ in diameter are produced by an atomizer nozzle, the flowrate of liquid benzene to be injected into the incoming stream is 0.3 gpm.

It can be seen that the equations for predicting fogging in pipes under turbulent and laminar flow, as derived by Amelin, can be applied to practical condenser design for preliminary calculations. Testing is recommended for final design.

References

1. Perry, Robert H. and Chilton, Cecil H., "Chemical Engineers' Handbook," fifth ed., McGraw-Hill, New York.
2. Branan, C., "The Process Engineer's Pocket Handbook," 1978, Gulf Pub. Co., P.O. Box 2608, Houston, TX 77001.
3. Amelin, A. G., "Theory of Fog Condensation," 1967, Daniel Davey and Co., Inc., P.O. Box 6088, Hartford, CT 06106.

The author

Lidia Llamas LoPinto is a technical writing consultant specializing in process operating manuals and software documentation, 41 Travers Ave., New York, NY 10705, tel: 914-963-3695. She has worked in process-design and environmental and safety departments in a variety of assignments for Stauffer Chemical Co., Chemical Air Pollution Control Co. and American Cyanamid Co. She holds an M.Ch.E. and B.Ch.E. from Manhattan College. She is an Engineer in Training, a member of the Soc. of Women Engineers and an Industrial Advisor for the Junior Engineering Technical Soc.

Condensate recovery systems

**Recovering the energy in hot steam
condensate saves fuel. Some problems
can arise, but they are easily solved.**

Elmer S. Monroe, E. I. du Pont de Nemours & Co.

☐ Most engineers realize that good steam trapping can
save 10% of the cost of fuel for producing steam. But not
all realize that another 10% can be saved by condensate
recovery.

This article will consider:
- The fuel savings possible with condensate recovery.
- Problems that can arise when you are using condensate recovery.
- Engineering solutions to these problems.

Fig. 1 shows the elements of a basic industrial steam-
plant system that will affect the economics of a condensate recovery system. Steam produced in the boiler is
delivered to either process use, heating and ventilating
duty, or the deaerating feedwater heater. The amount
of steam used in heating the feedwater is a function of
the amount and temperature of the condensate that is
recovered and returned to the feedwater heater for
reuse.

As shown in Fig. 2, without any condensate recovery
the amount of steam that must go to the feedwater
heater is about one pound for every five pounds that is
exported. This represents the maximum amount of
steam that can be saved by condensate recovery.

Fig. 2 also shows the amount of recovered condensate
that can be theoretically returned to the powerhouse—
for two basic operating conditions—to provide the necessary heat to replace boiler steam. The upper group of
curves is for the case in which atmospheric flash tanks are
used. For this condition, recovering about 60 to 80% of
the condensate will result in reducing the boiler steam
about 10%.

When atmospheric flash tanks are not used (i.e., direct
return of condensate and flash steam), the amount of
condensate required to eliminate boiler steam to the
deaerating feedwater heater is from 45 to 65%, as shown
by the lower group of curves in Fig. 2. Thus, for the
simple steam plant shown in Fig. 1, the amount of
condensate that can be effectively recovered in a deaerator is limited by the elimination of atmospheric tanks.

Looked at from another viewpoint, Fig. 2 can tell us
that the effectiveness of recovering condensate is least
when we have atmospheric flash tanks, and greatest
when we do not have them.

Fig. 3 expresses these facts in another way. It shows
the approximate boiler fuel saved for the two types of

systems with varying amounts of recovered condensate.
To achieve a 10% reduction in boiler steam for a 200-psi
system, one must collect only 30% of the condensate for
a system without flash tanks. If the same system has
atmospheric flash tanks, it requires twice as much condensate recovery to achieve the same reduction in boiler
fuel. Stated simply, you cannot recover steam if you vent
it to the atmosphere from a flash tank!

While the simple exercises presented by Fig. 2 and 3
show the potential value of condensate recovery, the
situation at some plants may be more complicated. Col-
lecting condensate must be analyzed thoroughly from a
system viewpoint. Some of the complicating factors are:
- Use of steam-driven auxiliaries in the powerhouse.
- Deaerator difficulties.
- Use of surplus heat not recoverable in the deaerator.

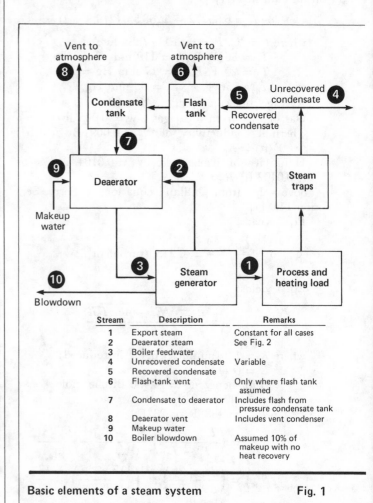

Stream	Description	Remarks
1	Export steam	Constant for all cases
2	Deaerator steam	See Fig. 2
3	Boiler feedwater	
4	Unrecovered condensate	Variable
5	Recovered condensate	
6	Flash-tank vent	Only where flash tank assumed
7	Condensate to deaerator	Includes flash from pressure condensate tank
8	Deaerator vent	Includes vent condenser
9	Makeup water	
10	Boiler blowdown	Assumed 10% of makeup with no heat recovery

Basic elements of a steam system   **Fig. 1**

Originally published June 13, 1983

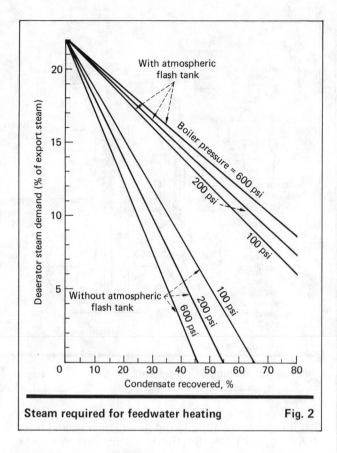

Steam required for feedwater heating   Fig. 2

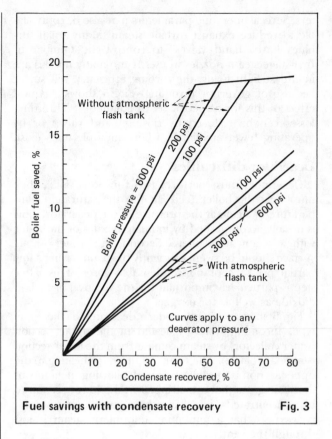

Fuel savings with condensate recovery   Fig. 3

Steam-driven auxiliaries

When steam must be supplied to the deaerator from the boiler, it is feasible to drive auxiliaries—such as fans and pumps—that exhaust at the deaerator pressure. This is cogeneration, and is widely practiced. The turbine exhaust steam must be used in the deaerator if cogeneration is to be economical. This imposes a reduction on the amount of condensate that can be recovered without exceeding the deaerator heat demand. Table I shows some typical values. It should be recognized that the performance of auxiliary steam turbines varies widely, so Table I is, at best, typical.

When auxiliary steam drives limit the amount of recov-

ered condensate heat that can be used in the deaerator, there are a number of options available to reduce the limitation. They are: use of electrically driven auxiliaries; use of more-efficient steam auxiliaries; and better care in operation of existing steam turbines.

Electrical auxiliaries are always preferable, from an energy conservation standpoint, when exhaust steam is being blown to the atmosphere. Steam rates for turbines vary widely, depending on many factors, but increasing both the number of stages and the speed has beneficial results for small turbines.

Fig. 4 shows the general benefits that can be achieved for a specific set of steam conditions and wheel diame-

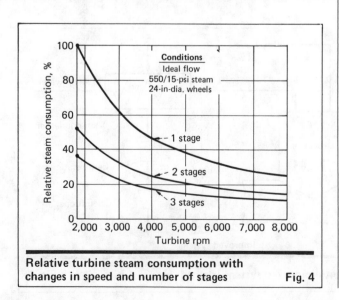

Relative turbine steam consumption with
changes in speed and number of stages   Fig. 4

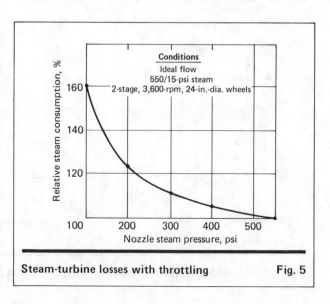

Steam-turbine losses with throttling   Fig. 5

ters. Several operating parameters may also be controllable to reduce exhaust turbine steam. Many small turbines have "hand valves" to control the number of first-stage steam nozzles in use. If too many nozzles are in use at light loads, the turbine efficiency will suffer because of governor throttling. Fig. 5 shows a typical effect of this for a specific set of conditions. Throttling losses can be reduced by closing hand valves or by operating fewer auxiliaries, where multiple units exist.

Deaerator difficulties

Deaerators must perform two functions. First, they must heat the boiler feedwater to the saturation temperature of steam at the pressure of the deaerator. This is usually accomplished by intimate mixing of the water with a steam atmosphere. Second, the heater water-surface should be swept efficiently by steam having a low partial pressure of noncondensable gases. This latter step is particularly important for the removal of carbon dioxide as well as for oxygen.

Fig. 6 shows a typical modern deaerator of the spray type. Incoming water is heated in the upper section (water collector) by steam coming from the lower section (stationary baffle). The steam required for heating in the upper section is mostly condensed, allowing the noncondensable gases that are stripped in the lower section to be concentrated—i.e., have their partial pressure increased so that a minimum amount of steam is lost through the vent.

This works well for units with high makeup water (i.e., low condensate return). If condensate return is high at temperatures that require little or no heating of the incoming water, the steam flow to the upper section is reduced and the stripping function of the deaerator suffers. It may be necessary to increase the venting steam flow appreciably to compensate for this, and the loss of steam to the atmosphere can be very wasteful. Our hoped-for thermal economies in recovering condensate may be dissipated in excess steam blown to the atmosphere in order to keep our deaerating feedwater heaters functioning properly.

Fig. 7 shows a solution that has worked for the author. If a vent condenser is added to the vent steam line, so

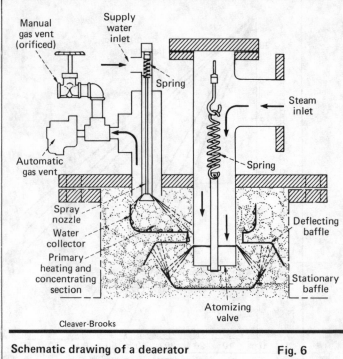

Schematic drawing of a deaerator   **Fig. 6**

that the incoming water to the deaerator is heated prior to entering the deaerator, the deaerating function will be reestablished.

This may seem to be a paradox, since the basic problem was too much heat in the incoming water to start with. In fact, the heating function of the deaerator is transferred to the vent condenser, and the heating section of the deaerator becomes an additional scrubbing section for the removal of the noncondensable gases. Every deaerator modified this way by the author has passed its deaerating performance tests, even though the tests were usually conducted by skeptical engineers.

It should be noted that the vent condenser is kept pressurized by the thermostatic trap, which should discard its condensate to waste, since it contains the very noncondensable gases that we are trying to get rid of.

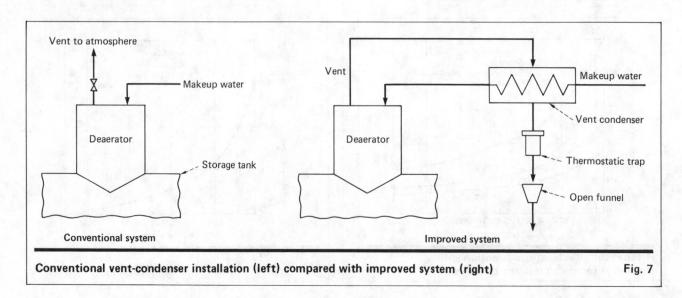

Conventional vent-condenser installation (left) compared with improved system (right)   **Fig. 7**

Where makeup is high in carbonates or CO_2, it may be desirable to use two separate vent condensers and two separate deaerator spray nozzles, as recommended in Ref. *3*.

Surplus heat usage

It is entirely possible that even with the solutions already proposed, there will be excess flash steam at the powerhouse when condensate recovery is high. This is a fortunate problem for most plants to have but, in the interest of energy conservation, should not be accepted.

Excess steam can be utilized in several ways:
- Use of low-pressure steam for either heating or air conditioning.
- Domestic hot-water heating.

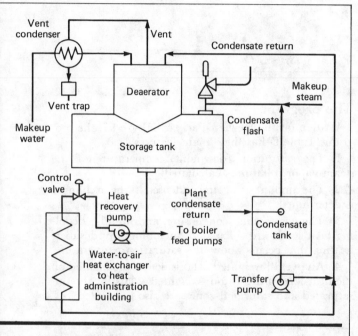

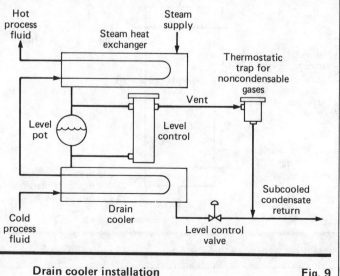

Hot-water heating with deaerator water   **Fig. 8**

Drain cooler installation   **Fig. 9**

- Preheating fuel or combustion air for boilers or other fired processes.

Ingenuity may be required to exploit these uses, but they can often be accomplished at minimum cost. For example, Fig. 8 shows an example where the steam heating system for an office building adjacent to a powerhouse was converted to hot-water heating, using deaerator water as the heating medium. Conversion was accomplished with a minimum of piping changes (that included removing the steam traps from the heating and ventilating coils).

The problem is not a new one to the marine industry, and we can turn to it for a solution. Marine practice has always included the use of "drain coolers." A typical drain cooler is shown in Fig. 9. It serves the purpose of preheating the process fluid through subcooling the condensate from the primary steam heat exchanger. One or more such units on the largest steam users can usually eliminate all blow-to-air steam, with even the highest rates of condensate recovery. Unless one has steam that is totally free of carbon dioxide, one should not attempt to accomplish the same subcooling by adding subcooling traps to the primary heat exchanger. And, even with totally CO_2-free steam, it is not a good idea if uniform process-heating temperatures are required.

Finally, one should always have a good steam-trapping program to ensure that worn steam traps are not leaking excessive amounts of steam, and so upsetting the powerhouse heat balance [1,2].

Precautions

When using condensate recovery, do not upset the deaerator performance or the powerhouse heat balance. Possible precautions include: Elimination of atmospheric flash tanks; use of electric auxiliary drives; use of more-efficient steam auxiliary drives; employment of low-pressure steam for heating and ventilating, or refrigeration; installing condensate drain coolers on large steam users; having a good steamtrap program.

References

1. Vallery, S. J., Are Your Steam Traps Wasting Energy?, *Chem. Eng.*, Feb. 9, 1981, p. 84, or Reprint No. 051.
2. Monroe, E. S., All About Steam Traps, *Chem. Eng.*, *Reprint* No. 248. Reprint includes: *Chem. Eng.*, Sept. 1, 1975 (How To Test Steam Traps); Jan. 5, 1981 (Select the Right Steam Trap); Apr. 12, 1981 (How To Size and Rate Steam Traps); and May 10, 1981 (Install Steam Traps Correctly).
3. Monroe, E. S., Effects of CO_2 in Steam Systems, *Chem. Eng.*, Mar. 23, 1981, p. 209.

The author

Elmer S. Monroe is a principal Power Consultant in the Engineering Dept., Louviers Bldg., E. I. du Pont de Nemours & Co., Wilmington, DE 19898, where he is currently working on a test for measuring the air-handling capacity of steam traps. He is a graduate of Virginia Polytechnic Institute and Cornell University, and is Vice Chairman of the ASME Performance Test Code Committee 39, on condensate handling, and a member of ISO TC 153/SC 4, on steam traps.

Calculator program for a steam condenser

This program calculates the weighted mean-temperature-difference between the fluids inside and outside the tubes of a steam condenser and the saturation temperature of the steam-vapor mixture as well as the various heat loads.

Larry J. Haydu, Kennecott Engineering Systems Co.

☐ There are three loads in a heat exchanger used for condensing steam out of a noncondensable vapor—gas cooling, condensing, and liquid subcooling.

In order to select the correct cooler/condenser for a particular job, it is necessary to determine the heat load distribution and the overall weighted mean-temperature-difference.

$$\text{Wtd. MTD} = Q_t / [\Sigma(Q_i/\Delta T_i)]$$

where Q_t = total·heat load

Q_i = incremental heat load over a selected interval

ΔT_i = log MTD for that interval

The program

Written for the Hewlett-Packard HP-67/97, the program (Table I) has these main features:

1. The saturation (dewpoint) temperature of the steam-vapor mixture is calculated.

2. The amount of steam condensed in the exchanger is determined.

3. The gas-cooling, condensing, and liquid-subcooling heat loads are calculated, including any desuperheating that occurs above the saturation temperature.

4. An overall weighted MTD is determined, based on the heat-load distribution. Weighted MTDs for the unsaturated and saturated zones are also calculated.

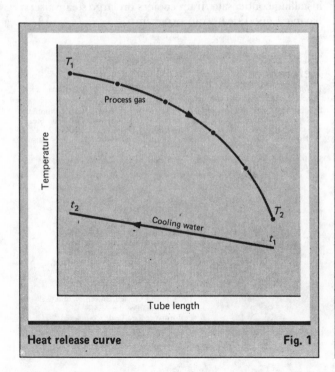

Heat release curve   **Fig. 1**

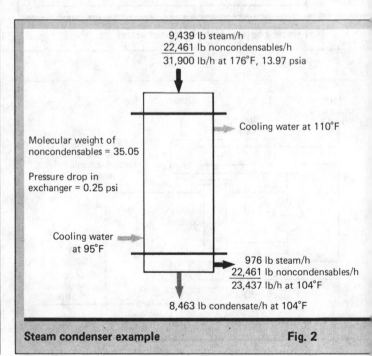

9,439 lb steam/h
22,461 lb noncondensables/h
31,900 lb/h at 176°F, 13.97 psia

Cooling water at 110°F

Molecular weight of noncondensables = 35.05

Pressure drop in exchanger = 0.25 psi

Cooling water at 95°F

976 lb steam/h
22,461 lb noncondensables/h
23,437 lb/h at 104°F

8,463 lb condensate/h at 104°F

Steam condenser example   **Fig. 2**

Originally published February 9, 1981

Calculator program for a steam condenser

Step	Key	Code	Step	Key	Code	Step	Key	Code	Step	Key	Code	Step	Key	Code
	Part I		035	3	03				107	÷	-24	141	2	02
	Side 1		036	8	08				108	RCLA	36 11	142	÷	-24
001	*LBLA	21 11	037	1	01	073	STOE	35 15	109	X≷Y	-41	143	RCLE	36 15
002	RCL9	36 09	038	.	-62	074	*LBLB	21 12	110	÷	-24	144	+	-55
003	5	05	039	3	03	075	2	02	111	RCLA	36 11	145	4	04
004	÷	-24	040	-	-45	076	STOI	35 46	112	-	-45	146	6	06
005	STO9	35 09	041	STOE	35 15	077	*LBLC	21 13		Part I		147	0	00
006	RCL1	36 01	042	PSE	16 51	078	RCL4	36 04		Side 2		148	+	-55
007	RCL2	36 02	043	RCL5	36 05	079	RCL9	36 09	113	1	01	149	1	01
008	÷	-24	044	X≠Y?	16-35	080	-	-45	114	8	08	150	1	01
009	STOA	35 11	045	GTOb	22 16 12	081	STO4	35 04	115	X	-35	151	6	06
010	RCL0	36 00	046	-	-45	082	RCLE	36 15	116	RCL0	36 00	152	5	05
011	STOB	35 12	047	CHS	-22	083	CHS	-22	117	X≷Y	-41	153	.	-62
012	1	01	048	STOD	35 14	084	3	03	118	STO6	35 06	154	1	01
013	8	08	049	GSBa	23 16 11	085	8	08	119	-	-45	155	X≷Y	-41
014	÷	-24	050	ST+1	35-55 01	086	1	01	120	STOC	35 13	156	-	-45
015	+	-55	051	P≷S	16-51	087	.	-62	121	2	02	157	4	04
016	LSTX	16-63	052	RCLE	36 15	088	3	03	122	÷	-24	158	9	09
017	X≷Y	-41	053	RCL6	36 06	089	-	-45	123	P≷S	16-51	159	3	03
018	÷	-24	054	-	-45	090	3	03	124	RCL0	36 00	160	.	-62
019	RCL4	36 04	055	5	05	091	0	00	125	+	-55	161	1	01
020	X	-35	056	÷	-24	092	2	02	126	RCLD	36 14	162	÷	-24
021	LOG	16 32	057	STOD	35 14	093	7	07	127	X	-35	163	.	-62
022	CHS	-22	058	RCLE	36 15	094	X≷Y	-41	128	ST+9	35-55 09	164	3	03
023	6	06	059	X≷Y	-41	095	÷	-24	129	ST+i	35-55 45	165	8	08
024	.	-62	060	-	-45	096	6	06	130	RCLC	36 13	166	Y^x	31
025	2	02	061	STOE	35 15	097	.	-62	131	ST+0	35-55 00	167	9	09
026	6	06	062	GTOB	22 12	098	2	02	132	P≷S	16-51	168	7	07
027	7	07	063	*LBLb	21 16 12	099	6	06	133	2	02	169	0	00
028	+	-55	064	RCL5	36 05	100	7	07	134	÷	-24	170	.	-62
029	3	03	065	RCL6	36 06	101	+	-55	135	RCL0	36 00	171	3	03
030	0	00	066	-	-45	102	10^x	16 33	136	+	-55	172	X	-35
031	2	02	067	5	05	103	RCL4	36 04	137	STOB	35 12	173	RCLC	36 13
032	7	07	068	÷	-24	104	X≷Y	-41	138	GSBa	23 16 11	174	X	-35
033	X≷Y	-41	069	STOD	35 14	105	-	-45	139	ST+i	35-55 45	175	ST+8	35-55 08
034	÷	-24	070	RCL5	36 05	106	RCL4	36 04	140	RCLD	36 14	176	ST+i	35-55 45
			071	X≷Y	-41									
			072	-	-45									

Calculation of the weighted MTD depends on the shape of the heat-release curve [1,2]. This curve indicates the amount of heat transfer at a given temperature as the gas is cooled and steam condenses out. A greater percentage of steam condenses just below the dewpoint than at lower temperatures, so the heat load per unit temperature drop is higher, giving the figure its curved shape (Fig. 1).

For calculation purposes, the heat-release curve is broken into several zones, so that when straight lines are drawn between the different points they will approximate the curve. The weighted MTD is then obtained by computing the logarithmic mean-temperature-difference between the vapor and coolant temperatures over the intervals of the condensing range. The equations are shown in Table II. It is assumed for the sake of this article that the vapor and coolant always flow countercurrently.

The program is in two parts and stored on two cards.

Part I calculates the heat load over the condensing temperature range in five successive steps. For each temperature interval, the gas-cooling, condensing, and liquid-subcooling heat duties are computed and accumulated in separate memory registers. If the steam is unsaturated at the inlet, a sixth temperature interval is

Nomenclature

Symbol	Item	Units
HS	Saturation humidity	lb steam/lb noncondensables
HV	Latent heat of vaporization	Btu/lb
M_a	Molecular weight of steam	eighteen lb/lb mole
M_b	Molecular weight of noncondensables	lb/lb mole
MTD	Mean-temperature-difference	°F
P_a	Partial pressure of steam	psi
P_t	Total pressure	psia
Q_c	Heat of condensation	Btu/h
Q_g	Heat of gas cooling	Btu/h
Q_i	Incremental heat load	Btu/h
Q_l	Heat of liquid subcooling	Btu/h
Q_t	Total heat load	Btu/h
T	Temperature of vapor	°F
t	Temperature of coolant	°F
T_c	Critical temperature of water	1,165.1°R
ΔT_i	Log mean-temperature-difference	°F
VP	Vapor pressure of steam	psi

Table I

Step	Key	Code	Step	Key	Code	Step	Key	Code	Step	Key	Code	Step	Key	Code
	Side 2 (cont'd)		211	ST+7	35-55 07	032	RCL1	36 01	063	ISZI	16 26 46	104	LN	32
177	RCLE	36 15	212	RTN	24	033	RCLE	36 15	069	RCL2	36 02	105	÷	-24
178	RCLD	36 14	213	R/S	51	034	x	-35	070	STO1	35 01	106	F2?	16 23 02
179	-	-45		Part II Side 1		035	P≠S	16-51	071	RCLD	36 14	107	STOC	35 13
180	STOE	35 15	001	*LBLA	21 11	036	RCL8	36 08	072	-	-45	108	P≠S	16-51
181	P≠S	16-51	002	P≠S	16-51	037	X≠Y	-41	073	STO2	35 02	109	RCLi	36 45
182	ISZI	16 26 46	003	RCL7	36 07	038	-	-45	074	RCL4	36 04	110	X≠Y	-41
183	6	06	004	RCL8	36 08	039	STO4	35 04	075	STO3	35 03	111	÷	-24
184	RCLI	36 46	005	+	-55	040	GSBD	23 14	076	6	06	112	ST+0	35-55 00
185	X≠Y?	16-35	006	RCL9	36 09	041	SF2	16 21 02	077	RCLI	36 46		Part II Side 2	
186	GTOC	22 13	007	+	-55	042	RCLB	36 12	078	X≠Y?	16-35	113	P≠S	16-51
187	RCL0	36 00	008	STOA	35 11	043	STO1	35 01	079	GTOC	22 13	114	RTN	24
188	RCL1	35 01	009	P≠S	16-51	044	2	02	080	P≠S	16-51	115	*LBLE	21 15
189	+	-55	010	RCL8	36 08	045	STOI	35 46	081	RCLA	36 11	116	RCLC	36 13
190	P≠S	16-51	011	STO3	35 03	046	GTOC	22 13	082	RCL0	36 00	117	÷	-24
191	RCL7	36 07	012	RCL7	36 07	047	*LBLB	21 12	083	÷	-24	118	RCL0	36 00
192	RCL8	36 08	013	-	-45	048	STO1	35 01	084	STOE	35 15	119	X≠Y	-41
193	+	-55	014	X≠Y	-41	049	2	02	085	STOD	35 14	120	-	-45
194	RCL9	36 09	015	÷	-24	050	STOI	35 46	086	RCL1	36 01	121	RCLA	36 11
195	+	-55	016	STOE	35 15	051	0	00	087	X>0?	16-44	122	RCL1	36 01
196	P≠S	16-51	017	1	01	052	STOC	35 13	088	GSBE	23 15	123	-	-45
197	RTN	24	018	STOI	35 46	053	*LBLC	21 13	089	RCLE	36 15	124	X≠Y	-41
198	*LBLa	21 16 11	019	RCLD	36 14	054	RCL1	36 01	090	RTN	24	125	÷	-24
199	RCL1	36 01	020	5	05	055	RCLD	36 14	091	*LBLD	21 14	126	STOD	35 14
200	RCL3	36 03	021	x	-35	056	-	-45	092	RCL1	36 01	127	RTN	24
201	x	-35	022	RCL6	36 06	057	STO2	35 02	093	RCL4	36 04	128	R/S	51
202	RCLB	36 12	023	+	-55	058	P≠S	16-51	094	-	-45			
203	.	-62	024	STOB	35 12	059	RCLi	36 45	095	STO0	35 00			
204	4	04	025	RCL5	36 05	060	RCLE	36 15	096	RCL2	36 02			
205	5	05	026	X≠Y?	16-35	061	x	-35	097	RCL3	36 03			
206	x	-35	027	GTOB	22 12	062	P≠S	16-51	098	-	-45			
207	+	-55	028	STO1	35 01	063	RCL3	36 03	099	STO9	35 09			
208	RCLD	36 14	029	X≠Y	-41	064	X≠Y	-41	100	-	-45			
209	x	-35	030	STO2	35 02	065	-	-45	101	RCL0	36 00			
210	P≠S	16-51	031	P≠S	16-51	066	STO4	35 04	102	RCL9	36 09			
						067	GSBD	23 14	103	÷	-24			

Equations used in the calculations

Table II

		Ref.
Antoine equation for vapor pressure of steam	$\log VP = C_1/(T + C_2) + C_3$	(4)
	Constants $C_1 = -3{,}027$ $C_2 = 381.3$ $C_3 = 6.267$	
Watson equation for heat of vaporization	$HV_2 = HV_1[(T_c - T_2)/(T_c - T_1)]^{0.38}$	(3)
Saturation humidity	$HS = \dfrac{M_a P_a}{M_b (P_t - P_a)}$	(5)
Weighted mean-temperature-difference	$\text{Wtd MTD} = \dfrac{Q_t}{\Sigma Q_i/\Delta T_i}$	(1,2)
Log mean-temperature-difference	$\Delta T_i = \dfrac{(T_1 - t_2)-(T_2 - t_1)}{\ln[(T_1 - t_2)/(T_2 - t_1)]}$	(1,2)
Total heat load	$Q_t = Q_g + Q_c + Q_l$	(1,2)
Specific heat of steam	0.45 Btu/lb °F	
Specific heat of water	1.0 Btu/lb °F	

Temperatures in the Watson equation are in °R; all other temperatures are in °F.

User's instructions and sample calculations   Table III

Step	Value	Example	Key
1. Clear primary and secondary registers			CL REG
			P⇌S
			CL REG
2. Store input data in primary registers			
	Steam inlet rate, lb/h	9,439	STO 0
	Noncondensable rate, lb/h	22,461	STO 1
	Molecular weight noncond., lb/lb mole	35.05	STO 2
	Specific heat inerts, Btu/lb °F	0.22	STO 3
	Operating pressure, absolute psia	13.97	STO 4
	Inlet gas temperature, °F	176	STO 5
	Outlet gas temperature, °F	104	STO 6
	Coolant inlet temperature, °F	95	STO 7
	Coolant outlet temperature, °F	110	STO 8
	Pressure drop of vapor, psi	0.25	STO 9
3. Load program Part I, sides 1 and 2			
4. Begin computations			A
5. Program pauses (briefly) to display saturation temperature, °F		172.23	
6. Recall output from primary and secondary registers			
	Q (total), Btu/h	9,463,302.32	Display
	Steam rate out, lb/h	976.26	RCL 0
			P⇌S
	Steam condensed, lb/h	8,462.74	RCL 0
	Q (gas cool) incl. desuperheat, Btu/h	483,665.53	RCL 7
	Q (condensed), Btu/h	8,584,213.88	RCL 8
	Q (liquid subcool), Btu/h	395,422.92	RCL 9
	Q (desuperheat), Btu/h	34,659.50	RCL 1
7. Return to primary registers			P⇌S
8. Load program Part II, sides 1 and 2			
9. Begin computations			A
10. Recall output			
	Overall weighted MTD, °F	37.60	Display
	Wtd. MTD saturation zone, °F	37.55	RCL D
	Wtd. MTD unsaturated zone, °F	57.97	RCL C
	Saturation temperature, °F	172.23	RCL B

used to figure the heat of desuperheating the steam-vapor mixture. The Watson analogy [3] is used to calculate the heat of vaporization for water (Table II). The Antoine equation [4] is used to predict the vapor pressure of steam.

Part II calculates the weighted MTD from the information in Part I. Individual MTDs are also figured for the desuperheated and saturated zones.

An example is illustrated in Fig. 2, and a step-by-step procedure for using the program is detailed in Table III.

For TI-58/59 users

The TI version of the program has two parts, program A and program B, (for listings, see Tables IV and V). Table VI contains user instructions and the same example that was given for the HP version.

Listing for TI version—program A   Table IV

Step	Code	Key	Step	Code	Key	Step	Code	Key
000	76	LBL	015	42	STO	030	43	RCL
001	11	A	016	20	20	031	00	00
002	43	RCL	017	43	RCL	032	55	÷
003	09	09	018	00	00	033	01	1
004	55	÷	019	42	STO	034	08	8
005	05	5	020	21	21	035	85	+
006	95	=	021	53	(	036	43	RCL
007	42	STO	022	53	(	037	20	20
008	09	09	023	43	RCL	038	54	)
009	43	RCL	024	00	00	039	65	×
010	01	01	025	55	÷	040	43	RCL
011	55	÷	026	01	1	041	04	04
012	43	RCL	027	08	8	042	54	)
013	02	02	028	55	÷	043	28	LOG
014	95	=	029	53	(	044	75	−

Step	Code	Key	Step	Code	Key	Step	Code	Key	Step	Code	Key	Step	Code	Key
045	06	6	108	43	RCL	171	75	-	234	42	STO	297	44	SUM
046	93	.	109	05	05	172	43	RCL	235	21	21	298	25	25
047	02	2	110	75	-	173	36	36	236	71	SBR	299	43	RCL
048	06	6	111	43	RCL	174	54	)	237	23	LNX	300	25	25
049	07	7	112	06	06	175	35	1/X	238	43	RCL	301	32	X:T
050	54	)	113	95	=	176	65	×	239	39	39	302	01	1
051	55	÷	114	55	÷	177	43	RCL	240	74	SM*	303	06	6
052	03	3	115	05	5	178	20	20	241	25	25	304	77	GE
053	00	0	116	95	=	179	65	×	242	43	RCL	305	13	C
054	02	2	117	42	STO	180	43	RCL	243	23	23	306	43	RCL
055	07	7	118	23	23	181	04	04	244	55	÷	307	17	17
056	94	+/-	119	43	RCL	182	75	-	245	02	2	308	85	+
057	95	=	120	05	05	183	43	RCL	246	85	+	309	43	RCL
058	35	1/X	121	75	-	184	20	20	247	43	RCL	310	18	18
059	75	-	122	43	RCL	185	95	=	248	24	24	311	85	+
060	03	3	123	23	23	186	65	×	249	85	+	312	43	RCL
061	08	8	124	95	=	187	01	1	250	04	4	313	19	19
062	01	1	125	42	STO	188	08	8	251	06	6	314	95	=
063	93	.	126	24	24	189	95	=	252	00	0	315	99	PRT
064	03	3	127	76	LBL	190	42	STO	253	95	=	316	43	RCL
065	95	=	128	12	B	191	40	40	254	94	+/-	317	00	00
066	42	STO	129	01	1	192	43	RCL	255	85	+	318	99	PRT
067	24	24	130	02	2	193	00	00	256	01	1	319	43	RCL
068	99	PRT	131	42	STO	194	75	-	257	01	1	320	10	10
069	43	RCL	132	25	25	195	43	RCL	258	06	6	321	99	PRT
070	05	05	133	76	LBL	196	40	40	259	05	5	322	43	RCL
071	32	X:T	134	13	C	197	95	=	260	93	.	323	17	17
072	43	RCL	135	43	RCL	198	42	STO	261	01	1	324	99	PRT
073	24	24	136	09	09	199	22	22	262	95	=	325	43	RCL
074	77	GE	137	22	INV	200	43	RCL	263	55	÷	326	18	18
075	22	INV	138	44	SUM	201	40	40	264	04	4	327	99	PRT
076	32	X:T	139	04	04	202	42	STO	265	09	9	328	43	RCL
077	75	-	140	03	3	203	00	00	266	03	3	329	19	19
078	43	RCL	141	00	0	204	43	RCL	267	93	.	330	99	PRT
079	24	24	142	02	2	205	22	22	268	01	1	331	43	RCL
080	95	=	143	07	7	206	55	÷	269	95	=	332	11	11
081	42	STO	144	94	+/-	207	02	2	270	45	Y×	333	99	PRT
082	23	23	145	55	÷	208	85	+	271	93	.	334	91	R/S
083	71	SBR	146	53	(	209	43	RCL	272	03	3	335	76	LBL
084	23	LNX	147	43	RCL	210	10	10	273	08	8	336	23	LNX
085	43	RCL	148	24	24	211	95	=	274	95	=	337	43	RCL
086	39	39	149	85	+	212	65	×	275	65	×	338	01	01
087	44	SUM	150	03	3	213	43	RCL	276	43	RCL	339	65	×
088	11	11	151	08	8	214	23	23	277	22	22	340	43	RCL
089	43	RCL	152	01	1	215	95	=	278	65	×	341	03	03
090	24	24	153	93	.	216	42	STO	279	09	9	342	85	+
091	75	-	154	03	3	217	38	38	280	07	7	343	53	(
092	43	RCL	155	54	)	218	44	SUM	281	00	0	344	43	RCL
093	06	06	156	85	+	219	19	19	282	93	.	345	21	21
094	95	=	157	06	6	220	74	SM*	283	03	3	346	65	×
095	55	÷	158	93	.	221	25	25	284	95	=	347	93	.
096	05	5	159	02	2	222	43	RCL	285	42	STO	348	04	4
097	95	=	160	06	6	223	22	22	286	37	37	349	05	5
098	42	STO	161	07	7	224	44	SUM	287	44	SUM	350	54	)
099	23	23	162	54	)	225	10	10	288	18	18	351	95	=
100	43	RCL	163	95	=	226	43	RCL	289	74	SM*	352	65	×
101	23	23	164	22	INV	227	22	22	290	25	25	353	43	RCL
102	22	INV	165	28	LOG	228	55	÷	291	43	RCL	354	23	23
103	44	SUM	166	42	STO	229	02	2	292	23	23	355	95	=
104	24	24	167	36	36	230	85	+	293	22	INV	356	42	STO
105	12	B	168	53	(	231	43	RCL	294	44	SUM	357	39	39
106	76	LBL	169	43	RCL	232	00	00	295	24	24	358	44	SUM
107	22	INV	170	04	04	233	95	=	296	01	1	359	17	17
												360	92	RTN

Listing for TI version—program B Table V

Step	Code	Key	Step	Code	Key	Step	Code	Key	Step	Code	Key	Step	Code	Key
000	76	LBL	056	21	21	111	75	-	166	77	GE	221	35	35
001	11	A	057	42	STO	112	53	(	167	33	X²	222	22	INV
002	43	RCL	058	02	02	113	73	RC*	168	76	LBL	223	87	IFF
003	17	17	059	43	RCL	114	25	25	169	32	X:T	224	02	02
004	85	+	060	08	08	115	65	×	170	43	RCL	225	35	1/X
005	43	RCL	061	75	-	116	43	RCL	171	24	24	226	43	RCL
006	18	18	062	53	(	117	24	24	172	99	PRT	227	35	35
007	85	+	063	43	RCL	118	54	)	173	43	RCL	228	42	STO
008	43	RCL	064	11	11	119	95	=	174	23	23	229	22	22
009	19	19	065	65	×	120	42	STO	175	99	PRT	230	22	INV
010	95	=	066	43	RCL	121	04	04	176	43	RCL	231	86	STF
011	42	STO	067	24	24	122	71	SBR	177	22	22	232	02	02
012	20	20	068	54	)	123	23	LNX	178	99	PRT	233	76	LBL
013	43	RCL	069	95	=	124	01	1	179	43	RCL	234	35	1/X
014	08	08	070	42	STO	125	44	SUM	180	21	21	235	73	RC*
015	42	STO	071	04	04	126	25	25	181	99	PRT	236	25	25
016	03	03	072	71	SBR	127	43	RCL	182	91	R/S	237	55	÷
017	53	(	073	23	LNX	128	02	02	183	76	LBL	238	43	RCL
018	43	RCL	074	86	STF	129	42	STO	184	23	LNX	239	35	35
019	08	08	075	02	02	130	01	01	185	43	RCL	240	85	+
020	75	-	076	43	RCL	131	43	RCL	186	01	01	241	43	RCL
021	43	RCL	077	21	21	132	01	01	187	75	-	242	10	10
022	07	07	078	42	STO	133	75	-	188	43	RCL	243	95	=
023	54	)	079	01	01	134	43	RCL	189	04	04	244	42	STO
024	55	÷	080	01	1	135	23	23	190	95	=	245	10	10
025	43	RCL	081	02	2	136	95	=	191	42	STO	246	92	RTN
026	20	20	082	42	STO	137	42	STO	192	00	00	247	76	LBL
027	95	=	083	25	25	138	02	02	193	43	RCL	248	33	X²
028	42	STO	084	61	GTO	139	43	RCL	194	02	02	249	53	(
029	24	24	085	24	CE	140	04	04	195	75	-	250	53	(
030	01	1	086	76	LBL	141	42	STO	196	43	RCL	251	43	RCL
031	01	1	087	22	INV	142	03	03	197	03	03	252	20	20
032	42	STO	088	43	RCL	143	43	RCL	198	95	=	253	75	-
033	25	25	089	05	05	144	25	25	199	42	STO	254	43	RCL
034	43	RCL	090	42	STO	145	32	X:T	200	09	09	255	11	11
035	23	23	091	01	01	146	01	1	201	53	(	256	54	)
036	65	×	092	01	1	147	06	6	202	43	RCL	257	55	÷
037	05	5	093	02	2	148	77	GE	203	00	00	258	53	(
038	85	+	094	42	STO	149	24	CE	204	75	-	259	43	RCL
039	43	RCL	095	25	25	150	43	RCL	205	43	RCL	260	10	10
040	06	06	096	00	0	151	20	20	206	09	09	261	75	-
041	95	=	097	42	STO	152	55	÷	207	54	)	262	53	(
042	42	STO	098	22	22	153	43	RCL	208	55	÷	263	43	RCL
043	21	21	099	76	LBL	154	10	10	209	53	(	264	11	11
044	43	RCL	100	24	CE	155	95	=	210	53	(	265	55	÷
045	05	05	101	43	RCL	156	42	STO	211	43	RCL	266	43	RCL
046	32	X:T	102	01	01	157	24	24	212	00	00	267	22	22
047	43	RCL	103	75	-	158	42	STO	213	55	÷	268	54	)
048	21	21	104	43	RCL	159	23	23	214	43	RCL	269	54	)
049	77	GE	105	23	23	160	00	0	215	09	09	270	54	)
050	22	INV	106	95	=	161	32	X:T	216	54	)	271	95	=
051	43	RCL	107	42	STO	162	43	RCL	217	23	LNX	272	42	STO
052	05	05	108	02	02	163	11	11	218	54	)	273	23	23
053	42	STO	109	43	RCL	164	67	EQ	219	95	=	274	61	GTO
054	01	01	110	03	03	165	32	X:T	220	42	STO	275	32	X:T
055	43	RCL												

Superheated vapor condensation in heat exchanger design

Some exchanger transfer surface will be dry if the entering superheated vapor is hot enough. However, accounting properly for this dry-wall heat transfer could actually result in smaller surface area, and less-expensive tube material.

D. H. Foxall and H. R. Chappell, IMI Yorkshire Imperial Ltd.

☐ Condensation heat transfer from pure saturated vapors is well understood and usually can be evaluated accurately.

If the vapor is superheated at the heat exchanger inlet, the heat flux might be only marginally increased. In such a case there will be little error in evaluating the duty as if the vapor were saturated, except for also taking into account the additional sensible heat that must be transferred.

If, however, the superheat temperature is sufficiently high that some of the heat-transfer surface will be dry, the saturated-vapor assumption may not be satisfactory. This problem is not always dealt with suitably in heat-exchanger design, there being at least the following two incorrect assumptions that may be made:

1. A high superheat temperature must be accompanied by a high tube-wall temperature. If the duty is accurately evaluated, however, it may be found that a tube material that was thought unacceptable based on temperature considerations would actually be satisfactory. Such a situation may arise with steam-heated calorifiers, for which copper alloy tubes have sometimes been specified when copper could have been used. In duties in which condensation begins after the vapor has lost some of its superheat, the tube surface temperature is higher in the dry than in the wet region, but the difference may not be substantial.

2. Heat transfer in the dry-wall region will be so poor as to require a disproportionately large surface area. However, because the dry-wall heat flux is greater than the saturation flux, taking this factor into account may make it possible to reduce heat-exchange surface area significantly. This may be the case with, for example, hydrocarbon condensers, in which the mass velocity

Originally published December 29, 1980

Nomenclature

A Surface area per unit length of tube, ft²/ft
  A_c Coolant side
  A_v Vapor side
  A_w Effective mean-wall-area
h Film heat-transfer coefficient, Btu/(h)(ft²)(°F)
  h_c Across cooling-fluid boundary layer
  h_L Saturated-vapor condensing coefficient
  h_v Across vapor boundary layer (h_t at the transition temperature, T_t)
  h_s Superheated-vapor true condensing coefficient
  h_{sv} Superheated-vapor apparent condensing coefficient (allowing for sensible heat transferred)
K_w Thermal conductivity of tube wall, Btu/(h)(ft²)(°F)/ft
L Length of tube, ft
q Rate of heat transfer, Btu/h
r Fouling factor, (h)(ft²)(°F)/Btu
  r_c Cooling fluid
  r_v Condensing fluid
T Temperature, °F
  T_c Cooling fluid
  T_L Saturated vapor
  T_v Superheated vapor (T_i vapor inlet, T_t at onset of condensation)

T_b At vapor-surface interface (estimated from dry wall calculation, $T_b = T_v - \Delta T_v = T_c + \Delta T_{ov} - \Delta T_v$)
U Overall heat-transfer coefficient (related to vapor-side surface area), Btu/(h)(ft²)(°F)
U' Partial overall coefficient, excluding resistances across vapor boundary layer and condensate film
  U_L, U'_L With saturated vapor
  U_v, U'_v Superheated vapor, dry wall (U_t, U'_t at the transition temperature, T_t)
  U_s, U'_s Superheated vapor, wet wall; note that definition of U_s, Eq. (7) and (8), excludes resistance across vapor boundary layer
x Tube wall thickness, ft
α $[h_s T_s / h_L T_L]^{1/3}$
ΔT Temperature difference, °F
  ΔT_c Across cooling-fluid boundary layer
  ΔT_L Across condensate film (saturated vapor)
  ΔT_s Across condensate film (superheated vapor)
  ΔT_v Across vapor boundary layer
  ΔT_{oL} Temperature difference between saturated vapor and cooling fluid ($T_L - T_c$)
  ΔT_{ov} Temperature difference between superheated vapor and cooling fluid ($T_v - T_c$)

may be high, and the condensing coefficient much lower than for steam.

Outlined in this article is a procedure for estimating the effect of superheat. Calculated first is the saturated-vapor heat transfer; next, the heat transfer at the vapor inlet temperature, assuming that no condensation takes place—which result establishes whether a dry region will actually occur.

If it is found that the surface will be dry at the inlet temperature, another calculation will determine the transition vapor temperature, T_t, below which condensation will occur, so that the dry and wet surface portions of the duty can be distinguished. As the vapor is cooled from T_t to T_L (the saturated vapor temperature), the mean flux remains higher than the saturation flux.

A method will be given in this article for estimating the flux in this superheated-vapor-to-wet-surface range. The procedure can be applied to any type of duty for which saturated-vapor and dry-wall calculations can be carried out. It is first illustrated by a simple example and later by more-complex but perhaps more-realistic calculations.

Reference to flux (heat transfer per unit surface area) in all cases relates to the vapor-side surface area. Flux at the coolant surface is in general different because, while the total amount of heat flowing is the same, the amount of surface area is not. The difference in flux at the two surfaces is substantial if the tube is finned.

The overall heat-transfer coefficient, U, is also defined so that it relates to the vapor-side surface (whether this is internal or external). This involves a minor difference from the usual convention of relating U always to the same surface (most frequently, the external surface).

Saturated-vapor coefficient defined

With the foregoing definition of U, the general form of equation for heat transfer with saturated vapor can be written:

$$\frac{1}{U_L} = \left(\frac{1}{h_c}\right)\left(\frac{A_v}{A_c}\right) + r_c\left(\frac{A_v}{A_c}\right) + \left(\frac{x}{K_w}\right)\left(\frac{A_v}{A_w}\right) + r_v + \frac{1}{h_L} \quad (1)$$

The analysis that follows later can be more easily presented if a partial overall heat-transfer coefficient (U' in general, or U'_L for saturated vapor) is defined. In this article, U' is defined as the overall coefficient that would be obtained from the resistances across the coolant boundary layer, tube wall and any fouling deposits, but excluding the resistance across the condensate film and (in the case of superheated vapor) across the vapor boundary layer.

With this definition, the overall coefficient for saturated vapor is:

$$\frac{1}{U'_L} = \left(\frac{1}{h_c}\right)\left(\frac{A_v}{A_c}\right) + r_c\left(\frac{A_v}{A_c}\right) + \left(\frac{x}{K_w}\right)\left(\frac{A_v}{A_w}\right) + r_v$$
$$= (1/U_L) - (1/h_L) \quad (1a)$$

Perhaps the first to use this type of approach were Colburn and Hougen in dealing with condensation from a mixture containing noncondensable gas [1].

However, they excluded only the resistance across the vapor boundary layer, having included the condensate film resistance in the determination of U'.

The overall temperature difference (ΔT_{oL}, in the case of saturated vapor) is distributed across the individual resistances in proportion to their magnitudes. The main concern here is to separate ΔT_{oL} into ΔT_L (across the condensate film) and into $\Delta T_{oL} - \Delta T_L$ (which is distributed across the other resistances).

Considering continuity of heat flow:

$$U_L \Delta T_{oL} = U'_L(\Delta T_{oL} - \Delta T_L) = h_L \Delta T_L \quad (2)$$

Any of these expressions gives the flux at the condensing surface.

Heat transfer per unit length of tube is obtained by multiplying this flux by A_v, the vapor-side surface area per unit length:

$$U_L A_v \Delta T_{oL} = U'_L A_v(\Delta T_{oL} - \Delta T_L) = h_L A_v \Delta T_L \quad (3)$$

Superheated vapor with dry wall

The means for the determination of heat transfer is provided by the fundamental Eq. (1), but it is suggested that the subscript L in the terms be changed to v to emphasize that the vapor-side heat-transfer mechanism is different:

$$\frac{1}{U_v} = \left(\frac{1}{h_c}\right)\left(\frac{A_v}{A_c}\right) + r_c\left(\frac{A_v}{A_c}\right) + \left(\frac{x}{K_w}\right)\left(\frac{A_v}{A_w}\right) + r_v + \frac{1}{h_v} \quad (4)$$

The overall temperature difference, ΔT_{ov}, which is greater than the saturation temperature difference, ΔT_{oL}, is also changed.

With the definition of partial overall coefficient already given:

$$\frac{1}{U'_v} = \left(\frac{1}{h_c}\right)\left(\frac{A_v}{A_c}\right) + r_c\left(\frac{A_v}{A_c}\right) + \left(\frac{x}{K_w}\right)\left(\frac{A_v}{A_w}\right) + r_v \quad (4a)$$

$$= (1/U_v) - (1/h_v)$$

U'_v and U'_L are determined from the same set of resistances, and in practice the difference may be negligible. However, the change in temperature may alter the value of h_c significantly between the two calculations, and U' cannot necessarily be assumed constant. The possible importance of this point will be shown in the examples that follow.

Should the surface be dry, the equation for flux corresponds to Eq. (2):

$$U_v \Delta T_{ov} = U'_v(\Delta T_{ov} - \Delta T_v) = h_v \Delta T_v \quad (5)$$

The surface is dry if its calculated temperature, T_b, is above saturation point, T_L, and it is wet if T_b is below T_L—i.e.,

Dry: $T_b > T_L$
Wet: $T_b < T_L$
But $T_b = T_c + (\Delta T_{ov} - \Delta T_v)$
And $T_L = T_c + \Delta T_{oL}$
The criterion can thus be expressed:
Dry: $\Delta T_{ov} - \Delta T_v > \Delta T_{oL}$
Wet: $\Delta T_{ov} - \Delta T_v < \Delta T_{oL}$
The solution to Eq. (4) and (5) must be examined by this criterion.

If a dry surface is confirmed, the calculation is valid and gives the correct flux and surface temperature.

A calculation that shows that the surface will be wet is correct in this respect but otherwise invalid. The flux and surface temperatures obtained should not be used in design calculations.

Before illustrating this point numerically, it is convenient to define the transition vapor temperature, T_t, at the boundary between the two regions.

It will be clear from the foregoing that when the vapor temperature is T_t:

$$T_b = T_L \text{ and } \Delta T_{ov} - \Delta T_v = \Delta T_{oL}$$

Using the subscript t to denote also the particular values of the heat-transfer coefficient obtained at vapor temperature T_t, Eq. (5) now yields:

$$U_t(T_t - T_c) = U'_t \Delta T_{oL} = h_t(T_t - T_L) \quad (5a)$$

If the vapor is sufficiently superheated at the inlet temperature for the surface to be dry, a simple calculation will then give a vapor temperature near to T_t. A second dry-wall calculation using Eq. (4) at this vapor temperature will enable T_t to be estimated accurately.

This appears to be the most appropriate point at which to illustrate the procedure with Example 1.

1: Steam condensing inside tube

This example presents the numerical procedure in its most fundamental form, without the complications usual in actual practice.

A 1-in.-dia. horizontal plain round tube is immersed in water at 200°F. Steam at atmospheric pressure flows through the tube, its velocity being 100 ft/s at its inlet temperature of 500°F. Heat transfer to the water is by natural convection.

Conditions are clean and the tube-wall thickness is assumed negligible so that its resistance, and the difference between external and internal surface areas, can be ignored.

For heat transfer at saturation temperature, Eq. (1) reduces to:

$$(1/U_L) = (1/h_c) + (1/h_L)$$

The partial overall coefficient, U'_L, simply equals the value of h_c determined at these conditions. The film coefficient h_c is based on the usual natural-convection assumptions [2] and h_L is calculated by the method recommended by Kern (i.e., calculating as for external condensation on a horizontal tube, except that the condensate loading is taken to be twice the calculated mean value) [3].

The saturated-vapor calculation provides the first entry in Table I: $U_L = 150.3$ Btu/(h)(°F)(ft²), $\Delta T_{oL} = 12$°F, and the product of U_L and $\Delta T_{oL} = 1,804$ Btu/(h)(ft²), the saturation flux.

Assuming the surface to be dry at the inlet temperature of 500°F, h_v is estimated as for a dry gas [4]. In this example, the Reynolds number (approximately 20,000) is well into the turbulent region. At the inlet temperature, $\Delta T_{ov} - \Delta T_v$ is found to be 21.35°F. As this is greater than ΔT_{oL} (12°F), it is established that the surface will be dry at the vapor inlet. The flux is therefore correctly estimated to be 3,932 Btu/(h)(ft²).

Two simple approximations are now made to find a vapor temperature near to T_t:

1. At T_t, the temperature difference associated with the partial overall coefficient is ΔT_{oL} [Eq. (5a)]. It is assumed for the moment that the variation in U' over the wet-wall range is small enough for the value U'_L already obtained to be treated as a constant. An approximation to the transition flux is then given by $U'_L \Delta T_{oL}$ [in Example 1, $155.4 \times 21 = 1,865$ Btu/(h)(ft²)].

The error in this approximate result is usually fairly small.

2. Flux in the dry-wall region is taken to be proportional to $(T_v - T_c)$, again an assumption that is accurate enough for the immediate purpose. The flux estimate from the first approximation gives a value of T_v fairly near the transition point:

$$\frac{T_v - T_c}{T_i - T_c} = \frac{\text{Approximate transition flux}}{\text{Inlet flux}} = \frac{U'_L \Delta T_{oL}}{U_v \Delta T_{ov}}$$

Here, T_i is the vapor inlet temperature. In Example 1, this gives:

$$T_v = 200 + (500 - 200)(1,865/3,932) = 342.3°F$$

The second dry-wall calculation is carried out at this temperature, which proves to be below the transition temperature, as the calculated $\Delta T_{ov} - \Delta T_v$ ($11.48°F$) is less than ΔT_{oL} ($12°F$). Although the calculation, therefore, should not be used for design purposes, it enables T_t to be estimated by interpolation.

The temperature T_t is that at which $\Delta T_{ov} - \Delta T_v = \Delta T_{oL}$. Therefore, by interpolation:

$$\frac{T_t - 342.3}{500 - 342.3} = \frac{12 - 11.48}{21.35 - 11.48}$$

This expression is given only numerically to avoid introducing additional subscripts that might cause unnecessary confusion. The procedure is obvious.

For Example 1, T_t is estimated in this way to be $350.6°F$. It is unlikely that there will be a significant error in this figure; but, should there be any doubt, its accuracy can be checked by means of another dry-wall calculation.

The transition flux also can now be corrected. The procedure recommended is to interpolate for h_t between the two values of h_v already obtained [14.11 and 13.61 Btu/(h)(ft²)(°F), at $T_v = 500°F$ and $342.3°F$, respectively]. From these figures, h_t (at $350.6°F$) is estimated to be 13.64 Btu/(h)(ft²)(°F); and, from Eq. (5a):

Transition flux $= h_t(T_t - T_L) = 13.64 \times 138.6 = 1,891$ Btu/(h)(ft²).

Although the second dry-wall calculation proved invalid, the value of h_v obtained was the vapor-side coefficient, assuming the surface to be dry; so it is useful for obtaining h_t without a more detailed calculation. U'_t, which will be useful if the wet-wall region is to be further investigated, is obtained from Eq. (5a).

The magnitude of the error in assuming the transition flux to be equal to $U'_L \Delta T_{oL}$ is now evident. In this example, h_c, being a natural-convection coefficient, is sensitive to changes in ΔT_c and therefore varies appreciably. The error in the approximation is essentially that h_c is taken to be constant; however, as will be seen, the effect is not significant.

In many cases (as in Example 2, which follows), h_c does not vary significantly, so $U'_L \Delta T_{oL}$ represents the transition flux accurately.

In all cases, the tube wall temperature found at the vapor inlet determines the suitability, as to strength, of a material that is preferred in other respects (e.g., on corrosion-resistance grounds). If the wall is dry, it will be above saturation temperature but the excess is not usually substantial.

The effective mean flux in the dry-wall region is given with little error by the logarithmic mean of the flux at inlet and transition temperatures [2,788 Btu/(h)(ft²) in Example 1]. Strictly, perhaps, the mean flux should be determined at the caloric temperature, as shown by Colburn, who used a graphical method that may not have significantly improved the accuracy of the result [5].

2: Shell-side condensing, plain tube

In this example, propane is cooled from 146°F and condensed at 98°F in a shell-and-tube exchanger. Water at a mean temperature of 82°F flows inside the tubes at 5 ft/s. The tubes are ¾-in. O.D. × 0.65-in. plain wall of 70/30 copper-nickel. Fouling factors of 0.001 (outside) and 0.002 (inside) are specified.

The mean vapor mass-velocity is found to be 122,000 lb/(h)(ft²).

Again, the saturated vapor is considered first. Here, it will be seen that the largest resistance is the one across the condensate film, and that this uses more than half the overall temperature difference (Table II).

The dry-wall calculation at the inlet temperature yields $\Delta T_{ov} - \Delta T_v = 20.57°F$, which is greater than ΔT_{oL} ($16°F$) and establishes that the surface will be dry.

Following the same procedure, an approximate estimate of 3,325 Btu/(h)(ft²) is obtained for the transition flux. In the present case, there is little error in this determination, because h_c (and consequently U') is almost constant. To show the effect of minor variations in h_c, the calculations have allowed for the small change in coolant film temperature between the two parts of the duty, but usually this would be ignored and h_c assumed constant.

The temperature $T_v = 131.8°F$ is then found for use in the second dry-wall calculation. At this temperature, the calculated $\Delta T_{ov} - \Delta T_v$ is $15.80°F$, i.e., below ΔT_{oL}, and so it is in the wet-wall region but close to T_t, which is now estimated by interpolation to be $132.4°F$.

The vapor convection heat-transfer coefficient varies little in the dry-wall region and can be interpolated between the two figures already calculated for h_v, giving $h_t = 96.7$ Btu/(h)(ft²)(°F). And from Eq. (5a):

Transition flux $= h_t(T_t - T_L) = 96.7 \times 34.4$
$= 3,327$ Btu/(h)(ft²).

This result confirms the accuracy of the original approximation.

The logarithmic mean flux over the dry-wall region is 3,784 Btu/(h)(ft²).

In this example, the flux at the transition temperature is more than twice the saturation flux, and the manner in which the flux changes between the two temperatures may be of interest.

Superheated vapor with wet wall

The following derivation applies at vapor temperatures between T_t and T_L.

Condensing calculations for pure vapors generally assume that heat transfer takes place by conduction across the condensate film, and that with saturated vapors the flux is provided essentially by latent heat. With condensation from a superheated vapor, sensible heat must flow to the liquid surface and across the condensate film, because there is a temperature difference between the vapor and the vapor-to-liquid interface.

Because part, but not all, of the heat received by the coolant flows across the vapor boundary layer, it is convenient to deal with this situation in much the same way as for saturated vapor. An effective film coefficient, h_{sv}, can be defined for heat flow across the condensate film from the sum of the two components of heat flux:

$$h_{sv}\,\Delta T_s = h_s\,\Delta T_s + h_v\,\Delta T_v \qquad (6)$$

Whereas h_s is the true condensing coefficient related to the flow of latent heat, h_{sv} can be regarded as an apparent condensing coefficient, adjusted to include the sensible heat flux.

With this definition, the overall coefficient for superheated vapor is:

$$\frac{1}{U_s} = \left(\frac{1}{h_c}\right)\left(\frac{A_v}{A_c}\right) + r_c\left(\frac{A_v}{A_c}\right) + \left(\frac{x}{K_w}\right)\left(\frac{A_v}{A_w}\right) + r_v + \frac{1}{h_{sv}} \qquad (7)$$

Eq. (7) corresponds to Eq. (1) for saturated vapor, giving an overall coefficient U_s comparable to U_L in that it applies with the temperature difference ΔT_{oL} between the saturated vapor and coolant. Because the individual resistances in Eq. (7) all transfer the same amount of heat, ΔT_{oL} is distributed proportionately among them in the usual way. The remaining resistance $1/h_v$ transfers a smaller amount of heat, using the remaining temperature difference, ΔT_v.

The partial overall coefficient as defined excludes the resistances $1/h_{sv}$ and $1/h_v$:

$$1/U'_s = 1/U_s - 1/h_{sv} \qquad (7a)$$

And, corresponding to Eq. (2):

$$U_s\,\Delta T_{oL} = U'_s(\Delta T_{oL} - \Delta T_s) = h_{sv}\,\Delta T_s \qquad (8)$$

Saturated-vapor condensing coefficients are usually evaluated from an equation such as:

$$h_L = CW^{-1/3}$$

Here, W is the rate of condensation per unit surface area. For convenience, it is assumed that the effect of the physical properties of the condensate film has been evaluated and absorbed into the constant, C.

For a saturated vapor: $\qquad W = \left(\dfrac{h_L\,\Delta T_L}{\lambda}\right)^{-1/3}$

Yielding: $\qquad h_L = C\left(\dfrac{h_L\,\Delta T_L}{\lambda}\right)^{-1/3}$

In these equations, λ is the latent heat of condensation, Btu/lb.

If superheat is present, the corresponding condensing equation gives:

$$h_{sv} = CW^{-1/3}$$

Because the effect of superheat on the condensate film temperature is small, the value of C can be assumed equal to that for saturated vapor. However, W is reduced because some of the heat flowing across the liquid film is sensible heat and does not contribute to the rate of condensation.

With superheated vapor:

$$W = (h_s\,\Delta T_s)/\lambda$$

Yielding:

$$h_{sv} = C[(h_s\,\Delta T_s)/\lambda]^{-1/3}$$

Eliminating C and λ:

$$h_{sv} = h_L[(h_L\,\Delta T_L)/(h_s\,\Delta T_s)]^{1/3} = h_L/\alpha \qquad (9)$$

Where:

$$\alpha = [(h_s\,\Delta T_s)/(h_L\,\Delta T_L)]^{1/3} \qquad (10)$$

For some purposes, a more useful form of Eq. (9) is obtained if both sides are multiplied by ΔT_s:

$$\Delta T_s = (\alpha/h_L)h_{sv}\,\Delta T_s \qquad (9a)$$

An iterative solution from the foregoing analysis can be obtained for the general case in which h_c is variable, and a simpler alternative procedure can be used when h_c is essentially constant.

Case 1. Variable h_c—Example 1 is appropriate for illustrating this case, and a calculation for $T_v = 300°F$ is presented.

By extrapolation from the dry-wall calculations of Table I, h_v is estimated to be 13.48 Btu/(h)(ft²)(°F); $\Delta T_v = 300 - 212 = 88°F$, and so the sensible heat flux $h_v\,\Delta T_v$ is estimated to be 1,187 Btu/(h)(ft²). Also from Table I, $h_L = 4,617$ Btu/(h)(ft²)(°F), $\Delta T_{oL} = 12°F$ and $\Delta T_L = 0.39°F$.

The first step is to assume a value of $h_{sv}\,\Delta T_s$, for instance 1,840 Btu/(h)(ft²). The required value will be somewhat less than that obtained by linear interpolation between 1,804 and 1,891 Btu/(h)(ft²).

Via Eq. (6), estimate
$\quad h_s\,\Delta T_s = 1,840 - 1,187 = 653$
Via Eq. (10), estimate
$\quad \alpha = (653/1,804)^{1/3} = 0.713$
Via Eq. (9a), estimate
$\quad \Delta T_s = (0.713 \times 1,840)/4,617 = 0.284$

The iteration is fundamentally a matter of "balancing" Eq. (8), while making use of the fact that U' varies little over the wet-wall range.

Between U'_L (saturated vapor) = 155.4, $\Delta T_{oL} - \Delta T_L = 11.61$, and U'_t (transition) = 157.6, $\Delta T_{oL} = 12$, U'_s can be interpolated with negligible error at $\Delta T_{oL} - \Delta T_s = 12 - 0.284 = 11.716°F$, giving $U'_s = 156.0$ at $T_v = 300°F$. However, if U'_s can be interpolated in this way, so can the flux, and the arithmetic is reduced:

Estimated flux at $T_v = 300°F$:

$\quad 1,804 + (1,891 - 1,804)[(0.39 - 0.284)/0.39]$
$\quad = 1,828$ Btu/(h)(ft²)

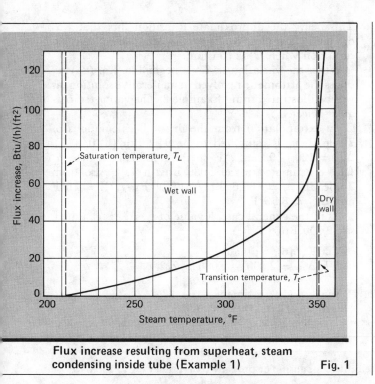

Flux increase resulting from superheat, steam condensing inside tube (Example 1)   **Fig. 1**

This result is the same as from $U'_s(\Delta T_{oL} - \Delta T_s) = 156.0 \times 11.716$.

The calculation requires only the foregoing estimate of ΔT_s (i.e., 0.284) and information in Table I. The assumed total flux of 1,840 Btu/(h)(ft^2) is too high, and a second calculation confirms the derived figure of 1,828 Btu/(h)(ft^2).

The flux variation in the wet-wall region is small in this duty and, to show the form of curvature obtained on a reasonable scale, Fig. 1 gives the incremental values above the saturation flux.

Case 2. Constant h_c—If the variation in h_c is negligible, so is the variation in U'. Then U'_L, the value determined at saturation temperature, can be taken to apply to the whole duty. This is essentially true in Example 2, as will be seen from Table II: U' varies only between 207.8 and 208.1 Btu/(h)(ft^2)($^\circ$F).

In such a case, Eq. (8) can be written:

$$h_{sv}\,\Delta T_s = U'_L(\Delta T_{oL} - \Delta T_s)$$

Eq. (9a) is now used to eliminate ΔT_s from the right-hand side:

$$h_{sv}\,\Delta T_s = U'_L\,\Delta T_{oL} - \alpha(U'_L/h_L)h_{sv}\,\Delta T_s$$

Saturated-vapor and dry-wall calculations for Example 1: steam condensing inside tube						Table I
Saturated vapor:						
$T_c = 200$	$h_c = 155.4$	$1/h_c = 0.006435$	$\Delta T_{oL} - \Delta T_L = 11.61$	$U'_L = 155.4$	Saturation flux:	Approximate transition flux:
$T_L = 212$	$h_L = 4{,}617$	$1/h_L = 0.000217$	$\Delta T_L = 0.39$	$h_L = 4{,}617$		
$\Delta T_{oL} = 12$		$1/U_L = 0.006652$	$\Delta T_{oL} = 12.00$	$U_L = 150.3$	1,804	$U'_L\,\Delta T_{oL} = 1{,}865$
Dry wall (1st calculation):						
$T_c = 200$	$h_c = 184.2$	$1/h_c = 0.00543$	$\Delta T_{ov} - \Delta T_v = 21.35$	$U'_v = 184.2$	Inlet flux:	T_v for dry wall (2nd calculation)
$T_v = 500$	$h_v = 14.11$	$1/h_v = 0.07087$	$\Delta T_v = 278.65$	$h_v = 14.11$		$\dfrac{T_v - 200}{500 - 200} =$
$\Delta T_{ov} = 300$		$1/U_v = 0.07630$	$\Delta T_{ov} = 300.00$	$U_v = 13.11$	3,932	$\dfrac{1{,}865}{3{,}932}$
						$T_v = 342.3$
Dry wall (2nd calculation):						
$T_c = 200$	$h_c = 155.05$	$1/h_c = 0.00645$	$\Delta T_{ov} - \Delta T_v = 11.48$			Transition temperature:
$T_v = 342.3$	$h_v = 13.61$	$1/h_v = 0.07348$	Wet: $< \Delta T_{oL}$			$\dfrac{T_t - 342.3}{500 - 342.3} =$
$\Delta T_{ov} = 142.3$		$1/U_v = 0.07993$				$\dfrac{12 - 11.48}{21.35 - 11.48}$
						$T_t = 350.6$
Transition temperature:						
$T_c = 200$			$\Delta T_{oL} = 12$	$U'_t = 157.6$	Transition flux:	
$T_t = 350.6$			$T_t - T_L = 138.6$	$h_t = 13.64$		
$T_t - T_c = 150.6$			$T_t - T_c = 150.6$		1,891	

But from Eq. (2):

$$U'_L = \frac{U_L \Delta T_{oL}}{\Delta T_{oL} - \Delta T_L}, \text{ and } \frac{U'_L}{h_L} = \frac{\Delta T_L}{\Delta T_{oL} - \Delta T_L}$$

Eliminating U'_L and h_L, Eq. (8) becomes:

$$h_{sv} \Delta T_s = U_L \Delta T_{oL} \left[\frac{\Delta T_{oL}}{\Delta T_{oL} - (1 - \alpha) \Delta T_L} \right] \quad (8a)$$

When the vapor is saturated, $\alpha = 1$ (the heat transferred is entirely latent), and Eq. (8a) simplifies to:

$$h_{sv} \Delta T_s = U_L \Delta T_{oL}$$

This confirms that the total flux equals the saturation flux.

When the vapor is at transition temperature, $\alpha = 0$ (the heat transferred then being entirely sensible), and Eq. (8a) becomes:

$$h_{sv} \Delta T_s = U_L \Delta T_{oL}[\Delta T_{oL}/(\Delta T_{oL} - \Delta T_L)]$$
$$= U'_L \Delta T_{oL}$$

This being the approximate transition flux that was originally assumed confirms the accuracy of the assumption when U' is constant.

When the variation in U' is negligible, Eq. (8a) reduces the arithmetic involved in an iterative solution, as will be illustrated using Example 2. The vapor temperature selected is 120°F.

By extrapolation from the dry-wall calculations of Table II, h_v is estimated to be 94.9 Btu/(h)(ft^2)(°F), $\Delta T_v = 120 - 98 = 22$°F, so the sensible-heat·flux, $h_v \Delta T_v$, is estimated to be 2,088 Btu/(h)(ft^2). Also from Table II, $h_L = 180.6$ Btu/(h)(ft^2)(°F), $\Delta T_{oL} = 16$°F and $\Delta T_L = 8.56$°F.

As in the general case, a value of $h_{sv} \Delta T_s$ is assumed, say 2,300 Btu/(h)(ft^2).

Via Eq. (6), estimate
$h_s \Delta T_s = 2,300 - 2,088 = 212$

Via Eq. (10), estimate
$\alpha = (212/1,546)^{1/3} = 0.516$

Saturated-vapor and dry-wall calculations for Example 2: shell-side condensing of propane (plain tube)						Table II
Saturated vapor:						
$T_c = 82$ $T_L = 98$ $\Delta T_{oL} = 16$	$h_c = 1,238$ $h_L = 180.6$	$(1/h_c)(A_v/A_c) = 0.000977$ $r_c(A_v/A_c) = 0.002420$ $(x/K_w)(A_v/A_w) = 0.000416$ $r_v = 0.001000$ $1/h_L = 0.005538$ $1/U_L = \underline{0.010351}$	$\Delta T_{oL} - \Delta T_L = $ 7.44 $\Delta T_L = 8.56$ $\Delta T_{oL} = 16.00$	$U'_L = 207.8$ $h_L = 180.6$ $U_L = 96.63$	Saturation flux: 1,546	Approximate transition flux: $U'_L \Delta T_{oL} = $ 3,325
Dry wall (1st calculation):						
$T_c = 82$ $T_v = 146$ $\Delta T_{ov} = 64$	$h_c = 1,252$ $h_v = 98.6$	$(1/h_c)(A_v/A_c) = 0.000967$ $r_v(A_v/A_c) = 0.002420$ $(x/K_w)(A_v/A_w) = 0.000416$ $r_v = 0.001000$ $1/h_v = 0.010142$ $1/U_v = \underline{0.014944}$	$\Delta T_{ov} - \Delta T_v = $ 20.57 Dry: $> \Delta T_{oL}$ $\Delta T_v = 43.43$ $\Delta T_{ov} = 64.00$	$U'_v = 208.1$ $h_v = 98.6$ $U_v = 66.9$	Inlet·flux: 4,282	T_v for dry wall (2nd calculation): $\dfrac{T_v - 82}{146 - 82} = $ $\dfrac{3,325}{4,282}$ $T_v = 131.8$
Dry wall (2nd calculation):						
$T_c = 82$ $T_v = 131.8$ $\Delta T_{ov} = 49.8$	$h_c = 1,248$ $h_v = 96.6$	$(1/h_c)(A_v/A_c) = 0.000970$ $r_c(A_v/A_c) = 0.002420$ $(x/K_w)(A_v/A_w) = 0.000416$ $r_v = 0.00100$ $1/h_v = 0.010350$ $1/U_v = \underline{0.015156}$	$\Delta T_{ov} - \Delta T_v = $ 15.80 Wet: $< \Delta T_{oL}$ $\Delta T_{ov} = 49.80$			Transition temperature: $\dfrac{T_t - 131.8}{146 - 131.8} = $ $\dfrac{16 - 15.80}{20.57 - 15.80}$ $T_t = 132.4$
Transition temperature:						
$T_c = 82$ $T_t = 132.4$ $T_t - T_c = 50.4$			$\Delta T_{oL} = 16$ $T_t - T_L = 34.4$ $T_t - T_c = 50.4$	$U'_t = 208.0$ $h_t = 96.7$	Transition flux: 3,327	·

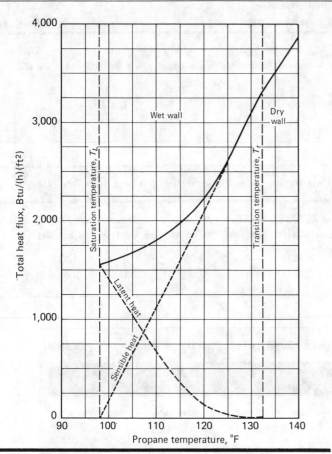

Effect of superheat, shell-side condensing of propane (Example 2, plain tube)   **Fig. 2**

Via Eq. (8a), estimate
$$h_{sv} \Delta T_s = 1,546\{16/[16 - (0.484 \times 8.56)]\}$$
$$= 2,084$$

A typical sequence of approximations is given in Table III, which shows that the result "balances" at a total flux of 2,216 Btu/(h)(ft²).

The accuracy of this alternative procedure depends on the variation in h_c being small; and, of course, the general procedure (Case 1, variable h_c) still also yields the required result. However, convergence to the final figure will appear relatively slow in Example 2, in which there is a large difference between transition and saturation flux; and thus the simplified arithmetic of the alternative procedure (Case 2, constant h_c) perhaps provides a worthwhile advantage.

Wet-wall calculations are invalid if based on a vapor temperature actually above T_t. The error becomes obvious, however, as the calculated sensible-heat flux, $h_v \Delta T_v$, is higher than the transition flux, and the latent heat required is negative.

Immediately below T_t, the latent-heat flux, $h_s \Delta T_s$, is negligible. An upper limit to its value can be found by considering unit increments, i.e., $h_s \Delta T_s = 0, 1$, etc. The difference between the calculated and the assumed total flux will change sign when the assumed $h_s \Delta T_s$ exceeds the actual value.

For example, when $T_v = 130°F$, h_v is estimated from Table II to be 96.3 Btu/(h)(ft²)(°F), and $\Delta T_v = 32°F$, and

so the estimated sensible-heat flux is 3,081 Btu/(h)(ft²). The calculations in Table IV show that $h_s \Delta T_s$ is less than 1 Btu/(h)(ft²) at $T_v = 130°F$.

Fig. 2 presents the calculated total flux for the wet-wall region in Example 2, and also the sensible-heat and latent-heat components. The flux is, of course, entirely of sensible heat until the vapor cools to T_t; beyond this point, there is a gradual replacement of sensible heat by latent heat.

As with Example 1, a smooth curve is predicted from the steep, almost linear, relationship in the dry-wall region, becoming relatively flat as the saturation flux is approached. A "straight line" assumption would overestimate the mean flux in the superheated-vapor-to-wet-wall region. If the increment is to be taken into account in the heat-exchanger rating, it appears advisable to assume half the linear value; i.e., in Example 2, the mean flux assumed for vapor cooling from T_t to T_L would be $1,546 + 0.25(3,327 - 1,546) = 1,991$ Btu/(h)(ft²).

Mean flux for several duties

The amount of heat-transfer surface required for the removal of superheat may be substantially less than would be calculated by using the saturation flux, although in a unit that desuperheats and condenses, the sensible heat is usually a fairly small proportion of the total (about 16% in Example 2) and the effect tends to be obscured. Even so, the reduction in transfer surface may be worthwhile.

When a duty can be divided into two or more distinct parts, the overall mean flux, by definition, is $\Sigma q / \Sigma A_L$—i.e., the total amount of heat transferred per unit time, divided by the total surface area. Using the data already obtained, the mean flux can be determined, without necessarily calculating the individual surface areas for the different parts of the duty, from:

$$\cfrac{MC_{p1}(T_i - T_t) + MC_{p2}(T_t - T_L) + M\lambda}{\underbrace{MC_{p1}(T_i - T_t)}_{\substack{\text{Dry-superheat}\\\text{mean flux}}} + \underbrace{MC_{p2}(T_t - T_L)}_{\substack{\text{Wet-superheat}\\\text{mean flux}}} + \underbrace{M\lambda}_{\substack{\text{Saturation}\\\text{flux}}}}$$

M, the mass flowrate, cancels out. C_{p1} and C_{p2} are the mean specific heats over the two vapor cooling-ranges and λ is the latent heat of condensation (corrected if necessary to allow for the mean subcooling of the condensate film). For Example 2, this yields:

$$\cfrac{(0.591 \times 13.6) + (0.571 \times 34.4) + 141.5}{\left(\cfrac{0.591 \times 13.6}{3,784}\right) + \left(\cfrac{0.571 \times 34.4}{1,991}\right) + \left(\cfrac{141.5}{1,546}\right)} =$$

$$1,635 \text{ Btu/(h)(ft}^2)$$

This is between 5 and 6% above the saturation flux. Strictly, a small proportion of the latent heat is transferred in the superheated-vapor-to-wet-wall region—i.e., at a higher flux than is assumed in this calculation. The result is, therefore, slightly conservative.

3: Shell-side condensing, low-fin tubing

It may be noticed from the data of Table II that the duty is one in which the use of low-fin tubing should be considered. This duty has, in fact, been selected partly for the purpose of showing the effect of changing the type of surface.

The data are based on an actual application; plain tube was originally specified but after closer analysis it was subsequently accepted that the low-fin tube would be more economical.

In Example 3, only the tube specification has been changed, so that the effect of this factor can be clearly shown. In other respects, the duty is as specified for Example 2 (including the water velocity of 5 ft/s). The tube is now 19 fins/in. low fin, with $\frac{3}{4}$-in. fin dia., $\frac{5}{8}$-in. root dia., and 0.065-in. residual wall.

The main calculations, following the suggested procedure, are given in Table V.

When changing from plain to low-fin tube, the external resistance becomes relatively small because of the greater surface area per unit length. Therefore, the external surface temperature is closer to the external fluid temperature, and a smaller temperature difference is needed to drive the heat across this resistance.

Sequence of approximations "balancing" total flux   Table III

$h_{sv}\,\Delta T_s$, assumed	$h_v\,\Delta T_v$, calculated	$h_s\,\Delta T_s$, via Eq. (6)	α, via Eq. (10)	$h_{sv}\,\Delta T_s$, via Eq. (8a)
2,300	2,088	212	0.516	2,084
2,200	2,088	112	0.417	2,249
2,215	2,088	127	0.435	2,217

Determining the upper limit of latent heat flux Table IV

$h_v\,\Delta T_v$, calculated	$h_s\,\Delta T_s$, assumed	$h_{sv}\,\Delta T_s$, via Eq. (6)	α, via Eq. (10)	$h_{sv}\,\Delta T_s$, via Eq. (8a)
3,081	0	3,081	0	3,325
3,081	1	3,082	0.0865	3,024

Saturated-vapor and dry-wall calculations for Example 3: shell-side condensing of propane (low-fin tube)   Table V

Saturated vapor:

$T_c = 82$	$h_c = 1,303$	$(1/h_c)(A_v/A_c) = 0.002947$	$\Delta T_{oL} - \Delta T_L =$ 13.18	$U'_L = 77.2$	Saturation flux:	Approximate transition flux: $U'_L T_{oL} = 1,235$
$T_L = 98$		$r_c(A_v/A_c) = 0.007680$				
$\Delta T_{oL} = 16$		$(x/k_w)(A_v/A_w) = 0.001320$				
		$r_v = 0.001000$	$\Delta T_L = 2.82$	$h_L = 361.9$		
	$h_L = 361.9$	$1/h_L = 0.002763$	$\Delta T_{oL} = 16.00$	$U_L = 63.65$		
		$1/U_L = 0.015710$			1,018	

Dry wall (1st calculation):

$T_c = 82$	$h_c = 1,321$	$(1/h_c)(A_v/A_c) = 0.002907$	$\Delta T_{ov} - \Delta T_v =$ 32.63	$U'_v = 77.5$	Inlet flux:	T_v for dry wall (2nd calculation):
		$r_c(A_v/A_c) = 0.007680$	Dry: $> T_{oL}$			$\dfrac{T_v - 82}{146 - 82} = \dfrac{1,235}{2,529}$
$\Delta T_{ov} = 64$		$(x/K_w)(A_v/A_w) = 0.001320$				
		$r_v = 0.00100$	$\Delta T_v = 31.37$	$h_v = 80.6$		
	$h_v = 80.6$	$1/h_v = 0.012407$	$\Delta T_{ov} = 64.00$	$U_v = 39.51$		
		$1/U_v = 0.025314$			2,529	$T_v = 113.3$

Dry wall (2nd calculation):

$T_c = 82$	$h_c = 130.5$	$(1/h_c)(A_v/A_c) = 0.002943$	$\Delta T_{ov} - \Delta T_v =$ 15.67			Transition temperature:
$T_v = 113.3$		$r_c(A_v/A_c) = 0.007680$				$\dfrac{T_t - 113.3}{146 - 113.3} =$
$\Delta T_{ov} = 31.3$		$(x/K_w)(A_v/A_w) = 0.001320$	Wet: $< \Delta T_{oL}$			$\dfrac{16 - 15.67}{32.63 - 15.67}$
		$r_v = 0.00100$				
	$h_v = 77.5$	$1/h_v = 0.012903$				$T_t = 113.95$
		$1/U_v = 0.025846$	$\Delta T_{ov} = 31.30$			

Transition temperature:

$T_c = 82$			$\Delta T_{oL} = 16$	$U'_t = 77.3$	Transition flux:
$T_t = 113.95$			$T_t - T = 15.95$ $h_t = 77.6$		
$T_t - T_c = 31.95$			$T_t - T_c = 31.95$		1,238

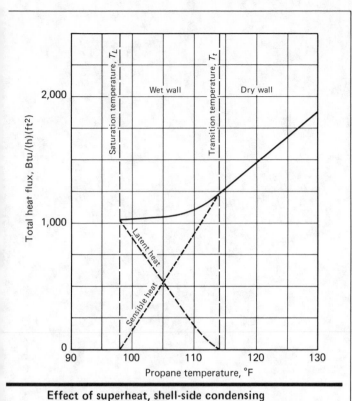

Effect of superheat, shell-side condensing of propane (Example 3, low-fin tube)   Fig. 3

quired. Referring to Eq. (3), this requires the mean flux to be multiplied by A_v (0.196 ft²/ft for ¾-in.-O.D. plain tube, and 0.496 ft²/ft for the corresponding size of low-fin tube):

Plain tube: $1,635 \times 0.196 = 320.5$ Btu/h per foot length

Low-fin tube: $1,073 \times 0.496 = 532.2$ Btu/h per foot length

So low-fin tube transfers about 66% more heat than plain tube.

Summary

For condensation from a superheated vapor, a thermal rating based on the flux at saturation temperature will be conservative, but in many cases the error will be small. The rate of heat transfer used for calculations must take into account the sensible heat flux from the vapor as it cools to the saturation temperature.

A calculation assuming the surface to be dry at the vapor inlet will determine whether a dry surface will occur. If a dry region is confirmed, this calculation will give the maximum heat-transfer surface temperature and the maximum heat flux. A method of approximation can then be used to find a temperature near the transition point (at which condensation begins).

A second dry-wall calculation enables the transition temperature to be estimated accurately and also gives the flux at this temperature. If the heat transferred to a dry surface is a substantial portion of the total, the thermal rating should take into account the higher flux that occurs in this part of the duty.

A procedure has been given for estimating the flux in the superheated-vapor-to-wet-wall region, which is also greater than the saturation flux. For practical purposes, it is suggested that the mean flux increase in this part of the duty should be taken to be 25% of the increase at transition temperature. The suggested procedure can be applied to any situation in which saturated-vapor and dry-wall heat transfer can be calculated.

Thus, in the present case, the surface temperature is found to be higher at a given vapor temperature with low-fin than with plain tube, and so the surface remains dry over a larger range. The maximum surface temperature, determined at inlet conditions, is of course increased.

Because h_c, and therefore U', are still almost constant, the first approximation to the transition flux, 1,235 Btu/(h)(ft²), obtained directly from the saturation flux is found to be accurate.

The complete flux plot for the low-fin duty, including the superheated-vapor-to-wet-wall range, is given in Fig. 3. As will be seen, the flux increment up to the transition point is relatively small with the low-fin tube. Nevertheless, the mean flux for the entire duty, 1,073 Btu/(h)(ft²), calculated as shown before for plain tube, is still 5 to 6% above saturation flux, 1,018 Btu/(h)(ft²).

To estimate the overall advantage of the low-fin tube, the mean heat transfer per unit length of tube is re-

References

1. Colburn, A. P., and Hougen, O. A., *Ind. Eng. Chem.*, Vol. 26, 1934.
2. McAdams, W. H., "Heat Transmission," 3rd ed., McGraw-Hill, 1954, p. 177.
3. Kern, D. Q., "Process Heat Transfer," McGraw-Hill, 1950, p. 269.
4. McAdams, W. H., op. cit., p. 219.
5. Colburn, A. P., *Ind. Eng. Chem.*, Vol. 25, 1933.

The authors

Hugh R. Chappell is Senior Heat Transfer Engineer with IMI Yorkshire Imperial Ltd., P.O. Box 166, Leeds LS1 1RD, England, working mainly on sales development of finned tubes. He was previously a senior mechanical test engineer and a technical inspector with IMI Yorkshire. He has served on a number of British standards committees concerned with the preparation of standards for methods on the mechanical testing of metals and on heat exchangers.

D. H. Foxall, Manager of Heat Transfer Services for IMI Yorkshire Imperial Ltd., is concerned with product development, sales promotion and technical service on heat-transfer tubes. He was educated at King Edward VI School, Birmingham, U.K., and at Birmingham University, where he obtained a B.Sc. (1st Class Honours, Mathematics), followed by two years of graduate study in engineering production. He had previously worked for British Motor Corp. as a development engineer, and for Imperial Chemical Industries on product development.

Section V
Dryers

Selection of industrial dryers

Here is a discussion of the types of batch and continuous dryers that are available, and a procedure for choosing the most suitable one for a particular process.

C. M. van 't Land, *Akzo Chemie Nederland bv*

☐ Drying is the removal, by heat, of volatile substances (moisture) from a mixture, to yield a solid product [1]. Although this definition includes the drying of such materials as pottery, timber and particulate materials, this article will discuss drying only for the chemical, pharmaceutical and food industries.

Dried material to be sold to customers must meet agreed-upon specifications. To attain these specifications, it may be necessary to assess the conditions of:
- The dryer.
- The equipment preceding the dryer.
- The equipment following the dryer.

The choice of continuous versus batch drying also depends on the nature of the equipment preceding and following the dryer. (Production capacities exceeding 100 kg/h often require use of a continuous dryer.)

Table I contains a list of possible criteria for judging a dried particulate material; Table II outlines the data that may have to be collected before one can start work on selecting a dryer.

Alternatives to full drying

Before drying is attempted, it is worthwhile to consider the possibility of alternatives:

Flaking instead of drying—Materials having a relatively low melting point can be flaked instead of dried. Upon heating of the slurry, the solid phase melts, and a liquid-liquid separation may be made using a disk centrifuge.

Selling wet filter cake—It is possible that the customer may be able to use wet filter cake instead of a dry product. (If the solid dissolves in the liquid only to a very small degree, the cake remains loose; otherwise it may set up and become hard to handle.)

In-product drying—A wet centrifuge- or filter-cake may be dried by admixing it with a material that forms a hydrate, or a higher hydrate, with the water in the wet cake. Heat evolution accompanies the hydration.

Selling product that is not bone-dry—The customer may be willing to accept product that still contains some moisture. The removal of the last quantity of moisture is relatively expensive compared with the removal of the bulk of liquid.

Some drying truisms

If it is possible to remove moisture mechanically, this will always be more economical than removing it by evaporation.

Steam-tube rotary dryers and direct-heat rotary dryers are universally applicable. They can be chosen if there is a limited quantity of experimental data, or insufficient time to go through the selection procedure suggested later in this article. However, the tailor-made solution, using the selection procedures, will often lead to a less expensive solution to the drying problem.

The drying of certain products is almost universally associated with a particular type of dryer that has been found especially suitable. Examples are the use of tunnel dryers for prunes or continuous bin dryers for grain.

Selection schemes

Fig. 1 and 2 are flowcharts for the selection of batch dryers (Fig. 1) and continuous dryers (Fig. 2), both for particulate materials.

Batch dryers

Here are some comments on Fig. 1 (for selection of batch dryers):

Vacuum dryers—If the maximum product temperature is lower than or equal to 30°C, it is worthwhile to look at a vacuum dryer. A good driving force for evaporation can be created while keeping the temperature low. The vacuum tray dryer is the simplest, but the product must usually be sieved to break down any agglomerates (the breakdown may be aided mechanically).

The capacity of the vacuum tray dryer is rather low. It may be economic to consider an agitated vacuum dryer (Fig. 3) in which the contents are moved mechanically. Such dryers are widely used.

If the product is oxidized by air during drying, consider either vacuum drying or inert-gas drying.

If either the product or the removed liquid is toxic, the equipment must be kept closed as much as possible. Again, a vacuum dryer can render good service. (In addition, dust formation is avoided.)

Fluidized-bed dryers—If the average particle size is about 0.1 mm, or larger, fluidized-bed drying (Fig. 4) may be considered. (If smaller particles must be dealt with, the equipment required to handle them may be too large to be feasible.) Inert gas may be used if there is the possibility of explosion of either the vapor or dust in air.

If such a dryer is being considered, it is easy to carry out tests in a small fluid-bed dryer.

Other dryers—As Fig. 1 shows, the remaining possibilities are the tray dryer and the agitated pan dryer.

Originally published March 5, 1984

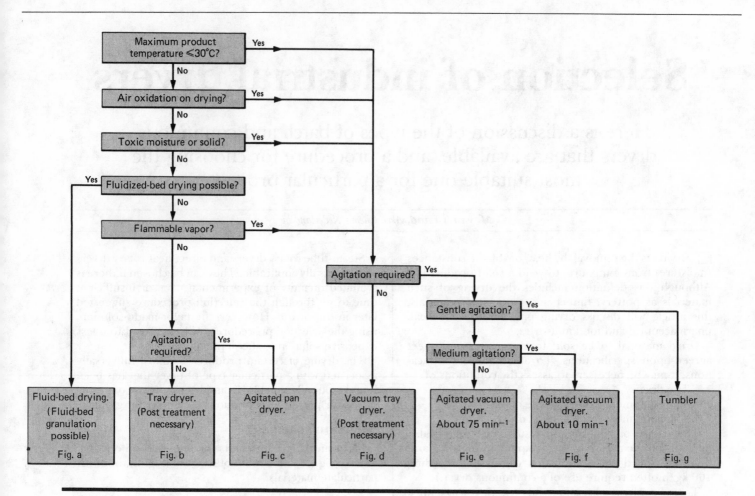

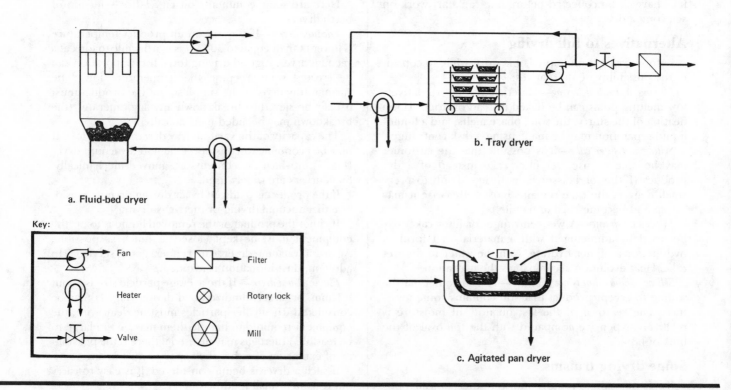

Key:

→ Fan		□ → Filter	
Heater		⊗ Rotary lock	
Valve		⊕ Mill	

a. Fluid-bed dryer

b. Tray dryer

c. Agitated pan dryer

Decision tree for the selection of a batch dryer suitable for any particular process need together with sketches of the

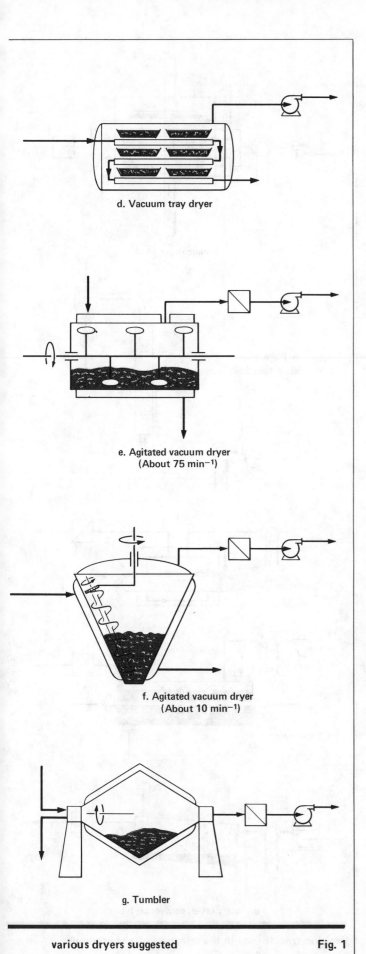

d. Vacuum tray dryer

e. Agitated vacuum dryer
(About 75 min⁻¹)

f. Agitated vacuum dryer
(About 10 min⁻¹)

g. Tumbler

various dryers suggested   Fig. 1

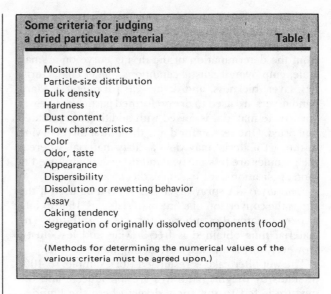

Some criteria for judging a dried particulate material	Table I

Moisture content
Particle-size distribution
Bulk density
Hardness
Dust content
Flow characteristics
Color
Odor, taste
Appearance
Dispersibility
Dissolution or rewetting behavior
Assay
Caking tendency
Segregation of originally dissolved components (food)

(Methods for determining the numerical values of the various criteria must be agreed upon.)

Continuous dryers

Solvent evaporation—In continuous drying, if a solvent must be evaporated and then recovered, it is usually not optimum to choose a convection dryer. Since solvent must be condensed from a large carrier-gas flow, the condenser and other equipment become rather large.

Milling/drying—If it is necessary to decrease particle size, in addition to drying, the two operations may be advantageously combined. The wet particulate solid is transported by warm or hot gas into a mill. Gas and particulate solid leave the mill, fly through a line and are separated. The comminution often greatly helps the drying by exposing internal moisture. This type of drying is encountered in cases where the fineness is of great importance to the application. Examples are cases where a rapid and complete dispersion (or dissolution) or a high activity (m²/g) are being aimed for.

Band (belt) dryers—A band dryer (Fig. 5) is preferable if the particles are rather coarse (i.e., over 5 to 10 mm). The particles are evenly spread onto a slowly moving, e.g., 5 mm/s, perforated belt. The belt moves into a drying cabinet and warm gas passes downward through the layer. This type of dryer is chosen when it is not possible to suspend the particles in the drying gas. The

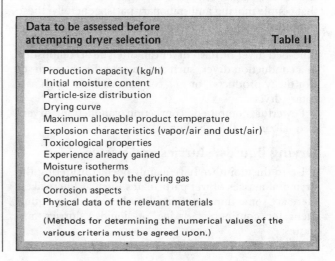

Data to be assessed before attempting dryer selection	Table II

Production capacity (kg/h)
Initial moisture content
Particle-size distribution
Drying curve
Maximum allowable product temperature
Explosion characteristics (vapor/air and dust/air)
Toxicological properties
Experience already gained
Moisture isotherms
Contamination by the drying gas
Corrosion aspects
Physical data of the relevant materials

(Methods for determining the numerical values of the various criteria must be agreed upon.)

dryer must offer a residence time, say 15 min, because bound moisture must diffuse through the pellet.

The performance of such a dryer can be predicted from the determination of the drying curve on a small scale, employing realistic conditions (pellet characteristics, layer thickness, and drying-gas parameters). Many band dryers are used to dry preformed particles. The wet particulate material is mixed with additives, granulated and dried. One reason for doing this is that direct drying of the wet material may yield a dusty material, whereas the granules are less dusty. Band dryers are also used in food applications, for example, diced carrots.

Spray dryers—A spray dryer (Fig. 6) can be used if the aim is the conversion of a fine material (e.g., 15 μm) into a coarser material of spherical form (e.g., 150 μm). The material thus obtained is free-flowing and less dusty. However, this is an expensive drying method.

The wet filter cake is reslurried (for example, to 40% of solids by weight), additives are introduced, and the mixture is fed to the spray dryer, where the liquid is evaporated. To keep the size of the equipment reasonable, a minimum inlet-gas temperature is required (perhaps 200°C) to produce a solids outlet temperature that exceeds 75°C.

Flash dryers—The flash dryer is the workhorse of industry (Fig. 7). However, because drying must take place within 10 s, the removal of bound moisture is difficult. Since the dryer is essentially a vertical line, drying and vertical transport can be combined.

Fluid-bed dryers—Use of a fluid-bed dryer is a possibility if the particle size exceeds 0.1 mm. A round piece of equipment holding a thick product layer is one option. The holdup must be large, as the composition of the dryer contents equals the outlet composition. The thick product layer means that much fan power is needed to push the drying gas through it. Because caking will not easily occur, the construction can be stationary. Such construction can allow high drying-gas temperatures—up to 500 – 600°C.

A rectangular-shaped dryer will permit plug flow. Fig. 8 shows a stationary type with a thick product layer; a shallow layer may require vibration for transport and to prevent caking. However, a vibrated construction (Fig. 9) cannot withstand high temperatures. A realistic maximum drying-gas inlet temperature is 300°C. Moreover, the hot drying gas must pass through the flexible devices that couple moving and stationary parts; generally, these cannot withstand high temperatures either.

Miscellaneous dryers—Jobs that cannot be handled by the fluid-bed dryer or flash dryer can often be accomplished in a conduction dryer, such as a steam-tube rotary dryer (for dusty products), or in a convection dryer, such as a rotary dryer.

Powerful combined convection/conduction dryers also fall under this heading.

Drying liquids, slurries and pastes

Up to this point we have mainly been considering the drying of masses of wet particulates, such as filter cakes. Here are some things to consider when choosing equipment for continuous drying of liquids, slurries and pastes:

Spray dryers—These may be chosen if the isolation of a

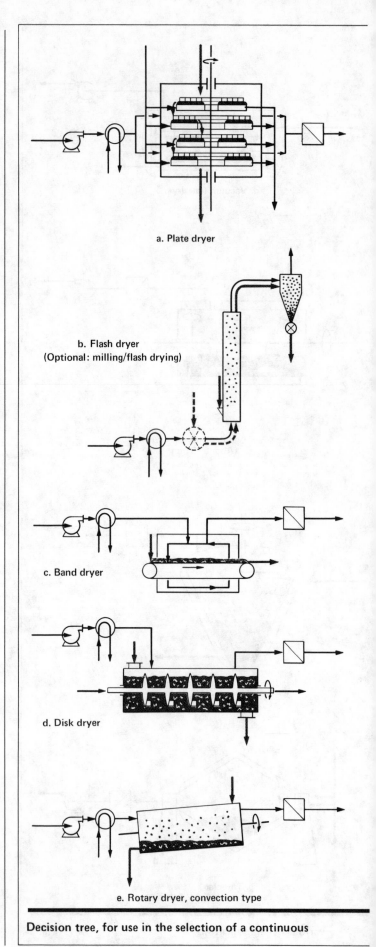

a. Plate dryer

b. Flash dryer
(Optional: milling/flash drying)

c. Band dryer

d. Disk dryer

e. Rotary dryer, convection type

Decision tree, for use in the selection of a continuous

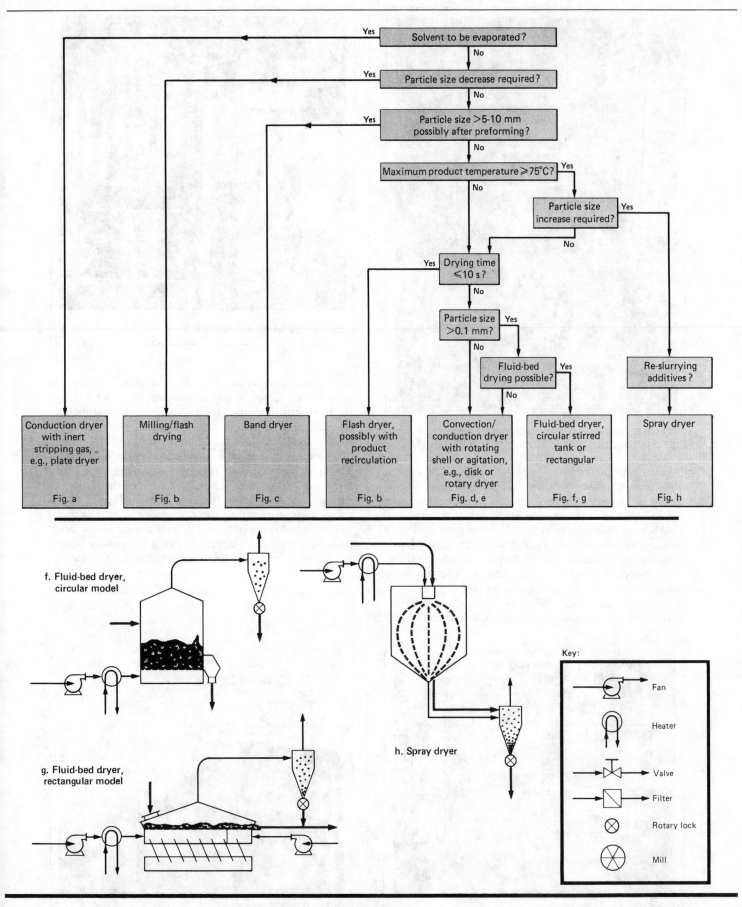

f. Fluid-bed dryer, circular model

g. Fluid-bed dryer, rectangular model

h. Spray dryer

Key:

Fan

Heater

Valve

Filter

Rotary lock

Mill

dryer, leads the user to one of the dryers shown in the sketches   Fig. 2

Hosokawa Nauta

Agitated vacuum dryer with solvent recovery   Fig. 3

Batch type of fluid-bed dryer   Fig. 4

Glatt

solid from a solution or slurry via conventional crystallization and liquid/solid separation is either impossible or too complicated. Sometimes, spray drying is chosen if the characteristic spherical particle shape is desired (as in making instant coffee). Typically, the average particle size (on a weight basis) falls in the range of 50 to 200 μm.

This dryer's short residence time is a plus for heat-sensitive products, e.g., milk powder and organic salts.

Film drum dryers—As with spray dryers, drum dryers may be chosen in cases where crystallization and liquid/

solid separation are not feasible. This type of dryer can be used for pastes, and it can be placed under vacuum for heat-sensitive products. Examples of its use include the flaking of mashed potatoes, and drying of other instant foods.

Cylindrical scraped-surface evaporator/crystallizer/dryers— These dryers resemble the equipment used for other thin-film techniques. The fluid to be dried passes through in plug flow. Between the feed point (of a solution or suspension) and the product outlet (a more or less free-flowing powder), there is a zone where much power is consumed (per unit wall area), as a viscous paste is converted into particles. In principle, the equipment is

Babcock-BSH

Band dryer uses a slowly
moving perforated belt   Fig. 5

Spray dryer produces a
spherical-shaped product   Fig. 6

Niro Atomizer

suitable for pastes as well as liquids and slurries. Vacuum may be applied for heat-sensitive products, and solvent recovery is possible.

For batchwise drying of liquids, slurries or pastes, consider:

Agitated pan dryers or agitated vacuum dryers (Fig. 10)—In both of these dryers, solvent recovery is possible. In the case of liquids and slurries, as the drying proceeds, the power consumption rises to a maximum and then decreases. (As noted above, a liquid or slurry is converted into a more or less free-flowing powder via a viscous paste.) Hence, the motor, gearing and stirrer must be adequately sized for the maximum power consumption.

Special drying techniques

Infrared drying—This is generally confined to surface drying, as of newly painted automobile bodies.

Freeze drying—In this technique, ice is sublimed from a frozen product while under vacuum. Drying temperatures typically range from -30 to $-10°C$, so the process is ideal for very heat-sensitive products such as those of a biological nature.

Microwave drying—This process is used for relatively large and expensive articles that contain little moisture. It has also recently been applied to the drying of pharmaceutical products under vacuum.

Some additional comments

Conduction vs. convection dryers—For both batch and continuous dryers, a distinction can be made between conduction and convection drying. Convection dryers (e.g., flash and spray dryers) often require relatively large solid/gas separation equipment. Thermal efficiency of convection dryers increases with increasing inlet temperature of the drying gas. There is no such effect with conduction dryers. Furthermore, conduction dryers may be preferred for dusty products.

A different aspect is the relationship between product temperature and heating-medium temperature. In convection dryers, the product usually adopts the adiabatic saturation temperature. But in conduction dryers, product in contact with hot metal will attain the metal's temperature.

Finally, the residence time in conduction dryers can be up to several hours. But the residence time in flash and spray dryers can be quite short—ten seconds or less. Both time and temperature are important in regard to thermal degradation.

In the classification we have used, the plate dryer (Fig. 11) is a hybrid type. This type of dryer can be useful

Flash dryer—one of the most common types used in industry   Fig. 7

Babcock-BSH

for solvent recovery or dusty products. The raking action counteracts caking.

Combining dryer types—A combination of two different types of dryers is sometimes optimum. One example is the spray drying of milk, followed by a post-treatment in a fluid-bed dryer.

Dust explosions—If there is a possibility of a dust explosion in a convection dryer, special precautions must be taken. These may involve use of inert gas, provision of explosion panels, triggered injection of suppression agents, and the like.

Scaling up—Some of the dryers thus far discussed have a limited possibility for scaling up. In comparing a drum dryer with a spray dryer, it is clear that the spray dryer can cope with larger evaporation loads.

Continuous, stationary fluid-bed dryer   Fig. 8

Escher Wyss

Escher Wyss

Continuous, vibrated fluid-bed dryer   Fig. 9

There is practically no limit to scaling up a rotary dryer. Convection-type rotary dryers are well established in the mining industry.

Vacuum drying—Continuous vacuum drying is not often found; it is difficult to feed and extract the material under vacuum. However, there are some special cases where a liquid, slurry or pumpable paste is fed and a dried solid (with relatively good handling characteristics) is removed.

Babcock-BSH

**Agitated vacuum dryer for batch
drying of liquids, slurries and pastes**   Fig. 10

Testing on small-scale dryers

For testing, it is usually wise to seek the cooperation of a limited number of reputable drying-equipment manufacturers. It is often possible to ship samples of the material to be dried and have them tested in the manufacturer's small-scale units. This is relatively simple and straightforward, but there are risks:

■ The material may have changed in character, due to chemical or physical changes between the time the wet product was tapped off the process stream and the time of drying.

■ Because the quantity of sample shipped is limited, it is not possible to check long-run performance. For some desired process results, such as proper transport of the material through the dryer, and absence of caking or dusting, it is essential to carry out experiments over a long period.

■ The performance of the dryer is not checked in relationship to the process-plant infrastructure—for example, the behavior of the dried material in the plant's solids-handling equipment.

Consequently, if the experiments done by the equipment supplier are successful, it is good practice to install a small-scale dryer in the process plant, where it can be fed from a side stream. If a new product or process is being dealt with, the dryer should be in the pilot plant. If there is no pilot plant, representative material must be made or obtained and then dried at the equipment manufacturer's facility. If one wishes to see full-scale equipment in operation, the manufacturer may be able to arrange an introduction to a facility using such dryers.

Before purchase, it is necessary to obtain guarantees from the dryer manufacturer and to agree on the method of analysis that will determine whether the guarantees are met.

In cases where several makes or types of dryers can do the job suitably, the variable and fixed costs for the equipment should be calculated. Then, the qualitative pros and cons can be listed. By studying this material, it should be possible to reach a purchase decision.

The literature on the selection of industrial dryers is sparse—a situation that is outlined by van Brakel [3]. Further progress can be made by developing simple relationships between the evaporation load and a characteristic main diameter of a dryer type.

Additionally, the dryer size could be correlated with a cost (investment) figure. This would enable the engineer to visualize the consequences of any particular choice.

Noden [4] approaches the issue in this way. However, this article is almost 15 years old, and updating is required. To complete the selection picture, one would also have to factor in personnel costs and energy consumption.

Examples of dryer selection

Example 1—In an existing plant, the rotary dryers being used for salt (NaCl) dated back to the 30s. These continuously functioning dryers were worn out and needed replacement. After a successful drying test at a manufacturer's facility, it was decided to install a flash dryer. The choice was based on the following:
- Water was to be evaporated.
- There was no need to change particle size.
- The average particle size was about 0.4 mm.
- The inorganic salt was not temperature-sensitive.
- Only surface moisture was to be removed.

The flash dryer was installed, but was not successful owing to the formation of a fine product mist from the equipment. Two reasons for this were thought of: 1) the high velocities in the flash dryer abrade the crystal, producing fine particles, and 2) the rapid evaporation of the surface moisture causes nucleation therein (the surface moisture being a saturated solution of salt in water).

The problem disappeared when the flash dryer was replaced by a trough-like vibrated fluid-bed dryer having a shallow product layer. Residence time is now minutes instead of seconds, and velocities are much lower.

Example 2—A plant was built for producing about 40 kg/h (dry basis) of a solid organic peroxide. It was decided to install a batch dryer, for these reasons:
- Maximum product temperature was 40°C.
- Water was to be evaporated (about 25% by weight)
- The solid was not very toxic.
- Oxidation by air did not occur.
- A dust explosion was possible.
- Average particle size was about 500 μm.

Testing established the feasibility of fluid-bed drying, using a recycled inert gas. The drying gas is warmed, indirectly, by warm water to 40°C. The diameter of the supporting plate is 0.92 m.

The process is controlled by measuring the temperature of the leaving gas. When it reaches a specified temperature, the warm water supply is switched off, and the batch is cooled by an inert gas for 10 min.

Example 3—A case of drying of organo-tin compounds is described in the literature [2]. Many of these compounds are more or less toxic. Between 10 and 30% of moisture (a mixture of various solvents, which may include water) is to be evaporated. The maximum allowable product temperature is between 50 and 90°C. Vacuum used is between 30 and 50 torr. Batch size is 1,500 to 2,000 kg, and the bulk density of the dry product is about 500 kg/m³.

Originally, a tumbler dryer was used, but it was later replaced—in order to increase capacity—by a proprietary dryer comprising a conical mixer with a screw-type mixing element rotating along the circumference of the cone and around its own axis. Both dryers were of stainless steel. Their capacities were:

	Tumbler	Conical mixer
Drying time, h	20 – 40	10 – 30
Volume, L	6,800	4,000

Plate dryer is useful for handling dusty products   **Fig. 11**

Krauss-Maffei

Hourly production is about 100 kg/h, so the choice of a batch dryer seems logical. A vacuum dryer was selected because of the toxicity of many of the products and the flammability of the vapors. The reasons for replacing the tumbler dryer by the vacuum dryer with medium agitation were:
- Shorter drying time because of better heat transfer (no crusts).
- Fixed provision for charging and discharging.
- Homogeneous product—no manual cleaning because of the absence of crusts.
- Low maintenance costs. No need to convey utilities from stationary parts to rotary parts (and vice versa) as in the tumbler.

According to the reference, the drying cost per unit of product was reduced by 40% by replacing the tumbler.

References

1. Keey, R. B., "Drying Principles and Practice," Pergamon Press, Oxford, 1972.
2. Schaake, P., and Stigter, H., *Aufbereitungstechnik*, Vol. 1, p. 27 (1975).
3. Van Brakel, J., *Chem. Eng. (London)*, July 1979, p. 493.
4. Noden, D., *Chem. Process. Eng. (London)*, Oct. 1969, p. 67.

The author

C. M. van 't Land is Process Development Manager, Research Centre Deventer, Akzo Chemie Nederland bv, Emmastraat 33, 7411 EK Deventer, Holland. He has specialized in evaporation, crystallization, liquid/solid separation, and drying. He holds an M.S. in chemical engineering from the Twente University of Technology, and is a member of Koninklijke Nederlandse Chemische Vereniging (Royal Dutch Chemical Soc.).

Indirect drying of solids

Here is a guide to the use of this technique, and to the wide variety of dryer designs—for example, shelf, drum, tubular, disc, and paddle styles—available.

William L. Root, III, Komline-Sanderson Engineering Corp.

☐ Drying can be defined as the process whereby, through heating, moisture is evaporated from a solid to produce a relatively dry substance. This contrasts with moisture removal by mechanical means, such as via filtration, centrifugation or decantation.

Drying is one of the oldest and most important of all unit operations. However, while other operations have been reduced to predictable mathematical expressions, drying is still largely an art. The only reliable way to check a commercial-scale design is to pretest one's product in the specific type of dryer to be used in the process. Mechanical features in each manufacturer's design make it difficult to project data from one design to another. Each dryer reacts differently. In addition, upstream processing can create chemical and physical variations—such as in particle size, level of impurities, and capillary size—that can have a distinct effect on the drying characteristics of each material.

Many attempts have been made to classify drying equipment—some from the user's viewpoint and some from the manufacturer's. McCormick in the "Chemical Engineers' Handbook" [1] tried to categorize units from the user's perspective. Sloan [2] identified 20 types of dryers, and Lapple and Clark [3] discussed 39 designs. Dittman [4] observed that previous classifications were difficult to follow, and reduced the principal classes to two by focusing on the method of heat transfer: indirect or direct.

Indirect dryers, also called nonadiabatic units, separate the heat-transfer medium from the product to be dried by a metal wall. Heat-transfer fluids may be of either the condensing type (e.g., steam, and diphenyl fluids, such as Dowtherm A) or the liquid type (such as hot water, and glycol solutions). Seldom does one use a noncondensing gaseous medium, because of the low film coefficients associated with such systems. When high temperatures—say, over 660°F—must be achieved, indirect firing may be the only possibility.

Little or no carrier gas generally is required to remove the vapors released from the solids during drying. (An exception to this occurs when drying to a very low moisture content: carrier gas then may be necessary to reduce the partial pressure of liquid's vapor in the environs surrounding the particle to be dried below that of the equilibrium pressure—thus allowing drying to progress.)

In contrast, direct (or adiabatic) dryers use the sensible heat of a gas that contacts the solid to provide the heat of vaporization of the liquid. (This kind of dryer can be further broken down into two types: suspended

Originally published May 2, 1983

particle, and bed units. Flash, spray and fluid-bed designs typify the suspended-particle approach, while bed units 'include cabinet, tray, belt, tunnel, and rotary styles.) The adiabatic dryer is more closely related to the humidification process than is the indirect one.

The major difference between humidification and drying is that humidification is evaporation from pure liquid, while drying involves vaporizing of a liquid dispersed in or on a solid. Thus, the direction of heat transfer in drying is opposite to that of mass transfer. A resistance to drying above that found in humidification must therefore be considered. This resistance comes from the passage or movement of the liquid and its vapor through the solid to the exterior vapor/solid interface, and accounts for the falling rate encountered during drying, as opposed to the constant rate found in humidification.

The advantages of indirect drying

Indirect dryers boast several distinctive operating features.

First of all, since the product does not contact the heat-transfer medium, the risk of cross-contamination is avoided.

Because of the limited amount of noncondensable gas encountered, solvent recovery is easier than in an adiabatic dryer, where there is a large volume of gas to be cooled to recover the solvent.

Similarly, dusting is minimized because of the small volume of vapors involved in indirect drying.

The dryers also allow operation under vacuum or closely controlled atmospheres, which can avoid product degradation.

Likewise, explosion hazards may be easier to control in such units.

When one compares the heat input to the heat required to evaporate the liquid, indirect dryers offer a high thermal efficiency. (To illustrate this, Fig. 1 shows the overall heat input to evaporate water in a typical unit using no carrier gas.) The temperature of evaporation in the unit approximates the boiling point of the liquid under the operating conditions in the dryer.

Typically, the indirect dryer is used for small- or medium-size production. The product from such a unit has a higher bulk density than the same material produced, for example, in a spray dryer. Particle-size degradation usually can be minimized by proper selection of agitator speed or design.

In most cases, mechanical separation—via filtration, centrifugation, decanting, etc.—should precede the drying step. Such mechanical means normally use less energy than drying and cause no thermal degradation, and so are an economic first step. However, they cannot achieve the low moisture levels attainable by drying.

If small batches of material are to be processed, such a preliminary mechanical step may not be desirable, because it actually may impose a penalty in labor costs and in materials-handling losses.

Sometimes, to avoid hazards in materials handling, it makes sense to remove all moisture in one step in a dryer. While this is usually at the sacrifice of thermal efficiency, the results may be a lower overall capital cost, and increased safety.

Sources of moisture and heat

Before going into more detail on indirect units, some general background on drying may be helpful.

Liquid that is to be removed by the drying process may occur as surface, mechanically bound, or chemically bound moisture. (The words liquid and solvent are used interchangeably in this article, with solvent chosen in some contexts only to indicate that the liquid may be worth recovering.)

Surface or free moisture refers to the liquid that wets the outer surface of the solid. As one might expect, this kind of moisture is removed rather rapidly by the drying process.

Mechanically bound moisture occupies the interstices of the solid, and, as such, moves to the surface by dif-

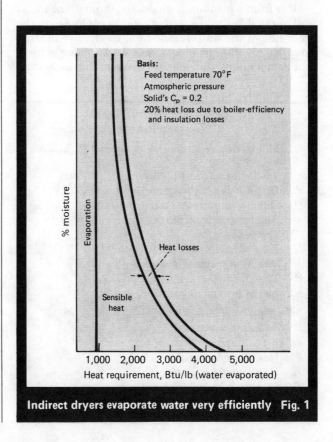

Indirect dryers evaporate water very efficiently   Fig. 1

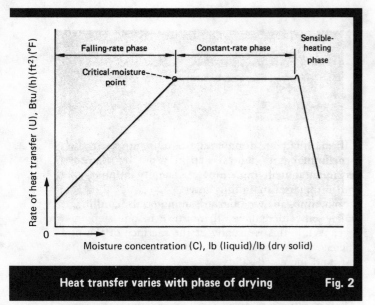

Heat transfer varies with phase of drying   Fig. 2

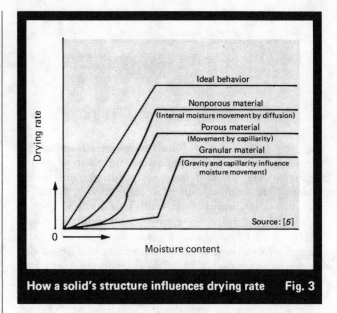

How a solid's structure influences drying rate   Fig. 3

fusion, capillary action, pressure gradients, etc. This moisture will never keep the outer surface of the solid completely wetted.

Chemically bound moisture often appears as water of hydration. It may also be produced by the structural reorientation of a chemical or by reaction. The heat required to release this moisture is, in addition to the heat of vaporization, the heat of crystallization or of reaction.

Heat to remove moisture can be provided by conduction, convection or radiation. However, in indirect dryers, most of the heat transfer occurs via conduction.

Heat from a medium is conducted through the metal wall of the dryer to the wetted solid in contact with the other side of the wall. In addition, conductive heating takes place within the process mass itself as long as there is a temperature differential between the individual

particles. Agitation can improve conduction by more rapidly bringing particles into contact with the heat-transfer surface.

Convection, although not a major factor during most of the drying cycle, aids in heat transfer during the early stages. As the temperature of the solid/liquid phase begins to increase, the vapor pressure of the liquid rises, and thus some of the liquid changes to vapor. The vapor begins to migrate and condense on the colder wetted solids, thereby transferring some heat by convection.

In most indirect dryers, radiation plays little or no part in heat transfer. The temperatures encountered are just too low to effect radiant heating. This is not the case, though, for some indirectly fired units. In fact, in such devices, radiation may be the major means of conveying heat for evaporation.

The stages of drying

Regardless of how heat is provided, the drying cycle consists of three steps: 1—Before any evaporation can take place, sensible heat must be added to the drying mass until the boiling point of the liquid under the given operating conditions is reached. 2—Once the boiling point is reached, evaporation takes place at a rate related to the moisture level in the solid; this rate normally is constant over a certain moisture range. 3—At some point, though, the critical moisture point is reached, and the drying rate begins to fall. This overall sequence is represented by a typical drying curve, as shown in Fig. 2.

(During drying, conditions must be such as to keep the vapor pressure of evaporated liquid in the environs of the dryer less than the vapor pressure of the remaining liquid. This provides a pressure differential to remove the evaporating liquid from the solid's surface.)

The sensible-heat input to the process mass is often forgotten in the theoretical presentation of the drying process. Nevertheless, the temperature of the process mass must be raised to the boiling point of the liquid under the operating conditions of the dryer. During this

Nomenclature

C Moisture concentration of drying mass, lb (liquids)/ lb (dry solids)

c Specific heat, Btu/(lb)(°F)

d Distance particle has traveled in dryer, ft

Q Summation of heat flowing, Btu/h

q Rate of heat flow, Btu/h

T Temperature, °F

U Rate of heat transfer, Btu/(h)(ft²)(°F)

W Weight of material flowing, lb/h

Θ Time, h

λ Latent heat of vaporization, Btu/lb

Subscripts

B Bed temperature

$d1, d2$, etc.   Specific location 1, 2, etc.

e Heat for evaporation

H Heat-transfer medium

$L1, L2$, etc.   Liquid 1, liquid 2, etc., in multi-liquid

$S1, S2$, etc.   Solid 1, solid 2, etc., in multi-solid

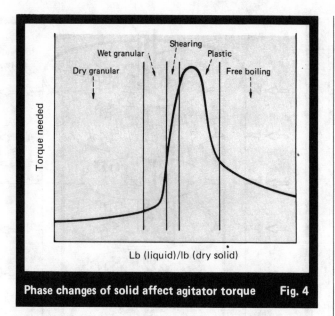

Phase changes of solid affect agitator torque   Fig. 4

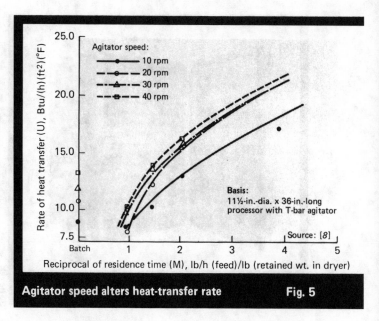

Agitator speed alters heat-transfer rate   Fig. 5

time, as the bed temperature rises, some evaporation takes place to satisfy the equilibrium vapor-pressure conditions between the gaseous phase surrounding the solid and the liquid on the solid. The vapors generated have a higher temperature than some of the surrounding solid/liquid mass, and thus some condensation takes place on the solids. For this reason, the rate of heat transfer during sensible heating varies as both the temperature increases and the moisture content is lowered somewhat.

Constant-rate drying is the first stage of drying that is normally taught in the classical approach. The rate of heat input/(unit area)(unit time)(degree temperature difference)—or of moisture release/(unit area)(unit time)—remains constant. Drying occurs at or near the boiling point of the liquid. There is no resistance to vaporization, since the moisture appears on the surface of the solid, completely wetting the outer surface. The constant drying rate is proportional to the difference between the vapor pressure of the liquid wetting the surface of the solid and that of the vapor surrounding the wetted solid. It has been found that the vapor pressure of the wetting liquid is very close to that of the pure liquid at the same temperature.

At the critical moisture content of the solid, the drying rate begins to fall, because the moisture that had completely wetted the solid has now disappeared. The reduced amount of moisture being evaporated comes from the interstices of the solid through the porous structure; this is not sufficient to completely wet the exterior surface of the solid.

The falling-rate drying is controlled by the physical properties of the liquid and solid. The rate of movement of the liquid and its vapor depends on capillary size, glazing of the solid, pressure gradients between trapped liquid and vapor and the environs of the solid, as well as on cracking, checking, etc. At the same time, the heat-transfer rate to the interior of the solid is being slowed because of the receding boundary of the liquid-wetted portion of the particle. This boundary movement increases the resistance to heat flow because it reduces the

thermal conductivity within the solid. Cracking and checking disrupt the paths of heat transfer, further decreasing the rate.

Coates and Pressburg [5] discussed the effect of drying-rate changes caused by particle structure, and depicted them graphically, as in Fig. 3. One can notice the nonlinear variations on the rate of drying, caused by glazing, cracking and checking of the surface of the solid particles. The linear falling rates are evident when the mass transfer of the liquid or vapor is purely by diffusion.

Drying rates are reported in several forms—weight of liquid removed/(unit area)(unit time); weight of solid processed/(unit area)(unit time); and heat-flow rate/(unit area)(unit time)(degree temperature difference). The first two methods require identical conditions of temperature, pressure and moisture level for meaningful comparisons between systems. The third approach avoids these restrictions.

If one observes the condition of the drying bed as the moisture level is reduced, five distinct phase changes will be seen: free boiling, plastic, shearing, wet granular and dry granular. A representative plot of horsepower required versus moisture content of the mass within a batch dryer appears as Fig. 4, with the various phase changes noted. During free boiling, the vapors sometimes evolve in bubble form from an almost-liquid matrix. When the plastic stage is reached, the mass being dried shows a tendency to agglomerate, and horsepower requirements peak. With continued loss of moisture, the mass begins to shear or break apart due to the conflicting movements of the agitated particles. Further evaporation causes the mass to take on the appearance of a wet granular bed. Continued drying leads to the formation of a free-flowing, dry powdery mass.

The benefit of agitation

Agitation in an indirect dryer provides an improvement in the rate of heat transfer, and thus the rate of drying. As the degree of agitation increases, dependency on the thermal conductivity of the process mass as a

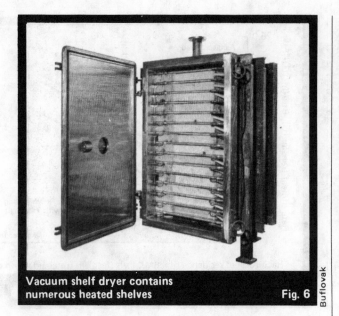

Vacuum shelf dryer contains numerous heated shelves   Fig. 6

Buflovak

whole is reduced. More individual particles of the mass are exposed to the heat-transfer surface, and, therefore, the bed temperatures within the dryer become more uniform.

This improvement is not gained without some tradeoffs. For instance, residence-time distribution within the dryer is modified because of the effect of backmixing.

The agitator design and its speed play an important part in enhancing the heat-transfer and drying rates. Speed raises the rate of heat transfer up to the point that mechanical fluidization takes place. At this point, the particles in the bed lose their intimate contact with each other, and thus reduce the effect of conductivity—so, the heat-transfer rate begins to fall.

However, when using a cylindrical vessel, this rate improves if the agitator speed is such that the material is held against the wall in a relatively thin film by centrifugal force. Then, one notices that the heat-transfer rate starts to improve rather rapidly.

This freeze dryer allows internal stoppering of product   Fig. 7

Pennwalt Stokes

Plate dryer allows temperature control of each plate   Fig. 8

Krauss Maffei

Gunes and Schlunder [6] investigated the effect of speed in a simple drying system, and confirmed the above observations. Kasatin et al. [7] reported on the effect of speed and production rate in a hollow-screw heat exchanger: They showed that, as production rate increased and the speed changed, there was an improvement in the heat-transfer rate. Porter [8] studied the impact of agitator speed and throughput on the performance of a cylindrical dryer having a T-bar agitator. His data are reported graphically in Fig. 5. Uhl and Root [9] summarized several other authors' findings, and presented some new data on this theme.

At first, agitators were modified to improve the rate of raking of the process mass from the heat-transfer surface and of fresh material to that surface. In more-recent designs, attempts have been made to promote localized mixing within the unit—via discontinuous paddles, breaker bars, scrapers, and the like. These units, however, cannot in any way be considered as blenders, since the mixing is not designed in most cases to promote a homogeneous mass.

There are a few tumbling and rotary blenders used for drying, but usually at a sacrifice of process time when compared with units specifically designed for moisture removal.

The first types of indirect dryers

The most-long-lived dryer designs include shelf (tray), pan, drum, and horizontal (rotary vacuum) units, and the steam tube.

The simplest, although the most labor intensive, type is the *shelf* or *tray dryer*, as shown in Fig. 6. Product to be dried is placed on trays that are then put on heated shelves within the dryer proper. The chamber pressure

Jacketed pan dryer employs top-entry agitator   Fig. 9

Bethlehem

Atmospheric double-drum dryer applies liquid at pinch at top   Fig. 10

Buflovak

is reduced to the desired operating level. The shelves are then heated to the appropriate dehydration temperature. The product is gently dried by heat being conducted from the shelves into the lower portion of the retaining trays, and by heat radiating from the underside of the shelf above. Extremely high vacuum can be readily obtained. The feed is usually a solid, such as a filter cake. The shelf dryer often is used to process extremely dusty and/or expensive material where small quantities must be carefully handled.

Today, the same basic principles are applied to *freeze dryers.* Fig. 7 illustrates a typical system. Here, the design

accommodates a frozen feed, and water is removed by sublimation. The chambers are often modified to provide the ability to stopper ampules or bottles within the chamber, under the operating conditions or selected inert atmospheres. Such dryers frequently find use in the pharmaceutical and food industries.

A more-recent adaptation of the tray dryer, which cuts labor intensiveness and allows continuous operation, is the *plate dryer* (see Fig. 8). In this device, the solids are fed onto the top plate (tray), and gently raked from plate to plate down the unit toward the discharge. Heat is thus conveyed by conduction through the relatively thin bed of material on each plate. The raking provides good mixing, mitigating risks of localized overheating or overdrying of particles, while maintaining close to plug flow. Each plate can be controlled to a different temperature level if desired.

The *pan dryer* (Fig. 9) is a jacketed, flat-bottom, verti-

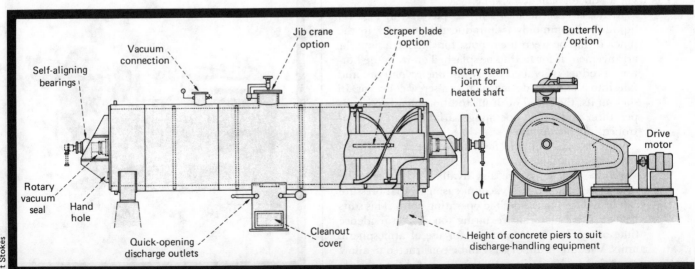

Self-aligning bearings

Vacuum connection

Jib crane option

Scraper blade option

Butterfly option

Rotary steam joint for heated shaft

Drive motor

Rotary vacuum seal

Hand hole

Quick-opening discharge outlets

Cleanout cover

Out

Height of concrete piers to suit discharge-handling equipment

Pennwalt Stokes

This horizontal (rotary vacuum) dryer has double-spiral agitator that is chain driven   Fig 11

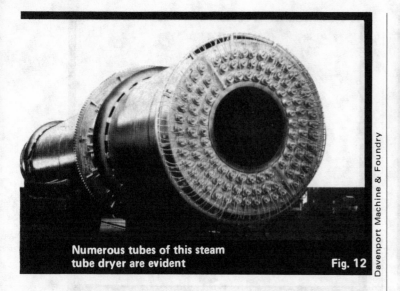

Numerous tubes of this steam tube dryer are evident   **Fig. 12**

Davenport Machine & Foundry

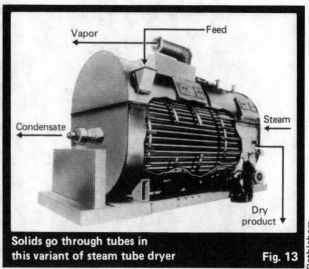

Solids go through tubes in this variant of steam tube dryer   **Fig. 13**

Bethlehem

cal-cylinder-style unit normally having a side discharge and a top-entering scraper/agitator. The jacketed bottom and side walls are heated by steam or circulating heat-transfer oil. The vessel interior may be operated under atmospheric conditions or vacuum (down to 10 mm Hg absolute). A typical agitation system will scrape all of the heat-transfer surfaces, and provide a positive circulation of product outward across the bottom of the dryer toward the side wall but inward from the side wall to the center shaft near the top surface of the bed. Another design incorporates breaker bars to reduce the tendency of product to ball or lump. Feed may be a solid or a slurry. A pan dryer is often used when different products must be handled by the same dryer, since it is particularly easy to clean.

Drum dryers place a thin film of a slurry feed on the surface of an internally heated drum. There are a number of versions, and methods of operation, depending upon application. Variables include: the number of feed-application rolls, the direction of rotation of the rolls, the method of feeding the rolls, film thickness, and type of enclosure. Product is removed from the drum via a "doctor" blade or knife.

In the *atmospheric drum dryer,* as pictured in Fig. 10, liquid feed commonly is introduced at the top, in the pinch or nip between the drums; for slurries, a popular arrangement is for feed to be splashed on from the bottom. Product is scraped off within one revolution, and falls into a removal device, such as a screw conveyor. In such units, the speed of drum rotation, the drum temperature, and other operating variables can be controlled independently.

This dryer is suitable for handling extremely-heat-sensitive materials.

Vacuum drum dryers are also available. In them, the drums are enclosed in a vacuum casing so that pressure can be reduced to the desired operating point. This suits heat-sensitive materials requiring too short a residence time or too low a temperature for use of atmospheric units. Vacuum drum dryers can be constructed to allow operation under sterile conditions, e.g., for the production of pharmaceuticals. (Industrial-sludge drying is an important current application for enclosed- and vac-

uum-type units because they permit complete solvent removal and recovery; dried product serves as a nonpolluting landfill.)

The *horizontal* or *rotary vacuum dryer* (Fig. 11) consists of a cylindrical, jacketed process vessel within which some form of mechanical agitation is provided. Suitable for handling solids or slurries, this unit can be operated either batchwise or continuously.

The agitator can come in several shapes. Its shaft may be hollow, to increase heat-transfer surface. From the shaft, supporting arms project radially. These arms also may be heated. At the ends of the arms, devices such as ribbons, plows or scrapers are attached. These attachments are canted to guide material movement, even if only toward the discharge door. On batch units, this door normally is centrally located.

Fig. 12 illustrates a *steam tube dryer*—basically a cylindrical body in which a number of hollow tubes are arranged symmetrically around the perimeter of the dryer shell. The body is mounted on a set of trunnions in such

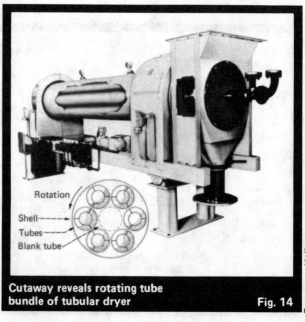

Rotation
Shell
Tubes
Blank tube

Cutaway reveals rotating tube bundle of tubular dryer   **Fig. 14**

Patterson-Kelley

Twin Shell dryer is noted
for blending capability

Fig. 15

Patterson-Kelley

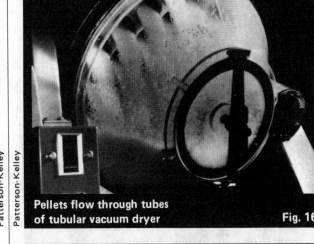

Patterson-Kelley

Pellets flow through tubes
of tubular vacuum dryer

Fig. 16

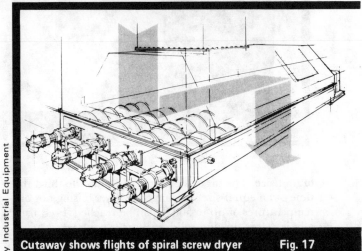

Joy Industrial Equipment

Cutaway shows flights of spiral screw dryer   Fig. 17

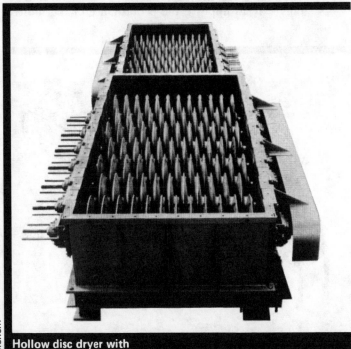

Bethlehem

Hollow disc dryer with
centerline-parallel mass flow

Fig. 18

a way that the unit is sloped, aiding transport of material through it. The body rotates on the trunnions, providing a tumbling action that continuously exposes fresh material to the heat-transfer surface. Free-flowing feed enters the high end of the body, and tumbles around the steam-heated tubes. Moisture is released continuously to the atmosphere as hot vapor. This type of dryer is often gas-swept to improve the rate of vapor removal and to control the relative humidity of the exhaust gas.

Recently, a different type of steam-tube dryer has appeared on the market—in it, pictured in Fig. 13, material flows through the tubes rather than outside of them. This permits controlled, gentle drying of a wide range of flowing solids. This unit is provided with a limited quantity of sweep air.

The *tubular dryer* (Fig. 14) is almost as old as the conventional steam-tube unit. The shell of this dryer is fixed, and a tube bundle (tube reel) is rotated within it. The bundle is equipped with lifters to elevate the solids fed into the device. The feed thus tumbles down among the tubes, as in the typical steam-tube dryer. The shell is shaped so as to provide a domed vapor space, allowing particle separation without the need for sophisticated dust-removal auxiliaries. The stationary shell eliminates large rotating seals, substituting instead small-diameter shaft seals. Many of these units are used in the starch and brewing industries.

A later development is the *cone* or *twin shell dryer* (Fig. 15). In this batch unit, a jacketed shell rotates about a set of trunnions so as to tumble the solids feed. The vapors generated during the drying cycle are removed through one of the trunnion ends, while the heat-transfer medium enters and exits through the other end. Widely regarded for its blending capabilities, this device is easily cleaned, and finds favor for drying a wide range of heat-sensitive, fragile and coarse materials.

The *tubular vacuum dryer* represents a variation on the cone design. As seen in Fig. 16, this unit essentially consists of a shell-and-tube heat exchanger located between two conical end-sections of the typical cone dryer. The tubes provide a constant heat-transfer-surface/process-volume (s/v) ratio, so drying time is constant regardless

This hollow disc unit
features axial mass movement   Fig. 19

Bepex

dryer is by increasing its s/v ratio. Probably the first unit developed in response to this was the *spiral screw dryer* (Fig. 17). The screw flights are hollow, so that heat-transfer fluid can flow through them. The screw troughs are fashioned so that one, two or four screw agitators can be put in a single one, and include a vaulted area to aid in the disengagement of solid particles from the vapors generated.

A little later, the *hollow disc dryer* appeared in the marketplace. Two different versions emerged—one (Fig. 18) in which the mass flow is parallel to the centerline of the heated disc or perpendicular to the axis of the shaft on which the discs are mounted; the other (Fig. 19) in which mass flow is parallel to the axis of the shaft. In both cases, it is common to employ some sort of clip at the periphery of the disc to transport the process mass, and to incorporate a vaulted area in the troughs. Either design is able to accommodate feed as a slurry or a solid.

In the first design, the discs are meshed, with a number of shafts being used in a single unit. In the second design, only one shaft is used per trough, and breaker bars are installed to increase the mixing of the solids between the discs. The first variant allows a more complete flooding of the heat-transfer surface.

The *paddle dryer* (Fig. 20) was designed to provide complete submergence of the heat-transfer surface in the material being dried. At the same time, the degree of local mixing around the heat-transfer surface was increased, resulting in an overall improvement in the heat-transfer rate. The paddles are hollow and are heated, as is the shaft. They are arranged on the shaft so as to provide a discontinuous screw profile, to enhance movement. One, two or four agitators are incorporated in each shell. The trough contains a vaulted area for solids disengagement.

Three other designs round out this compilation:

The *mechanically fluidized dryer* (Fig. 21) consists of a horizontal, jacketed, cylindrical shell containing a

of unit size. The process cycle takes about one-third the time of an equal-size vacuum cone dryer. Using hot oil, temperatures of up to 500°F can be reached. Feed usually is a free-flowing solid, such as a polymer.

Improved surface/volume-ratio designs

Twenty years ago, Horzella [10] pointed out that the only way to improve the performance of the indirect

Paddle dryer has twin agitators
in omega-shaped trough   Fig. 20

Komline-Sanderson Engineering

Mechanically fluidized dryer
employs adjustable paddles   Fig. 21

Bepex

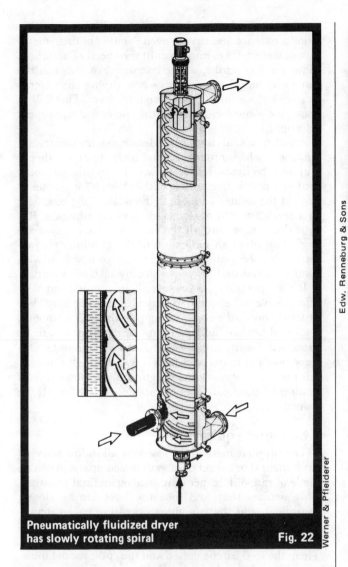

Pneumatically fluidized dryer
has slowly rotating spiral   Fig. 22

Werner & Pfleiderer

Edw. Renneburg & Sons

Electrically heated calciner
can reach high temperatures   Fig. 23

to dry feedstock prior to its being sent for reduction or calcination.

The importance of pilot testing

Probably the most crucial phase of the indirect-dryer selection process is pilot testing. Indeed, most sales (perhaps eighty-five percent) take place only after such trials. For this reason, manufacturers usually have a laboratory available for demonstrating their equipment with a potential client's feed material. In addition, most can provide rental equipment for actual plant tests. The latter method allows a prospective customer to check the dryer's performance against everyday variation in feed by using a slipstream from the production line.

high-speed agitator. This agitator has narrow, flat, pitched blades, and runs at three to ten times the velocity at which the gravity force on the particles is in balance with centrifugal force. The solid feed, thus, is in a thin layer as it moves across the heated shell. Residence time is relatively short. Often this unit is used with a heated purge-gas stream, thereby combining indirect with direct heat transfer.

Also available is a vertical system that uses conveying air to create a fluidizing condition. In the *pneumatically conveyed dryer* (Fig. 22), material is deposited in a thin film on a vertical, jacketed, cylindrical shell. A heated center shaft, on which air guides are mounted, slowly rotates to ensure good distribution of solids over the heat-transfer surface, preventing dead zones from forming. Feeds usually are fine to powdery solids.

Both of these designs provide a spiral pattern of flow as material advances through them.

An *indirectly heated calciner/dryer* is shown in Fig. 23. It contains a rotating shell housed within a heated chamber. The shell is set between breachings, and rotary seals are used. The chamber is heated either by direct fire or electrically, and temperatures in the thousands of degrees can be reached. So, this unit provides the means to remove water of hydration from many chemicals, and

Checklist of key data for dryer evaluation

	System component		
Data	Solid	Wetting liquid	Product
Moisture content	R		R
Specific gravity	D	D	D
Bulk density	D		D
Specific heat	D	D	D
Melting point	R		R
Softening point	R		R
Explosive limits	R	R	R
Boiling point		R	
Heat of fusion	A		
Heat of hydration	A		
pH	D	D	D
Sensitivity to heat	R		R
Abrasiveness	R		R
Corrosiveness	R	R	R
Screen analysis	D		R
Temperature of feed	R	R	
Temperature of discharge			R
Key: A = if applicable D = desired R = required			

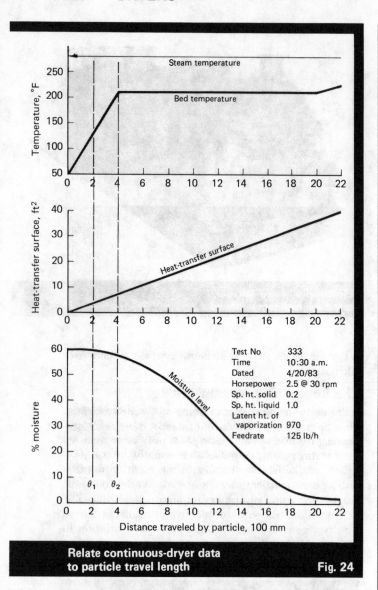

Test No   333
Time   10:30 a.m.
Dated   4/20/83
Horsepower   2.5 @ 30 rpm
Sp. ht. solid   0.2
Sp. ht. liquid   1.0
Latent ht. of
vaporization   970
Feedrate   125 lb/h

**Relate continuous-dryer data
to particle travel length**   **Fig. 24**

For simpler and older dryer designs (such as horizontal or rotary vacuum, pan, plate, or drum units), data obtained are almost always transferable from one supplier's equipment to another's. However, for more-complex and proprietary devices, it is difficult, if not impossible, for one manufacturer to use another's laboratory data for scaleup.

Before embarking on any test program, one should compile certain key information about the product to be dried, as summarized in the Table.

It is also well worthwhile to consider the following:
- How is the moisture held—free, mechanically or chemically bound?
- Are there any phase changes during the drying process?
- Has the material been dried previously—with what success, and using what kind of equipment?
- Are there any limitations in space, in materials of construction, or in source or type of heat?
- What type of enclosure, vapor-handling system, and feed-and-discharge systems are envisioned?

A crude plot of the drying curve of the material can be helpful for drawing preliminary conclusions. Data

can be obtained either from an infrared analyzer or by using a balance and drying oven. Neither of these methods, of course, takes into account the effects of agitation. However, the results can be compared to tests run the same way on other materials whose drying characteristics are known for a given design of dryer. This will at least give some feel for a starting point for equipment selection.

Another useful test is to determine the material's stickiness, which often is a good measure of whether it will foul the heat-transfer surface. To do this, put some feed in a plastic bag, and knead it. Watch for any tendency of the solids to adhere to the walls of the bag. Mix in a small amount of dry solids, and knead again. Repeat these steps until all the various phase changes are seen. This gives an indication of the handling characteristics of the product, and permits comparison with some material that you have already successfully dried.

If the operating parameters allow, a crude but very informative technique is to "dry it in a frying pan." For instance, heat an electric frying pan that has a smooth, uncoated bottom, and add some of the feedstock to be processed. Gently mix the material with a spatula. Observe how the material acts as it goes through the various phase changes—some insight into the possibility of fouling of the heat-transfer surface can frequently be gained.

Evaluating equipment

The simplest method for assessing all indirect-drying equipment is to collect data so that a comparison can be made of rate of feed per unit area versus final moisture. This assumes that feed moisture level can be closely controlled, and that the dryer operating parameters—such as speed of rotation, heat-transfer-surface temperature, and process-side pressure—are held constant. Then, the feed can be varied and the end-product moisture determined. Thus, by varying a value at a time, one can make a plot of a family of curves; the optimal operating conditions can then be selected by trial and error. Dryers typically rated this way include drum, steam tube, and indirectly fired types.

When it is possible to get a sufficient number of samples from a dryer to allow moisture level to be correlated with temperature, a more sophisticated approach can be used. In the case of a batch dryer, one would plot time as the abscissa; for a continuous unit, distance from the feed point. (One can readily see the interrelationship of time in the batch unit with distance traveled in the continuous one.)

The object is to develop a curve like the one shown in Fig. 24. This involves collection of the following data:
- Bed samples for moisture analysis.
- Bed temperature at the exact point each sample was taken.
- Temperature of the heat-transfer medium.
- Agitator speed at the time of sampling.
- Horsepower reading at the time of sampling.
- Process-side operating pressure of the dryer.
- Quantity and pressure of sweep gas (if applicable).

Batch testing usually suffices even for evaluation of continuous units, and provides conservative figures for design—inasmuch as many studies have shown that the calculated and operating heat-transfer coefficients in a continuous unit will be higher than those found in batch equipment. (As Porter [8] reports, the continuous mass flow helps improve the mixing around the heat-transfer surface beyond what is normally observed due solely to agitator speed, and thus increases the rate of heat transfer.)

The data from the above plot must now be turned into the classical-style drying curve. Because of the flexibility of the heat-transfer coefficient expressed as heat/(unit area)(unit time)(degree temperature difference), our interpretation will be made in that direction, and the drying curve will take on a shape as shown in Fig. 25.

The data points for this curve are arrived at by a stepwise calculation technique. To illustrate, let us study the information over specific time increments.

The average moisture concentration, C, of zone (d1 to d2) is arrived at as follows (since the moisture data are normally reported on a wet basis from the laboratory):

$$C = \left[\frac{\%\ \text{moisture}_{d1}}{100 - \%\ \text{moisture}_{d1}} + \frac{\%\ \text{moisture}_{d2}}{100 - \%\ \text{moisture}_{d2}} \right] / 2 \quad (1)$$

For convenience, let us assume that our system is a simple one, involving a single solid and a single wetting liquid. However, if a system contains more than two components, the calculations would have to be extended to cover all the constituents—and would be done identically.

The heat required by the process from time Θ_1 to Θ_2 is found from the following expressions:

First, for sensible heat:

$$q_{S1} = W_{S1} \times C_{p_{S1}} \times (T_{B_{d2}} - T_{B_{d1}}) \quad (2)$$

$$q_{L1} = W_{L1} \times C_{p_{L1}} \times (T_{B_{d2}} - T_{B_{d1}}) \quad (3)$$

To figure the evaporative heat, we use:

$$q_{e_{L1}} = (W_{L_{d1}} - W_{L_{d2}}) \times \lambda \quad (4)$$

For the case that we are considering, there is no heat of crystallization or reaction with which to contend. If heat is put in the process for these or any other reasons, it has to be counted to find the total heat load:

$$Q = q_{S1} + q_{L1} + q_{e_{L1}} + \cdots \quad (5)$$

The wetted surface area during the time period must now be determined. Often during tests, it may be impossible to collect enough feed to fill the dryer. In these cases, it is important to use only the truly wetted or submerged heat-transfer surface.

For instance, where the heat-transfer surface of the agitator is only partially submerged in a free-boiling mass, one should use the entire heat-transfer area. As the surface lifts from the free-boiling mass, it, in all probability, will be wetted, and evaporation will take place from the surface even though it is not submerged. By similar reasoning, if the agitator is only partially submerged in a dry granular bed, one should use only the submerged surface in the calculations.

The temperature difference is normally calculated by using the logarithmic mean temperature difference. (However, if the differential time or length is kept small,

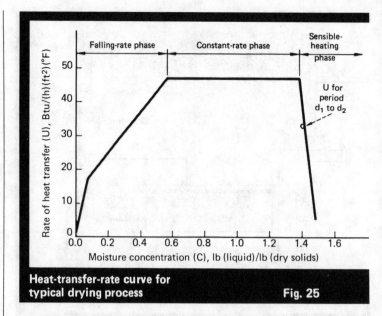

Heat-transfer-rate curve for typical drying process   Fig. 25

the arithmetic average will be almost as accurate.) The *LMTD* can be calculated by:

$$LMTD = \frac{(T_{H_{d2}} - T_{B_{d2}}) - (T_{H_{d1}} - T_{B_{d1}})}{ln\left[(T_{H_{d2}} - T_{B_{d2}})/(T_{H_{d1}} - T_{B_{d1}})\right]} \quad (6)$$

Now, having determined the total heat transferred during the time period, the wetted area, and the temperature difference, one can calculate the overall heat-transfer coefficient:

$$U = Q/(A)(LMTD) \quad (7)$$

This stepwise analysis of the operating curve should be continued until a sufficient number of points are determined to complete the curve of drying rate versus moisture level, as shown in Fig. 25.

After the drying curves are completed, and the various test runs are compared, one can begin to select the parameters for designing a production unit. By regression analysis, the size of that unit can be selected. Since the feed moisture-levels will undoubtedly change, judgment must be used in adjusting the curves for a heat-transfer value during the sensible-heating phase.

With the parameters now better defined, usually only three or four steps—such as sensible heating/drying, constant-rate drying, and falling-rate drying—are needed to calculate the drying surface.

Observations should be made on the physical changes of the process mass during drying, the tendency of the solids to foul the heating surface, the dust loading of the exhaust gases, the effect of agitator speed, of feedrate, etc. These observations may influence the choice of unit as much as the heat-transfer rate.

No attempt has been made to provide specific heat-transfer coefficients for the various pieces of equipment described, since this can be misleading.

All the designs could provide the same heat-transfer rate at a specific stage in the drying process. However, depending upon the moisture level at another point, the thermal conductivity of the bed, the degree of agitation, etc., the rate could vary drastically with unit type.

Each dryer manufacturer should be able to provide

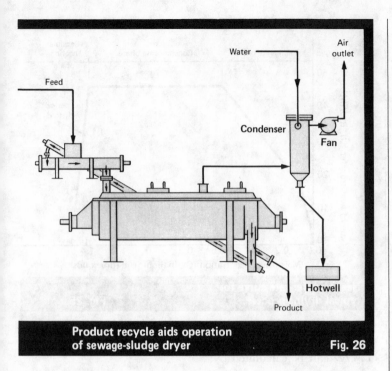

**Product recycle aids operation
of sewage-sludge dryer**   **Fig. 26**

some guidelines after the specific system's parameters have been defined. In nearly all cases, any recommendations should be verified by testing in the specific piece of equipment that one intends to buy.

To avoid confusion, evaluate dryer costs per job function, not per square foot. Often, the cost of housing the unit must be taken into consideration, so it is advisable to use a unit with a good s/v ratio. Also, keep in mind the desirability of keeping this ratio close to that of the test unit on which the evaluation studies were done.

Assuring successful operation

Once an appropriate dryer has been selected, installation requires various pieces of ancillary equipment. Commonly, such auxiliaries include feeders, vapor-handling equipment, discharge devices, and instrumentation. Guidance about this equipment is beyond the scope of this article.

However, a few words are in order concerning instrumentation—particularly about monitoring the temperature of the solids being processed: In indirect dryers, most products do not show an increase in their bed temperature until the cooling effect of the vaporized liquid is lost, usually when moisture level falls below five per-

cent. So, it is generally impossible to control the product moisture level at higher values by using temperature as the control factor. Below this point, temperature can be measured to infer moisture level. There are moisture analyzers that enable monitoring over a fuller range of values.

Another design factor that frequently poses problems is the arrangement of the vapor ducting. Surprisingly often, this ducting is fabricated so that any vapor condensed actually drains back into the dryer proper. To prevent this refluxing, the duct system should provide a run that has a negative slope from the horizontal and is closely coupled to the dryer.

Occasionally during drying, the process mass has a tendency to foul the heat-transfer surface. The material most often will be going through the plastic stage of the phase changes described earlier. Many solids that demonstrate this fouling problem can be handled simply by backmixing some of the previously dried material with the feed. This produces a pseudo-feed having a moisture content that is well within the shearing phase of the material.

Attention should be paid to the mixing action during the blending of the feed and the previously dried material—if a homogeneous mass is not prepared, balls with undried centers may be produced by the dryer. Relying on internal backmixing within the dryer of feed and recycle may result in a poorly homogenized mix, and thus poor drying results. Fig. 26 illustrates a recycle system successfully used on sewage sludge.

References

1. Perry, R. H., and Chilton, C. H., "Chemical Engineers' Handbook," 5th ed., p. 20–18, McGraw-Hill, New York, 1973.
2. Sloan, C. E., Drying Systems and Equipment, *Chem. Eng.,* June 19, 1967, p. 167.
3. Lapple, W. C., and Clark, W. E., Drying Methods and Equipment, *Chem. Eng.,* Oct. 1955, p. 191, and Nov. 1955, p. 177.
4. Dittman, F. W., How to Classify a Drying Process, *Chem. Eng.,* Jan. 17, 1977, p. 106.
5. Coates, J., and Pressburg, B. S., Drying Uses Mass and Heat Transfer, *Chem. Eng.,* July 24, 1961, p. 151.
6. Gunes, S., and Schlunder, E. U., Influence of Mechanical Stirring on Drying Rates in Contact Drying of Coarse Granulation, *Verfahrenstechnik,* Vol. 14, No. 6, p. 387 (1980).
7. Kasatin, A. C., Lekas, V. M., and Elkin, L. M., Investigation of Operation of Screw Conveyor Heat Exchanger Reactors, *Intl. Chem. Eng.,* Vol. 4, No. 1, p. 85 (1964).
8. Porter, W. L., "A Study on Heating a Bed of Free Flowing Solids in a Continuous, Horizontal Processor" (Report No. 175), Bethlehem Corp., Bethlehem, Pa., 1965.
9. Uhl, V. W., and Root, W. L., Indirect Drying in Agitated Units, *Chem. Eng. Prog.,* June 1962, p. 37.
10. Horzella, T., Practical Indirect Heating of Solids, *Chem. Eng. Prog.,* Mar. 1963, p. 90.

The author

William L. Root, III, is a Product Manager for Komline-Sanderson Engineering Corp. (P.O. Box 257, Peapack, NJ 07977; telephone: (201) 234-1000). He has more than thirty years' experience with indirect drying, and has authored a number of articles and presented a number of papers on the subject. He holds several patents on paddle-style dryers. He received a B.S. in chemical engineering from Drexel University, and is a member of AIChE, ACS and the Water Pollution Control Federation.

Finding thermal dryer capabilities

David A. Lee*

☐ When investigating the drying requirements of a new or different material, an engineer must often determine whether the particular convection-type unit under consideration will be suitable from a production standpoint, before getting involved with detailed design calculations. This situation might arise when a firm is considering building its own new unit, buying a used or reconditioned one, or revamping an existing system. Here is a method, based on heat and mass balances, to make such an initial determination.

In such situations, the following information about the system, shown schematically in the figure, is generally available, or can easily be derived or estimated: exhaust-air volumetric flowrate, Z (actual ft^3/min); exhaust air temperature, T_{ex} (°F); rating of inlet air heater, Q_r (Btu/h); amount of volatiles in feed, v_f (%); amount of volatiles remaining in dried product, v_r (%); feed temperature, T_f (°F); and ambient temperature, T_a (°F). In addition, it is generally assumed that the material residence time will be sufficient for the desired degree of drying.

Assuming that volumetric flowrate of the exhaust air, Z, and its temperature, T_{ex}, are known, with X being the rate at which solvent is evaporated (lb/h) and Y the flowrate of the dry inlet air (lb/h), and that any air leakage into or out of the unit is negligible, the following mass balance is obtained:

$$Z = (1/60)[X/\rho_{x(vapor)} + y/\rho_{y(gas)}] \tag{1}$$

where ρ_x is the density of the volatile solvent vapor and ρ_y that of the drying gas, both at the operating temperature and pressure. Eq. (1) can be rearranged to:

$$Y = \rho_y[60Z - (X/\rho_x)] \tag{2}$$

where X and Y are the only unknowns.

Assuming that the product is dried to bone dryness and that there are no significant heats of reaction and/or solution, the heat balance around the dryer is:

$$1.05\left[X(H_g - H_l) + X\left(\frac{100 - V_f}{V_f}\right)(C_{ps})(T_d - T_f)\right] = \\ YC_{pg}(T_{in} - T_{ex}) \tag{3}$$

where H_g = the enthalpy of the evaporated solvent vapor (Btu/lb); H_l = the enthalpy of the liquid solvent in the feed (Btu/lb); C_{ps} = the specific heat of the dried solids, generally not higher than 0.5 Btu/(lb)(°F); T_{in} = the inlet temperature of the drying air (°F); T_d = the temperature of the dried product (which can reasonably be assumed to be the same as that of the exhaust air); and C_{pg} = the specific heat of the drying gas (Btu/(lb)(°F)). The 1.05 factor includes a 5% contingency for miscellaneous heat losses. Substituting T_{ex} for T_d and solving for Y yields:

*Business Development Consultant, 5 Riverside Dr., New York, NY 10023. Telephone: (212) 574-7362

Originally published December 12, 1983

$$Y = \frac{1.05X\left[(H_g - H_l) + \left(\frac{100 - V_f}{V_f}\right)(C_{ps})(T_{ex} - T_{in})\right]}{C_{pg}(T_{in} - T_{ex})} \tag{4}$$

Eqs. (2) and (4) can now be solved for X as:

$$X = \frac{60Z}{\dfrac{1.05}{\rho_y}\left[\dfrac{(H_g - H_l) + \left(\frac{100 - V_f}{V_f}\right)(C_{ps})(T_{ex} - T_f)}{C_{pg}(T_{in} - T_{ex})}\right] + \dfrac{1}{\rho_x}} \tag{5}$$

Example

A used convection (flash, fluid-bed, rotary, spray, etc.) dryer is being considered for a new material that contains 30% solids and 70% water and is to be dried to bone dryness. Pilot-scale tests indicate that it can be dried using an inlet air temperature of 400°F and an exhaust gas temperature of 150°F. The ambient air temperature is assumed to be 60°F. The dryer's exhaust-gas volumetric flow rating is 3,600 acfm at one atmosphere, and the gas-fired heater supplied with the unit has a rating of 3 million Btu/h. Plant personnel must determine: (a) what production capacity the unit is capable of, and (b) whether the gas-fired heater will be able to handle the heat duty.

Solution. The following data are needed: T_{ex} = 150°F; ρ_y (at 150°F) = 0.065 lb/ft^3; Z = 3,600 acfm; H_g = 1,126 Btu/lb; H_l = 28 Btu/lb; C_{ps} = 0.5 Btu/(lb)(°F); T_f = 60°F; C_{pg} = 0.25 Btu/(lb)(°F); $\rho_{water\ vapor}$ (at 150°F) = 0.0103 lb/ft^3; and T_{in} = 400°F.

These numbers can be used in Eq. (5) to find the amount of water that can be evaporated, X = 560 lb/h. Multiplying this by 30/70 gives the capacity as 240 lb/h of dried product.

To determine whether the air heater is adequate, the amount of air, Y, needed to dry the material is calculated, from either Eq. (2) or (4), to be 10,506 lb/h. The heat transferred from the hot air to the product is Q = $YC_{pg}(T_{in} - T_a)$, Q = (10,506)(0.25)(400 − 60) = 893,101 Btu/h. Thus, the air heater's capacity is sufficient.

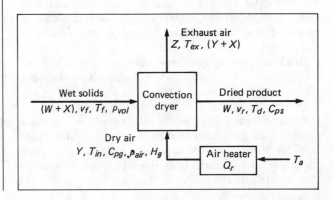

Consider microwave drying

Already well established in food processing, the technique provides quick heating throughout a material— not just of its surface. Because of this, microwave dryers promise energy savings along with increased throughput.

Preston E. Hubble, Dow Chemical U.S.A.

☐ Though little known in many of the chemical process industries, microwave drying has been making steady inroads in food processing since the early 1960s. Pasta producers have led the way, installing three or four new units annually in the U.S. in the past few years.

Yet, the features of the technique that have made it attractive to food processors should also be appealing for many other CPI applications.

Compared to conventional vacuum/steam or hot-air dryers, microwave units may provide:

■ Greater energy efficiency, often enabling energy savings of one-third or more.
■ Lower temperature and shorter residence time in the dryer.
■ Greater throughput.
■ Lessened floor-space requirements.
■ Equal-or-better product quality, including such factors as color and texture.

Principle of operation

As its name implies, the technique relies on microwave radiation to achieve heating. Microwaves are located between infrared and radio frequencies in the electromagnetic spectrum; the two common frequencies allotted for industrial uses are 2,450 MHz and 915 MHz—corresponding to wavelengths of about 5 in. and 13 in., respectively.

The microwaves penetrate the material to be dried, and are absorbed by water or other polar solvents, causing molecular vibrations and thus heating simultaneously throughout the material. Typically, the solid is only heated to the vaporization temperature of water or the solvent.

Attenuation occurs as the waves pass through a material. However, the surface, even though it receives the most energy, is generally cooler than the interior of the material. This is due to (1) evaporative cooling, and (2) the rate of energy input to the interior, which is usually greater than the rate of heat transfer to the outside surface for dissipation. As a result, there is a positive vapor-pressure gradient from the interior to the surface that accelerates the moisture transfer.

Microwave unit for macaroni drying employs eight passes. Generator (with ducting) is on right.

Originally published October 4, 1982

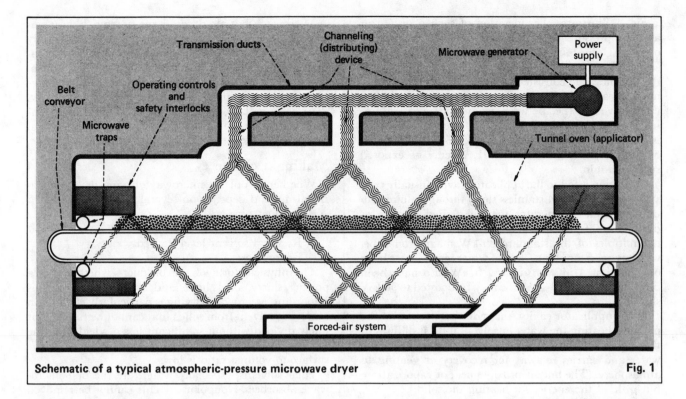

Schematic of a typical atmospheric-pressure microwave dryer   **Fig. 1**

Microwave drying is not suitable for all materials. Some solids damp the microwave-induced vibrations; others will continue to absorb energy even in the absence of polar solvents, making the materials susceptible to overheating.

Contrast with conventional dryers

Other dryers heat the surface of the solid, and then rely on conduction to transfer the heat to the interior.

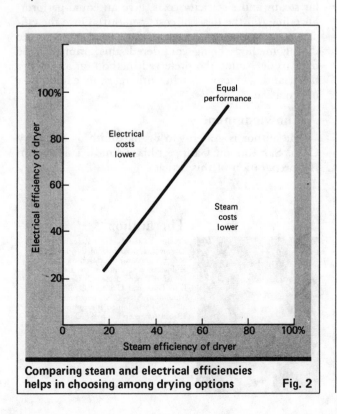

Comparing steam and electrical efficiencies helps in choosing among drying options   Fig. 2

Conduction is comparatively slow; so, a high temperature must be applied to the solid's surface to achieve a practical heating time. And a post-heat equilibration step may be necessary, since the temperature otherwise might vary considerably across the cross-section of the solid.

In comparison, microwave drying provides faster energy transfer. This generally permits quicker drying at lower temperature. Temperature gradients are smaller across the solid because the entire material has received the energy input, not just the exposed surface. And the net result is that a considerably smaller unit is required, as little as one-third the size of a conventional dryer. Plus, further floor-space savings are possible since dryer sections can be stacked.

The lower temperature and smaller size of the microwave dryer cuts down on the heat loss through the dryer walls, and hence boosts energy efficiency.

A look at an actual system vividly demonstrates this effect: for drying ceramic blocks, 56% of the total energy input was lost through the walls of a conventional hot-air dryer; wall losses for the microwave dryer amounted to only 21% of the total energy input. In this case, the microwave unit boasted an overall energy efficiency of 32.9%, versus 24.8% for the conventional dryer.

Pasta makers using microwave dryers report annual energy savings of 40 to 52% over previously used hot-air systems.

The hardware

Microwave dryers may operate at atmospheric pressure or under vacuum to effect evaporation at lower temperatures for heat-sensitive products. Hot-air systems employ a belt conveyor, while vacuum systems come in belt-conveyor and rotary-drum versions.

Fig. 1 illustrates the operation of such a dryer, an ambient-pressure unit. Powdered material is metered onto an adjustable-speed continuous conveyor belt (made of a suitable dielectric, such as fiber glass), with the thickness and consistency of the cake controlled by a mechanical blade. The material then passes through a microwave trap (i.e., between two plastic pipes containing flowing water) into the applicator cavity. The trap is designed to keep microwave leakage well below the current OSHA standard of 0.01 W/cm² of exposed surface/6 min.

Radiation is fed to the applicator cavity (usually constructed of Type 304 stainless steel) through aluminum ducting from a generator that, if desired, can be as distant as 50 ft or so. Generators typically are available in multiples of the standard 30-kW magnetron tube, which is good for about 75 lb of water removal per hour of operation. Units as small as 5 kW also have been built. (The standard-size tube may be boosted to 50 kW shortly.)

In the applicator cavity, the radiation penetrates the product, heating the water or solvent until it diffuses to the surface of the cake. Hot air then evaporates the water and carries it away for recovery or venting to atmosphere. (The hot air provides heat of vaporization, but little of the energy for heating the solid.)

Dried product leaves the cavity through another trap, and drops from the end of the conveyor belt. It then passes to further processing or packaging.

Pelletized products or large, regular solids ($>\frac{1}{4}$-in. dia.) are even easier to dry, as meshed conveyor belts can be used, allowing more surface exposure to the hot-air flow.

Straightforward maintenance

For resistance to chemical corrosion, the microwave application cavity, conveyor belt, and air ducts can be constructed from a variety of suitable materials. Protective coatings can be used, since the applicator surfaces are subject to very little erosive wear.

Electrically, the dryer is not a complex device. The circuitry is based on standard relays, circuit breakers, and other components familiar to plant-maintenance electricians.

The microwave generator consists of a power supply with a transformer to convert line voltage to high-voltage d.c. to operate the magnetron tube. It is a self-excited oscillator that, given proper voltage, filament power and magnetic field, requires no adjustments or tuning.

The magnetron does have a finite life. Failure generally results from the gradual erosion and, finally, breaking of the filament, just as with an incandescent lightbulb. Replacement only requires disconnecting the filament leads, cooling water, and cooling air, removing three hold-down screws, then taking out the failed tube, and installing a new one.

The 30-kW magnetron currently has an average life of about 5,000 h, with some tubes reaching 12,000 to 15,000 h. The magnetron has a prorated warranty of 2,500 h, and currently sells for approximately $5,000. So, the warranted cost is about $2/h, and the expected actual cost is about $1/h.

Microwave hot-air dryers have a main blower, an exhaust blower, and a single conveyor belt as the only moving parts. The conveyor, depending upon its construction material, may require a belt-tracking system. The mechanical components require regular maintenance and lubrication, but nothing out of the ordinary.

At pasta makers, who probably have the most extensive maintenance history of any microwave-system users, all plants use their normal maintenance people.

Making a choice

Whether to opt for microwave drying or a conventional method depends on several factors:
- Product type.
- Solvent to be removed.
- Residual solvent level desired.
- Energy costs.

The physical form of the product—whether it is a powder, slurry, fixed-shape solid, or some other type—is important. Microwave drying is particularly effective in removing solvent from solids that cannot be tumbled or finely divided without significant loss of yield. Also the advantage here goes to the microwave system when the surface-to-volume ratio is low.

Polar solvents, especially water, are the best microwave absorbers. Nonpolar solvents cannot be removed efficiently by microwaves.

Removing the last few tenths of a percent of a solvent can dramatically increase costs for any drying operation. Depending upon the product, a microwave unit may still be the least expensive alternative down to at least 0.1% residual solvent.

Of course, a paramount factor in the decision is energy cost. To assess the relative costs for microwave drying versus other options, construct a graph similar to the one shown in Fig. 2. The curves, drawn for particular steam and electricity costs, give an equal-performance line (on the basis of costs per Btu). Once the efficiencies of steam and electrical usage are defined for the drying methods being considered, this graph can be used to determine the preferred method on an energy basis, and to factor this in with relevant capital-cost information.

Acknowledgment

The author is grateful to Frank Smith of Microdry Corp., San Ramon, Calif., for his technical assistance in the preparation of this article.

The author

Preston E. Hubble is a Research Engineer, specializing in the design, construction and operation of pilot-plant and semiworks units, at the Eastern Division Process Research Center of Dow Chemical U.S.A., P.O. Box 520, Magnolia, AR 71753; telephone: 501-235-2264. He joined the company in 1978 after receiving a B.S.Ch.E. from the University of Virginia, and transferred to the Magnolia operation in 1981.

Section VI
Other Equipment

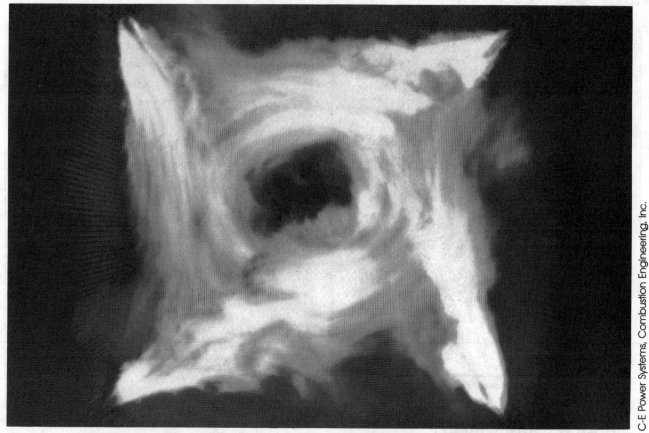

CONTROLLING FIRED HEATERS

Furnaces operated with low excess-air are fuel-efficient
but unstable. Careful instrumentation and control
schemes are necessary to prevent explosion hazards.

Vincent G. Gomes, McGill University

There are two primary objectives in furnace operation—fuel efficiency and safety while maintaining the desired flow, temperature and pressure conditions. The complex interactions between the process variables, and the multiple simultaneous manipulations needed to meet the objectives, make automatic control essential. Split-second decisions required to prevent an explosion hazard in a fuel-efficient furnace (low excess-air operation) are best left to automatic protection systems. Human response-time is inadequate for controlling continuous furnace operations.

Brief review of fundamentals

The factors that determine heater efficiency are:
1. Fluegas exit temperature.
2. Excess-air for combustion.
3. Type of fuel.
4. Heater casing loss.

Improvement in heater efficiency is usually realized by incorporating: a heat-recovery system; improved instrumentation and control; more-efficient burners; improved insulation; efficient soot blowers; reduced air leaks.

Additional heat-recovery equipment has come to be widely used with fired heaters. Design of the heat-recovery system is based on fluegas temperature, dewpoint, and the temperature of the stream that picks up the waste heat. The decision to use such equipment should be based on a technoeconomic feasibility study.

Broadly speaking, heat-recovery systems can include:
1. Process-stream heating in convection section.
2. Steam generation.
3. Air-preheating system for combustion air. Air preheaters are the most widely used heat-recovery system for fired heaters, and are instrumental in boosting the efficiency to about 90%. This article will concentrate on the control and instrumentation of such a system. The important process variables for control of a fired heater with air preheating are:
- Fuel flowrate.
- Air flowrate.
- Operating excess-air.
- Process fluid flowrate.
- Process fluid temperature.
- Furnace draft.
- Flame condition.
- Combustibles.

Originally published January 7, 1985

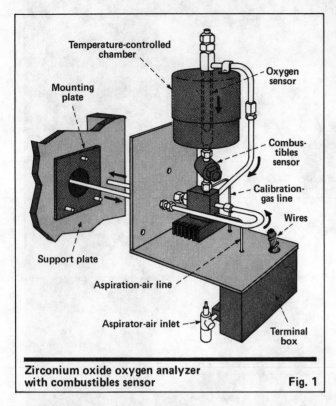

Zirconium oxide oxygen analyzer with combustibles sensor   Fig. 1

Labels: Temperature-controlled chamber; Oxygen sensor; Mounting plate; Combustibles sensor; Calibration-gas line; Wires; Support plate; Aspiration-air line; Aspirator-air inlet; Terminal box

Annubar unrecovered pressure loss (typical)   Table I

Pipe size (nominal)	% of differential pressure unrecovered (i.e., pressure loss)*
4	11
5	9
6	21
8	16
10	13
12	11
14	10
16	9
18	8
20	7
24	6
30	5
36	4
42	3
48	3
60	2

*By comparison, the total permanent pressure loss for orifice plates is approximately 60% of the differential pressure (rule of thumb)

Choosing a transducer

Good control requires good measuring devices. The focus here will be mainly on primary measuring elements.

Flow measurement

Fuel-oil, and process-fluid, flowrate measurement can be satisfactorily achieved by orifice meters. The fuel-gas flowrate can be corrected by using a densitometer. Air-flow measurement, however, poses some problem, since an orifice produces a high, permanent, pressure loss. The venturimeter yields a low pressure drop, but is comparatively expensive, and may require a duct transition from a rectangular to a circular cross-section. The Annubar element has been found suitable for this job.

The Annubar is an averaging type of flow element. Essentially, an Annubar element has characteristics similar to the pitot tube, but with vastly improved accuracy, ranging within 1% of value based on 95% of test points. It is further reported to have a 0.1%-of-value repeatability, based on an average of various differential-pressure readings. The permanent pressure loss in the Annubar is comparable to that of the venturi tube; hence, it is much less energy-intensive compared with the orifice plate (Table I). Further, the Annubar is relatively insensitive to surface wear or abrasion on edges or sensor parts, ensuring long-term accuracy. However, the overall system accuracy and flow range are limited by the differential-pressure secondary instrumentation (transmitters, meters, etc.). Hence, care must be exercised in the selection of the secondary instruments.

Excess-air estimation

The indicators used to control low-excess-air trim are oxygen and carbon monoxide meters. Controversy exists as to which is preferable, but oxygen analyzers are cheaper and are more widely used. The zirconium-oxide ceramic sensing element has come to be widely preferred. It offers several advantages over the other existing types—such as reduced maintenance requirements, minimum sample-conditioning needs, ability to handle dirty fluegases, and greater resolution at low oxygen content—and is readily adaptable for use as a probe type or extractive type. The extractive type can be used in combination with other suitable monitoring devices such as the combustibles sensor. A portion of the aspirated sample is fed in a closed loop to the sensor (which is housed in a temperature-controlled chamber) and is discharged back to the furnace (Fig. 1).

The sensor output signal is determined with respect to the oxygen content of a reference gas such as air. The electromotive force, E, produced by the cell is given by:

$$E \propto T \log \frac{[O_2]_{\text{Ref. gas}}}{[O_2]_{\text{Sample gas}}}$$

$[O_2]_{\text{Reg. gas}}$ = Concentration of oxygen in reference gas = 0.209 for air; $[O_2]_{\text{Sample gas}}$ = Concentration of oxygen in sample gas; T = Temperature (absolute)

Because its output voltage is temperature-dependent, the cell has to be maintained at a constant temperature. Since the resulting output signal is inversely proportional to the logarithm of the O_2 concentration of the sampled gas, the signal strength is higher at lower concentration. Therefore, greater accuracy, reliability and resolution are obtainable at the lower range of operating excess-air (see Table II).

It is advisable to locate the analyzer installation at the heater breeching, where errors due to air leakage are

Typical oxygen-analyzer specifications — Table II

Accuracy: 1% of excess O_2
Repeatability: 0.2% of measured value
Response time: 5 s (approximate)
Sample temperature: 3,200°F (max.), with ceramic probe
Sample flowrate: 0.1–120 std. ft^3/h
Sample pressure: 2 psig
Aspirator air requirements: 10 to 20 std ft^3/h at 15 to 100 psi.
Combustibles monitor: Catalytic detector independent of O_2 sensor

Typical flame-scanner specifications — Table III

Response range: 190–270 nm (wavelength)
Flame-off delay: 1–3 s (preset)
Power consumption: 15 VA
Temperature range (operating): 0–60°F
Field of view: 3 deg
Purge/cooling air: 1 ft^3/min
"Flame-on" sensitivity: 1 μW/cm^2
"Fault" sensitivity: 100 μW/cm^2

expected to be minimum. If the analyzer is mounted at the exit of the convection section or in the fluegas duct, a leak analysis is recommended to determine the required correction to analyzer readings.

Combined O_2 and CO analyzers can be used to trim excess air, if the extra cost is justifiable.

Temperature measurement

Thermocouple temperature measurement is adequate in a heater environment. For better estimation of fluegas temperature, a velocity thermocouple is recommended [1], because of its superior sampling technique.

Fired-heater control

General description

The main objectives of the control system for a furnace with air preheating are to:
1. Meter fuel according to load demand.
2. Proportion air and fuel for complete combustion.
3. Optimize excess-air for fuel efficiency.
4. Initiate protective measures in the event of a flameout or a fan failure.
5. Maintain optimum draft conditions.
6. Monitor fluegas combustibles and air-preheater cold-end temperature.
7. Monitor process-stream conditions.

There are many possible variations of the control scheme that depend on the particular fired-heater system and the philosophy regarding component failure. A representative scheme will be examined, in the following

Flame scanning

Use of forced-draft burners with electrical ignition requires monitoring for flame failure. The ultraviolet flame scanner offers an excellent solution for multiple-burner heaters having combination firing. Typically, the UV cell comprises a pair of highly polished molybdenum (or sometimes tungsten) electrodes positioned at a certain distance from each other inside a helium-filled glass bulb. The gas between the excited electrodes becomes ionized upon being struck by UV photons. The resultant pulse frequency is a direct measure of the radiation intensity received, thus providing a realizable means for discrimination between the main flame and its neighbors.

However, the UV radiation fields are not evenly distributed within a flame envelope; hence, a proper viewing angle across the plane of the flame is required. Besides, it is advisable to eliminate interference from neighboring flames by proper positioning. Usually, each burner is integrally fitted with a UV scanner, factory-set for optimum viewing plane and angle.

The life expectancy of a UV scanner drops drastically with higher operating temperature. Hence, proper housing and use of cooling air, in addition to an air purge for cleaning the optics, are necessary. Table III provides a brief summary of typical specification figures.

In addition to the detecting-tube type, solid-state scanners also are available. However, solid-state devices (diode/transistor) are very temperature-sensitive and require installation at a cool, remote location. Therefore, fiber-optic bundles normally are used for transmitting the light signal to the solid-state detector, which is usually placed not more than 6 ft away. Commercially available fiber-optic bundles exhibit a high attenuation of the UV spectrum, but are considerably better at visible and lower frequencies. Hence, red, far-infrared or far-violet spectra are selected for operation.

Modern flame detectors are designed and located for sensing multiple characteristics of a flame before the presence of the flame is acknowledged. Thus, it is not unusual to find a burner unit fitted with more than one detector head and detector unit logic, including self-checking features, to take care of sensor failure.

discussion, over which modifications can be effected to suit a particular system. The controllers most widely used for the analog control system to be described are the parameter-optimized proportional-integral (PI) or the proportional-integral-derivative (PID) type.

Fuel-air control

The firing-rate-demand signal is used for regulating the fuel and air flowrates. This signal is derived from the deviation of the process-fluid outlet condition from that desired (setpoint). The process-fluid outlet condition is usually determined from the fluid temperature. In case of vaporization within a very narrow temperature range, the fluid pressure is used as the feedback. The simultaneous fuel and air control (Fig. 2) employs a cross-limit control system. It ensures that fuel demand does not exceed measured airflow (plus tolerance) and that the airflow does not drop below measured fuel flow (plus tolerance). The firing-rate-demand signal is sent to a pair of signal-selector relays—high- and low-signal selectors.

The high-signal selector compares the firing-rate demand against the operating total-fuel-flow signal. The latter is obtained from a "summer" of the conditioned fuel-gas and fuel-oil flowrate signals. The high-signal selector includes a small negative bias applied to the total-fuel-flow signal. This permits faster response to load changes by the fuel/air controllers within the limits of the bias. The high-signal selector: causes the air to lead the fuel during the increasing firing-rate-demand mode; and causes the air to lag the fuel during the decreasing firing-rate-demand mode.

The output of the high-signal selector is the setpoint for the airflow controller. The feedback signal to the latter is the oxygen-trimmed airflow signal. This signal is derived from the airflow transmitter and subsequently adjusted by a multiplication factor determined by the oxygen trim controller. The feedback to the oxygen trim controller is the oxygen analyzer signal. The output of the oxygen controller, as a safety precaution, is filtered by high and low limiters to a narrow range of 0.8 to 1.2 (typical). In the automatic mode, the output of the airflow controller is sent to the forced-draft (FD) fan inlet-vane positioner through the manual/automatic control station and the flame safety interlocks. In the manual mode, the airflow control signal is interrupted, and a manually generated signal is substituted. In case of

flameoff, a purge signal is activated and an override signal for startup is substituted for the control signal.

The low-signal selector compares the firing-rate-demand signal against the oxygen-trimmed airflow signal. A small positive bias is applied to the airflow signal to permit a certain initial response to load changes, though only within the limit of the bias. The normal functions of the low-signal selector are: cause the fuel to lead the air on a decreasing firing-rate-demand mode: and cause the fuel to lag the air with increasing firing-rate demand.

The output of the low-signal selector is the setpoint for the fuel flow controller, the feedback signal being the total fuel flowrate. For combination firing, the fuel-gas flowrate signal is corrected for density variations, and the fuel-oil flowrate signal is adjusted for equivalent Btu with respect to the fuel gas. Although fuel oil and fuel gas can be controlled simultaneously in case of combination firing, the controller action is usually arranged for maximum firing of the cheaper fuel. The fuel-oil atomizing steam is controlled by a differential pressure controller. The process-fluid flowrate is controlled by a flow controller. A feedforward control loop, to anticipate load changes due to fluctuations in the fluid flow, can be used for the fuel-air control system.

Fan control

The fired heater is operated under a balanced-draft condition. The FD fan output is controlled (Fig. 3) by the airflow controller, as described above. However, high pressure at the heater arch can damage the heater struc-

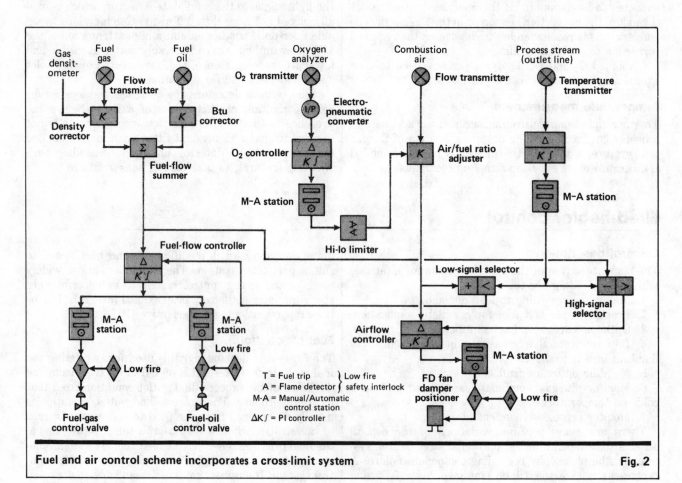

Fuel and air control scheme incorporates a cross-limit system   Fig. 2

ture. On the other hand, excessive low pressure may cause furnace implosion. The furnace draft is controlled for close to −1.0 mm water-gage pressure at the arch by adjusting the induced-draft (ID) fan inlet-vane positioner. The feedback to the draft controller is the pressure at the heater arch, via one or more highly sensitive pressure transmitter(s).

In the event of either FD or ID fan failures, certain corrective or protective steps are recommended, depending on the standby philosophy adopted. A system may accommodate either dropout doors on the burner-air plenum or a spare FD fan. The dropout doors are simply fully-open or fully-closed dampers operated by pneumatic cylinders. The solenoid valves activate the switching from forced to natural draft operation. The stack damper is configured similarly, with the option of manually adjusting the degree of opening. In the event of FD

fan failure the following course of action is prescribed:

1. The dropout doors are to open, or the standby FD fan is to switch on, within a specified time limit.

2. If a dropout-door system is used, the stack damper is to open within a specified time, and the ID fan trip.

3. If (1) and (2) (if applicable) do not occur within a specified time, the fuel to the furnace is to be cut off and the ID fan is to trip, in case of standby FD fan failure. Furnace purge action is to be initiated.

In the event of ID fan failure:

1. Stack damper is to open within a set time interval, otherwise the FD fan and fuel are to trip, with purge initiation.

2. If the furnace is designed to operate below a certain limit with natural draft, the heater load is to be adjusted likewise.

Fan failures are detected by line-pressure switches

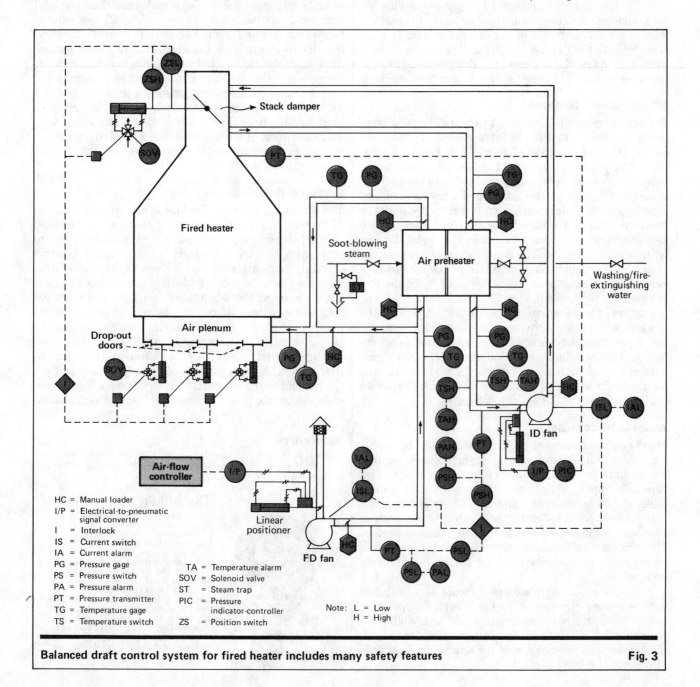

HC = Manual loader
I/P = Electrical-to-pneumatic signal converter
I = Interlock
IS = Current switch
IA = Current alarm
PG = Pressure gage
PS = Pressure switch
PA = Pressure alarm
PT = Pressure transmitter
TG = Temperature gage
TS = Temperature switch

TA = Temperature alarm
SOV = Solenoid valve
ST = Steam trap
PIC = Pressure indicator-controller
ZS = Position switch

Note: L = Low
      H = High

Balanced draft control system for fired heater includes many safety features   Fig. 3

(low at FD fan outlet and high at ID fan inlet) and low-motor-current switches. In the event of excessive high or low pressures at the furnace arch, fuel and fan trip action should take place, to protect the furnace. During a fuel trip, the FD fan inlet vanes are held in the last position, until the operator switches to manual control.

Air-preheater instrumentation

A few instruments are exclusively used for monitoring the air preheater performance (Fig. 3).

Isolation dampers are provided on the inlet and outlet ducts for air and fluegas. Suitable panel and field-mounted pressure and temperature indicators serve as quick reference to performance quality. Further, a low-temperature alarm on the fluegas outlet duct serves to indicate the operating limit for preventing cold-end corrosion. The high-temperature alarm also serves to indicate the possibility of fire hazard owing to combustible accumulation, or low heat-transfer due to fouling. Alarms are also activated in case of large pressure excursions for both the FD and the ID fan. The response lag of the fan dampers, the flexing of the damper linkages, and the wearing-out of bearings should be carefully checked.

Startup safety features

A prefiring purge is essential, to ensure that any combustibles accumulated in the furnace are completely removed prior to initiation of firing. This can be accomplished by passing air through the furnace at a minimum rate, or by passing low-pressure steam into the firebox, in case there is a high fire hazard under hot furnace conditions. The 30% minimum airflow is maintained until the heater reaches 30% of rated capacity. Initial firing is accomplished with a group of ignitors that light the fuel with an electric spark. The flame detectors are brought online. If the combustion control drops the air flowrate below the minimum permissible (typically 30%), the fuel is automatically tripped.

Further, the safeguard system does not permit the startup unless safety sequences are followed. For example, a prefiring purge must be carried out for a definite length of time to allow a specified number of furnace gas-volume displacements or the ignitors should be activated for a certain length of time. The interlock is bypassed during the testing and startup schedules.

Aspects to consider

Many equipment-related problems are caused by selecting underdesigned components or improperly locating components for demanding service requirements. This especially applies to field-mounted sensors. Such components should be weather-resistant with respect to moisture (at least NEMA-3 rating) and ambient temperature. Further considerations should include mounting-surface temperature, and possible vibration-induced operating problems of components installed on the fan or firing equipment.

The sensor-location philosophy should also be based on: obtaining a representative signal over the operating range, freedom from contamination, and accessibility for maintenance, calibration, etc. Transmitters for individual sensing taps, and the need for sensing-line purging, also should be kept in mind.

Specific component problems should be considered during procurement—for example, setpoint drifting, deadband (inability to reset on signal reversal) problems, and service life of switching elements.

Tuning the control scheme described above can be a challenging task by itself. However, proper testing, analysis and documentation should lead, finally, to a tuned system. The basic tuning method involves the use of input perturbations such as step or frequency tests to determine the controller gain, response time and stability. Also, standard computations by Ziegler-Nichols or Cohen-Coon methods [2] provide preliminary controller parameters before final onstream tuning. One should test controllers by simulation, prior to actual operations.

Also it is important to consider the sequence of tuning controllers, and the interaction between them. For example, the airflow controller and the heater-pressure (draft) controller interact (with respect to their outputs) to a certain extent. In such cases, the draft controller is tuned first and its response tested to changes in airflow; the airflow controller is tuned separately—upon placing them onstream, an oscillating response may be obtained, because the airflow controller's response is more rapid during simultaneous controller action than when tested by itself. In this case the furnace-pressure controller must be able to deal with any airflow disturbance with minimum upset of furnace pressure. Hence, the optimum furnace-pressure controller must be retained and the air-flow-controller sensitivity reduced.

Some final thoughts

The increasing demands for higher efficiency and safety of heaters may justify the use of microprocessor-based control. Considering the increasing capabilities of microprocessors, and the downward trend of their price, this certainly appears attractive. Even if implementation costs happen to be comparable, it is worthwhile to take a rapid glance at the advantages of digital control when applied to heaters—flexibility in logic implementation, process deadtime compensation, synchronization between primary and secondary control loops, use of sophisticated control logic (e.g., adaptive control) elimination of switching transients, and bumpless transfer from the manual to the automatic mode. Complex digital control systems are probably already in use on fired heaters.

References

1. Reed, R. D., "Furnace Operations," Gulf Pub. Co., Texas, 1981, p. 50.
2. Smith, C. L., "Digital Computer Process Control," International Textbook Co., 1972.

The author

Vincent G. Gomes is presently a research assistant in the Dept. of Chemical Engineering, McGill University, 3480 University St., Montreal, Quebec H3A 2A7, Canada, specializing in the areas of process dynamics and control application. He holds a B. Tech. degree from the Indian Institute of Technology, Kharagpur, India. He has also had five years of experience as a process engineer in the Heat and Mass Transfer Div. of Engineers India Ltd. (New Delhi), working on process design of heat-transfer equipment.

Consider the plate heat exchanger

Certain features make it a better choice than the shell-and-tube unit for some applications. Among the advantages are a high rate of heat transfer, compactness, and ease of maintenance.

K. S. N. Raju and *Jagdish Chand*, Panjab University

☐ Capital and operating costs, maintenance requirements, weight and space limitations, temperature approach, and pressure and temperature levels play critical roles in the choice between a plate exchanger and a tubular exchanger.

Capital and operating costs

A plate exchanger becomes attractive when an expensive material of construction is required. (When mild-steel construction is acceptable, a shell-and-tube exchanger is often more economical.) A plate unit may also be cheaper when heat must be transferred among three or more fluids. It also need not be insulated, and (for the same duty) can be supported on a less-expensive foundation than a shell-and-tube one. Fig. 1 compares capital costs of plate and tubular exchangers.

Also, less energy is required for pumping fluids in plate exchangers, which lowers operating cost:

Easier maintenance in less space

Plate exchangers offer complete accessibility to all parts for easy inspection, cleaning and replacement, and do not require extra space for maintenance. Cleaning in place with chemicals is comparatively easy because plate channels harbor no dead spaces and promote turbulent flow.

Because of this accessibility and ease of cleaning, plate exchangers are particularly suited for handling fouling fluids and for hygienic services demanding frequent cleaning.

And spare-parts stocking is simpler, because plate-exchanger components are generally standard.

Further, an empty plate exchanger weighs much less than a shell-and-tube unit for the same duty, and takes up far less space.

Temperature and pressure

Temperature approaches of the plate exchanger can be as low as 1°C, making it ideal for high recovery of energy, whereas the practical limit for the shell-and-tube exchanger is about 5°C.

Plate heat-transfer coefficients are also much higher, because of the highly turbulent flow in the plate passages, which is attained with comparatively moderate pressure loss. As a result, the plate exchanger is capable of 90% heat recovery, whereas only 50% recuperation is economically feasible with the tubular exchanger. Because the plate exchanger's heat-transfer area can be altered by adding or removing plates, the exchanger is more flexible as well.

Plate exchangers are not suitable for pressures exceeding 25 kg/cm^2 and temperatures over 250°C.

The trend in plate-exchanger development has been toward larger capacity, and higher working temperature and pressure. These exchangers are now available in many sizes of plates having a variety of patterns in either corrugated or embossed form.

Basic features of the plate exchanger

A plate exchanger consists of a frame and corrugated or embossed metal plates. The frame includes a fixed plate, pressure plate, pressing arrangement and connecting ports (Fig. 2) [1]. Plates are pressed together in the frame. The end-plates do not transfer heat.

Adjoining plates are spaced by gaskets to form a narrow uninterrupted passage through which liquids flow in contact with the corrugated surfaces of the two plates. The corrugation imparts turbulence. Hot and cold streams flow in alternate spaces between the plates. Several arrangements of the fluid streams are possible.

Turbulence is attained in plate exchangers at Reynolds numbers from 10 to 500 (in smooth-pipe laminar flow, Reynolds numbers range up to 2,100) because the corrugation breaks down the stagnant insulating film at the heat-transfer surface. This, combined with the thin plate and lower fouling factor, contributes to the plate exchanger's higher heat-transfer coefficients than those of the shell-and-tube exchanger at the same Reynolds-number flowrate. Heat-transfer coefficients of 2,000 to 5,000 kcal/(h) (m^2) (°C) can be obtained in plate exchangers. The number and size of heat-transfer plates chosen depends on the flowrate, temperature changes, allowable pressure drop and physical properties of the fluids.

Originally published August 11, 1980

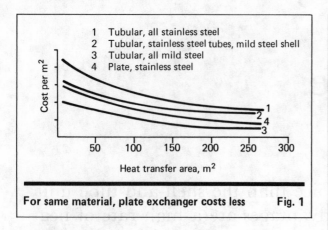

1 Tubular, all stainless steel
2 Tubular, stainless steel tubes, mild steel shell
3 Tubular, all mild steel
4 Plate, stainless steel

For same material, plate exchanger costs less   Fig. 1

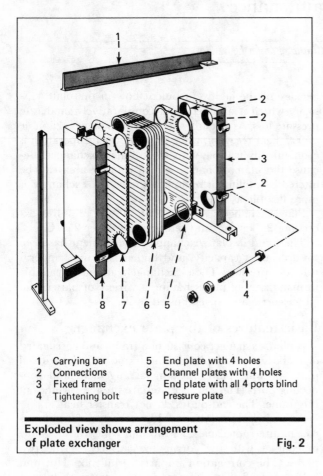

1 Carrying bar	5 End plate with 4 holes
2 Connections	6 Channel plates with 4 holes
3 Fixed frame	7 End plate with all 4 ports blind
4 Tightening bolt	8 Pressure plate

**Exploded view shows arrangement
of plate exchanger   Fig. 2**

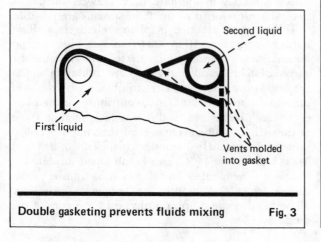

Double gasketing prevents fluids mixing   Fig. 3

Plate arrangements

A series of plates are clamped together so that the corrugations interlock to form narrow flow channels, which produce turbulence even at low velocities. The corrugation also increases the rigidity of the thin plate, enabling it to better withstand distortion from high pressure. Having a large number of support points also minimizes pressure deflection.

By means of gaskets, the two fluids can be arranged in countercurrent flow, and flow volumes can be divided into a number of parallel streams. Gaskets seal the plates at their outer edges and around the ports, which are designed so that the inlet port can be at the top or bottom. Gaskets provide a double seal between the liquid streams, making mixing impossible.

The interspace between the seals is vented to atmosphere, giving a visual indication of leakage and an escape path for the fluid (Fig. 3). A variety of hole combinations are possible, so the plate pack can be adjusted for different services.

Manufacturers have evolved proprietary corrugations and embossed patterns for optimizing heat-transfer and pressure-drop characteristics. Plate size and thickness, and frame design, are therefore related to the particular type of corrugated or embossed pattern. At present, there are no common plate-exchanger design standards such as those of TEMA for shell-and-tube exchangers.

Plates can be made from material that can be cold worked, regardless of its welding characteristics. Common plate materials are stainless steel, titanium, nickel, Monel, Incoloy 825, Hastelloy C, phosphor bronze and cupro-nickel. Titanium-stabilized materials provide greater corrosion resistance.

Plate thicknesses range from 0.5 to 3 mm. Average gaps between the plates run from 1.5 to 5 mm. Plate sizes vary from 0.03 to about 1.5 m². Surface areas extend from 0.03 to about 1,500 m². The largest plate exchanger available handles flows to 2,500 m³/h.

Frame-and-plate design

The frame of a plate exchanger consists of a strong upright end-plate with horizontally projecting upper and lower carrying bars. The plates are suspended from the upper bar and compressed between the end-plate and the thick pressure plate, which is itself carried on this upper bar.

Frame design or plate thickness limits the service pressure. The plate pack is compressed by means of a tightening device, which has a scale to indicate the pressure exerted, so as to prevent under or over compression. Larger units may be tightened hydraulically.

Fig. 4 shows how the connecting plates split the plate pack into several sections, each capable of performing a different function [1]. The connecting plates and frames, not being in contact with liquids, are made of weldable carbon steel, or sometimes of cast iron. They are given anti-rust coating. For stringent sanitary demands, their surfaces are clad with stainless steel.

Gasket selection is critical

Each plate has a gasket, whose functions are to effect an overall seal and close off the flow path of one of the

Nomenclature

A_p	Area of a plate, m^2 (2 A_p = area per channel, bounded by two plates)
A_t	Total area of the thermal plates ($A_p N$), m^2
b	Distance between plates, m
C	Constant in Eq. (3)
C_p	Heat capacity of the fluid, kcal/(kg)(°C)
D_e	Equivalent diameter of the flow channel, m
d	Diameter of the particle, mm
d_f	Fouling or dirt factor, (h)(m^2)(°C)/kcal
F	Correction factor for LMTD, dimensionless
f	Friction factor, dimensionless
G	Mass flowrate of the fluid, kg/(h)(m^2)
g	Gravitational constant, m/h^2
h	Film coefficient of heat transfer, kcal/(h)(m^2)(°C)
k	Thermal conductivity, (kcal)(m)/(h)(m^2)(°C)
L	Length of flow passage, m
l	Distance between plates, mm
m	Exponent in Eq. (3)
n	Exponent in Eq. (3), or number of substreams
N	Number of thermal plates
NTU	Number of Transfer Units, or performance factor or thermal length, θ, or temperature ratio, TR, dimensionless
Nu	Nusselt number, $(h)(D_e)/k$, dimensionless
Pr	Prandtl number, $(C_p)(\mu)/k$, dimensionless
ΔP	Pressure drop, kg/m^2
q	Heat-transfer rate, kcal/h
Re	Reynolds number, $(D_e)(G)/\mu$, dimensionless
t	Temperature, °C
t_i	Channel inlet temperature of the process fluid, °C
t_o	Channel outlet temperature of the process fluid, °C
Δt	Temperature difference, °C
Δt_m	Log mean temperature difference (LMTD) between a channel and immediately adjacent channels, °C
U	Overall heat-transfer coefficient, kcal/(h)(m^2)(°C)
W	Width of the plate, m
w	Mass flowrate of the fluid, kg/h
x	Exponent in Eq. (3), or plate thickness, m
ε	Heat-transfer effectiveness, defined by Eq. (15)
θ	Performance factor or thermal length, or NTU
μ	Fluid viscosity, kg/(h)(m)
μ_w	Fluid viscosity at wall temperature, kg/(h)(m)
ρ	Fluid density, kg/m^3

Subscripts

av	Average
c	Cold
ci	Cold fluid, inlet
co	Cold fluid, outlet
h	Hot
hi	Hot fluid, inlet
ho	Hot fluid, outlet
max	Maximum
min	Minimum
p	Plate

process streams. The gasket is held in a deep depression around the plate perimeter and bonded to the plate. Gasket deterioration is restrained by having a minimum of its surface exposed to the liquids.

Different gasket materials and their applications are listed in Table I [2,3,4]. For high temperatures, a compressed-asbestos-fiber gasket may be necessary. This will resist mixtures of organic chemicals, and raise the service temperature of a plate exchanger to about 200°C. Because such a gasket is virtually inelastic (compared with rubber ones), plates and frames must be designed for the greater compression forces required to seal it.

Factors in plate-exchanger applications

Viscous fluids that might flow laminarly in a shell-and-tube exchanger are likely to flow turbulently in a plate exchanger. When a liquid contains suspended solids, the difference between the space separating the plates and the diameter of the largest particle should not be less than 0.5 mm.

The application of the plate exchanger in many of the chemical process industries is curtailed by the upper limit on its size, which is set by the presses available for stamping out the plates from the sheet metal. Exchangers larger than 1,500 m^2 are not normally available.

Maximum operating pressure also limits the plate

Characteristics of some gasket materials			Table I
Gasket material	**Temperature, °C**	**Service**	**Remarks**
Styrene-butadiene rubber	To 85	General-purpose for aqueous systems	
Acrylonitrile butadiene rubber	To 140	Excellent resistance to aqueous systems, fats and aliphatic hydrocarbons	
Ethylene-propylene terpolymer	To 150	High-temperature resistance for a wide range of chemicals	
Resin-cured butyl rubber	To 150	Excellent resistance to wide range of chemicals, including acids, alkalis, ketones and amines	Poor resistance to fats
Silicone rubber	Low	Limited to sodium hypochlorite and low temperatures generally	
Fluoro-elastomer	Above 150	Excellent for oils	Use limited by high cost
Compressed asbestos fiber	To 250	Resistant to organic chemical mixtures	Plates and frames must be stronger because of poor elasticity
Klingerite	To 250		

exchanger. Although it is possible to design and build one capable of operating up to 25 kg/cm², the normal service pressure is about 10 kg/cm². Available gasket materials limit the operating temperature of plate exchangers to about 250°C.

Plate exchangers are not suitable for gases. Highly viscous fluids also present problems in flow distribution, especially when being cooled. Flow velocities below 0.1 m/s result in low heat-transfer efficiency.

Plate exchangers are not well adaptable for condensing (particularly, vapors under vacuum) because the narrow plate gaps and induced turbulence cause appreciable pressure drop. However, there are now specially designed exchangers available for evaporation and condensation.

Varied flow patterns possible

Some of the possible flow patterns in plate exchangers are illustrated in Fig. 5: (a) series flow, in which a continuous stream changes direction after each vertical path; (b) parallel flow, in which the stream divides and then reconverges; (c) a loop system, in which both streams flow in parallel; (d) and (e) other complex flow patterns.

The number of parallel passages is set by the exchanger output and the allowable pressure drop. Of course, the larger the number of parallel passages, the lower the pressure drop. The number of series passages is determined by plate efficiency and heat-exchange requirements. If a liquid is cooled into viscous flow, the number of passages can be reduced to increase velocity.

Whereas the plate exchanger can take advantage of up to 82% of the theoretical log-mean-temperature difference (LMTD), the shell-and-tube exchanger can use only 50% of it, because of the baffled cross-flow.

Flow distribution

The distribution of flow through the plates in a pass is usually assumed to be uniform. This may not be the case when fluids are viscous, throughputs are high, and plate packs are long. Calculating the actual flow distribution is not easy.

The flow distribution through the plates in a pack is determined by the pressure profiles in the two manifolds (i.e., the inlet and exit manifolds for each stream). In the U arrangement, the inlet and outlet ports for the two streams are on the same end-plate, but they are on the opposite end-plates in the Z arrangement. The pressure profile in manifolds is determined by two factors: (1) the fluid friction and (2) momentum changes resulting from the change of fluid velocity (fluid velocity decreases as it flows out of the inlet manifold and increases as it flows into the exit manifold).

The variations in pressure due to these two factors

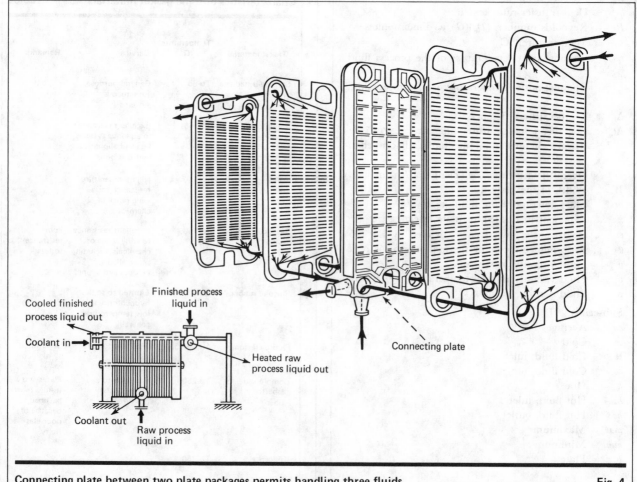

Cooled finished process liquid out

Coolant in

Finished process liquid in

Heated raw process liquid out

Coolant out

Raw process liquid in

Connecting plate

Connecting plate between two plate packages permits handling three fluids   **Fig. 4**

can be in the same or in the opposite direction, depending on whether the manifold is an inlet or exit one and on the flow arrangement (i.e., whether U or Z) [6].

In the U arrangement, the pressure difference between the manifolds, and hence the plate flowrate, decreases from the inlet and outlet faces toward the closed end (Fig. 6).

In the Z arrangement, the pressure differences between the manifolds do not vary as much as in the U arrangement. The variation in the plate flowrate is much less with the Z than with the U arrangement (Fig. 6).

Heat-transfer mechanism

As shown in Fig. 7, the fluid in one channel may receive heat simultaneously from two other streams flowing in the opposite direction. As plates are added to increase the heat-transfer area, the flow pattern becomes more complex and many types of configuration are possible (Fig. 5). Because of the nature of flow patterns in plate exchangers, the log-mean-temperature difference must be corrected by a factor.

Plates are generally termed "soft" or "hard." Low heat-transfer coefficients and small pressure loss per pass characterize the first, and the opposite the second. Hard plates are more complex, are long and narrow, and have deep corrugations and smaller gaps between plates. Soft plates, on the other hand, are short and wide.

For deciding which type of plate will be most suitable for a service, use is made of the Number of Transfer Units (NTU):

$$NTU = |t_i - t_o|/\Delta t_m = (2A_pU)/(wC_p) \quad (1)$$

NTU, also known as the performance factor, or thermal length θ, or temperature ratio TR, can be defined as the total temperature change for the process fluid, divided by the Δt_m for the exchanger, or:

$$NTU = (U_{av}A_t)/(wC_p) \quad (2)$$

A good design is one in which the exchanger surface area matches the thermal duty, and uses the available pressure efficiently. Hard plates are most suitable for difficult duties that require high NTU values (such as regeneration, in which end-temperature differences are small) because the plates achieve high heat recovery.

Soft plates are better for easier duties requiring low NTU values (approximately below 1).

A high NTU duty calls for a relatively high-pressure-loss plate, i.e., one that is long and narrow, and a low NTU duty the opposite.

The number of transfer units depends on the plate configuration, as well as its length. A typical plate exchanger (water to water duty) will have 2 to 2.5 NTU per pass, and a shell-and-tube about 0.5 NTU per pass. For very high duties (NTU = 9), an exchanger may be designed with three passes of plates (each NTU = 3), with the passes in series. An unequal number of passes is usually adopted when the flowrates and permissible pressure drop for the two liquids are different.

The rating of plate exchangers—that is, the calculations for deciding basic plate size and quantity, and the flow pattern—is complex. Recommendations from manufacturers should be sought for specific problems. However, a plate size can be approximated for liquid-to-liquid duties via generalized correlations. Basically, the exchanger area for a given duty is calculated by assuming that all the available pressure drop is consumed and that any size of exchanger is available for providing the needed surface area.

Achieving the required thermal duty while also using all the available pressure drop is difficult. One way of solving this complex problem is to increase the gap between the plates. There are, however, disadvantages to this. Because a plate is generally thin, its strength depends on many contact points with adjacent plates to

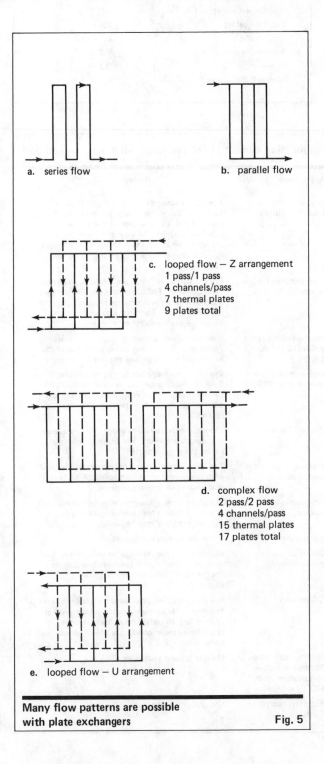

a. series flow

b. parallel flow

c. looped flow – Z arrangement
   1 pass/1 pass
   4 channels/pass
   7 thermal plates
   9 plates total

d. complex flow
   2 pass/2 pass
   4 channels/pass
   15 thermal plates
   17 plates total

e. looped flow – U arrangement

Many flow patterns are possible with plate exchangers   Fig. 5

Fouling factors for plate exchangers   Table II

Fluid	Fouling factor, $d_f \times 10^5$, (h) $(m^2)(°C)$/kcal
Water	
Demineralized or distilled	0.2
Soft	0.4
Hard	1.0
Cooling water (treated)	0.8
Sea (coastal)	1.0
Sea (ocean)	0.6
River, canal	1.0
Engine jacket	1.2
Oils, lubricating	0.4 to 1.0
Oils, vegetable	0.4 to 1.2
Solvents, organic	0.2 to 0.6
Steam	0.2
Process fluids, general	0.2 to 1.2

Application guide to heat-exchanger selection   Table III

Application	Remarks
Low-viscosity fluids (less than 10 cP)	The plate exchanger requires the smallest surface area. For fluids at high temperature or pressure and for noncorrosive fluids, use the tubular exchanger. When elastomeric gaskets are not suitable, use the plate with compressed asbestos fiber gaskets or a spiral or lamellar exchanger.
Low-viscosity liquids to steam	For noncorrosive fluids, use a tubular exchanger of carbon steel. For corrosive or hygienic duties and when steam pressures are moderate, use the plate exchanger. For large volumetric steam flowrates, use the spiral or lamellar.
Medium-viscosity fluids (10 to 100 cP)	With such fluids on both sides, use the plate exchanger. If gasketing causes problems or the solids content is high, use the spiral exchanger.
High-viscosity fluids (greater than 100 cP)	The plate is suitable because of its turbulent flow. Plates have been used up to 50,000 cP in certain cases. For extreme viscosities, the spiral is preferred. Spirals have been used for viscosities up to 500,000 cP.
Fouling liquids	The spiral or plate may be used. For easy access, the plate is preferred. For fibrous suspended solids, the spiral is better.
Slurries, suspensions and pulps	The spiral is best, having been successfully used with ore slurries up to 50% solids. A plate may also be used in certain situations.
Heat-sensitive liquids	The plate is best; however, in certain cases, such as when a fluid is heat-sensitive and highly viscous, the spiral may be useful.
Air cooling	Extended surface exchangers.
Gas or air under pressure	Within limits, the plate can be used. Otherwise, select the tubular (extended surface on the gas side), or the lamellar or spiral exchanger.
Condensing	For noncorrosive duties, select a tubular of carbon steel. For corrosive duties, pick the spiral or lamellar. In certain cases, such as those requiring hygienic conditions, the plate may be considered.
High-pressure (over 35 atm) or high-temperature (over 500°C) service	Use the tubular exchanger.
Extremely corrosive fluids	Use graphite exchangers.

withstand high pressure. If the gap is wide, the normal support points will no longer make contact, and alternative support points must be fitted. Another solution for this problem is to use several basic channel-types, but this requires a series of costly pressing tools.

Still another solution is to install, in the same plate pack, channels having different θ values [7,8]. Such mixing can produce any desired θ between the highest and lowest channel values. A high θ plate is corrugated with obtuse-angled chevrons (Fig. 8a), and so yields a comparatively large pressure drop, whereas a low θ plate has acute-angled chevrons (Fig. 8b) and offer less flow resistance. Fig. 8c, d and e represent three different types of channel combinations.

Effective log-mean-temperature difference

Plate exchanger LMTD is determined in the same manner as for shell-and-tube exchangers. The LMTD correction factor, F, for plate exchangers depends on the number of fluid passes. When the flow ratio between the fluids falls between 0.66 and 1.5, it is usually possible to have an equal number of passes on both sides of the exchanger. When both fluids take the same number of

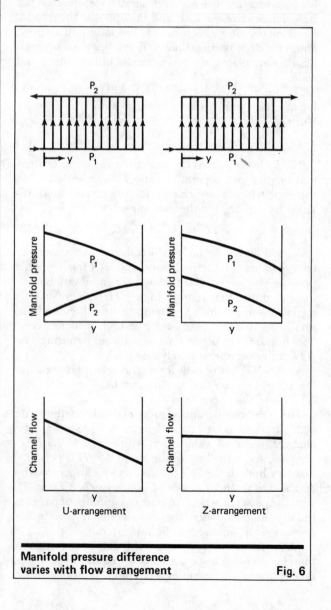

Manifold pressure difference varies with flow arrangement   Fig. 6

passes, the LMTD correction factor is usually quite large. This factor is particularly important when a close, or even relatively close, approach is required.

When flow ratios vary widely, a multipass system with an unequal number of passes is used. For such a system, the correction factor can be quite low, though not as low as the corresponding ones for multipass tubular exchangers. Fig. 9 gives approximate values for the correction factor, F, for various pass systems at NTU up to 11 [9].

Fouling is lessened

Fouling is restrained in plate exchangers by highly turbulent flow (which keeps solids in suspension), smooth plate surface, and the absence of low-velocity zones (such as are present on the shell side of tubular exchangers). Corrosion-resistant plate material also reduces fouling tendencies in a plate exchanger, because deposits of corrosion products to which fouling can adhere are absent.

Fouling factors for designing plate exchangers are much less than for shell-and-tube ones, with resistance values of the first being about 10% to 20% of the second. Simplicity of cleaning, whether chemical or mechanical, also allows lower design fouling-factors. Factors recommended for plate-exchanger design are given in Table II, assuming operation at the economic pressure drop of about 0.3 kg/cm² [9].

Curbing corrosion and erosion

Because plates are so thin compared with tubes, corrosion allowances recommended in standard reference books for process equipment are almost meaningless. As a general guide, the maximum permissible corrosion rate for a plate exchanger is 2 mils/yr.

For a particular corrosive environment, a change from a shell-and-tube to a plate exchanger may call for an upgrading of alloy. For example, whereas Type 316 stainless steel is specified for a tubular exchanger cooling sulfuric acid, a plate exchanger may require a 25 Ni, 20 Cr, 4 Mo, 2 Cu alloy, or even Incoloy 825 (40 Ni, 25 Cr, 3 Mo, 2 Cu). The design engineer must depend on the equipment supplier to recommend a suitable material for a specific duty.

Although a plate exchanger may require a more expensive material of construction, the thin gage of plates, with its high heat-transfer coefficient, frequently means that the plate exchanger's cost/unit of heat transfer is lower.

Plate geometry, in addition to promoting turbulent flow, also intensifies erosion. Therefore, materials such as cupro-nickel alloys used in tubular exchangers are not suitable for plate exchangers.

For these exchangers, materials, such as Monel or titanium, that combine excellent corrosion resistance to the chloride ion with immunity to erosion-corrosion should be specified [10].

Heat transfer and pressure-drop correlations

The film coefficients of heat transfer for plate exchangers are usually correlated by an equation such as:

$$Nu = CRe^n Pr^m (\mu_{av}/\mu_w)^x \qquad (3)$$

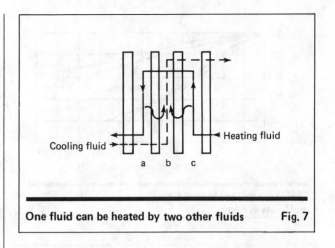

One fluid can be heated by two other fluids   Fig. 7

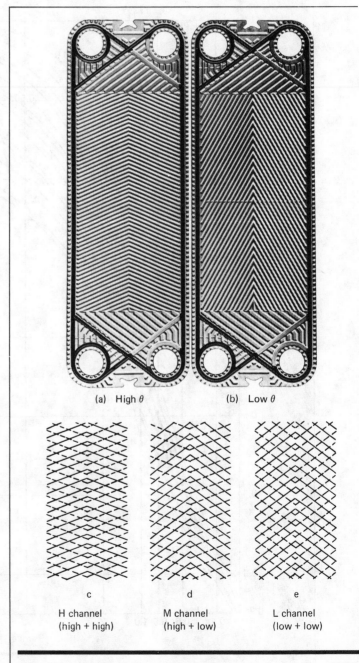

(a) High θ   (b) Low θ

c   d   e

H channel   M channel   L channel
(high + high)   (high + low)   (low + low)

High and low θ plates can be combined   Fig. 8

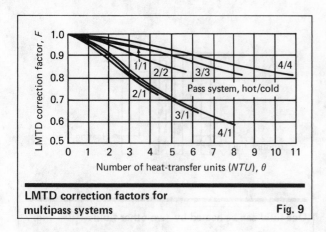

LMTD correction factors for multipass systems

Fig. 9

The constants and exponents are determined empirically and are valid only for a particular plate design. Typical reported values for turbulent flow are [9]: $C = 0.15$ to 0.40; $n = 0.65$ to 0.85; $m = 0.30$ to 0.45 (usually 0.333); and $x = 0.05$ to 0.20.

A widely adopted correlation for estimating film coefficients for turbulent flow in plate exchangers is [11]:

$$h = 0.2536(k/D_e)(Re_{av})^{0.65}(Pr_{av})^{0.4} \qquad (4)$$

The equivalent diameter, D_e, is defined as four times the cross-sectional area of the channel, divided by the wetted perimeter of the channel, or:

$$D_e = (4Wb)/(2W + 2b) \qquad (5)$$

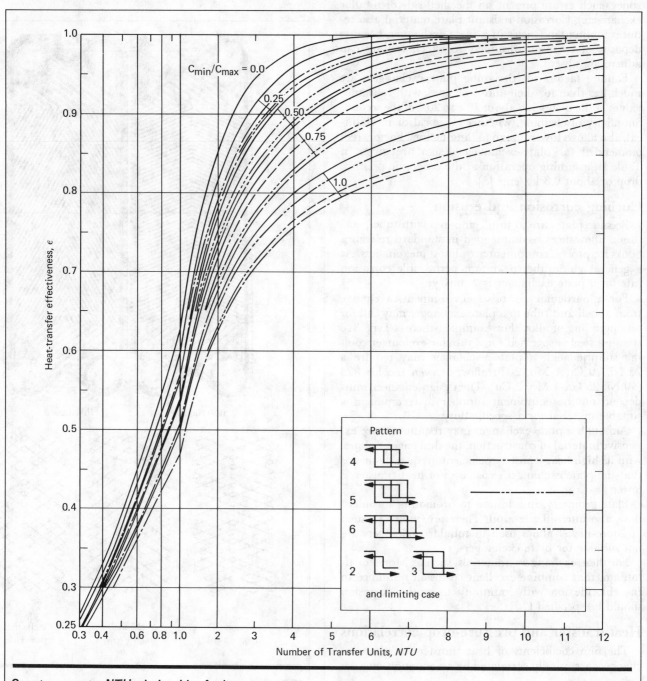

Countercurrent ϵ-NTU relationships for loop patterns

Fig. 10

In Eq. (5), D_e equals approximately $2b$, because b is negligible in comparison with W.

For laminar flow (Re < 400), Jackson, et al., proposed the equation [12]:

$$h = 0.742 C_p G (Re_{av})^{-0.62} \times$$
$$(Pr_{av})^{-0.667} (\mu_{av}/\mu_w)^{0.14} \quad (6)$$

Flow is normally laminar in plate exchangers handling highly viscous and polymeric materials.

Pressure drop in a plate exchanger can be estimated by the equations recommended by Cooper [13]:

$$\Delta P = (2 f G^2 L)/(g D_e \rho) \quad (7)$$

In Eq. (7), $f = 2.5/Re^{0.3}$.

These equations represent highly simplified situations. Any accurate estimation of pressure drop should take into consideration the plate geometry, as well as the pressure losses in the ports.

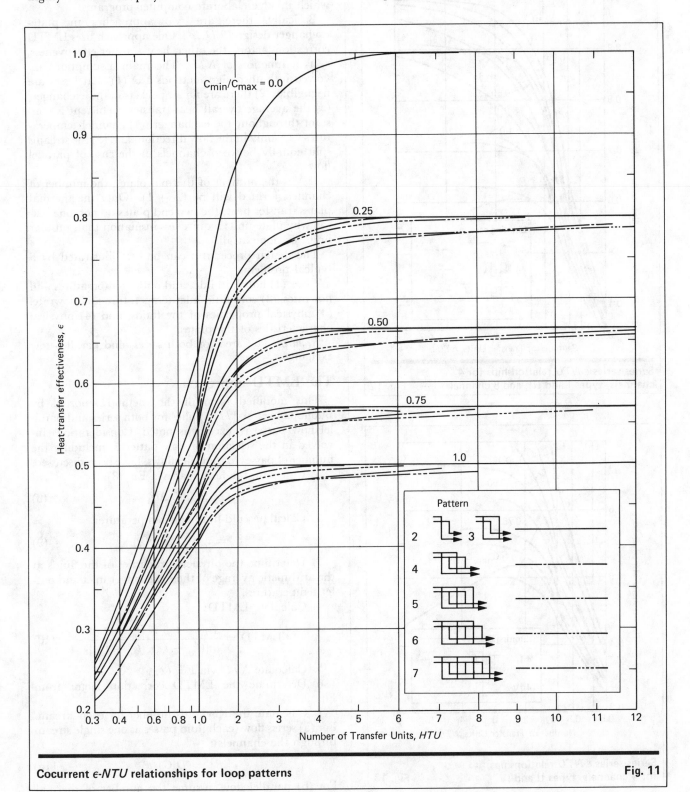

Cocurrent ϵ-NTU relationships for loop patterns   Fig. 11

Average velocities in plate exchangers are lower than in tubular exchangers. Velocities typically range from 0.5 to 0.8 m/s in plate heat exchangers, and 1 m/s in tubular ones. However, because of the highly turbulent flow in plate exchangers, heat-transfer coefficients are much higher than in tubular exchangers [e.g., for water, 2,500–3,500 kcal/(h)(m²)(°C) compared with 1,000–1,500 kcal/(h)(m²)(°C)]. The required heat transfer can be achieved with fewer passes in the plate than in the tubular exchanger.

Design procedures

The complexities of plate designs and flow configurations limit the application of available information (which is scanty) on plate exchangers. Manufacturers have their own design procedures for their exchangers, which involve elaborate computer programs.

Basically, there are two approaches in plate-exchanger design [9,11,14]. One approach uses LMTD correction factors; the other, heat-transfer effectiveness, ε, as a function of NTU. The main assumptions involved in both design methods are: (1) heat losses are negligible; (2) there are no air pockets in the exchanger; (3) the average overall heat-transfer coefficient is constant throughout the exchanger; (4) channel temperatures vary only in the flow direction; and (5) the streams split equally between channels in the case of parallel flow.

If N is the number of thermal plates, the number of channels formed will be $(N + 1)$. (Only the thermal plates transfer heat; the two end plates do not, one reason why plate units require less insulation than tubular units, or none at all.)

The design procedures can be best illustrated by a typical problem.

Given: (1) hot-fluid inlet and outlet temperatures and flowrates, (2) cold-fluid inlet temperature and flowrate, (3) physical properties of the fluids, and (4) physical characteristics of the plates.

Required: The area for both series- and parallel-flow exchangers.

The LMTD approach

This modified form of the method reported by Buonopane et al. [11] applies for both series and parallel flow patterns, except as noted. Considerations involved in the selection of flow patterns, including the number of passes for parallel flows, have been discussed.

1. Calculate the heat load via Eq. (8):

$$q = (wC_p\Delta t)_h \qquad (8)$$

2. Calculate cold-fluid exit temperature:

$$t_{co} = t_{ci} + (q/(wC_p)_c) \qquad (9)$$

3. Determine the physical properties of the fluids at the arithmetic average of their exchanger inlet and outlet temperatures.

4. Calculate LMTD:

$$\text{LMTD} = \frac{(t_{hi} - t_{co}) - (t_{ho} - t_{ci})}{\ln[(t_{hi} - t_{co})/(t_{ho} - t_{ci})]} \qquad (10)$$

5. Calculate NTU via Eq. (2).

6. Determine the LMTD correction factor from Fig. 9.

7. Calculate the Reynolds number for each stream. For (a) series flow (each fluid passes as one single stream through the channels):

$$Re = (D_eG)/\mu \qquad (11a)$$

For (b) parallel flow, assume the number of thermal

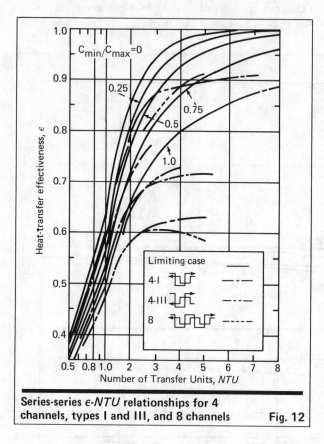

Series-series ϵ-NTU relationships for 4 channels, types I and III, and 8 channels   **Fig. 12**

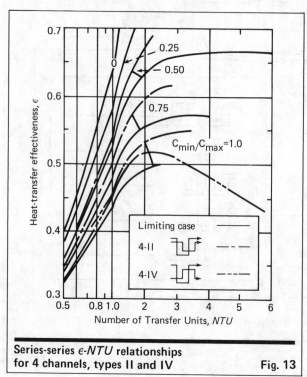

Series-series ϵ-NTU relationships for 4 channels, types II and IV   **Fig. 13**

plates and determine the number of substreams, n_c and n_h, into which each fluid is divided:

$$Re = [D_e(G/n)]/\mu \qquad (11b)$$

8. Calculate heat-transfer coefficients from either Eq. (4) or (6), depending upon whether Re is, respectively, greater or less than 400.

9. Calculate the overall heat-transfer coefficient:

$$U_{av} = 1/[(1/h_h) + (x/k)_p + (1/h_c) + d_{f_h} + d_{f_c}] \qquad (12)$$

10. Calculate total heat-transfer area:

$$A_t = q/(U_{av}\text{LMTD } F) \qquad (13)$$

11. Calculate the number of thermal plates:

$$N = A_t/A_p \qquad (14)$$

12. For parallel flow, from the number of thermal plates calculated in Step 11, determine n for the hot and cold streams. If N is an odd number, n_h and n_c will be equal. If N is an even number, n_h and n_c will be unequal, and one of the fluids will have one more substream than the other (e.g., if $N = 4$ then $n_c = 3$ and $n_h = 2$, or $n_c = 2$ and $n_h = 3$).

13. Compare values of n_c and n_h determined in Step 12 with the corresponding values assumed in Step 7b. If the calculated values do not agree with the assumed values, repeat Steps 7b through 13, replacing assumed values with the values calculated in Step 12, until the values agree.

Steps 1 to 11 are common for both series and parallel flows. Steps 12 and 13 apply only to parallel flow.

Effectiveness-NTU approach

Jackson and Troupe have reported a design procedure that does not involve LMTD correction factors [14]. The concepts of heat-transfer effectiveness, NTU and fluid heat-capacity ratio are instead made use of in the development of the method that can be applied to plate exchangers having different plate and flow configurations.

Although the procedure is suitable for computer programming, the results can also be presented in four diagrams (Fig. 10, 11, 12 and 13) that give ε-NTU relationships as a function of flow patterns and fluid heat-capacity ratios. The procedure can be illustrated in the following steps:

1. Calculate the heat load via Eq. (8).
2. Calculate cold-fluid exit temperature via Eq. (9).
3. Determine the physical properties of the fluids at the arithmetic average of their inlet and outlet temperatures.
4. Calculate heat-transfer effectiveness:

$$\varepsilon = \frac{(wC_p)_h(t_{hi} - t_{ho})}{(wC_p)_{min}(t_{hi} - t_{ci})}$$
$$= \frac{(wC_p)_c(t_{co} - t_{ci})}{(wC_p)_{min}(t_{hi} - t_{ci})} \qquad (15)$$

5. Calculate heat-capacity ratio, $(wC_p)_{min}/(wC_p)_{max}$.
6. Assume an exchanger containing an infinite number of channels and find the required NTU, using the appropriate ε-NTU relationship (Fig. 10, 11, 12 and 13).
7. Calculate the Reynolds number for each stream.

For series flow, use Eq. (11a). For parallel flow, assume the number of thermal plates and find the number of substreams, n_c and n_h, into which each fluid is divided, using Eq. (11b).

8. Calculate heat-transfer coefficients with Eq. (4) or (6), depending upon whether Re is, respectively, greater or less than 400.

9. Calculate the overall heat-transfer coefficient, using Eq. (12).

10. Calculate the approximate number of thermal plates:

$$N = NTU(wC_p)_{min}/(U_{av}A_p) \qquad (16)$$

11. Assume an exchanger of $(N + 1)$ channels and

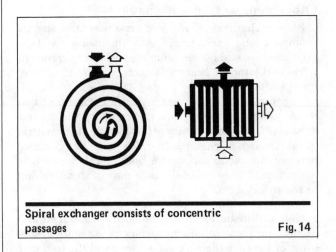

Spiral exchanger consists of concentric passages   **Fig. 14**

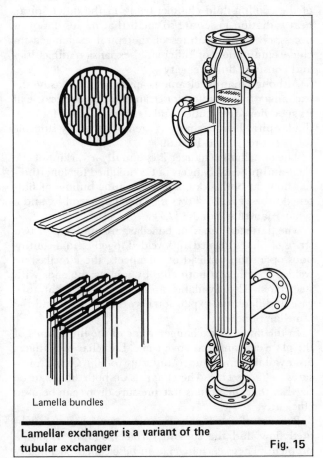

Lamella bundles

Lamellar exchanger is a variant of the tubular exchanger   **Fig. 15**

find the required *NTU* from the appropriate curve (Fig. 10, 11, 12 and 13).

12. For (a) series flow, recalculate *N* with Eq. (16). Repeat steps 11 and 12a, until the *N* calculated in Step 12a matches that assumed in Step 11. For (b) parallel flow, repeat the calculations described in Steps 7b through 12b, until the *N* calculated in Step 12b matches the one assumed in Step 11.

Steps 1 to 11 are common for both series and parallel flows. Steps 7a and 12a apply to series flow, and 7b and 12b to parallel flow.

In the two design approaches presented, the calculation of heat load, stream temperatures and overall heat-transfer coefficients is the same.

Other compact heat exchangers

Plate exchangers belong to a class normally termed "compact heat exchangers." Any discussion on these exchangers would not be complete without reference to spiral and lamellar heat exchangers, which are coming into prominence.

The spiral exchanger's heating surface consists of two relatively long strips of plate, spaced apart and wound around an open, split center to form a pair of concentric spiral passages or channels (Fig. 14) [1,7]. The distance between the metal surfaces in both spiral channels is maintained by means of distance pins or studs welded to the metal sheet.

Fouling or scaling in spiral exchangers is low, because good flow distribution and turbulence are obtained in a single long pass without bypassing or stagnation. Because of the exchanger's compactness, and the cold end of the cooling-fluid channel being in the outer spiral, heat radiation losses are so small that no insulation is necessary. Other features of the spiral exchanger include elimination of differential-expansion difficulties and constant fluid velocity.

Although manual cleaning cannot be done easily, the exchanger is well suited for cleaning in place because of its good flow distribution and its single channel for each fluid. Spiral heat exchangers are not generally suitable for pressures above 15 atm.

The lamellar exchanger is essentially a variant of the shell-and-tube exchanger, but is designed for longitudinal flow on both sides, and has a tube bundle of flattened tubes. Fig. 15 shows the lamellar assembly, and a typical lamellar bundle [1,7].

The flattened tubes, or lamellas, are made up of two strips of plates formed and welded together in a continuous operation. Instead of tubesheets, the lamellas are welded together at both ends by joining the ends with steel bars. Also available are constructions that take care of differential expansion between the shell and the lamellar bundle.

In the lamellar exchanger, some of the limitations of the plate exchanger are overcome. Lamellar exchangers are available for temperatures up to 600°C and pressures to 35 kg/cm². The channels on both sides can be sized so that the permissible pressure drops can be used efficiently.

Because of the highly turbulent, uniformly distributed flow and the smooth surfaces, fouling is not a problem. These exchangers can be cleaned chemically because cleaning fluids that normally attack gaskets can be used in the all-welded channel system inside the lamellas.

The flexibility with respect to cleaning, the variation of heat-transfer surface, and the flow arrangements that are possible with a plate exchanger are not available with the lamellar exchanger. Though the lamellar offers ease of accessibility to all its parts, a single lamella cannot be replaced without cutting the bundle apart.

Table III gives some general guidelines on the selection of heat exchangers [1,7].

Acknowledgments

The authors wish to recognize the assistance of M/S Alfa-Laval AB and M/S APV International Ltd., which supplied information and photographs on the different types of compact heat exchangers.

References

1. "Thermal Handbook," Alfa-Laval AB, Sweden, 1969.
2. Cowan, C. T., *Chem. Eng.*, July 7, 1975.
3. Lane, D. E., Heat Transfer Survey, *Chem. Proc. Eng. (Bombay)*, Vol. 47, Aug. 1966, p. 127.
4. "Paraflow Seminar, Principles of Plate Heat Transfer," The APV Co., Crawley, U.K.
5. "Paracool, Closed Circuit Cooling," The APV Co., Crawley, U.K.
6. Wilkinson, W. L., *The Chem. Engr.*, Vol. 285, May 1974, p. 289.
7. "Heat Exchanger Guide," Alfa Laval AB, 2nd ed., Sweden.
8. Clark, D. F., *The Chem. Engr.*, Vol. 285, May 1974, p. 275.
9. Marriott, J., *Chem. Eng.*, Apr. 5, 1971, p. 127.
10. Cowan, C. T., *Chem. Eng.*, Vol. 81, June 9, 1975, p. 100.
11. Buonopane, R. A., Troupe, R. A., and Morgan, J. C., *Chem. Eng. Prog.*, Vol. 59, No. 7, 1963, p. 57.
12. Jackson, B. W., and Troupe, R. A., *Chem. Eng. Prog.*, Vol. 60, No. 7, 1964, p. 62.
13. Cooper, A., *The Chem. Engr.*, Vol. 285, May 1974, p. 280.
14. Jackson, B. W., and Troupe, R. A., *Chem. Eng. Prog.*, Symposium Series, Vol. 62, No. 64, 1966, p. 185.
15. Wood, R., *Proc. Eng.*, Nov. 1976, p. 75.

The authors

K. S. N. Raju is a professor of chemical engineering at Panjab University (Chandigarh 160014, India), where for 20 years he has contributed to the development of the chemical engineering department. His special interests include phase equilibrium, heat pipes and pollution control. A member of the Indian Institute of Chemical Engineers, he has published over 30 articles and a book on applied chemistry. He received his first degree in chemical engineering from Andhra University (Waltair, India), his second from the Indian Institute of Technology (Kharagpur), and his Ph.D. from Panjab University.

Jagdish Chand, professor of chemical engineering at Panjab University, has been with the school for 15 years. Previously, he was a design engineer with D. C. M. Chemical Works in New Delhi, India. His special interests include heat transfer and process design. A member of the Indian Institute of Chemical Engineers, he has published five articles. He received his first degree in chemical engineering from Jadavpur University (Calcutta), his M.S. from the Illinois Institute of Technology, and his Ph.D. from Panjab University.

Indirect heat transfer in CPI fluidized beds

Here is the way that these fluidized beds operate, and a sampling of some of the processes for which they are used.

Robert A. Rossi, Consultant

☐ The use of gas-fluidized solids systems within the chemical process industries (CPI) has experienced significant growth since its inception over 60 years ago. The majority of these systems have been direct fluid beds (DFB), which transfer significant amounts of heat and mass between solids and gases that are in intimate, direct contact.

Over the last thirty years, and particularly within the last fifteen years, there has been increasing interest in fluid beds that exchange heat between immersed transfer surfaces and the fluidized solids. In such beds, heat is indirectly transferred to and from the bed, while mass transfer is still direct between gas and solids. Such a system is known as an indirect fluid bed (IFB).

The basics of various gas-fluidized-solids systems have been well analyzed and the reader is referred to Ref. 1–4 for a thorough understanding of the subject. For a recent and quicker overview, Zenz [5] is a good source.

This article is intended to impart the basic concept of indirect heat transfer in fluidized beds and its applications for manufacturing particulate chemicals.

Technical background

Fluid-bed processing is excellent for thermal processing of particulate materials. The technique of levitating solids in an upward-flowing gas stream transforms the solids into a highly mobile state that creates an intimate gas/solids contact. At the proper gas rate, the levitated solids behave surprisingly like a liquid; i.e., they will conform to the container's shape, seek their own levels, and discharge through any orifice.

Fluidization is largely dependent on solids characteristics, primarily size distribution, density, shape and surface condition. Gas properties such as density and viscosity are also contributory, while the gas distributor is another variable.

Fig. 1 is a simplified drawing of a typical fluid-bed system. The main vessel is usually cylindrical, but can be rectangular. Regardless, the vessel is divided into three distinct zones: the plenum, the fluidized bed and the freeboard. The plenum and bed are separated by the gas distributor, which uniformly introduces fluidization gas into the solids bed. Fluidization quality is monitored by

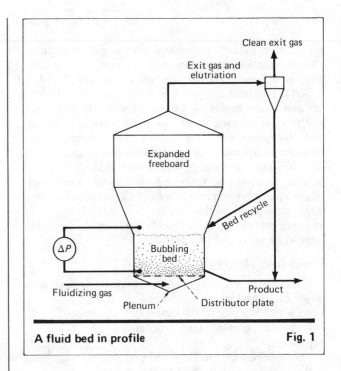

A fluid bed in profile   **Fig. 1**

measuring ΔP between the bed's top surface and a point directly above the gas distributor.

Fig. 2 depicts a typical fluidized regime by plotting normalized gas rate versus bed ΔP. As gas velocity is increased from zero, its ΔP across the bed will reach a maximum. At this point, friction between gas and solids equals the solids gravitational force, and the solids separate. This is defined as incipient fluidization. The noticeable pressure-drop peak and the short decline that are sometimes encountered after this point on the curve are generally attributed to asymmetrically shaped particles that assume a more stable configuration at maximum separation.

Increasing gas flow does not greatly increase interparticle distance, but does create void spaces starting at the distributor plate and rising vertically through the bed. These voids are analogous to "bubbles" in a boiling liquid phase and are created via coalescence of smaller

Originally published October 15, 1984

253

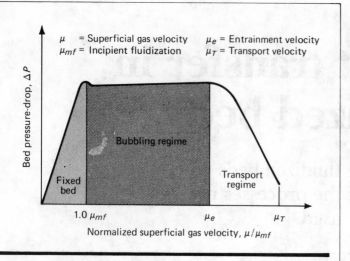

μ = Superficial gas velocity μ_e = Entrainment velocity
μ_{mf} = Incipient fluidization μ_T = Transport velocity

Fluidization diagram, showing the various regimes   Fig. 2

gas bubbles initially introduced via the gas distributor. This fluidization regime is often referred to as a bubbling bed. Zenz [5] covers this point in greater detail.

Bed pressure at any point in this regime is the weight per unit area of bed being supported above the point and is, therefore, analogous to hydraulic head in a liquid. The fluidized-solids bulk density that influences this pressure is a function of interparticle voidage, ϵ, and particle shape, and can be easily measured.

After interparticle space has developed, substantial relative particle movement, as well as general bulk movement, is induced by the bubbles. The net effect is very intimate gas/solids contact yielding extremely favorable interphase mass and heat transfer.

Fluid-bed classes

Commercial fluidized bed operations fall into four major categories as illustrated by Fig. 3 [6], which is a plot of the superficial gas velocity versus bed particulate expansion.

At low velocities, the Type "A" bed is the previously described bubbling bed. This bed has a defined fluidized-solids-bed/freeboard interface that is readily visible. A very high proportion of product solids is discharged from the bed via a gravity overflow or underflow technique. The Type "A" bed was one of the first applications of fluid-bed technology in industry.

Higher gas velocity decreases bubble size, with a higher rate of particle elutriation (carryover). Also the fluidized-solids-bed/freeboard interface becomes less defined. This is a Type "B" bed, often called a turbulent-layer bed, with a more equal ratio of elutriation to bed product when compared with the Type "A" bed.

The Type "A" and "B" beds are

still the most widely used commercial fluid beds for such coarse-particle applications as: ore roasting, calcination and reduction, polymer drying and cooling, organic waste incineration, coal combustion, catalytic reactions, and particle agglomeration at high temperatures.

With both beds, there is an increasing difference between mean gas velocity and mean solids velocity, as the gas velocity increases. This differential is known as the slip velocity.

Because of the resultant dense solids concentration and lower metal-erosion tendencies, the Type "A" and "B" beds are usually employed as IFBs. Further references to indirect heat transfer in fluidized beds are limited to these beds.

The Type "C" bed differs from "A" and "B" in that much higher gas velocities and solids concentrations are used. Extensive product recycling permits the high-solids density, while also creating a condition known as particle "clustering." These clusters have much higher transport velocities than the unclustered particles, which explains the greater slip velocities at increasing solids throughput. The recycling also permits better temperature and residence-time control.

This type of bed is rather new (since the early 1960s), and is finding increasing usage in high-tonnage, fine-particle applications. These are usually very strong endothermic and exothermic reactive processes that either evolve, or require, significant gas quantities. Typical applications are aluminum trihydrate, phosphate-rock and dolomite calcination, hematite reduction and, more recently, coal combustion.

Type "A" or "B" bed diameters would be very large for such high-gas-load fine-particle applications, in order to maintain the requisite slip velocities. The larger diameters and lower velocities also complicate fuel introduction and retard complete combustion at the economically preferred low excess-combustion-air rates. Bed pressure-drops for dense-phase Type "A" and "B" beds, as seen from Fig. 2, are also higher than for Type "C,"

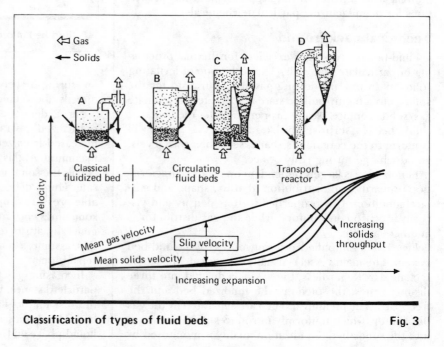

Classification of types of fluid beds   Fig. 3

an important consideration for high-gas-flow systems.

Due to its very high and erosive gas velocities, however, the Type "C" bed is rarely used with in-bed metallic heat-transfer devices, fragile particles, or systems that require long particle-residence times.

The last category, Type "D," is the classic transport reactor as primarily used in fluid catalytic cracking (FCC) and dry-process cement preheating.

Unlike Type "C," no product recycling is used, so that at the higher velocities and solids loading of the Type "D" bed, the clusters break down, with resultant solids elutriation. Again, because of high velocity and a more dilute solids phase, the Type "D" bed is not used as an IFB.

More detailed explanations of the Type "C" and "D" beds can be obtained by reading Ref. 7 and 8.

IFB basics

The Type "D" bed, as with Type "C," rarely uses in-bed heat-transfer surfaces. Therefore, how does a heat-transfer surface, when used with a Type "A" or "B" bed, yield an IFB application?

Fig. 4 depicts an immersed heat-transfer surface in a Type "A" or "B" bubbling bed. Also a generalized relationship between superficial gas velocity and the bed-side heat-transfer coefficient, h_w, is presented.

While there is much discussion about the exact mechanism that creates h_w, it is generally accepted that significant bulk movement of fluidized solids to and from the surface is the major factor. When particles contact the wall, they gain or release enthalpy via conduction, and then return to the bed for further enthalpy exchange. Fresh solids then replace the displaced solids, and the process continues. The frequency of particle/surface

Fluid-bed indirect heat transfer variables			Table I
Gas properties	**Solids properties**	**Flow conditions**	**Geometric properties**
Conductivity	Diameter	Velocity	Bed height
Viscosity	Shape	Bed voidage	Transfer surface
Density	Density		Bed diameter
Specific heat	Specific heat		Distributor design
	Conductivity		

contact time and gas conductivity are considered to be major influences on h_w.

Much experimental work (especially in coal combustion) is being done to develop empirical relationships that are suitable for scale-up purposes. This work is complicated by the many possible gas/solids systems being considered and the numerous variables that affect h_w. Table I lists most known major and minor variables.

Fig. 4 further illustrates the significant effect of gas velocity on h_w. At low velocities, i.e., fixed bed or no particle motion, the stagnant particles cannot move heat to and from the surface; consequently, any coefficient developed is very low. Once minimum fluidization velocity is attained, however, a significant rapid rise in h_w begins. As gas velocity increases through the bubbling bed regime, the h_w will stabilize and then gradually decline. This decline is due to higher bed voidage and bubble population that inhibit particle/wall contact, thereby creating an insulating effect. At transport velocity, the solids concentration is very dilute, which decreases the probability for wall contact. Further, the particle will most likely leave the bed rather than move enthalpy to and from it. Accordingly, h_w will decrease rapidly in the transport regime.

Fig. 5 [9] indicates the impact of bed geometry (L/D ratio) on h_w in a bubbling bed. As can be seen, the taller bed (L/D = 6.2) has a lower h_w than a squat bed (L/D = 2.0) at the same velocity. The lower h_w for the higher-L/D bed is related to larger bubble growth in tall fluid beds, which, as with Fig. 4, reduces particle/wall contact. At high enough velocities, the h_w's coincide at a common peak and then mutually decline.

Note that most work to date has primarily investigated gas, solid, and flow conditions, while geometric properties have been less explored. A good review of past work that attempts to reconcile many of these factors is by Grewal and Saxena [10]. Botterill [11] has been a principal investigator in the overall area of fluid-bed heat transfer, and should be consulted also.

Because of the specific and limited nature of past empirical results, they should be used only for estimation purposes and not final design. Laboratory and pilot-plant work with the anticipated process and heat-transfer surface is still the best route to a reliable IFB design.

Process advantages

The decision to employ IFB heating or cooling requires a thorough knowledge of the desired process. There are many cases where indirect heating/cooling is not justi-

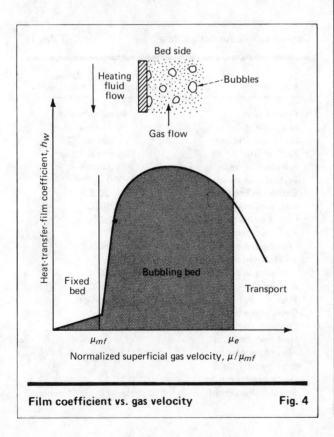

Film coefficient vs. gas velocity   **Fig. 4**

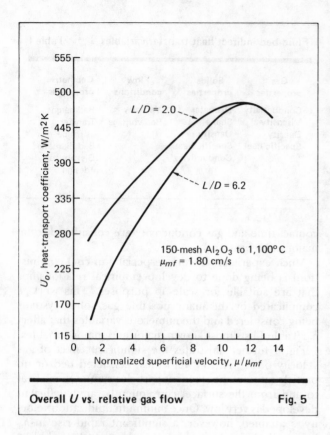

Overall *U* vs. relative gas flow   Fig. 5

fied, due to a combination of high particle-fluidization velocities, high bed temperature, absence of a controlled gas composition, or "sticky" particle characteristics. Most of these processes can be satisfactorily processed in DFBs. Where applicable, however, an IFB offers the following advantages:

■ The high indirect heat-transfer rate significantly reduces gas-flow requirements. Possibly 50−90% of a process's heat load can be transferred indirectly. This decreases the amount of gas needed for good fluidization or, as in drying, an optimum psychrometric ratio. Hence, lost exit-gas enthalpy is significantly reduced, resulting in reduced equipment size and operating cost.

■ The economic turndown ratio of an IFB is usually much greater than that of a DFB. Since much of the heat generation is unrelated to gas flow, it can be completely stopped without affecting bed fluidization conditions. Therefore, it is possible to run an IFB at 50 to 90% turndown (depending on the indirect heat loads) and still maintain 100%-rate energy economics. A Type "A" DFB can only turndown 10 to 25% before changing fluidization conditions adversely affect both operability and economics.

■ The tremendous leverage gained by the multiple of the heat-transfer coefficient, LMTD, and heat-transfer surface density permits very high heat inputs into low-temperature, heat-sensitive applications. This multiple more than overcomes the hindrance of not being able to use high gas inlet temperatures.

■ Heat-transfer fluids used can be common ones such as water, steam, diphenyls, molten salts, and combustion products. Electricity, depending on the process and site economics, is also very attractive as a high-temperature heat source.

■ Indirect heat transfer significantly reduces fluid-bed size when the solids fluidization velocity is low, in a high-tonnage, endothermic or exothermic process. Often, a single-train indirect-heated fluid bed will do the job of two or three direct-heated units. Better yet, lower-fluidization-velocity particles are typically smaller-diameter ones that yield high heat-transfer coefficients.

■ When a plugflow, rectangular IFB of low height profile is used, the solids will flow counter to the thermal fluid, resulting in a very favorable LMTD. In other words, the fluid bed can now be constructed like a counter-current heat exchanger with all its attendant benefits.

■ Many gas/solids fluid-bed chemical reactions require a specific stoichiometric ratio to proceed. By isolating the heat source from the gas supply, a significant degree of reaction variability and control is obtained. This is especially relevant with very-fine-particle systems, i.e., 100% under 44 μm, or 325 Tyler mesh.

The aforementioned advantages can be best illustrated with a comparative calculation between a direct- and an indirect-heated bed for the same application.

Comparative example

A high-tonnage, moderate-temperature cleaning process to remove trace impurities from a fine, fragile solid provides the simplest comparison to illustrate these advantages. Table II lists the process design data for both an IFB and a DFB for the same application. Fig. 6 and 7 are schematics of each process, while Table III summarizes the significant differences.

The DFB, Fig. 6, is a simple process wherein the solids are heated with externally heated fluidizing gas. This gas not only provides process enthalpy but also fluidizes the mass to ensure the gas/solids contact needed for heat

Comparative case design data		Table II
Variable	**Indirect fluid bed**	**Direct fluid bed**
Process type	Simple solids heating	
Feed rate	90.7 m.t./h	
Solids specific heat	0.2 cal/gm-°C	
Particle size	100% minus 75 μm	
Fluidization velocity	0.50 m/s	
Settled bulk density	960 kg/m³	
Expanded bed voidage	30%	
Inlet solids temperature	10°C	
Outlet solids temperature	260°C	
Fluidization gas inlet temperature	650°C	
Fluidization gas outlet temperature	260°C	
Ambient air temperature	25°C	
Fuel	Natural gas	
Overall pipe heat-transfer coefficient	285 W/m²K	N.A.
Heat-transfer fluid inlet temperature	400°C	N.A.
Heat-transfer fluid outlet temperature	370°C	N.A.
Fluidized-bed depth	1.83 meter	0.61 meter

Comparative case conclusions		Table III
Result	Indirect fluid bed	Direct fluid bed
Net process heat load	4,536,290 Kcal/h	
Gross heat input	5,326,882 Kcal/h	5,757,309 Kcal/h
Thermal efficiency	85.2%	78.8%
Heat input split		
Direct	0%	100%
Indirect	100%	0%
Power requirement	71 kW	150 kW
Bed fluidization gas	4,387 Nm³/h	24,153 Nm³/h
Heater exit gas	6,165 Nm³/h	0
Total exit gas	10,552 Nm³/h	24,153 Nm³/h
Bed area	6.04m²	25.1 m²

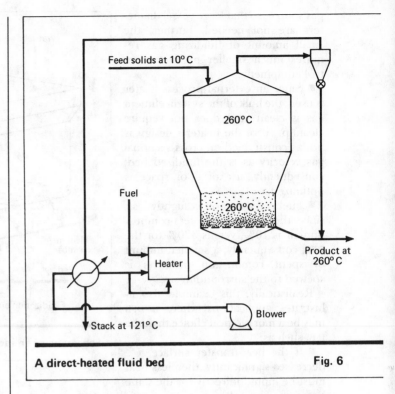

A direct-heated fluid bed   **Fig. 6**

transfer. Heat economy is realized by indirectly preheating incoming fluidizing air with bed exit gases. A cylindrical bed is used since it is mechanically better suited to handle the 650°C inlet plenum gas. While this process is effective, there are two major constraints that affect operating and capital cost:

■ Since we are limited to 650°C inlet gas temperature, the system ΔT is small, and a large gas quantity is required. This gas passes directly through the bed and it must be totally cleaned by subsequent, and sometimes costly, pollution-control equipment.

■ The particles' low fluidization-velocity range requires a large-diameter vessel to compensate for gas flow. This will affect capital cost.

The IFB, Fig. 7, is somewhat more complicated but not without its benefits. Unlike the DFB, it uses a rectangular bed to optimize the solids-flow/heat-transfer-fluid LMTD effect. Also, the plenum side-inlet gas is at a low temperature, precluding any mechanical constraints.

In this process, an external direct-fired heater, operating at low excess-combustion-air, heats a diphenyl thermal fluid to a temperature above that of the bed, but less than the fluid's degradation temperature.

While this fluid's temperature is much less than 650°C (as with the DFB inlet gas), the high bed-side film coefficient yields tremendous heat influx according to the classical heat-transfer equation:

$$Q_{bed\ input} = U_o A_t \Delta T_{lm}$$

Where U_o is the overall heat-transfer coefficient of the in-bed tubular surface. (A diphenyl fluid at proper velocity will create a very high tubeside coefficient, much higher than the bed-side coefficient, such that the latter controls). For such a fine-particle system, a U_o of 285 W/(m)(K) is not unreasonable.

Of equal importance is the amount of heat-transfer surface area, A_t, that can be "loaded" into a bed without impeding fluidization.

The loading ratio can vary between 5 and 30 m² of transfer surface/m² bed surface, depending on the application and experience.

Add to these two factors the solids counterflow LMTD effect, and a significant multiple is developed that yields a very high enthalpy input within a confined vessel at low temperatures. This more than compensates for the lower thermal-fluid temperature when compared with the DFB's higher inlet air temperature.

Therefore, the IFB process yields several advantages over the DFB process for this comparison:

■ The heat source is "decoupled" from the fluidizing

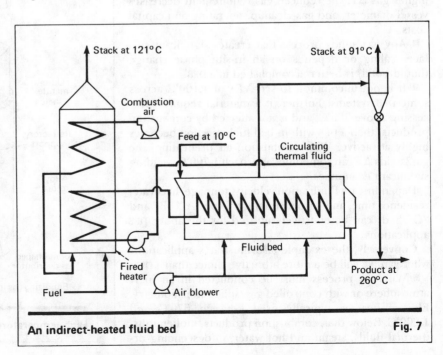

An indirect-heated fluid bed   **Fig. 7**

gas source, so that large vessel diameters are not needed. Further, the small amount of fluidizing gas requires much smaller pollution-control equipment.

■ Since an external process heater is used, the bulk of the system effluent gas is clean, and does not require cleanup. Also, the heater's design is not as constrained on cross-sectional gas velocity as is the fluidized bed; consequently, use of floor space is optimized.

■ Fuel consumption is slightly less, since the lower IFB excess-combustion-air is 10%, versus 320% for the DFB; consequently, a smaller amount of spent combustion products is spewed to the surroundings.

Realistically, this example also illustrates applications where a DFB may be a more logical choice than the IFB, such as:

■ If the heat-transfer surface, U_0, decreases significantly, then less heat must be input indirectly for the same bed area. This will require more fluidization-gas enthalpy input. At some point, a balance is reached where the thermal-fluid recirculating system becomes costly relative to its enthalpy contribution.

■ An increase in the minimum acceptable fluidization velocity, because of a large-particle system, will improve the DFB's capital cost, since much more enthalpy can be input to the bed with fluidization gas. This reduces both the bed diameter and the capital cost. As before, the thermal-fluid recirculating system also becomes more costly.

■ If a more heat-insensitive material is processed, the limitation on inlet gas temperature is relaxed. With higher gas ΔT, the reduced gas requirement decreases vessel diameter, and gas-cleanup, operating and capital costs.

■ Any gas/solids process that creates significant surface scaling or depends on an in-situ phase change (liquid/solid) is better accomplished in a DFB.

■ It is not uncommon to see ΔT's of 1,100°C across some DFB systems. Further, if a material requires processing above 760°C, and is not affected by combustion products, then a DFB with in-bed fuel injection becomes highly attractive. With combustion air preheating, the gas/solids ΔT can sometimes approach 1,700°C, yielding significant economies.

Depending on fluidization velocity, tonnage, heat rate, residence time, and particle fragility, Type "A," "B" and "C" beds can be effectively utilized for the above DFB applications.

Conversely, the example also illustrates applications where an IFB will be a more attractive choice than a DFB:

■ Where a process must be conducted in an inert atmosphere or with controlled gas/solids stoichiometry. Electric heat is typically used between 670°C and 1,200°C. Below that, combustion products, molten salts, thermal fluids, steam and hot water, in descending or-

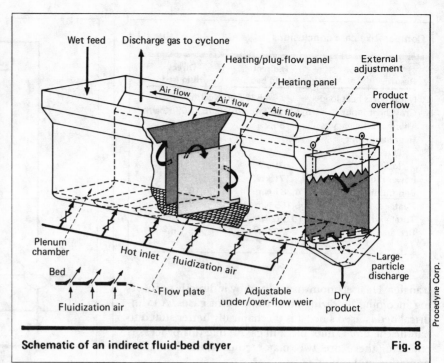

Schematic of an indirect fluid-bed dryer   **Fig. 8**

der, are all heat-source candidates. Typical applications may be: metal oxide reductions, catalyst preparation, metal powder annealing, inorganics halogenation, and volatiles stripping.

■ For fragile particle systems in which breakdown

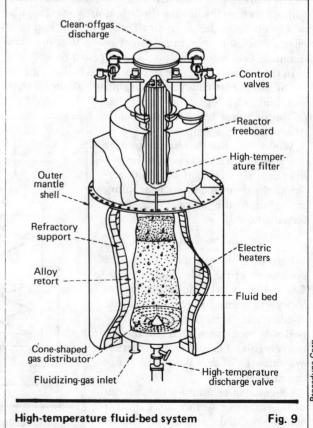

High-temperature fluid-bed system   **Fig. 9**

cannot be tolerated, but where the process does not require a controlled atmosphere, and in which product temperatures are up to 1,200°C and transport velocities less than 0.50 m/s. Some applications are catalyst preparation and regeneration, and fine-particle carbonate and hydroxide calcinations.

■ For systems that do not require a controlled atmosphere and in which product temperatures are below 670°C and transport velocities are less than 1.0 m/s. Some applications are inorganic salt drying, soda ash manufacturing, organic synthesis, and molecular sieve calcination.

■ When a high-tonnage, high-temperature particle system can indirectly yield its heat to a fluid or gas, which in turn provides clean, recovered heat for other process users (usually found in alumina, coke, and combustor ash cooling to preheat combustion air or boiler feedwater).

■ For drying of very heat-sensitive polymers that have large-constant-rate drying regimes and also require tightly defined residence times. Proven applications are polyvinyl chloride, polypropylene, polyethylene, acrylonitrile-butadiene-styrene copolymers, and polycarbonates. It is especially useful if solvents must be recovered from a recirculating gas loop.

■ For applications in places where electricity is relatively cheap compared with available fossil fuels and when the velocity is usually below 0.5 m/s, and with product temperature up to 1,200°C. Since no combustion products are emitted from an electrically heated system, its thermal efficiency may be two to five times that of a fossil-fuel system. Even when electricity costs three times as much on an equivalent Btu basis, the higher efficiency will favor an electric-heated system.

To better highlight these cases, let us look at some established IFB processes in the CPI.

Drying

IFB drying in processing water- and solvent-wet chemicals has been increasingly used over the last 15 years. Significant advantages over previously employed flash and rotary dryers—such as energy efficiency, high unit-capacity/capital-cost ratio and improved product quality—are well documented [12–14].

A simple laboratory batch fluid-bed dryer is used to develop design data by measuring and observing drying rate, fluidization characteristics and heat-transfer coefficients over the product's full drying range. The batch drying curve also yields the break between constant-rate and falling-rate drying.

Now let us examine an industrial dryer designed on this basis. Refer to Fig. 8.

The first fluidized bed compartment receives wet cake dispersed into a fluidized bed of semi-dried product. This is ensured by inputting all enthalpy via high-velocity fluidizing gas

since heat transfer panels would interfere with bed mixing. Compartment exhaust air is at saturation.

The semi-dried product then enters the second compartment, where it passes around alternate heat-transfer panels, yielding water as it proceeds. The heat-transfer coefficient used is that obtained during testwork.

Panel spacing is critical and is based on industrial experience with the particular material's fluidization characteristics. The panel's stress characteristic and its support and removal details are the most important mechanical design aspects of an IFB dryer.

Exit-air composition is controlled by directing the water-lean air from the falling-rate zone backwards, against product flow, and mixing with water-rich air from the constant-rate zone. In this manner, condensation over the first compartment is avoided.

Product directional flow toward the outlet weir is maintained with a gas-distributor plate that minimizes stagnant regimes and maximizes "plug-flow" or first-particle-in/first-particle-out flow patterns.

Continuous cooling

Hot solid products from industrial thermal processes such as rotary kilns, multiple hearths, and fluid beds have traditionally been cooled with direct and indirect rotary drum coolers, DFB, and indirect disc coolers. The most attractive application is that in which a heat-transfer fluid is indirectly heated for use by another thermal process system within the same plant. Clean air can also be preheated for combustion purposes.

Because of the high h_w, IFB's have been widely used for very hot (to 1,100°C) high-tonnage cooling applications. Also, as previously discussed, indirect cooling significantly reduces the air flow to that required only for fluidization and not heat removal. This reduces equipment size for fine-particle cooling applications such as alumina, FCC catalyst, combustor ashes, and coke fines.

Indirect cooling is accomplished via rugged pipe bundles arranged in a rectangular "plug-flow" fluid bed to

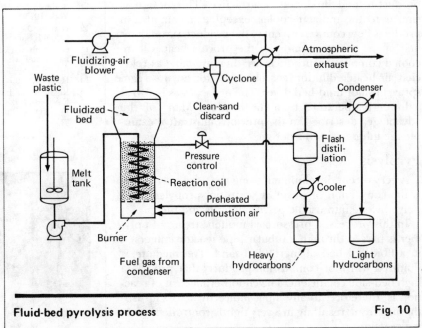

Fluid-bed pyrolysis process   **Fig. 10**

Procedyne Corp.

maximize countercurrent heat transfer. With this design, hot solids enter above the bed layer, are fluidized, and then indirectly contact the exiting heat-transfer fluid pumped through the pipe bundles.

Batch calcining

There are many chemical processes that require heating finely divided powders to elevated temperatures at tightly controlled residence times.

Many of these products have particle-size distributions between submicron and 44 μm. In addition, process temperatures up to 1,200°C are not uncommon. Further, they may require closely controlled gas compositions (hydrogen, halogens, steam/air, organics, etc.) to obtain a specific stoichiometric ratio. In many cases, the residence times or heat-up requirements are so narrow and specific that only a cyclic batch process will suffice.

For such high-temperature gas/solids processes an electrically heated IFB reactor is the only available choice [15]. Fig. 9 depicts such a calciner.

This batch system also includes two noteworthy technical innovations aside from electrical heating. They are a gas distributor specifically designed for fine particles and a high-temperature gas/solids filter. Both features are required to permit optimal utilization of indirect heating.

Continuous calcination

With continuous operations, similar processing principles are utilized; however, the geometry of the IFB is changed to accommodate any controlling residence-time distribution. This usually involves baffling, or a change in vessel shape to a rectangular bed, thereby minimizing bypassing and backmixing.

These are usually fine-particle applications which, unlike the batch calciner, have less-stringent residence-time requirements, but require greater throughputs. They often require a controlled gas atmosphere.

In applications where electricity is costly, or the product temperature is below 650°C, an indirect combustion-product-heated fluid bed is attractive. This design is similar to the indirect cooler, except that combustion products flow countercurrent to the product, through in-bed pipes, to supply most of the process heat. Clean, cooled combustion products are then available as recycled air-heater dilution gas or as gas for an upstream spray, flash or fluid-bed dryer. The bed exit gas is also at a lower temperature than the product. Many of the advantages described in the previous illustrative example are inherent in this design.

Pyrolysis

A very recent development using the IFB, Fig. 10 [16], is the conversion of waste or scrap polypropylene and polyethylene into usable petrochemicals and fuels.

In this process, a pressurized molten stream of polymer is passed through a tubular-pipe reactor immersed in a fluidized bed of inert silica-sand. The medium is fluidized with hot combustion products that provide a high bed-side coefficient. The endothermic heat of reaction is, therefore, totally input indirectly across an isothermal front, resulting in a very tightly controlled cracking reaction.

The pyrolyzate is separated into various gas and liquid fractions via downstream flash-separation and condensation operations.

Unlike direct-heated fluid-bed pyrolysis processes, this process is performed in the complete absence of air or other gases, thereby yielding little coke. Any coke that is developed is removed, without shutdown, via a patented air-oxidation technique. The exotherm generated is readily distributed back into the media bed.

This type of pyrolysis process may have other applications within the petrochemical industries.

Conclusion

The IFB is one more example of the versatility and economics that can be realized by using fluid beds for gas/solids operation. As more research is performed with various gas/solids heat-transfer systems and with wider use of established IFB processes, we can expect to see this technique finding wider application in the CPI.

Acknowledgment

The author wishes to thank Procedyne Corp., New Brunswick, N.J., for assistance in preparing this article.

References

1. Kunii, D., and Levenspiel, O., "Fluidization Engineering," John Wiley & Sons, New York, 1969.
2. Zenz, F. A., and Othmer, D. F., "Fluidization & Fluid Particle Systems," Reinhold, New York, 1960.
3. Leva, M., "Fluidization," McGraw-Hill, New York, 1959.
4. Davidson, J. F., and Harrison, D., "Fluidization," Academic Press, 1971.
5. Zenz, F. A., *Chem. Eng.*, Dec. 19, 1977, pp. 81 – 91.
6. Reh, L., Fluid Bed Combustion in Processing, Environmental Protection and Energy Supply, paper presented at the International Fluidized Bed Combustion Symposium of the American Flame Research Committee, Boston, Mass., April 1979.
7. Yerushalmi, J., and Cankurt, N. T., *Chemtech*, Vol. 9, pp. 564 – 572 (1978).
8. Reh, L., *Chem. Eng. Prog.*, Feb. 1971, pp. 58 – 63.
9. Staffin, H. K., and Rim, C., Calibration of Temperature Sensors Between 538°C and 1092°C in Air Fluidized Solids, paper presented at the Fifth Symposium on Temperature, Washington D.C., June 1971.
10. Grewal, N. S., and Saxena, S. C., *Int. J. Heat Mass Transfer*, 1980, Vol. 23, pp. 1505-1518.
11. Botterill, J. S. M., "Fluid Bed Heat Transfer," Academic Press, New York, 1975.
12. Herron, D., Polymer Drying Equipment Selection, paper presented at the 86th National AIChE Meeting, Houston, Tex., Apr. 1979.
13. Christiansen, O. B., Fluid Bed Drying and Cooling Systems with Internal Heat Exchange Surfaces, paper presented at the 86th National AIChE Meeting, Houston, Tex., Apr. 1979.
14. Hess, D., and Rossi, R. A., *Chem. Eng. Prog.*, Apr. 1983, pp. 43 – 50.
15. Rossi, R. A., Indirect Heat Transfer in Fluidized Bed Dryers and Calciners, *J. Powder Bulk Solids Technol.*, May 1984.
16. Bhatia, J., and Rossi, R. A., *Chem. Eng.*, Oct. 1982, pp. 58 – 59.

The author

Robert A. Rossi is a consultant, 7855 Blvd. East, North Bergen, NJ 07047, tel: (201) 662-1741. He has had over 13 years of experience in development, design, and marketing of high- and low-temperature gas/solids processing systems. He holds a B.S.Ch.E. degree from New Jersey Institute of Technology, and has done graduate work at Stevens Institute of Technology. He has been associated with Niro Atomizer, Inc., Dorr-Oliver Inc. and Procedyne Corp. He is a member of AIChE, ACS, American Institute of Mining, Metallurgical and Petroleum Engineers, and the Technical Assn. of the Pulp and Paper Industry.

Calculator design of multistage evaporators

A simple mathematical model is developed by making four modest assumptions about how multistage evaporators work. The resulting Hewlett-Packard program can be used to calculate the heat-exchange area and the other significant variables.

S. Esplugas and *J. Mata,* University of Barcelona

☐ To design multistage evaporators, we generally use iterative calculations. However, if we make the following assumptions, it is possible to achieve an analytical solution: the feed is at its boiling point; all heating surfaces are equal; sensible heats are negligible when compared with latent heats; boiling-point rise is negligible.

With these simplifications, we have developed a mathematical model and the program to solve it. The program (Table I) was run on a Hewlett-Packard 67, but it can be applied to any other HP calculator with the same type of magnetic card. The general applicability of the assumptions is discussed later on.

Mathematical model

Once the steady state is achieved, the mathematical model is composed of the following equations (see figure):

Material balance around Effect i:

$$L_{i-1} = L_i + V_i \tag{1}$$

$$L_{i-1}x_{i-1} = L_i x_i = L_0 x_0 = \text{constant} \tag{2}$$

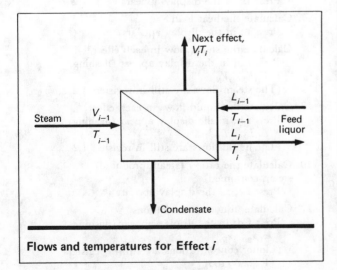

Flows and temperatures for Effect i

Heat balance around Effect i:

$$Q_i = V_{i-1}\lambda_{i-1} \tag{3}$$

$$Q_i = U_i A(T_{i-1} - T_i) \tag{4}$$

$$Q_i = V_i\lambda_i - L_{i-1}C_{pi}(T_{i-1} - T_i) \simeq V_i\lambda_i \tag{5}$$

Hence:

$$Q_i = V_i\lambda_i = V_{i-1}\lambda_{i-1} = V_0\lambda_0 = Q = \text{constant} \tag{6}$$

and then:

$$\sum_{i=1}^{N} V_i = V_0\lambda_0 \sum_{i=1}^{N} 1/U_i = L_0 - L_N \tag{7}$$

which expresses the overall mass-balance condition. From Eq. (4) and (6):

$$Q/A = \frac{T_{i-1} - T_i}{1/U_i} = \frac{T_0 - T_N}{\displaystyle\sum_{i=1}^{N} 1/U_i} \tag{8}$$

Thus:

$$T_i = T_{i-1} - \left(\frac{T_o - T_N}{\displaystyle\sum_{i=1}^{N} 1/U_i}\right)\left(\frac{1}{U_i}\right) \tag{9}$$

Knowing the temperatures, it is possible to evaluate the latent heat of steam, λ_i.

Applying Regnault's formula:

$$\lambda_i = 606.5 - 0.695\, T_i \tag{10}$$

where T_i is in °C, and λ_i is in kcal/kg.

Or:

$$\lambda_i = 1{,}114 - 0.695\, T_i \tag{10a}$$

if T_i is in °F and λ_i is in Btu/lb.

From the independent variables—T_0, T_N, L_0, x_0, x_N and U_i—are calculated the dependent variables, which are displayed or printed by the calculator.

Originally published February 7, 1983

Program can be used to calculate dependent variables for multistage evaporators

Step	Code	Key	Step	Code	Key	Step	Code	Key	Step	Code	Key	Step	Code	Key	Step	Code	Key
001	FIX	-11	038	GSB0	23 00	075	1/X	52	112	GSB6	23 06	149	ISZI	16 26 46	186	÷	-24
002	DSP2	-63 02	039	RCLi	36 45	076	ST+1	35-55 01	113	+	-55	150	GTOa	22 16 11	187	STOi	35 45
003	CLRG	16-53	040	P⇄S	16-51	077	P⇄S	16-51	114	RTN	24	151	*LBLb	21 16 12	188	PRTX	-14
004	P⇄S	16-51	041	ST0	35-55 00	078	RCLI	36 46	115	R/S	51	152	RCLC	36 13	189	ISZI	16 26 46
005	CLRG	16-53	042	P⇄S	16-51	079	1	01	116	*LBLD	21 14	153	STOA	35 11	190	GTOd	22 16 14
006	CF0	16 22 00	043	GTOB	22 12	080	-	-45	117	RCL0	36 00	154	1	01	191	R/S	51
007	1	01	044	R/S	51	081	STOI	35 46	118	x	-35	155	STOI	35 46	192	*LBL5	21 05
008	STOI	35 46	045	*LBL0	21 00	082	X≠0?	16-42	119	P⇄S	16-51	156	*LBL4	21 04	193	DSZI	16 25 46
009	R/S	51	046	RCLA	36 11	083	GTO2	22 02	120	RCL0	36 00	157	RCLA	36 11	194	RCLi	36 45
010	P⇄S	16-51	047	RCL0	36 00	084	RCLi	36 45	121	GSB3	23 03	158	RCLi	36 45	195	ISZI	16 26 46
011	STO2	35 02	048	-	-45	085	GSB3	23 03	122	x	-35	159	X=0?	16-43	196	R/S	51
012	P⇄S	16-51	049	P⇄S	16-51	086	P⇄S	16-51	123	RCL0	36 00	160	GTO5	22 05	197	*LBLe	21 16 15
013	R↓	-31	050	RCL0	36 00	087	RCL1	36 01	124	ENT↑	-21	161	-	-45	198	SF0	16 21 00
014	STO0	35 00	051	÷	-24	088	x	-35	125	RCLA	36 11	162	STOA	35 11	199	R/S	51
015	R↓	-31	052	STOD	35 14	089	1/X	52	126	-	-45	163	STOi	35 45	200	*LBL6	21 06
016	STOC	35 13	053	P⇄S	16-51	090	RCLC	36 13	127	÷	-24	164	PRTX	-14	201	CLX	-51
017	R/S	51	054	0	00	091	ENT↑	-21	128	R/S	51	165	ISZI	16 26 46	202	1	01
018	X⇄Y	-41	055	STOI	35 46	092	RCLB	36 12	129	*LBLE	21 15	166	GTO4	22 04	203	1	01
019	STOA	35 11	056	RCLi	36 45	093	-	-45	130	X⇄Y	-41	167	R/S	51	204	1	01
020	X⇄Y	-41	057	*LBL1	21 01	094	x	-35	131	RCL0	36 00	168	*LBLc	21 16 13	205	4	04
021	1/X	52	058	RCLi	36 45	095	ENT↑	-21	132	GSB3	23 03	169	1	01	206	RTN	24
022	RCLC	36 13	059	ISZI	16 26 46	096	STOE	35 15	133	x	-35	170	STOI	35 46	207	R/S	51
023	x	-35	060	RCLi	36 45	097	R/S	51	134	STOD	35 14	171	RCLC	36 13			
024	P⇄S	16-51	061	X=0?	16-43	098	*LBL3	21 03	135	1	01	172	RCLB	36 12			
025	RCL2	36 02	062	GTO5	22 05	099	ENT↑	-21	136	STOI	35 46	173	-	-45			
026	x	-35	063	RCLD	36 14	100	.	-62	137	RCLC	36 13	174	RCLE	36 15			
027	P⇄S	16-51	064	x	-35	101	6	06	138	RCLD	36 14	175	÷	-24			
028	STOB	35 12	065	+	-55	102	9	09	139	R/S	51	176	R/S	51			
029	R/S	51	066	STOi	35 45	103	5	05	140	*LBLa	21 16 11	177	*LBLd	21 16 14			
030	*LBLA	21 11	067	PRTX	-14	104	CHS	-22	141	RCLD	36 14	178	RCLC	36 13			
031	1/X	52	068	GTO1	22 01	105	x	-35	142	RCLi	36 45	179	P⇄S	16-51			
032	STOi	35 45	069	*LBLC	21 13	106	6	06	143	X=0?	16-43	180	RCL2	36 02			
033	ISZI	16 26 46	070	DSZI	16 25 46	107	0	00	144	GTO5	22 05	181	x	-35			
034	R/S	51	071	*LBL2	21 02	108	6	06	145	GSB3	23 03	182	P⇄S	16-51			
035	*LBLB	21 12	072	RCLi	36 45	109	.	-62	146	÷	-24	183	RCLi	36 45			
036	DSZI	16 25 46	073	GSB3	23 03	110	5	05	147	STOi	35 45	184	X=0?	16-43			
037	X=0?	16-43	074	P⇄S	16-51	111	F0?	16 23 00	148	PRTX	-14	185	GTO5	22 05			

How to use the program

With the calculator switch in RUN position, enter the magnetic card. Then follow the instructions given below, in the same order.

1. Initialize the calculator
   Press GTO. 000
   Press R/S; in the display appears   1.00

2. Enter general data
   Key in L_0, then press ENTER
   Key in T_0, then press ENTER
   Key in x_0
   Press R/S; in the display appears L_0
   Key in T_N, then press ENTER
   Display x_N
   Press R/S; in the display appears L_N

3. Enter heat-transfer coefficients
   Key in U_1, then press A; display $1/U_1$
   Key in U_2, then press A; display $1/U_2$
   . . . .
   Key in U_N, then press A; display $1/U_N$

4. Calculate effect temperatures
   Press B; in the display appear (flashing)
   $T_1, T_2, ..., T_N$
   (The temperatures are still in registers 1,2,...,N, too)

5. Calculate steam consumption (1st effect)
   Press C; in the display appears . . . V_0

6. Calculate the heating surface
   Press D; in the display appears . . A

7. Calculate the heat load
   Press E; in the display appears . . . Q

8. Calculate the steam flow in each effect
   Press f a; in the display appear (flashing)
   $V_1, V_2, ..., V_N$
   (The steam flows are still in registers 1,2,...,N)

9. Calculate the liquid flows in each effect
   Press f b; in the display appear (flashing)
   $L_1, L_2, ..., L_N$
   (The liquid flows are still in registers 1,2,...,N)

10. Calculate the ratio r (steam produced/ steam consumed)
    Press f c; in the display appears r

11. Calculate liquor concentrations
    Press f d; in the display appear (flashing)
    $x_1, x_2, ..., x_N$
    (Liquor concentrations are still in registers 1,2,...,N, too)

The program is initially devised for operating in metric units (see Nomenclature). If you wish to work in English units, then press f e; that converts the Regnault formula into these last units. This must be done at the beginning (i.e., before Key A is pressed).

Assumptions

The assumptions made for the mathematical analysis have the following limits:

The feed is at its boiling point—This is not really a limiting condition. It is only a general basis on which to compare results. If the feed is *not* at its boiling point (bp), the extra steam consumption to heat it to the boiling point can be estimated easily by hand calculations. Generally, in industrial situations, the condensed steam is used to preheat the feed in another heat exchanger, so the feed usually reaches its bp.

All the heating surfaces are equal—This is quite normal, especially in the design of industrial evaporators, due to economic considerations.

Sensible heats are negligible when compared with latent heats—This is true in the majority of the cases when the boiling-point rise (bpr) is negligible. The maximum possible error is about 20%—when the solutions are very concentrated and the bpr cannot be overlooked (see next point).

Boiling-point rise is negligible—This is the most limiting condition. It is true when the solution's molal concentration is not too high (according to Raoult's law), or when we deal with solutions of organic compounds of high molecular weights. On the other hand, when we deal with electrolytic compounds, and when the range of concentration is also very wide, the error of the estimated area can be as high as 25 to 35%.

This can be easily computed by adding the bpr of each stage (ΔT_e) and comparing this value with the thermic potential, ΔT, of the evaporator $(\Delta T = t_0 - t_N)$. If $\Sigma(\Delta T_e)_i$ is 30 or 40% of ΔT, then we can expect a high level of error (35%). This error can be 40% when the number of stages is very high (7 or more).

Nevertheless, this is not a serious problem because this program is only designed to find the approximate value of the evaporation area, the most important parameter in design considerations. It is also possible to correct the area as follows:

$$A_{\text{corrected}} = A_{\text{computed}} \frac{\Delta T}{\Delta T \Sigma(\Delta T_e)_i}$$

Example

Find the value of the variables for a three-effect evaporator system.

Data:

$L_0 = 1{,}000$ kg/h; $T_0 = 100\,°C$; $x_0 = 0.1$;
$T_N = 60\,°C$; $x_N = 0.2$
$U_1 = 200$ kcal/h m^2 °C
$U_2 = 400$
$U_3 = 800$

Results:

$L_N = 500$ kg/h;
$T_1 = 77.1$ °C; $T_2 = 65.7$ °C; $T_3 = 60.0$ °C;
$V_0 = 173.6$ kg/h;

Nomenclature

A	Heating surface, m^2 or ft^2
L_o	Feed flowrate, kg/h or lb/h
L_i	Liquor flowrate (Effect i), kg/h or lb/h
L_N	Product flowrate, kg/h or lb/h
N	Number of effects
Q	Heat load, kcal/h or Btu/h
r	Ratio between steam produced and steam consumed (dimensionless)
T_0	Steam temperature (first effect, condensation chamber), °C or °F
T_i	Liquor temperature (Effect i), °C or °F
T_N	Product temperature, °C or °F
U_i	Overall heat-transfer coefficient (Effect i), kcal/h m^2 °C or Btu/h ft^2 °F
V_0	Steam flow (first effect, condensation chamber), kg/h or lb/h
V_i	Steam flow (Effect i), kg/h or lb/h
x_0	Feed-liquor concentration (solids mass fraction), dimensionless
x_i	Liquor concentration (Effect i), dimensionless
x_N	Product concentration, dimensionless

$A = 20.4 \; m^2$
$Q = 93{,}243$ kcal/h
$V_1 = 168.6$ kg/h; $V_2 = 166.3$ kg/h; $V_3 = 165.1$ kg/h;
$L_1 = 831.3$ kg/h; $L_2 = 665.1$ kg/h; $L_3 = 500.0$ kg/h;
$r = 2.88$
$x_1 = 0.12$; $x_2 = 0.15$; $x_3 = 0.20$

For TI-58/59 users

The TI version of the program appears in Table II. User instructions are found in Table III, and the example is run in Table IV.

Program listing for TI version   Table II

Step	Code	Key	Step	Code	Key	Step	Code	Key
000	76	LBL	023	14	14	046	43	RCL
001	10	E'	024	91	R/S	047	00	00
002	86	STF	025	42	STD	048	42	STD
003	00	00	026	15	15	049	58	58
004	01	1	027	02	2	050	43	RCL
005	01	1	028	42	STD	051	19	19
006	01	1	029	00	00	052	55	÷
007	04	4	030	91	R/S	053	43	RCL
008	42	STD	031	76	LBL	054	15	15
009	59	59	032	12	B	055	65	×
010	91	R/S	033	35	1/X	056	43	RCL
011	76	LBL	034	72	ST*	057	13	13
012	11	A	035	00	00	058	95	=
013	42	STD	036	32	X:T	059	42	STD
014	19	19	037	01	1	060	16	16
015	91	R/S	038	44	SUM	061	98	ADV
016	42	STD	039	00	00	062	99	PRT
017	01	01	040	32	X:T	063	98	ADV
018	91	R/S	041	91	R/S	064	43	RCL
019	42	STD	042	76	LBL	065	58	58
020	13	13	043	13	C	066	32	X:T
021	91	R/S	044	58	FIX	067	02	2
022	42	STD	045	02	02	068	42	STD

Step	Code	Key	Step	Code	Key	Step	Code	Key	Step	Code	Key	Step	Code	Key
069	00	00	133	00	00	197	01	01	261	98	ADV	325	99	PRT
070	76	LBL	134	32	X:T	198	95	=	262	42	STO	326	01	1
071	22	INV	135	43	RCL	199	65	×	263	21	21	327	44	SUM
072	73	RC*	136	58	58	200	43	RCL	264	02	2	328	00	00
073	00	00	137	75	-	201	12	12	265	42	STO	329	43	RCL
074	44	SUM	138	01	1	202	95	=	266	00	00	330	00	00
075	17	17	139	95	=	203	35	1/X	267	43	RCL	331	67	EQ
076	01	1	140	77	GE	204	65	×	268	58	58	332	45	Y×
077	44	SUM	141	23	LNX	205	53	(	269	32	X:T	333	61	GTO
078	00	00	142	98	ADV	206	43	RCL	270	76	LBL	334	44	SUM
079	43	RCL	143	87	IFF	207	19	19	271	42	STO	335	76	LBL
080	00	00	144	00	00	208	75	-	272	43	RCL	336	45	Y×
081	67	EQ	145	34	ΓX	209	43	RCL	273	21	21	337	98	ADV
082	24	CE	146	06	6	210	16	16	274	55	÷	338	53	(
083	61	GTO	147	00	0	211	54	)	275	53	(	339	43	RCL
084	22	INV	148	06	6	212	95	=	276	43	RCL	340	19	19
085	76	LBL	149	93	.	213	99	PRT	277	59	59	341	75	-
086	24	CE	150	05	5	214	98	ADV	278	75	-	342	43	RCL
087	53	(	151	42	STO	215	42	STO	279	93	.	343	16	16
088	43	RCL	152	59	59	216	20	20	280	06	6	344	54	)
089	14	14	153	76	LBL	217	65	×	281	09	9	345	55	÷
090	75	-	154	34	ΓX	218	43	RCL	282	05	5	346	43	RCL
091	43	RCL	155	43	RCL	219	17	17	283	65	×	347	20	20
092	01	01	156	58	58	220	65	×	284	73	RC*	348	95	=
093	54	)	157	32	X:T	221	53	(	285	00	00	349	99	PRT
094	55	÷	158	02	2	222	43	RCL	286	54	)	350	98	ADV
095	43	RCL	159	42	STO	223	59	59	287	95	=	351	02	2
096	17	17	160	00	00	224	75	-	288	99	PRT	352	42	STO
097	95	=	161	76	LBL	225	93	.	289	72	ST*	353	00	00
098	42	STO	162	33	X²	226	06	6	290	00	00	354	76	LBL
099	18	18	163	43	RCL	227	09	9	291	01	1	355	52	EE
100	02	2	164	59	59	228	05	5	292	44	SUM	356	43	RCL
101	42	STO	165	75	-	229	65	×	293	00	00	357	19	19
102	00	00	166	93	.	230	43	RCL	294	43	RCL	358	65	×
103	76	LBL	167	06	6	231	01	01	295	00	00	359	43	RCL
104	23	LNX	168	09	9	232	54	)	296	67	EQ	360	13	13
105	73	RC*	169	05	5	233	55	÷	297	43	RCL	361	55	÷
106	00	00	170	65	×	234	53	(	298	61	GTO	362	73	RC*
107	65	×	171	73	RC*	235	43	RCL	299	42	STO	363	00	00
108	43	RCL	172	00	00	236	01	01	300	76	LBL	364	95	=
109	18	18	173	95	=	237	75	-	301	43	RCL	365	72	ST*
110	95	=	174	35	1/X	238	43	RCL	302	98	ADV	366	00	00
111	32	X:T	175	44	SUM	239	14	14	303	43	RCL	367	99	PRT
112	01	1	176	12	12	240	54	)	304	19	19	368	01	1
113	22	INV	177	01	1	241	95	=	305	42	STO	369	44	SUM
114	44	SUM	178	44	SUM	242	99	PRT	306	22	22	370	00	00
115	00	00	179	00	00	243	98	ADV	307	02	2	371	43	RCL
116	73	RC*	180	43	RCL	244	43	RCL	308	42	STO	372	00	00
117	00	00	181	00	00	245	20	20	309	00	00	373	67	EQ
118	85	+	182	67	EQ	246	65	×	310	43	RCL	374	53	(
119	32	X:T	183	35	1/X	247	53	(	311	58	58	375	61	GTO
120	95	=	184	61	GTO	248	43	RCL	312	32	X:T	376	52	EE
121	99	PRT	185	33	X²	249	59	59	313	76	LBL	377	76	LBL
122	32	X:T	186	76	LBL	250	75	-	314	44	SUM	378	53	(
123	01	1	187	35	1/X	251	93	.	315	43	RCL	379	58	FIX
124	44	SUM	188	43	RCL	252	06	6	316	22	22	380	09	09
125	00	00	189	59	59	253	09	9	317	75	-	381	22	INV
126	32	X:T	190	75	-	254	05	5	318	73	RC*	382	86	STF
127	72	ST*	191	93	.	255	65	×	319	00	00	383	00	00
128	00	00	192	06	6	256	43	RCL	320	95	=	384	98	ADV
129	01	1	193	09	9	257	01	01	321	72	ST*	385	91	R/S
130	44	SUM	194	05	5	258	54	)	322	00	00			
131	00	00	195	65	×	259	95	=	323	42	STO			
132	43	RCL	196	43	RCL	260	99	PRT	324	22	22			

User instructions for TI version

Table III

Step		Key	Output
1.	Clear registers	Press **CMs**	
2.	Enter data:		
	Feed flowrate, L_0	Press **A**	L_0
	Steam temperature, T_0	Press **R/S**	T_0
	Feed-liquor concentration, x_0	Press **R/S**	x_0
	Product temperature, T_N	Press **R/S**	T_N
	Product concentration, x_N	Press **R/S**	2
3.	Enter heat-transfer coefficients for each unit:		
	U_1	Press **B**	$1/U_1$
	U_2	Press **B**	$1/U_2$
Output will be:			
	Product flowrate		L_N
	Liquor temperature for each effect		T_i (i = 1 to n)
	Steam flowrate		V_0
	Heating surface		A
	Heat load		Q
	Steam flowrate for each effect		V_i (i = 1 to n)
	Liquor flowrate to each effect		L_i (i = 1 to n)
	Ratio of steam produced/consumed		r
	Liquor concentration for each effect		x_i (i = 1 to n)

Example for TI version

Table IV

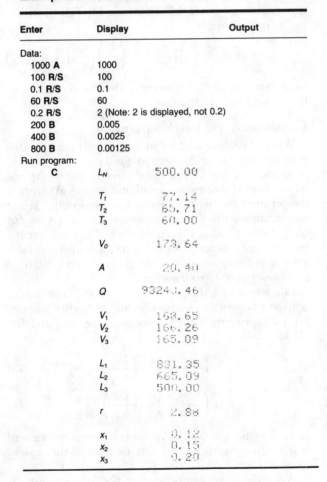

Enter	Display	Output
Data:		
1000 **A**	1000	
100 **R/S**	100	
0.1 **R/S**	0.1	
60 **R/S**	60	
0.2 **R/S**	2 (Note: 2 is displayed, not 0.2)	
200 **B**	0.005	
400 **B**	0.0025	
800 **B**	0.00125	
Run program:		
C	L_N	500.00
	T_1	77.14
	T_2	65.71
	T_3	60.00
	V_0	173.64
	A	20.40
	Q	93243.46
	V_1	168.65
	V_2	166.26
	V_3	165.09
	L_1	831.35
	L_2	665.09
	L_3	500.00
	r	2.88
	x_1	0.12
	x_2	0.15
	x_3	0.20

The authors

Santiago Esplugas Vidal is a professor of chemical engineering at the University of Barcelona. He holds B.S. and Ph.D. degrees in chemistry from the University of Barcelona. His research areas are simulation of chemical processes and photoreactor engineering.

Juan Mata Alvarez is a professor of chemical engineering at the University of Barcelona, Spain. He has B.S. and Ph.D. degrees from the University of Barcelona. His research focuses on simulation of chemical processes and biologic treatment of wastewater.

Heat transfer in agitated vessels

Easy-to-use procedures set forth how to calculate heat transfer in agitated vessels for continuous and batch operations. They show how to establish heat-transfer coefficients, specify heat-transfer areas for jackets and coils, and determine heating- and cooling-cycle periods.

Frederick Bondy and Shepherd Lippa, The Heyward-Robinson Co.

☐ Heat transfer in agitated vessels depends on the type of agitator and jacketing or internal coils. An agitator is selected on the basis of the properties of the material and the processing required. Some typical agitators are shown in Fig. 1. Some common jacketing and coil arrangements are depicted in Fig. 2.

In general, the heat transfer occurs as part of a processing operation, such as suspending or dissolving solids, or dispersing a gas, in a liquid; emulsifying immiscible liquids; or regulating chemical reactions. The type, size, location and speed of an agitator will usually be set by such a mixing requirement [1,2,3,4]. In most cases, the agitator and the power it requires are determined before heat-transfer aspects are considered.

When processing is controlled by heat transfer, such variables as log mean temperature difference and transfer surface area will usually predominate over the agitator variables. Mixing can only affect the inside film resistance, which is but one of a number of resistances that determine the overall heat-transfer coefficient.

General relationships

Heat transfer in an agitated vessel having an external jacket follows the relationship $Q = UA \, \Delta T$. The overall heat-transfer coefficient, U, is determined from a series of five resistances to the transfer of heat:

$$1/U = 1/h_i + ff_i + x/k + ff_j + 1/h_j \qquad (1)$$

Besides applying only to jacketed vessels, Eq. (1) also is only valid when the vessel's diameter is very large in comparison to its wall thickness—i.e., the inner and outer heat-transfer surfaces are very nearly equal—as will be the case with almost all jacketed vessels. When, however, heat is transferred via internal coils or tubular baffles, the difference between the inner and outer heat-transfer surfaces is significant.

The term h_i represents the heat-transfer coefficient on the process side—i.e., the *inside* wall of a vessel having an external jacket, but the *outside* wall of a coil inside a vessel. With internal coils, the coefficient U must be referred to the inner or outer coil surface. The outer surface coefficient, U_o, is also established from a series of five resistances, but is calculated via Eq. (2), which involves coil diameter:

$$\frac{1}{U_o} = \frac{1}{h_i} + ff_i + \left(\frac{x}{k}\right)\left(\frac{d_{co}}{d_{cm}}\right) + \left(\frac{1}{h_{ci}}\right)\left(\frac{d_{co}}{d_{ci}}\right) + ff_{ci} \qquad (2)$$

Note that in Eq. (2), h_i represents the film coefficient on the outside wall of the coils.

Continuous vs. batch operation

When the vessel and its jacket (or coil) are operated continuously under isothermal conditions, the general equation $Q = UA \, \Delta T$ is applied directly. When the vessel is operated continuously with its contents at constant temperature but with different inlet and outlet jacket temperatures, the general equation becomes $Q = UA \, \Delta T_{lm}$. In both cases, U is constant, but in the latter the temperature difference is the log mean temperature difference between the vessel temperature and the inlet and outlet jacket temperatures.

In the case of batch operation, with the vessel's contents at temperature t_1 initially, and at t_2 after θ hours, Eq. (3) represents the relationship for heating and Eq. (4) for cooling:

$$\ln\left(\frac{T - t_1}{T - t_2}\right) = \left(\frac{UA}{m \, c_p}\right)\theta \qquad (3)$$

$$\ln\left(\frac{t_1 - T}{t_2 - T}\right) = \left(\frac{UA}{m \, c_p}\right)\theta \qquad (4)$$

Here, T is the constant jacket (or coil) temperature, and m is the weight and c_p the specific heat of the vessel's contents.

When the jacket temperature is not constant, Eq. (3) and (4) can still be used if the difference between the jacket inlet and outlet temperatures is small compared with the ΔT_{lm} between the average temperature of the jacket and the temperature of the vessel's contents. However, to use the average jacket temperature for T, the change in the jacket's inlet and outlet temperatures should not be greater than 10% of the ΔT_{lm}. When the

Originally published April 4, 1983

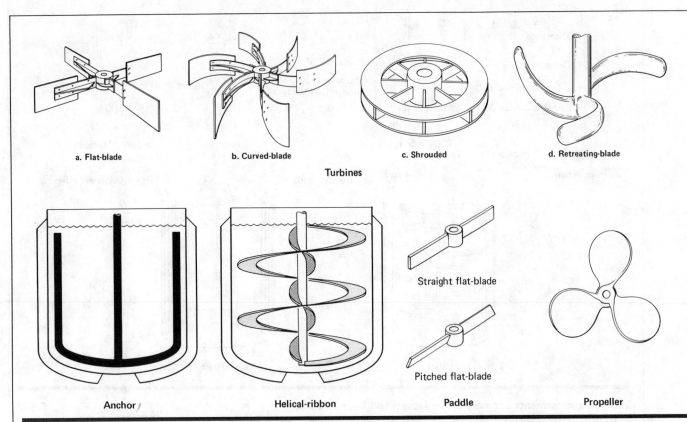

a. Flat-blade   b. Curved-blade   c. Shrouded   d. Retreating-blade

Turbines

Straight flat-blade

Pitched flat-blade

Anchor   Helical-ribbon   Paddle   Propeller

Flat-blade turbine (a) and propeller handle majority of mixing services; applications of other types are more specialized, such as: curved-blade (b) for fibrous materials; anchor and ribbon for viscous materials Fig. 1

change is greater, Eq. (5) should be used for heating and Eq. (6) for cooling:

$$\ln\left(\frac{T_1 - t_1}{T_1 - t_2}\right) = \left(\frac{WC}{m\,c_p}\right)\left(\frac{K-1}{K}\right)\theta \qquad (5)$$

$$\ln\left(\frac{t_1 - T_1}{t_2 - T_1}\right) = \left(\frac{WC}{m\,c_p}\right)\left(\frac{K-1}{K}\right)\theta \qquad (6)$$

Here, T_1 = jacket inlet temperature, °F, and $K = e^{(UA/WC)}$.

With Eq. (3), (4), (5) and (6), the coefficient U is assumed to be essentially constant [5]. If, during heating up or cooling down, the batch's temperature range is large and U varies significantly, the range must be divided into small increments, and the time it takes to achieve each temperature increment must be calculated separately [6].

Inside-film coefficients

In applying the following equations for calculating film coefficients in jacketed vessels, care should be taken in two primary regards:

1. The physical property data should be accurate, particularly those for thermal conductivity, k—which may not always be readily available. The value of k can have a major impact on the calculated film coefficient, and can vary widely: For example, at 86°F, it is 0.080 Btu/(h)(ft²)(°F/ft) for hexane, and 0.356 Btu/(h)(ft²)(°F/ft) for water.

2. The system being designed should, if possible, be similar geometrically to the agitated vessel for which the applicable equations were developed. All the equations presented are for dished-bottom vessels, with the exceptions of Eq. (23) and (24), which are for flat-bottomed vessels. However, these latter two equations could be used with dished-bottomed vessels without causing serious inaccuracy in the estimation of an inside-film coefficient.

Inside coefficients—jacketed vessels

Flat-blade turbine—For mixing at $N_{Re} > 400$ in a jacketed, baffled vessel, calculate the inside-film coefficient via Eq. (7) [6,7]:

$$N_{Nu} = 0.74(N_{Re})^{0.67}(N_{Pr})^{0.33}(\mu/\mu_w)^{0.14} \qquad (7)$$

Here, N_{Nu} = Nusselt number = $h_i D_T/k$; N_{Re} = Reynolds number = $D^2N\rho/\mu$; and N_{Pr} = Prandtl number = $c_p\mu/k$.

For mixing at $N_{Re} < 400$ in a baffled or unbaffled vessel, use Eq. (8) [6,7]:

$$h_i D_T/k = 0.54(N_{Re})^{0.67}(N_{Pr})^{0.33}(\mu/\mu_w)^{0.14} \qquad (8)$$

Eq. (7) and (8) were developed with 6-bladed turbines. Both equations apply to vessels of "standard" geometry, in which the ratio of liquid level (measured from the bottom of the dish) to tank diameter (Z/D_T) equals 1.0, and the ratio of impeller diameter to tank diameter (D/D_T) equals 1/3.

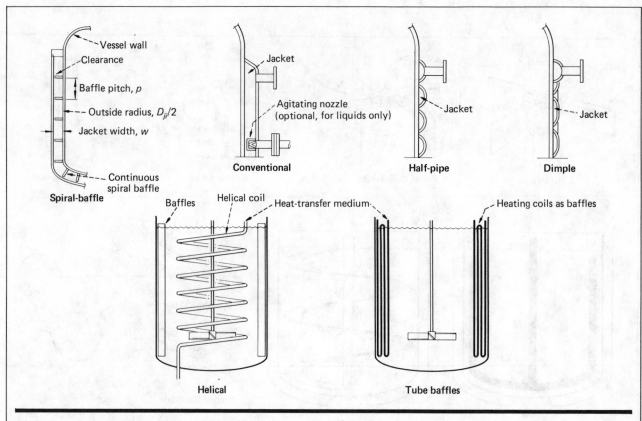

Common arrangements of heat-transfer jackets and internal coils   Fig. 2

A more general form of Eq. (7)—for $N_{Re} > 400$ and nonstandard geometry—is Eq. (9) [8]:

$$h_i D_T/k = 0.85(N_{Re})^{0.66}(N_{Pr})^{0.33} \times \\ (Z/D_T)^{-0.56}(D/D_T)^{0.13}(\mu/\mu_w)^{0.14} \quad (9)$$

The viscosity correction term, $(\mu/\mu_w)^{0.14}$, in Eq. (7), (8) and (9), as well as in the equations that follow, requires that the wall temperature be estimated. When a liquid is being heated, the term will be greater than 1.0 and h-corrected larger than h-uncorrected (because, of course, liquid viscosity decreases with increasing temperature). When a liquid is being cooled, the opposite will occur. Therefore, when a liquid whose viscosity varies significantly with temperature is being heated or cooled, it is advisable to apply the viscosity correction term, rather than assume that it equals 1.0.

The wall temperature, t_w, is estimated by trial-and-error via Eq. (10), which is based on equal heat flow through the jacket-side and vessel-side films and negligible temperature drop across the metal of the vessel wall ($T_w = t_w$):

$$t_w = T - \{(T - t)/[1 + (h_j A_o/h_i A_i)]\} \quad (10)$$

Here, A_o is the jacketed area based on the outside vessel diameter, and A_i is the area based on the inside diameter. There is a similar relationship for internal coils.

If the difference between A_o and A_i is negligible (usually the case with jacketed vessels), Eq. (10) simplifies to:

$$t_w = T - \{(T - t)/[1 + (h_j/h_i)]\} \quad (11)$$

In the first trial to estimate h_i, assume $(\mu/\mu_w)^{0.14} = 1.0$ when using the inside-film-coefficient equations. Unless the term varies greatly from 1.0, one iteration should be sufficient to establish t_w. The viscosity at the wall, μ_w, is then taken from viscosity data at t_w in order to calculate the correction term.

Retreating-blade turbine—When mixing in a jacketed, unbaffled vessel with a turbine having six retreating blades, calculate h_i via Eq. (12) [9,10]:

$$h_i D_T/k = 0.68(N_{Re})^{0.67}(N_{Pr})^{0.33}(\mu/\mu_w)^{0.14} \quad (12)$$

A discussion of the pertinent vessel geometry is provided by Holland and Chapman [6,11]. An equation that takes into account nonstandard geometry in terms of Z/D_T and D/D_T does not seem to be available for retreating-blade turbines.

For a jacketed, baffled vessel in which mixing is done by a glassed-steel impeller having three retreating blades, calculate h_i by means of Eq. (13) [9]:

$$h_i D_T/k = 0.33(N_{Re})^{0.67}(N_{Pr})^{0.33}(\mu/\mu_w)^{0.14} \quad (13)$$

With a similar, alloy, impeller:

$$h_i D_T/k = 0.37(N_{Re})^{0.67}(N_{Pr})^{0.33}(\mu/\mu_w)^{0.14} \quad (14)$$

The lower constant for the glassed-steel impeller is attributed to greater slippage around its curved surfaces than around the sharp corners of the alloy impeller [9].

Propeller—When mixing is done by a 45-deg.-pitched four-bladed impeller, calculate h_i via:

$$h_i D_T/k = 0.54(N_{Re})^{0.67}(N_{Pr})^{0.25}(\mu/\mu_w)^{0.14} \quad (15)$$

Nomenclature

A	Heat-transfer area, ft^2
A'	Effective heat-transfer area, ft^2
A_x	Cross-sectional flow area, ft^2
B	Number of baffles
C_j	Specific heat of fluid in jacket, Btu/(lb)(°F)
C_p	Specific heat of vessel contents, Btu/(lb)(°F)
D	Impeller dia., ft
D_c	Mean or centerline dia. of internal coil helix, ft
D_e	Equivalent dia. for heat transfer, ft
D'_e	Equivalent dia. for fluid flow, ft
D_{ji}	Inner dia. of annular jacket, ft
D_{jo}	Outer dia. of annular jacket, ft
D_T	Inner dia. of vessel, ft
d_{ci}	Inner dia. of pipe or coil, ft
d_{co}	Outer dia. of pipe or coil, ft
d_{cm}	Log mean dia. of pipe or coil, ft
e	Clearance, $(D_T - D)/2$, ft
F	Volumetric flowrate through jacket or coil, gal/min
f_f	Fanning friction factor, dimensionless
f_m	Moody friction factor, $f_m = 4f_f$, dimensionless
ff_i	Fouling factor, inside vessel, (h)(ft^2)(°F)/Btu
ff_j	Fouling factor, inside jacket, (h)(ft^2)(°F)/Btu
ff_{ci}	Fouling factor on coilside referred to inside coil area, (h)(ft^2)(°F)/Btu
ff_{co}	Fouling factor on coilside referred to outside coil area, (h)(ft^2)(°F)/Btu
g	Acceleration due to gravity, 4.17×10^8, ft/h^2
H_c	Height of total coil, ft
h	Film coefficient, Btu/(h)(ft^2)(°F)
h_{ci}	Coefficient on coilside referred to inside coil area, Btu/(h)(ft^2)(°F)
h_{co}	Coefficient on coilside referred to outside coil area, Btu/(h)(ft^2)(°F)
h_i	Coefficient on process side of heat-transfer area, i.e., inside surface of jacketed vessel or outside surface of internal coil, Btu/(h)(ft^2)(°F)
h_j	Coefficient on inside surface of jacket, Btu/(h)(ft^2)(°F)
i	Agitator ribbon pitch, ft
K	$e^{(UA/WC_j)}$, dimensionless
k	Thermal conductivity, Btu/(h)(ft^2)(°F/ft)
L	Length of coil or jacket passage, ft

m	Mass of material in vessel, lb
N	Agitator speed, rev/h
N_{Gr}	Grashof number, $D_e^3 \rho^2\, g\beta\Delta t_G/\mu^2$, dimensionless
N_{Nu}	Nusselt number, $h_i D_T/k$, dimensionless
N_{Pr}	Prandtl number, $c_p\mu/k$, dimensionless
N_{Re}	Reynolds number, $D^2 N\rho/\mu$ (in vessel), $D_e V\rho/\mu$ (in jacket), dimensionless
n	Number of coil turns/ft of coil height, 1/ft
ΔP	Pressure drop in straight pipe or passage, lb/in.^{2}
ΔP_c	Pressure drop in internal coil (total), lb/in.^{2}
ΔP_j	Pressure drop in jacket (total, including entrance and exit losses), lb/in.^{2}
p	Pitch of baffle spiral, ft
Q	Heat-transfer rate, Btu/h
T	Temperature in jacket or coil, °F
T_1	Inlet temperature in jacket or coil, °F
T_2	Outlet temperature in jacket or coil, °F
T_w	Wall or surface temperature in jacket or coil, °F
ΔT	Difference between t and T, °F
ΔT_{lm}	Log mean temperature difference, °F
t	Bulk temperature of vessel contents, °F
t_1	Initial temperature of vessel contents, °F
t_2	Final temperature of vessel contents, °F
t_w	Wall temperature inside vessel, °F
Δt_G	Difference between temperature of bulk liquid in jacket and vessel wall; $(T - T_w)$ for heating, and $(T_w - T)$ for cooling, °F
U	Overall heat-transfer coefficient in a jacketed vessel, Btu/(h)(ft^2)(°F)
U_o	Overall heat-transfer coefficient in an internal-coil vessel, Btu/(h)(ft^2)(°F)
V	Velocity in spiral coil or jacket, ft/h
W	Mass flowrate through jacket or coil, lb/h
W'	Effective mass flowrate through jacket, lb/h
w	Width of conventional or spiral jacket, ft
x	Wall thickness of vessel or coil, ft
Z	Liquid height (from bottom of lower head), ft
β	Coefficient of volumetric expansion (see Grashof number, N_{Gr}), 1/°F
θ	Time of heating or cooling cycle, h
μ	Viscosity at bulk temperature, lb/(ft)(h)
μ_w	Viscosity at wall surface, lb/(ft)(h)
η	Viscosity correction exponent, dimensionless
ρ	Density, lb/ft^3

Eq. (15) is based on limited data with regard to propeller pitch and vessel baffling [12]. For design purposes, divide the h_i obtained with this equation by a factor of about 1.3.

Paddle—At $N_{Re} > 4{,}000$ in both baffled and unbaffled jacketed vessels, use Eq. (16) [13]:

$$h_i D_T/k = 0.36(N_{Re})^{0.67}(N_{Pr})^{0.33}(\mu/\mu_w)^{0.14} \quad (16)$$

The pertinent vessel geometry is discussed by Holland and Chapman [11].

Under similar conditions, but with N_{Re} between 20 and 4,000, use Eq. (17) [14]:

$$h_i D_T/k = 0.415(N_{Re})^{0.67}(N_{Pr})^{0.33}(\mu/\mu_w)^{0.24} \quad (17)$$

The vessel geometry is similar to that in Holland and

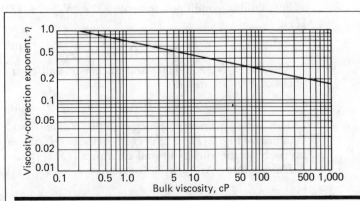

Viscosity correction for Eq. (23) applies to all tank sizes [17]   Fig. 3

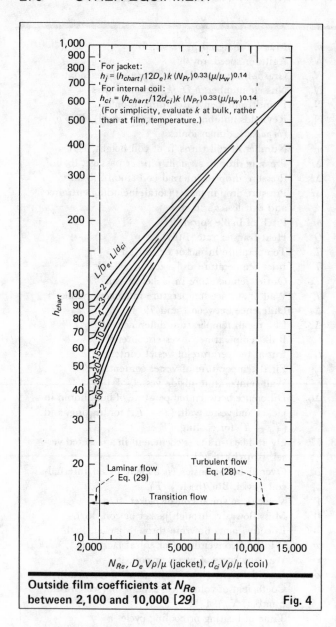

For jacket:
$$h_j = (h_{chart}/12D_e)k(N_{Pr})^{0.33}(\mu/\mu_w)^{0.14}$$
For internal coil:
$$h_{ci} = (h_{chart}/12d_{ci})k(N_{Pr})^{0.33}(\mu/\mu_w)^{0.14}$$
(For simplicity, evaluate k at bulk, rather than at film, temperature.)

Laminar flow
Eq. (29)

Turbulent flow
Eq. (28)

Transition flow

N_{Re}, $D_e V\rho/\mu$ (jacket), $d_{ci} V\rho/\mu$ (coil)

**Outside film coefficients at N_{Re}
between 2,100 and 10,000 [29]**   Fig. 4

Thermal conductivities for some common materials	Table I

Material	k, Btu/(h)(ft²)(°F/ft)
Carbon steel	26
Copper	218
Glass lining	0.58
Inconel	8.7
Monel (400, 404, R-405, 411)	14.3
Nickel (200, 201, 220, 225)	35
Stainless steel (304, 316, 321, 347)	9.1
Tantalum	32
Titanium	10.5

At $N_{Re} > 130$, calculate h_i by means of Eq. (22):

$$h_i D_T/k = 0.238(N_{Re})^{0.67}(N_{Pr})^{0.33}(\mu/\mu_w)^{0.14}(i/D)^{-0.25} \quad (22)$$

Inside coefficients—internal coils

Flat-blade turbine—When heating or cooling is through an internal helical coil, and mixing is by a turbine having six flat blades at $400 < N_{Re} < 1,500,000$ [17]:

$$h_i d_{co}/k = 0.17(N_{Re})^{0.67}(N_{Pr})^{0.37}(D/D_T)^{0.1}(d_{co}/D_T)^{0.5}(\mu/\mu_w)^\eta \quad (23)$$

Here, η is a function of μ and can be obtained from Fig. 3. The vessel geometry is depicted in Holland and Chapman's Fig. 5 [11]. Eq. (23) is considered applicable to all sizes of tanks with a six-flat-blade turbine, even those with wall baffles or an inside helical coil—if tube diameters are $0.018 \leq d_{co}/D_T \leq 0.036$—at a viscosity range probably to 10,000 cP. Tube spacing of coil wraps of 2–4 tube dia. showed no appreciable effect on the coefficient.

With a flat, four-bladed turbine and vertical tubes as baffles, and at $1,300 < N_{Re} < 2,000,000$ [18,19]:

$$h_i d_{co}/k = 0.09(N_{Re})^{0.65}(N_{Pr})^{0.3}(\mu/\mu_w)^{0.14}(D/D_T)^{0.33}(2/B)^{0.2} \quad (24)$$

Typical vessel geometry is shown in Fig. 7 of Ref. 11.

Retreating-blade turbine—With internal helical coils and an impeller of six retreating blades [9]:

$$h_i D_T/k = 1.40(N_{Re})^{0.62}(N_{Pr})^{0.33}(\mu/\mu_w)^{0.14} \quad (25)$$

For the vessel geometry, see Holland and Chapman's Fig. 6 [11].

No correlation seems to be available for the three-retreating-blade impeller.

Propeller—For internal-coil heating or cooling, and propeller mixing, calculate h_i via Eq. (26) [9]:

$$h_i d_{co}/k = 0.078(N_{Re})^{0.62}(N_{Pr})^{0.33}(\mu/\mu_w)^{0.14} \quad (26)$$

Eq. (26) is based on limited data with regard to propeller pitch and vessel baffling. When using it in design, divide the h_i obtained by a factor of about 1.3.

Paddle—For mixing with a paddle [13]:

$$h_i D_T/k = 0.87(N_{Re})^{0.62}(N_{Pr})^{0.33}(\mu/\mu_w)^{0.14} \quad (27)$$

For vessel geometry, see Fig. 4 of Ref. 11.

Chapman's Fig. 4 [11], except that no coils were used.

Anchor—At $30 < N_{Re} < 300$ and anchor-to-wall clearances of less than 1 in. [14], use:

$$h_i D_T/k = 1.0(N_{Re})^{0.67}(N_{Pr})^{0.33}(\mu/\mu_w)^{0.18} \quad (18)$$

For similar conditions, except $300 < N_{Re} < 4,000$ [14]:

$$h_i D_T/k = 0.38(N_{Re})^{0.67}(N_{Pr})^{0.33}(\mu/\mu_w)^{0.18} \quad (19)$$

At $4,000 < N_{Re} < 37,000$ and anchor-to-wall clearances of 1 to $5\frac{1}{8}$ in. [12]:

$$h_i D_T/k = 0.55(N_{Re})^{0.67}(N_{Pr})^{0.25}(\mu/\mu_w)^{0.14} \quad (20)$$

The vessel geometry is depicted in Holland and Chapman's Fig. 9 [11]. The overall coefficient, U, varies inversely to the anchor-to-wall clearance [15].

Helical ribbon—At $N_{Re} < 130$, calculate h_i via Eq. (21) [16]:

$$h_i D_T/k = 0.248(N_{Re})^{0.50} \times (N_{Pr})^{0.33}(\mu/\mu_w)^{0.14}(e/D)^{-0.22}(i/D)^{-0.28} \quad (21)$$

Excerpt of agitating-nozzle data for standard vessels of a fabricator								Table II
Vessel designation and capacity, gal.**		Agitating nozzles		Water at 70° F				
				Minimum for turbulence			ΔP = 20	
		Quantity	Size, in.	ΔP*, psi	F†	h_j‡	F†	h_j‡
P14:	5; 10	1	1¼	15	19	370	22	430
P20:	20; 30	1	1¼	15	19	430	22	490
P24:	50	1	1¼	17	21	300	22	315
P30:	100	1	1¼	12	17	420	22	540
ES32:	50; 75; 100	2	1½	4.0	30	550	68	950
EM40:	150; 200	2	1½	3.5	28	550	68	1,030
EL48:	200; 500	2	1½	8.9	45	350	68	470
ELL60:	500; 750	2	1½	7.2	41	370	68	520
XL66:	1,000	2	1½	6.2	37	400	68	620
XXL84:	1,500; 2,000	2	1½	6.8	39	380	68	565

*Pressure drop through jacket includes that for nozzle plus jacket †Total jacket flowrate, gpm ‡Jacket film coefficient, Btu/(h)(ft²)(°F)

**All vessels with same designation have same diameter. When more than one capacity is given (for example, 200 and 500 gal. for EL 48), the height differs for each capacity.

Source: Pfaudler Co.

No correlations seem to be available for internal-coil heating and mixing with an anchor or helical ribbon.

Fouling factors and wall resistances

Fouling factors (ff_i, ff_j, ff_{ci} and ff_{co}) for use in determining the overall heat-transfer coefficient, U, should be estimated from previous operating experience or judgment as to fouling severity. Inside and outside fouling factors may be selected from such sources as E. E. Ludwig's "Applied Process Design for Chemical and Petrochemical Plants" [30].

Wall resistances can be significant and should be calculated. The thermal conductivity of carbon steel should not be used for a stainless-steel shell. Listed in Table I are typical, conservative thermal conductivities, k, for various materials.

Outside coefficients—jacketed vessels

Annular jacket with spiral baffling—A heat-transfer liquid is circulated through this type of jacket. It is not used with condensing steam.

For heat-transfer purposes, this jacket can be considered a special case of a helical coil if certain factors are incorporated into equations for calculating outside-film coefficients. In the following equations, the equivalent heat-transfer diameter, D_e, for a rectangular cross-section is equal to $4w$ (w being the width of the annular space). Velocities are calculated from the actual cross-section of the flow area, pw (p being the pitch of the spiral baffle); and from the effective mass flowrate, W', through the passage.

The leakage around spiral baffles is considerable, amounting to 35–50% of the total mass flowrate, W [20]. To get a conservative outside-film coefficient and avoid making laborious trial-and-error pressure balances to establish precise flow and velocity in the spiral passage, the effective mass flowrate should be taken to be about 60% of the total mass flowrate to the jacket ($W' \approx 0.6W$).

At a given Reynolds number, heat-transfer coefficients of coils, particularly with turbulent flow, are higher than those of long, straight pipes, due to the greater friction. Flow through an annular jacket with spiral baffling acts similarly. Therefore, at $N_{Re} > 10,000$, the Sieder-Tate equation for straight pipe, multiplied by a turbulent-flow coil correction factor, $1 + 3.5 (D_e/D_c)$, can be used to calculate the outside-film coefficient [21]:

$$h_j D_e/k = 0.027(N_{Re})^{0.8}(N_{Pr})^{0.33}(\mu/\mu_w)^{0.14}[1 + 3.5(D_e/D_c)] \quad (28)$$

Here, $N_{Re} = (D_e V \rho/\mu)$.

At $N_{Re} < 2,100$, Eq. (29), also based on straight pipe, can be used:

$$h_j D_e/k = 1.86[(N_{Re})(N_{Pr})(D_e/L)]^{0.33}(\mu/\mu_w)^{0.14} \quad (29)$$

In the transition region, $2,100 < N_{Re} < 10,000$, obtain h_j from Fig. 4 [29]. Otherwise, use Eq. (28) and (29) for greater accuracy.

Annular jacket, no baffles—In the case of steam condensation, a film coefficient, h_j, of 1,000 Btu/(h)(ft²)(°F) is a safe assumption. In the case of liquid circulation, velocities will be very low because of the large cross-sectional flow area.

Outside coefficients for unbaffled jackets can also be obtained for turbulent, laminar and transition flow from, respectively, Eq. (28), Eq. (29) and Fig. 4—except that the turbulent-flow coil-correction factor, $1 + 3.5 (D_e/D_c)$, is not included in Eq. (28). In this case, the equivalent heat-transfer diameter, D_e, is found from Eq. (30), and the flow area from Eq. (31):

$$D_e = [(D_{jo})^2 - (D_{ji})^2]/D_{ji} \quad (30)$$

$$A_x = \pi[(D_{jo})^2 - (D_{ji})^2]/4 \quad (31)$$

At very low Reynolds numbers (laminar-flow region), natural convection will help heat transfer to a limited extent. Eq. (32), developed for laminar flow of water in annuli, gives approximate film coefficients [22]:

$$\frac{h_j D_e}{k} = 1.02(N_{Re})^{0.45}(N_{Pr})^{0.33} \times$$

$$\left(\frac{D_e}{L}\right)^{0.4}\left(\frac{D_{jo}}{D_{ji}}\right)^{0.8}\left(\frac{\mu}{\mu_w}\right)^{0.14}(N_{Gr})^{0.05} \quad (32)$$

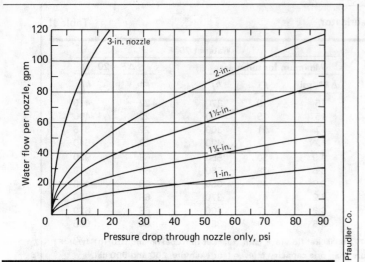

Capacity vs. pressure drop for representative agitating nozzles

Fig. 5

Source: Pfaudler Co.

The Grashof number, N_{Gr}, must be evaluated from fluid properties at the bulk temperature, T.

For Eq. (32) only, evaluate D_e from:

$$D_e = D_{jo} - D_{ji} \qquad (33)$$

The Eq. (33) D_e is only for laminar flow. For turbulent flow, obtain D_e from Eq. (30) for use in Eq. (28), as noted previously.

Other, more-complex correction factors for natural convection have been developed for vertical pipes [23]. However, it is recommended that neither the correction factors nor Eq. (32) be applied to jackets, internal coils or tubular baffles in which flow is laminar, because of their complexity, and so as to obtain more-conservative h_j values.

Furthermore, because film coefficients are very low with liquid circulating through an unbaffled jacket (even at reasonable flowrates), it is recommended that one or more agitating nozzles be installed at the inlet to the jacket. Film coefficients for annular jackets equipped with agitating nozzles can be obtained from vessel fabricators (see Table II excerpt).

To achieve an outside-film coefficient of at least 400, the nozzle for an annular jacket should be selected to provide a minimum of 0.01 total hydraulic horsepower per square foot of heat-transfer area. Hydraulic horsepower is given by HHP = (gpm)(psi)/1,715; gpm is the total flowrate through the jacket, and psi is the pressure drop across the nozzle (or nozzles), based on the flowrate through each nozzle.

For pressure drops and flow characteristics of various sizes of Pfaudler agitating nozzles, see Fig. 5. As an example, a jacket having 135 ft² of heat-transfer area requires hydraulic horsepower equal to 1.35. If the necessary jacket flowrate for heat transfer is 117 gpm (based on an allowable temperature rise of the jacket fluid), three 1½-in. nozzles with a 20-psi drop and a flow of 39 gpm (Fig. 5) are adequate, because HHP/ft² = (3)(39)(20)/(1715)(135) = 0.0101.

Half-pipe-coil jacket—Outside-film coefficients for jackets of this type can be determined from Eq. (28),

Eq. (29) and Fig. 4 for, respectively, turbulent, laminar and transition flow.

Pipe coils are made with a 180-deg. central angle (semicircular cross-section) or a 120-deg. central angle. For the first, $D_e = (\pi/2) d_{ci}$. For calculating velocity, $A_x = (\pi/8)(d_{ci})^2$ (with d_{ci} the pipe I.D.). For 120-deg. half-pipe coils, $D_e = 0.708 \, d_{ci}$, and $A_x = 0.154 \, (d_{ci})^2$.

Dimple jacket—Film coefficients in this case are also determined from Eq. (28), Eq. (29) and Fig. 4 for turbulent, laminar and transition flow, respectively. When using Eq. (28), omit the turbulent-flow-correction factor. Because of turbulence created by the dimples in the flow stream, the coefficients so obtained are not very accurate, probably erring on the low side. If available, coefficients based on experimental data should be used.

From typical dimensions of the fluid passages provided by a manufacturer, the equivalent diameter for heat transfer is approximately 0.66 in. [24]. The flow area is 1.98 in.²/ft of vessel circumference. A dimple-jacketed vessel of 6-ft, 0-in. dia., for example, would have a flow area equal to $(\pi)(6)(1.98)$, or 37.3 in.²

Outside coefficients—internal coils

Internal coils should be designed for turbulent flow. For special cases involving viscous heat-transfer fluids, laminar or transition flow may be unavoidable.

Use Eq. (28), Eq. (29) and Fig. 4, to calculate internal-coil outside-film coefficients, h_{ci}, for, respectively, turbulent, laminar and transition flow, as previously discussed, except as follows: Eq. (28)—substitute the ratio d_{ci}/D_c for D_e/D_c; Eq. (29)—substitute $(d_{ci}/D_c)^{1/2}$ for (D_e/L). For the special case of water, when the coil should always be designed for turbulent flow, Eq. (28) yields film coefficients that are not conservative, and Fig. 6 should be used instead.

In all cases, h_{ci} must be converted to h_{co} (film coefficient at outside of coil) before calculation of U_o.

Heat-transfer area

Surface area for heating and cooling agitated vessels can be provided by either external jacketing or internal coils (or tubular baffles). Jacketing is usually preferred because of: cheaper materials of construction, less tendency to foul, easier cleaning and maintenance, fewer problems in circulating catalysts and viscous fluids, and larger heat-transfer surface.

Coils should be considered only if jacketing alone will not provide sufficient heat-transfer area, if jacket pressure will exceed 150 psig, or if high-temperature vacuum processing is required. Under the latter two conditions, a coil offers the advantage of a higher overall film coefficient, because of its thinner walls.

Annular jacket—In this case, the effective heat-transfer area (with or without spiral baffling) is that wetted by both the vessel's contents and the heat-transfer fluid.

Half-pipe-coil jacket—These standard coils come in 2, 3 and 4-in. pipe sizes. Spacing between adjacent half-pipes is normally ¾ in.

The area between the half-pipes (not wetted by the heating medium) is not completely effective in heat transfer; however, the loss is small because the vessel wall transmits heat longitudinally. Multiplying the area between the half-pipes by 0.6 and adding the product to the area under the half-pipes yields conservative values for the total effective area. This method results in the following ratios of effective heat-transfer area to total heat-transfer area, A'/A, for the pipe diameters: 2 in.—0.90; 3 in.—0.93, and 4 in.—0.94.

Dimple jacket—For a $\frac{1}{2}$-in.-dia. dimple weld in a $2\frac{1}{2}$-in. square pattern, the A'/A ratio is 0.92.

Internal coils—The effective heat-transfer area is the total wetted area, based on the coil's outside surface. This outside transfer area, A_{co}, is calculated from:

$$A_{co} = \pi d_{co} H_c n[(\pi D_c)^2 + n^{-2}]^{1/2} \qquad (34)$$

Here, n is the number of coil turns per foot of coil height, and equals the inverse of the coil pitch, p.

Transfer-medium pressure drop

Pressure drops in jackets and coils can be significant when a high degree of turbulence (from high velocity) is required in order to improve the overall coefficient by increasing the h_j film coefficient of the heat-transfer medium. This may be the case when the inside-film coefficient, h_i, cannot be further improved. [See Eq. (28) and (29) for the effect of velocity on h_j.] Such an instance occurs when the pressure drop in a jacket or coil must be designed to operate within limits imposed by a once-through heat-transfer system serving multiple equipment and having a fixed pressure differential between the fluid supply and return headers.

To obtain the desired mass flow through a jacket or coil for carrying transferred heat within a pressure-drop limit, it may be necessary to divide the jacket or coil into a number of parallel zones (i.e., multipasses). Doing this will increase the total mass flow to the zones and, to a lesser extent, the jacket or coil velocity. If, however, the objective is to increase velocity to obtain a higher film coefficient, but accomplishing this via multipassing would result in an excessive volumetric flow, it may be advisable to resort to a single-pass jacket or coil, and to increase the fluid pressure and velocity via a booster pump.

Annular jacket with spiral baffling—This type of jacket can be considered similar to a helical coil having a rectangular, rather than a circular, cross-section.

Preferably, the system should be designed for turbulent flow of the heat-transfer fluid. The Fanning equation, Eq. (35), can be used to calculate pressure drop if: (1) an equivalent diameter for fluid in a rectangular cross-section is used when determining the Reynolds number and friction factor, and (2) a suitable multiplier is applied to the friction factor (or equivalent length is used) to account for the curvature of the helical flow passage.

$$\Delta P = 4f_f\left(\frac{V^2}{2g}\right)\left(\frac{L}{D'_e}\right)\left(\frac{\rho}{144}\right) = f_m\left(\frac{V^2}{2g}\right)\left(\frac{L}{D'_e}\right)\left(\frac{\rho}{144}\right) \qquad (35)$$

The equivalent diameter, D'_e, for fluid-flow calculations is given by:

$$D'_e = 4\left(\frac{\text{cross-sectional flow area}}{\text{wetted perimeter for flow}}\right)$$

$$= \left(\frac{4\,pw}{2p + 2w}\right) = \left(\frac{2\,pw}{p + w}\right) \qquad (36)$$

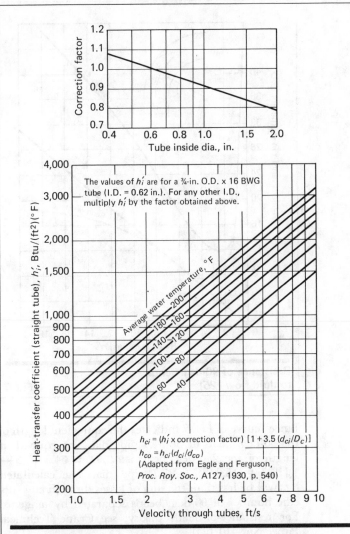

Film coefficients for water in tubes and coils   Fig. 6

Note that for the rectangular passage, D'_e differs from the equivalent diameter for heat-transfer calculations, D_e, which is equal to $4w$. For D_e, only the inner jacket wall is considered the wetted perimeter for heat transfer, because heat transfer is limited to this surface, and a more conservative h_j results from Eq. (28) and (29).

There are a number of reliable ways to account for the curvature of the helical flow passage [25,26,27]. One of the simplest and more conservative is given in the 1969 edition of Crane Co. Technical Paper No. 410 [25].

Fig. 7 gives the resistance of 90-deg. bends to the flow of fluids in terms of equivalent lengths of straight pipe. The resistance of bends greater than 90-deg. is found from:

$$L/D'_e = R_t + (n - 1)[R_l + (R_b/2)] \qquad (37)$$

Here, n = number of 90-deg. bends in coil; R_t = total resistance due to one 90-deg. bend in L/D'_e; R_l = resistance due to length of one 90-deg. bend in L/D'_e; and R_b = bend resistance due to one 90-deg. bend in L/D'_e.

As an example, determine the equivalent lengths in pipe diameters of a 90-deg. bend and a 270-deg. bend, both having a relative radius of 12. The "Total resis-

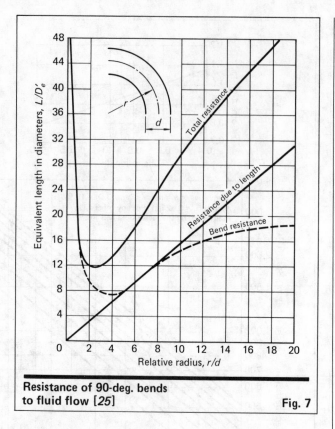

Resistance of 90-deg. bends to fluid flow [25]   **Fig. 7**

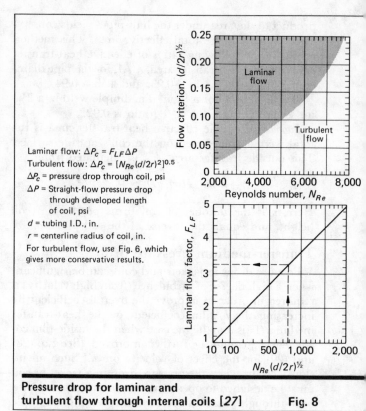

Laminar flow: $\Delta P_c = F_{LF}\Delta P$

Turbulent flow: $\Delta P_c = [N_{Re}(d/2r)^2]^{0.5}$

ΔP_c = pressure drop through coil, psi

ΔP = Straight-flow pressure drop through developed length of coil, psi

d = tubing I.D., in.

r = centerline radius of coil, in.

For turbulent flow, use Fig. 6, which gives more conservative results.

Pressure drop for laminar and turbulent flow through internal coils [27]   **Fig. 8**

tance" curve of Fig. 7 indicates the equivalent length of the 90-deg. bend to be 34.5 pipe dia. The equivalent length of the 270-deg. bend is: $L/D'_e = 34.5 + (3 - 1)[18.7 + (15.8/2)] = 87.7$ pipe dia. The calculated loss will be less than the sum of losses through any specific number of 90-deg. bends separated by tangents. For resistance-of-bends theory, see Crane Technical Paper No. 410 [25].

It is suggested that the outside radius of the vessel be substituted for the mean radius of the helical flow passage, r (Fig. 7), in using this method. Using the mean radius (which is equal to the outside radius of the vessel plus half the width of the annular jacket, because the equivalent length increases with the relative radius, as shown in Fig. 7) will yield more-conservative results.

Substituting Eq. (37) into Eq. (35), and including entrance and exit losses, yields:

$$\Delta P_j = f_m(V^2/2g)[R_t + (n - 1)(R_l + R_b/2)](\rho/144) + 2(V^2/2g)(\rho/144) \quad (38)$$

Here, a total of two velocity heads is assumed in the second term for entrance and exit losses (involving expansion, contraction and sharp turns of the fluid).

In the special case of laminar flow in the rectangular passages of the jacket, find the pressure drop from Fig. 8. (For turbulent flow, Fig. 6 gives more-conservative results.) The upper graph in Fig. 8 is used to determine whether flow is turbulent or laminar.

To find pressure drop through the developed length of coil for laminar straight flow, first use the following form of the Poiseuille equation:

$$\Delta P = 0.0167\mu LV/(D'_e)^2 \quad (39)$$

Next, multiply the ΔP from Eq. (39) by the factor, F_{LF},

obtained from the lower graph in Fig. 8 to obtain the curved-laminar-flow pressure drop.

Because of considerable leakage around spiral baffles (refer to earlier discussion on calculating outside-film coefficients for annular jackets with spiral baffles), it is recommended that the velocity in Eq. (35) and (38) also be based on the effective flowrate (i.e., 60% of the total flowrate). However, use the total flowrate to determine the pressure drops in the supply and return lines, control valve, fittings, etc.

Annular jacket, no baffles—Flow in this type of jacket can be considered approximately the same as that in the annulus of a double-pipe heat exchanger. The Fanning equation, Eq. (35), can be applied to this type of jacket (when not equipped with agitation nozzles), with a suitable equivalent diameter.

The equivalent diameter for calculating fluid flow, D'_e, can be found from:

$$D'_e = 4\left(\frac{\text{cross-sectional flow area}}{\text{wetted perimeter for flow}}\right)$$
$$= \frac{4(\pi/4)(D_{jo}^2 - D_{ji}^2)}{\pi(D_{jo} + D_{ji})} = D_{jo} - D_{ji} \quad (40)$$

Note that the equivalent diameter for calculating fluid flow, D'_e, differs from the equivalent diameter for calculating heat transfer, as given by Eq. (30). Only the inner jacket wall is used as the wetted perimeter in Eq. (30), because heat transfer is limited to this surface. Using the inner jacket wall also yields more-conservative values for film coefficients from Eq. (28) and (29).

As noted before, film coefficients will usually be low for this type of jacket without agitating nozzles. Pressure-drop data for agitating nozzles, such as shown in

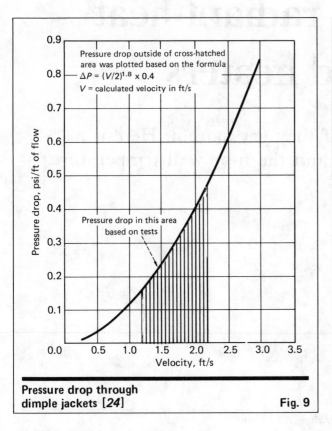

Pressure drop outside of cross-hatched area was plotted based on the formula

$\Delta P = (V/2)^{1.8} \times 0.4$

V = calculated velocity in ft/s

Pressure drop in this area based on tests

Pressure drop through dimple jackets [24]   **Fig. 9**

Internal coils—The procedure for calculating pressure drop through internal coils is similar to that described for annular jackets with spiral baffling and half-pipe-coil jackets. In this case, the equivalent diameters for heat transfer and for fluid flow are the same (i.e., equal to the coil's internal diameter).

References

1. Parker, N. H., *Chem. Eng.,* June 8, 1964, p. 165.
2. Hicks, E. W. *et al., Chem. Eng.,* Apr. 26, 1976, p. 102.
3. Gates, L. E., *et al., Chem. Eng.,* May 24, 1976, p. 144.
4. Hicks, R. W., and Gates, L. E., *Chem. Eng.,* July 19, 1976, p. 141.
5. Kern, D. Q., "Process Heat Transfer," McGraw-Hill, New York, 1950.
6. Holland, F. A., and Chapman, F. S., *Chem. Eng.,* Feb. 15, 1965, p. 175.
7. Brooks, G., and Su, G. D., *Chem. Eng. Prog.,* October 1959, p. 54.
8. Dickey, D., and Hicks, R. W., *Chem. Eng.,* Feb. 2, 1976, p. 93.
9. Ackley, E. J., *Chem. Eng.,* Aug. 22, 1960, p. 133.
10. Cummings, G. H., and West, A. S., *Ind. Eng. Chem.,* Vol. 42, 1950, p. 2303.
11. Holland, F. A., and Chapman, F. S., *Chem. Eng.,* Jan. 18, 1965, p. 153.
12. Brown, R. W., Scott, R., and Toyne, C., *Trans. IChE,* Vol. 25, 1947, p. 181.
13. Chilton, T. H., Drew, T. B., and Jebens, R. H., *Ind. Eng. Chem.,* Vol. 36, 1944, p. 510.
14. Uhl, V. W., *Chem. Eng. Prog. Symposium Series,* Vol. 51, 1954, p. 93.
15. Uhl, V. W., and Voznick, H. P., *Chem. Eng. Prog.,* March 1960, p. 172.
16. Blazinski, H., and Kuncewicz, C., *Intl. Chem. Eng.,* Vol. 21, 1981, p. 679.
17. Oldshue, J. Y., and Gretton, A. T., *Chem. Eng. Prog.,* Vol. 50, 1954, p. 615.
18. Rushton, J. H., Lichtman, R. S., and Mahoney, L. H., *Ind. Eng. Chem.,* Vol. 40, 1948, p. 1082.
19. Dunlap, I. R., and Rushton, J. H., *Chem. Eng. Prog. Symposium Series,* Vol. 49, 1953, p. 5.
20. Bolliger, D. H., *Chem. Eng.,* Sept. 20, 1982, p. 95.
21. Perry, R. H., and Chilton, C. H., (Ed.), "Chemical Engineers' Handbook," 5th ed., McGraw-Hill, New York, 1973, p. 10–17.
22. Chen, C. Y., Hawkins, G. A., and Solberg, H. L., *Trans. ASME,* Vol. 68, 1946, p. 99.
23. Martinelli, R. C., *et al., Trans. AIChE,* Vol. 38, 1942, p. 493.
24. Markovitz, R. E., *Chem. Eng.,* Nov. 15, 1971, p. 156.
25. Crane Co., "Flow of Fluids," Technical Paper No. 410, 1969.
26. *Op. cit.,* 1980.
27. *ASME Trans., J. of Basic Engineering,* Vol. 81, 1959, p. 126.
28. Kaferle, J. A., *Chem. Eng.,* Nov. 24, 1975, p. 86.
29. Sieder, E. N., and Tate, G. E., *Ind. Eng. Chem.,* Vol. 28, 1936, p. 1429.
30. Ludwig, E. E., "Applied Process Design for Chemical and Petrochemical Plants," Vol. 3, Gulf Pub. Co., Houston, Tex., 1965, pp. 57–58.

Fig. 5, can be obtained from manufacturers. Total jacket pressure drop = 1.25 × nozzle pressure drop (compare Fig. 5 with Table II).

Half-pipe-coil jacket—Flow and pressure-drop calculations for this type of jacket are similar to those described previously for an annular jacket with spiral baffling. This jacket should be designed for turbulent flow of the heat-transfer fluid.

The equivalent diameter for calculating fluid flow, D'_e, for 180-deg. half-pipe-coil jackets can be found from:

$$D'_e = 4\left(\frac{\text{cross-sectional area}}{\text{wetted flow perimeter}}\right)$$
$$= \frac{4(1/2)(\pi/4)(d_{ci})^2}{d_{ci} + \pi d_{ci}/2} = 0.611\, d_{ci} \qquad (41)$$

Note that the equivalent diameter for calculating fluid flow, D'_e, differs from that for calculating heat transfer, D_e, where the wetted perimeter is the inner jacket wall, or d_{ci} rather than $d_{ci} + \pi d_{ci}/2$.

For 120-deg. half-pipe coils, the equivalent diameter for fluid flow, D'_e, equals $0.321\, d_{ci}$. This differs from the equivalent diameter for heat transfer, D_e, which is equal to $0.708\, d_{ci}$.

Dimple jacket—Calculations of pressure drop are complex and may not yield accurate results. It is advisable to obtain such data from manufacturers. Fig. 9 provides a curve that is based on tests made by the Brighton Corp. [24]. When using Fig. 9, consider the velocity obtained (which is based on the cross-sectional flow area of the dimple jacket as defined previously) as the actual velocity. For predicting pressure drop, based on the Brighton Corp. findings, for baffling inside dimple jackets refer also to Kaferle [28].

The authors

Frederick Bondy is a process manager with The Heyward-Robinson Co. (One World Trade Center, New York, NY 10048), responsible for process engineering of refinery, chemical and polymer plants. His more than 20 years of experience includes operations engineering, process engineering, and supervision of process engineering. He previously had been with PPG Industries, The M. W. Kellogg Co., Chem Systems Inc., Hydrocarbon Research Inc. and Foster-Wheeler Energy Corp. He holds a B.Ch.E. from City College of New York and an M.Ch.E. from the University of Akron.

Shepherd Lippa is a process supervisor with The Heyward-Robinson Co., responsible for the process design of chemical and polymer plants. His thirty years of experience encompasses operations, process development, and process engineering. His previous employers have included Stauffer Chemical Co., B. F. Goodrich Co., and Crawford & Russell Inc. He holds a B.Ch.E. from City College of New York and an M.S.Ch.E. from the University of Michigan, is a member of AIChE and Tau Beta Pi, and is a registered professional engineer in Connecticut.

Program predicts radiant heat flux in direct-fired heaters

In these heaters, tubes can rupture if they get too hot. Here is a program to determine the heat flux and the tube wall temperature so that rupture can be avoided.

Tayseer A. Abdel-Halim, KTI Corp.

☐ One of the main causes for tube rupture in direct-fired heaters is the high tube-metal temperature (T_{TM})—or tube-skin temperature—experienced by some of the tubes in the furnace's radiant section.

Although it is customary for vendors of heaters to assume an average radiant heat flux in the unit's radiant section, such an assumption is incorrect. The radiant heat flux is a function of the tube metal temperature, and varies through the radiant coil according to the following equation, which is a modification of the Stefan-Boltzmann law:

$$q = K(T_{BW}^4 - T_{TM}^4) \qquad (1)$$

Here, $K = 654.4 \times 10^{-12}$ Btu/(h)(ft^2)(R^4). This value is based on assuming a radiant heat flux of 10,000 Btu/(h)(ft^2) for a T_{BW} of 1,620°F (2,080 R) and a T_{TM} of 880°F (1,340 R). This value of K is also for one row of tubes, and assumes that tubes are spaced on center lines equal to two times their nominal diameters, and are 1½ nominal dia. from the refractory surface.

Trial-and-error calculations

To obtain the radiant heat flux for the tubes, a two-way trial-and-error method is usually used:

1. A radiant heat flux, q, is assumed at the coil outlet.
2. The tube metal temperature is calculated:

$$T_{TM} = T_{FO} + \Delta T_{TOTAL} \qquad (2)$$

where:

$$\Delta T_{TOTAL} = \Delta T_{FILM} + \Delta T_{COKE} + \Delta T_{METAL} \qquad (3)$$

$$\Delta T_{TOTAL} = \frac{q}{h_i}\left(\frac{\text{O.D.}}{\text{I.D.} - 2t_c}\right) +$$
$$\frac{q \times t_c}{k_c}\left(\frac{\text{O.D.}}{\text{I.D.} - t_c}\right) + \frac{q \times t_a}{k_w}\left(\frac{\text{O.D.}}{\text{O.D.} - t_a}\right)$$
$$= q \times \text{O.D.}\left[\frac{1}{h_i(\text{I.D.} - 2t_c)} + \right.$$
$$\left. \frac{t_c}{k_c(\text{I.D.} - t_c)} + \frac{t_a}{k_w(\text{O.D.} - t_a)}\right] \qquad (4)$$

Film coefficients are calculated by a method given by API [1]. In Eq. (4), values of q_{ASSD} are first put in for q, then a value is calculated and is used.

3. The radiant heat flux at each tube outlet is then calculated from Eq. (1).

4. If the calculated heat flux is close enough to the assumed value, we move back one tube in the coil and repeat the same steps. If it is not, we go back to step 1 and assume a new heat flux.

5. The heat absorbed by each tube is calculated, based on the average of the heat fluxes at the inlet and outlet of the tube.

6. The total duty calculated for the whole radiant coil should be close to the heat absorbed in the radiant section. If not, start again at the coil outlet with a different guess for the radiant flux.

(text continues on p. 279)

Originally published December 17, 1979.

Program performs two-part trial-and-error calculations in predicting radiant heat fluxes   **Table I**

Location	Code	Key	Location	Code	Key	Location	Code	Key	Location	Code	Key	Location	Code	Key	Location	Code	Key
Setup			(Setup cont'd)			(Setup cont'd)			(Main-line program cont'd)			(Main-line program cont'd)			(Main-line program cont'd)					
000	76	LBL	062	55	÷	124	25	CLR	184	43	RCL	246	00	0	308	04	04
001	11	A	063	06	6	125	42	STO	185	00	00	247	00	0	309	43	RCL
002	42	STO	064	95	=	126	20	20	186	95	=	248	32	X⇄T	310	22	22
003	00	00	065	42	STO	127	03	3	187	65	×	249	43	RCL	311	69	OP
004	42	STO	066	27	27	128	00	0	188	43	RCL	250	02	02	312	06	06
005	02	02	067	43	RCL	129	42	STO	189	17	17	251	75	−	313	03	3
006	43	RCL	068	03	03	130	01	01	190	55	÷	252	43	RCL	314	05	5
007	14	14	069	65	×	131	76	LBL	191	02	2	253	00	00	315	01	1
008	42	STO	070	43	RCL	132	33	X²	192	95	=	254	95	=	316	03	3
009	40	40	071	07	07	133	03	3	193	42	STO	255	50	I×I	317	03	3
010	00	0	072	65	×	134	01	1	194	26	26	256	22	INV	318	07	7
011	42	STO	073	89	π	135	75	−	195	43	RCL	257	77	GE	319	01	1
012	01	01	074	65	×	136	43	RCL	196	14	14	258	16	A'	320	07	7
013	42	STO	075	43	RCL	137	01	01	197	75	−	259	43	RCL	321	69	OP
014	33	33	076	05	05	138	95	=	198	43	RCL	260	30	30	322	04	04
015	42	STO	077	55	÷	139	42	STO	199	26	26	261	32	X⇄T	323	43	RCL
016	30	30	078	01	1	140	29	29	200	55	÷	262	61	GTO	324	02	02
017	42	STO	079	02	2	141	73	RC*	201	43	RCL	263	13	C	325	69	OP
018	04	04	080	95	=	142	29	29	202	18	18	264	76	LBL	326	06	06
019	42	STO	081	42	STO	143	99	PRT	203	95	=	265	16	A'	327	98	ADV
020	08	08	082	08	08	144	97	DSZ	204	42	STO	266	43	RCL	328	43	RCL
021	42	STO	083	55	÷	145	01	01	205	21	21	267	01	01	329	21	21
022	17	17	084	43	RCL	146	33	X²	206	76	LBL	268	75	−	330	42	STO
023	42	STO	085	04	04	147	91	R/S	207	45	Yˣ	269	01	1	331	14	14
024	18	18	086	95	=				208	71	SBR	270	95	=	332	43	RCL
025	42	STO	087	42	STO	**Main-line program**			209	23	LNX	271	99	PRT	333	30	30
026	20	20	088	17	17	148	76	LBL	210	85	+	272	98	ADV	334	32	X⇄T
027	42	STO	089	43	RCL	149	15	E	211	43	RCL	273	01	1	335	97	DSZ
028	21	21	090	13	13	150	43	RCL	212	21	21	274	03	3	336	01	01
029	42	STO	091	55	÷	151	04	04	213	95	=	275	01	1	337	13	C
030	22	22	092	53	(	152	85	+	214	42	STO	276	04	4	338	43	RCL
031	42	STO	093	43	RCL	153	01	1	215	22	22	277	03	3	339	40	40
032	24	24	094	14	14	154	95	=	216	43	RCL	278	06	6	340	42	STO
033	42	STO	095	75	−	155	42	STO	217	16	16	279	69	OP	341	14	14
034	25	25	096	43	RCL	156	01	01	218	65	×	280	04	04	342	43	RCL
035	42	STO	097	15	15	157	42	STO	219	53	(	281	43	RCL	343	20	20
036	26	26	098	54	)	158	30	30	220	53	(	282	26	26	344	99	PRT
037	42	STO	099	95	=	159	32	X⇄T	221	43	RCL	283	69	OP	345	91	R/S
038	28	28	100	42	STO	160	76	LBL	222	23	23	284	06	06			
039	42	STO	101	18	18	161	13	C	223	85	+	285	43	RCL	**Subroutine del-tee**		
040	29	29	102	53	(	162	43	RCL	224	04	4	286	26	26	346	76	LBL
041	42	STO	103	43	RCL	163	02	02	225	06	6	287	44	SUM	347	23	LNX
042	35	35	104	11	11	164	85	+	226	00	0	288	20	20	348	43	RCL
043	42	STO	105	75	−	165	43	RCL	227	54	)	289	02	2	349	24	24
044	33	33	106	43	RCL	166	00	00	228	45	Yˣ	290	01	1	350	65	×
045	43	RCL	107	12	12	167	95	=	229	04	4	291	03	3	351	53	(
046	03	03	108	54	)	168	55	÷	230	75	−	292	02	2	352	43	RCL
047	55	÷	109	55	÷	169	02	2	231	53	(	293	03	3	353	21	21
048	43	RCL	110	53	(	170	85	+	232	43	RCL	294	07	7	354	75	−
049	09	09	111	43	RCL	171	01	1	233	22	22	295	69	OP	355	43	RCL
050	95	=	112	14	14	172	00	0	234	85	+	296	04	04	356	15	15
051	42	STO	113	75	−	173	00	0	235	04	4	297	43	RCL	357	54	)
052	04	04	114	43	RCL	174	95	=	236	06	6	298	21	21	358	85	+
053	43	RCL	115	15	15	175	42	STO	237	00	0	299	69	OP	359	43	RCL
054	05	05	116	54	)	176	00	00	238	54	)	300	06	06	360	12	12
055	55	÷	117	95	=	177	43	RCL	239	45	Yˣ	301	03	3	361	95	=
056	01	1	118	42	STO	178	01	01	240	04	4	302	07	7	362	42	STO
057	02	2	119	24	24	179	77	GE	241	54	)	303	03	3	363	25	25
058	95	=	120	43	RCL	180	45	Yˣ	242	95	=	304	00	0	364	01	1
059	75	−	121	14	14	181	43	RCL	243	42	STO	305	03	3	365	55	÷
060	43	RCL	122	42	STO	182	02	02	244	02	02	306	07	7	366	53	(
061	06	06	123	21	21	183	85	+	245	01	1	307	69	OP	367	43	RCL

Continued   Table I

Location (Subroutine del-tee cont'd)	Code	Key	Location (Subroutine del-tee cont'd)	Code	Key	Location (Subroutine del-tee cont'd)	Code	Key
368	25	25	395	19	19	422	95	=
369	65	×	396	55	÷	423	42	STO
370	53	(	397	01	1	424	35	35
371	43	RCL	398	02	2	425	43	RCL
372	27	27	399	54	)	426	00	00
373	75	−	400	54	)	427	65	×
374	43	RCL	401	95	=	428	43	RCL
375	19	19	402	42	STO	429	05	05
376	55	÷	403	34	34	430	55	÷
377	06	6	404	43	RCL	431	01	1
378	54	)	405	06	06	432	02	2
379	54	)	406	65	×	433	95	=
380	95	=	407	01	1	434	65	×
381	42	STO	408	02	2	435	53	(
382	33	33	409	55	÷	436	43	RCL
383	43	RCL	410	53	(	437	33	33
384	19	19	411	43	RCL	438	85	+
385	55	÷	412	10	10	439	43	RCL
386	53	(	413	65	×	440	34	34
387	03	3	414	53	(	441	85	+
388	06	6	415	43	RCL	442	43	RCL
389	65	×	416	05	05	443	35	35
390	53	(	417	75	−	444	54	)
391	43	RCL	418	43	RCL	445	95	=
392	27	27	419	06	06	446	92	RTN
393	75	−	420	54	)			
394	43	RCL	421	54	)			

Nomenclature

A	Radiant-coil total heat-transfer area, ft^2
A_T	Surface area per tube, ft^2
h_{io}	Inside film coeff. at coil outlet, Btu/(h)(ft^2)(°F)
h_{ii}	Inside film coeff. at coil inlet, Btu/(h)(ft^2)(°F)
I	Number of tubes per pass
I.D.	Tube inside dia., in.
K	Constant in Eq. (1), 6.454×10^{-10} Btu/(h)(ft^2)(R^4)
k_c	Coke thermal conductivity, Btu/(h)(ft^2)(°F/ft)
k_w	Tube-metal conductivity, Btu/(h)(ft^2)(°F/in.)
L	Effective tube length, ft
O.D.	Tube outside dia., in.
Q	Total heat absorbed in radiant coil, Btu/h
q_{ASSD}	Assumed heat flux, Btu/(h)(ft^2)
q_{CALC}	Calculated heat flux, Btu/(h)(ft^2)
T_{BW}	Bridge wall temperature, °F
T_{FI}	Fluid temperature at coil inlet, °F
T_{FO}	Fluid temperature at coil outlet, °F
T_{TM}	Tube-metal temperature, °F
t_a	Average tube-wall thickness, in.
t_c	Thickness of coke deposit, in.
ΔT_{TOTAL}	As defined by Eq. (3)
ΔT_{FILM} ΔT_{COKE} ΔT_{METAL}	Temperature differences across layers, °F

Storage information for calculating radiant flux. Values are given for a typical problem   Table II

Location	Data	Example
00	q_{ASSD}	
01	Index	
02	q_{CALC}	
03	No. of radiant tubes	64
04	I	
05	O.D.	6.625 in.
06	t_a	0.280 in.
07	L	45.5 ft
08	A	
09	No. of passes	8
10	k_w	200 Btu/(h)(ft^2)(°F/in)
11	h_{io}	152 Btu/(h)(ft^2)(°F)
12	h_{ii}	145 Btu/(h)(ft^2)(°F)
13	Q	58.522×10^6 Btu/h
14	T_{FO}	1,094°F
15	T_{FI}	730°F
16	K	6.454×10^{-10} Btu/(h)(ft^2)(R^4)
17	A_T	
18	Temp. gradient	
19	t_c	0.0625 in.
20	Sum of heat absorbed	
21	T_{FO} at each tube	
22	T_{MT}	
23	T_{BW}	1,728°F
24	Film coeff. gradient	
25	Film coeff. at each tube outlet	

Results from example. Value of T_{MT} at outlet is about 1,200°F   Table III

Outlet tube No.		Outlet tube No.	
Coil outlet 8.		3.	
0.	ABS	7452386.362	ABS
1094.	FOT	876.8735111	FOT
1200.128378	TMT	1005.570646	TMT
9889.452579	RATE	11814.17588	RATE
7.		2.	
6495718.392	ABS	7667922.84	ABS
1053.597391	FOT	829.1799272	FOT
1163.906945	TMT	962.2439578	TMT
10303.49122	RATE	12150.96661	RATE
6.		1.	
6745079.431	ABS	7871321.785	ABS
1011.643785	FOT	780.2212257	FOT
1126.582845	TMT	917.5026026	TMT
10702.11287	RATE	12467.90522	RATE
5.		Coil inlet 0.	
6989467.183	ABS	8061693.4	ABS
968.1701162	FOT	730.078435	FOT
1087.772944	TMT	871.4096137	TMT
11087.81582	RATE	12763.66857	RATE
4.		58509389.63	
7225800.239	ABS		
923.2264832	FOT		
1047.428127	TMT		
11459.16445	RATE		

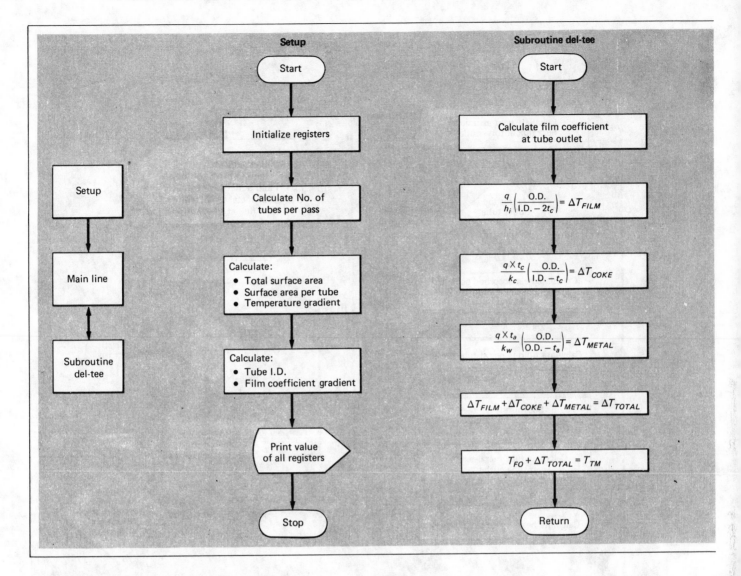

7. Repeat this two-way iterative process until the solution converges to a stable value.

A program is presented for use on the TI-59. This program is adaptable to the HP-97, or to any other programmable calculator.

The program is divided into three sections: setup, mainline and subroutine "del-tee." The logic diagram for the program appears in the figure. The program is shown in Table I, and program storage-allocation appears in Table II.

User's instructions

1. Initialize the following registers: 03, 05, 06, 07, 09, 10, 11, 12, 13, 14, 15, 16, 19 and 23.

2. Enter a first guess for the heat flux at the coil outlet and press **A**. The calculator will go through setup and print out the values of all its registers.

3. Press **E**. The calculator will give a printout of absorption, fluid temperature, tube metal temperature and heat flux at the outlet of each tube.

4. The calculator will also print the value of the calculated total heat absorbed in the coil. If this value is not close enough to the actual heat absorption, guess another value for the heat flux and press **A**, then press **E**.

Notes

The author has found out that a good first guess that forces the program to converge readily is to use a value somewhat less than that of the average heat flux for the whole coil.

A Newton-Raphson or interval-halving [2] subroutine could have been used to force the program to the next guess. Since this requires a substantial increase in both program size and execution time, and since by the above method of guessing we always know the correctness of the second assumption made, the addition of a convergence subroutine cannot be justified.

To obtain the maximum tube-metal temperatures (defined as the temperature of the front 60 deg of the tube), multiply the above TMTs by 1.8. This number, given by API [1], is for tubes arranged in single rows against a wall, and on center-to-center spacing equal to twice the nominal tube diameter.

Example

Data for an example are given in Table II. The heater has 64 tubes arranged in an 8-pass flow. The coil tubes are Sch. 40A pipe, with a thickness of 0.280 in. The thermal conductivity of coke—3.0 $Btu/(h)(ft^2)$

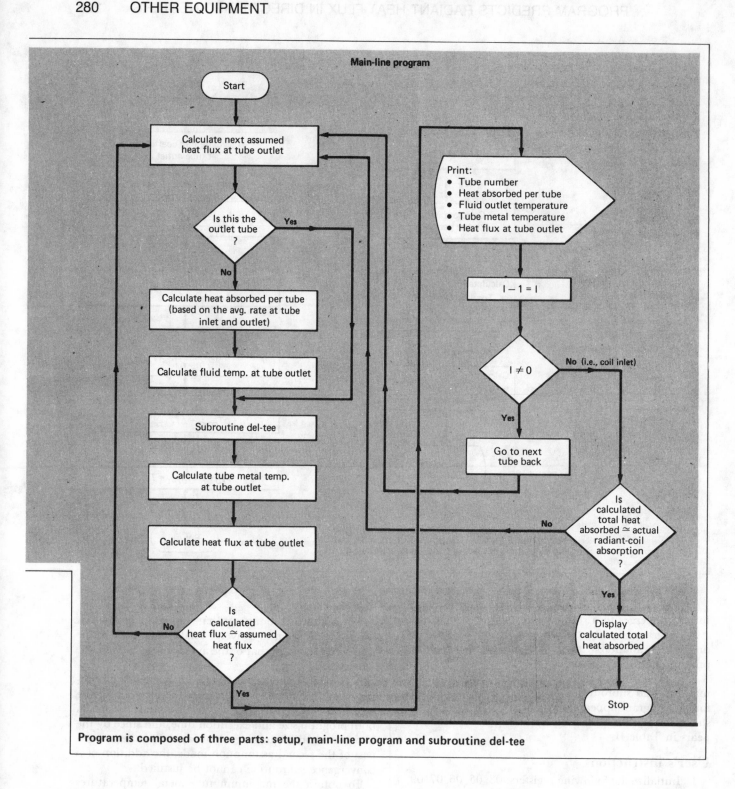

Main-line program

Start

Calculate next assumed
heat flux at tube outlet

Is this the
outlet tube
?   Yes

No

Calculate heat absorbed per tube
(based on the avg. rate at tube
inlet and outlet)

Calculate fluid temp. at tube outlet

Subroutine del-tee

Calculate tube metal temp.
at tube outlet

Calculate heat flux at tube outlet

Is
calculated
heat flux ≃ assumed
heat flux
?   No

Yes

Print:
• Tube number
• Heat absorbed per tube
• Fluid outlet temperature
• Tube metal temperature
• Heat flux at tube outlet

I − 1 = I

I ≠ 0   No (i.e., coil inlet)

Yes

Go to next
tube back

Is
calculated
total heat
absorbed ≃ actual
radiant-coil
absorption
?   No

Yes

Display
calculated total
heat absorbed

Stop

Program is composed of three parts: setup, main-line program and subroutine del-tee

(°F/ft)—is built into the program. Note T_{BW} is 1,728°F.

The heat flux was guessed to be 8,000 Btu/(h)(ft²). Results of the calculations are shown in Table III. ABS stands for heat absorbed per tube, in Btu/h. RATE is heat flux, Btu/(h)(ft²). Note that T_{MT} is symbolized by TMT, and T_{FO} is symbolized by FOT. Similar substitutions are made elsewhere.

For HP-67/97 users

Two programs are required to perform these calculations on the HP calculators. Table IV provides a listing of both programs A and B, and Table V gives the operating steps necessary to run the program. After entering the data, it is wise to record them on a separate magnetic card. Also, recording each of the programs on magnetic cards facilitates the running of the programs and preserves the programs for later use. As in the TI version, when KEY A is pressed, the program lists all data registers. When KEY E is pressed (Step 7), the tube number, heat absorbed, fluid temperature, tube-metal temperature and heat flux at the oulet of each tube are printed. Finally, the total heat absorbed in the coil is printed. If this value is not close enough to the actual value of heat absorbed, return to step 2 and continue. Data entry may be made via magnetic card. Table VI describes the user-defined keys, flags, and data storage locations for the HP version.

Program Listing for HP version

Table IV

Program A

Step	Key	Code
001	*LBLA	21 11
002	FREG	16-13
003	P⇄S	16-51
004	FREG	16-13
005	P⇄S	16-51
006	R/S	51
007	*LBLB	21 12
008	STO0	35 00
009	STO2	35 02
010	P⇄S	16-51
011	RCL4	36 04
012	P⇄S	16-51
013	STO1	35 01
014	RCL3	36 03
015	RCL9	36 09
016	÷	-24
017	STO4	35 04
018	RCL5	36 05
019	1	01
020	2	02
021	÷	-24
022	RCL6	36 06
023	6	06
024	÷	-24
025	-	-45
026	STO8	35 08
027	RCL3	36 03
028	RCL7	36 07
029	Pi	16-24
030	RCL5	36 05
031	x	-35
032	x	-35
033	x	-35
034	1	01
035	2	02
036	÷	-24
037	RCL4	36 04
038	-	-45
039	P⇄S	16-51
040	STO7	35 07
041	RCL3	36 03
042	RCL4	36 04
043	RCL5	36 05
044	-	-45
045	÷	-24
046	STO8	35 08
047	RCL1	36 01
048	RCL2	36 02
049	-	-45
050	RCL4	36 04
051	RCL5	36 05
052	-	-45
053	÷	-24
054	STOE	35 15
055	RCL4	36 04
056	STOB	35 12
057	CLX	-51
058	P⇄S	16-51
059	STO7	35 07
060	R/S	51

Program B

Step	Key	Code
001	*LBLE	21 15
002	P⇄S	16-51
003	0	00
004	STO1	35 01
005	P⇄S	16-51
006	RCL4	36 04
007	1	01
008	+	-55
009	STOI	35 46
010	STOA	35 11
011	*LBLC	21 13
012	RCL2	36 02
013	RCL0	36 00
014	+	-55
015	2	02
016	÷	-24
017	1	01
018	0	00
019	0	00
020	+	-55
021	STO0	35 00
022	RCLI	36 46
023	RCLA	36 11
024	X≤Y?	15-35
025	GTO1	22 01
026	*LBL3	21 03
027	RCL2	36 02
028	RCL0	36 00
029	+	-55
030	P⇄S	16-51
031	RCL7	36 07
032	x	-35
033	2	02
034	÷	-24
035	STO1	35 01
036	RCL4	36 04
037	RCL1	36 01
038	RCL8	36 08
039	÷	-24
040	-	-45
041	STOB	35 12
042	P⇄S	16-51
043	*LBL1	21 01
044	GSB2	23 02
045	P⇄S	16-51
046	RCLB	36 12
047	+	-55
048	STOC	35 13
049	RCL6	36 06
050	RCLD	36 14
051	4	04
052	6	06
053	0	00
054	+	-55
055	4	04
056	Y^x	31
057	RCLC	36 13
058	4	04
059	6	06
060	0	00
061	+	-55
062	4	04
063	Y^x	31
064	-	-45
065	x	-35
066	P⇄S	16-51
067	STO2	35 02
068	RCL0	36 00
069	-	-45
070	ABS	16 31
071	1	01
072	0	00
073	0	00
074	X⇄Y	-41
075	X≤Y?	16-35
076	GTOa	22 16 11
077	GTOC	22 13
078	*LBLa	21 16 11
079	RCLI	36 46
080	1	01
081	-	-45
082	PRTX	-14
083	SPC	16-11
084	P⇄S	16-51
085	RCL1	36 01
086	PRTX	-14
087	P⇄S	16-51
088	ST+7	35-55 07
089	RCLB	36 12
090	PRTX	-14
091	RCLC	36 13
092	PRTX	-14
093	RCL2	36 02
094	PRTX	-14
095	SPC	16-11
096	RCLB	36 12
097	P⇄S	16-51
098	STO4	35 04
099	P⇄S	16-51
100	DSZI	16 25 46
101	GTOC	22 13
102	RCL1	36 01
103	P⇄S	16-51
104	STO4	35 04
105	P⇄S	16-51
106	RCL7	36 07
107	PRTX	-14
108	SPC	16-11
109	SPC	16-11
110	R/S	51
111	*LBL2	21 02
112	RCLE	36 15
113	RCLB	36 12
114	P⇄S	16-51
115	RCL5	36 05
116	-	-45
117	x	-35
118	RCL2	36 02
119	+	-55
120	P⇄S	16-51
121	RCL8	36 08
122	P⇄S	16-51
123	RCL9	36 09
124	6	06
125	÷	-24
126	-	-45
127	x	-35
128	1	01
129	X⇄Y	-41
130	÷	-24
131	P⇄S	16-51
132	STO9	35 09
133	P⇄S	16-51
134	RCL9	36 09
135	P⇄S	16-51
136	RCL8	36 08
137	P⇄S	16-51
138	RCL9	36 09
139	1	01
140	2	02
141	÷	-24
142	-	-45
143	3	03
144	6	06
145	x	-35
146	÷	-24
147	P⇄S	16-51
148	STO3	35 03
149	RCL6	36 06
150	1	01
151	2	02
152	x	-35
153	P⇄S	16-51
154	RCL0	36 00
155	P⇄S	16-51
156	RCL5	36 05
157	RCL6	36 06
158	-	-45
159	x	-35
160	÷	-24
161	RCL0	36 00
162	RCL5	36 05
163	1	01
164	2	02
165	÷	-24
166	x	-35
167	RCL9	36 09
168	RCL3	36 03
169	+	-55
170	X⇄Y	-41
171	R↓	-31
172	+	-55
173	R↑	16-31
174	x	-35
175	RTN	24
176	R/S	51

Step	Procedure	Enter	Press	Display
	User instructions for HP version			**Table V**
1	Turn calculator on			
2	Load program A			
3	Enter data:			
	a) Number of radiant tubes	#	STO 3	#
	b) Outer diameter of tube, in.	O.D.	STO 5	O.D.
	c) Average tube wall thickness, in.	t_a	STO 6	t_a
	d) Effective tube length, in.	L	STO 7	L
	e) Number of passes	#	STO 9	#
			P≤S	
	f) Tube-metal conductivity, Btu/(h)(ft²)(°F/in.)	k_w	STO 0	k_w
	g) Inside film coefficient at coil outlet, Btu/(h)(ft²)(°F)	h_{io}	STO 1	h_{io}
	h) Inside film coefficient at coil inlet, Btu/(h)(ft²)(°F)	h_{ii}	STO 2	h_{ii}
	i) Total heat absorbed in radiant coil, Btu/h	Q	STO 3	Q
	j) Fluid temp. at coil outlet, °F	T_{FO}	STO 4	T_{FO}
	k) Fluid temp. at coil inlet, °F	T_{FI}	STO 4	T_{FI}
	l) Constant, Btu/(h)(ft²)(R⁴)	K	STO 6	K
	m) Thickness of coke deposit, in.	t_c	STO 9	t_c
	n) Bridgewall temp., °F	T_{BW}	STO D	T_{BW}
			P≤S	
4	List data registers if desired		A	
5	Enter heat flux guess and execute	Guess	B	
6	Load program B			
7	Execute		E	See text

Program information for HP version   **Table VI**

User defined keys

Program A

A - List data registers

B - Enter heat-flux guess and execute

Program B

E - Execute

Flags

None

Data registers

Primary				
0	q_{ASSD}		A	L
1	Used		B	T_{FO}
2	q_{CALC}		C	T_{MT}
3	Used		D	T_{BW}
4	I		E	Film coeff.
5	O.D.			gradient
6	t_a		I	Index
7	Used			
8	Used			
9	Used			

Secondary		
0	k_w	
1	Used	
2	h_{ii}	
3	Used	
4	T_{FO}	
5	T_{FI}	
6	K	
7	A_T	
8	Temp. gradient	
9	t_c	

References

1. Amer. Petroleum Inst., publication RP-530, "Recommended Practice for the Calculation of Heater Tube Thickness," 2nd ed., Dec. 1976.
2. Carnahan, Brice, et al., "Applied Numerical Methods," John Wiley & Sons, Inc., New York, 1969.

The author

Tayseer A. Abdel-Halim is a process engineer with the Process Plant Div. of KTI Corp., 221 East Walnut St., Pasadena, CA 91101. Tel: (213) 577-1600, x284. He is involved with the design of ethylene plants. He has also worked for Born, Inc., where he was responsible for the design and cost estimation of direct-fired heaters, and for Heat Research Corp., as a senior process engineer. Abdel-Halim obtained a B.Sc. degree in refinery engineering from the Egyptian High Inst. for Petroleum Engineering, and an M.Sc. degree in chemical engineering from the University of Tulsa. He is a member of AIChE.

Solar ponds collect sun's heat

So simple that it is hard to believe they will work, solar ponds are an effective means of trapping and storing useful amounts of energy inexpensively.

Robert K. Multer, Aidco Maine Corp.

☐ Although most chemical engineers are concerned about the cost and availability of energy, few have considered solar ponds as an energy source. One problem is that many Ch.E.s intuitively find it difficult to believe that they will work. But, they *do* work.

Making a solar pond is simple. You dig a wide hole in the ground, put some brine in it (sodium chloride brine, industrial waste, whichever is cheapest), float some fresh water on top, and stand back.

The pond gets warm. Of course it does—the sun is shining on it. But how warm? Could you calculate it? Can you guess?

In Israel, back around 1960, Harry Tabor and Rudolph Block saw one of theirs rise to a temperature of about 90°C. Not 90°F, as I first thought when I read their report, but 90°C (194°F) [1].

In 1980, Howard Bryant at the University of New Mexico boiled his pond for several weeks at 109°C (229°F), but of course that was a hot summer. Bryant [2] and Carl Nielsen [3] at Ohio State University in Columbus have been operating demonstration models of solar ponds since about 1975, and have shown that they can deliver useful amounts of heat year round in the northern U.S., even when covered with ice!

In 1980, a design study for a 30,000-ft² pond [4] intended to furnish industrial process heat to a chemicals factory in New Mexico concluded that the cost of the heat was competitive with that from local natural gas at 1980 prices.

How solar ponds work

How can something so ludicrously simple as a layer of fresh water floating on a layer of brine be a useful source of heat? Why do such ponds get hot at all? The obvious explanation might be hard to believe if based on theory alone. However, the theory [7] is confirmed by the many natural and constructed solar ponds that now exist.

Clear water, whether salty or not, is transparent to the visible part of sunlight. (Obviously it must be, else the water would not look clear.) When sunlight falls on and through a clear pond, most of the visible portion is absorbed near the bottom, especially if the bottom is dark and the pond about 8 ft deep. About half of the solar radiation that reaches the earth's surface is visible;

the rest is mostly infrared, and is absorbed in the upper 3 ft of the pond.

When the salt water in the bottom of a solar pond gets hot, why does it not rise to the surface and dump energy to the atmosphere? This does not occur, because the salinity is made high enough (typically 15 to 20% solids) so that even at the boiling temperature, the brine remains denser than the fresher, cooler water on top. A solar pond is a simple heat trap: a solar collector and energy store all in one, with convection entirely suppressed.

(The term "solar pond" has been used to describe a variety of covered and uncovered liquid pools in which solar energy is collected. The term is used in this article to describe only salt-gradient nonconvective solar ponds that are uncovered and that store heat as well as collect it.)

An ordinary pond never becomes much warmer than the air above it, because when the liquid is heated it rises and carries heat to the surface, where it is discharged to the atmosphere.

In a solar pond, hot, dense brine lies beneath cooler, less-dense liquid, so that convection is prevented. Heat losses by radiation and conduction are small enough that total heat loss is less than solar gain until temperatures rise to useful levels. Bottom temperatures can be near 95°C (200°F) at usefully high operating efficiencies. Heat storage is controllable by design, and so large as to avert the effects of fluctuations in daily and monthly sunshine.

Energy can be withdrawn by pumping hot liquid from the lowest level, or by an immersed heat exchanger (or by other means). Some chemical processes can proceed within a solar pond.

The gradients of salinity, density and temperature can be remarkably stable, and gradient maintenance needs no further study [8].

Size of solar ponds

One of the earliest solar ponds built in Israel (the one that reached 90°C) was only 3 ft deep, and had no appreciable thermal storage capacity except in the earth beneath. Such a pond might be used in a place where most seasons are sunny, and for applications in which energy demand can fluctuate during seasonal variations in sunshine. At the other extreme, some authors have suggested that large, multi-acre ponds (for winter district-heating of entire residential communities) might be as much as 20 to 30 ft deep, for immense thermal storage capacity.

Most solar ponds for year-round industrial processes are likely to be about 8 to 10 ft deep. Theoretical studies have not convincingly predicted how much heat may be usefully stored in the earth beneath a solar pond. It is known that significant heat is withdrawn from the earth as the pond is cooled below that temperature reached by the earth when the solar input greatly exceeds energy demand (during summer). It is also known that, even as far as 3 ft below a pond bottom, the earth becomes as warm as the pond brine.

Too little money has been available so far to observe earth temperature fluctuations at greater depths in real, operating ponds. However, tests may eventually show that liquid depths greater than 10 ft increase brine costs without adding to useful energy output.

Optimum pond depth will surely vary with site, local climate, and energy-demand pattern. In general, solar-pond area contributes to annual energy output, while seasonal storage capacity is a function of depth. Peak temperature depends on the combination of area and depth.

Temperature available

Solar ponds will be most cost-effective where heat can be used at low temperatures. Higher temperatures increase all of the heat losses from a pond, and diminish the percentage of input available for use. It seems certain that in the northern U.S., operating temperatures can rise to at least 180°F in late summer, and be kept above 120°F in winter. In southerly, sunny places, higher temperatures may be practical.

Widespread use of solar ponds might lead to modification of annual factory production schedules and operating temperatures, since the heat from a pond will probably cost less if process demands vary with the seasons, and provided that low-temperature processing is advantageous.

Since a wide range of temperatures will always exist from top to bottom in a solar pond, these versatile heat traps could find additional use as dumps for waste process heat. When Professor Nielsen, at Ohio State, first built a solar pond in 1975, he was searching for a way to store heat under a nonconvective brine layer; the idea of using solar heat came later.

The brine layer

The salinity required in the brine layer is determined by the operating temperature. Hotter ponds need a steeper density gradient and, therefore, more salt at the bottom of the pond. Ponds designed to operate at 120° to 180°F will probably have salt concentrations of 15 to 20%. Too low a salinity can permit instabilities in which a lower layer becomes less dense than liquid above—all nonconvective salt-gradient solar ponds require fresh water at the surface—so that a wave of hot water rises and unloads a substantial portion of the pond's stored heat to the atmosphere.

After such an event, ponds have been observed to regain their stability, and resume heating, without attention. A high enough salinity-density gradient will avoid such instabilities. Usually, brine specific gravity will not exceed 1.2. Temperatures approaching 300°F might be possible with higher densities attainable with very soluble salts, provided that heat input is great and losses small.

Theoretical predictions of density and needed salinity are difficult because of the surprising absence of published data on densities of all possible concentrations of soluble salts at the various possible temperatures. However, laboratory tests to learn the concentrations needed with any desired purchased salt, or mixture of waste salts, should not be difficult. Most solar ponds have been built with sodium or magnesium chloride, but any cheap salt will serve if it is sufficiently soluble.

Diffusion of the brine layer

There have been many theoretical studies of upward diffusion of salt, which could eventually destroy the salinity gradient. About 30 years ago, Tabor and others in Israel proposed the "falling pond" concept, in which fresh water would be added periodically to the top, while brine is withdrawn from the bottom. The withdrawn brine could be discarded and replaced by salt when this is cheap, since the amounts involved are small in relation to the value of the heat being withdrawn between recharges. Alternatively, the withdrawn brine could be recycled after concentration in a small auxiliary solar evaporation pond.

Bryant's demonstration that boiling of the lower layers in a solar pond is possible while the upper surface remains near atmospheric temperature offers the possibility of counteracting upward salt diffusion by an internal distillation. This could take place annually during a brief period at the end of summer. This should lift pure water from the depths to near the surface, while leaving the salt near the bottom.

In practice, ponds have operated with little attention to upward diffusion. Actually, solar ponds exist in nature. A 1902 paper [5] describes natural ones in Transylvania that for centuries have been at 71°C (160°F) every summer, and never cooler than 27°C (81°F).

Natural salt lakes in the southwestern U.S. are not solar ponds because the infrequent rainfall evaporates from the surface before the required gradients of density and temperature are established. Salt lakes in regions of very high rainfall might not become solar ponds if the land contours permit wide fluctuations in water level, with consequent flushing away of the gradient layer. In other climates, control of natural runoff appears to be advisable.

Solar ponds seem to need almost no maintenance if properly designed and built. The high salinity and absence of dissolved oxygen keep solar ponds optically clear and free of microorganisms. Airborne debris either sinks to the bottom, or floats until carried away by surplus rainfall. What sinks usually turns dark brown, but a pond deeper than about 5 ft, and of typical good transparency, will absorb most of the sunlight in the liquid optical path even if the bottom is reflective.

A solar pond bottom that is designed to be white, or

that becomes so from unusual, light-colored debris or precipitated salt, could generate gradient instabilities under some operating conditions.

Solar ponds for the CPI

It seems that solar ponds are more than ready for commercial application. The chemical process industries—with their engineers capable of exploiting and refining the technology, and their variety of available low-cost brines—could benefit from its early adoption on a large scale.

The CPI could obtain much of their low-temperature—65° to 95°C (150° to 200°F)—process heat from solar ponds built on space available at plantsites, and often unsuitable for other use. Pond costs could be reduced by the use of waste brines that are available at zero or negative price. Solar ponds can conveniently store waste heat available from the plants at a variety of temperatures.

As an incidental benefit, these ponds might provide a source of water for fire protection and emergency cooling. Often it will be possible to build solar ponds from waste ponds already in existence.

Costs of solar ponds

Solar ponds can be built with local labor and materials, using standard industrial pond methods. System costs depend on size and site, while energy output varies with location and energy demand.

In southwestern states, where very cheap salt is often at hand, systems can deliver up to 100,000 Btu/ft² annually, and costs may be as low as $1/ft² for large ponds. In the north, although ponds collect solar energy even during cloudy days (in contrast to most other solar heating systems), output may be nearer to 30,000 Btu/ft² annually.

With 10-year amortization, these costs can be translated into initial energy prices of less than $12/million Btu, and apparently often less than $5 [9].

Solar ponds should be practical wherever heat is needed at temperatures up to about 95°C (200°F), and with special techniques, perhaps to 150°C (300°F). They can yield any desired amount of uninterruptible "free" heat, year after year, regardless of weather, season, latitude or altitude, with very small expense for operation and maintenance, and with justifiable capital investment. With normal precautions against groundwater contamination, environmental impact is negligible or even beneficial.

Too good to be true?

Solar ponds seem almost too good to be true, and that may be one reason for the lack of exploitation since their discovery 80 years ago, and their first designed demonstration in the 1950s. The absence of any special "product" to sell may also have hindered development, but can now be a special advantage to industries that are able to design and build these systems for their own use.

If potential users of solar ponds begin planning them right now, it will be 1984 before the first are ready to scale up to valuable sizes. It will be 1990 before a plant can possibly draw the maximum possible energy from

its own full-sized solar pond(s), since a typical system will not reach equilibrium temperature until two years or more after startup.

Solar ponds have been shown to be more than three times more cost-effective than the next-best solar technology at temperatures below about 100°C (212°F).

They have existed untended, in nature, for centuries. Artificial demonstration models have operated for more than five years with no uncontrollable difficulties. They involve no unfamiliar technology, and require no new kinds of hardware or material.

Output can be predicted by methods understandable by any chemical engineer. Costs can be estimated as readily as for ordinary industrial ponds. There are no limits on size, and cost declines as size increases. It seems that pilot plants of modest size can deliver heat at prices competitive with traditional fuels now, so information needed for building large ponds can be gathered without large costs for research and development.

Now, in Israel, a 7,500-m² pond, only 2.5 m deep, delivers 150 kW in a pilot plant for a planned major power-generating station with eventual output of 2,000 MW from a 5,000-km² solar "lake." Under a mix of public and private funding, a comparable solar pond electric system is being considered for the Salton Sea in California [6].

Modest investments in solar pond technology seem like good insurance, and could turn out to be very profitable to those companies that first begin to rely heavily on the inexhaustible, low-cost heat from salt-gradient nonconvective solar ponds.

References

1. Tabor, H. Z., Solar ponds, *Solar Energy,* Vol. 7, No. 4, p. 189 (1963).
2. Zangrando, F., and Bryant, H. C., A Salt-Gradient Solar Pond, *Solar Age,* Apr. 1978, p. 21.
3. Nielsen, C. E., Experience With a Prototype Solar Pond for Space Heating, Proc. Joint Conference of American and Canadian Solar Energy Societies, Winnipeg, Canada, Vol. 5, pp. 169–182 (1976).
4. Multer, R. K., Solar Pond Energy System to Furnish Process Heat for Chemicals Factory in New Mexico (to be published in 1981 by Solar Energy Research Institute, Golden, Colo.).
5. Kalecsinski, A. V., *Annals der Physik,* Vol. 4, No. 7, p. 408 (1902).
6. Bronicki, Y. L., A Solar-Pond Power Plant, *IEEE Spectrum,* Feb. 1981, pp. 56–59.
7. Weinberger, H., The Physics of the Solar Pond., *Solar Energy,* Vol. 8, pp. 45–56 (1964).
8. Nielsen, C. E., Salt-Gradient Solar Pond Development, Proc. 3rd Annual Solar Heating and Cooling R&D Contractors' Meeting, EG-G-04-4155, Sept. 1978.
9. Multer, R. K., Solar Pond Energy Systems, *ASHRAE J.,* Vol. 22, No. 11, pp. 80–82 (1980).

The author

Robert K. Multer is a consultant for Aidco Maine Corp., Orr's Island, ME 04066, telephone (207) 833-6700, a company that he founded in 1974 to do solar engineering work. Three of his solar designs were funded by the U.S. government, and under contract with Solar Energy Research Institute/Dept. of Energy, he designed and evaluated the economics of a 30,000-ft² solar pond to provide process heat for a proposed chemicals factory in New Mexico. He holds a B.Ch.E. from Cornell and is a member of AIChE, American Soc. for Testing and Materials, American Soc. of Heating, Refrigeration and Air Conditioning Engineers, and others.

Assessing heat transfer in process-vessel jackets

Several process-vessel jacket designs are available,
and selecting the best one for an application
involves predicting jacket heat-transfer coefficients.
Here are ways to do that for complex configurations.

Donald H. Bolliger, Monsanto Chemical Intermediates Co.

☐ Jacketing a process vessel provides excellent heat
transfer in terms of efficiency, control and product
quality. However, a jacketed vessel usually costs more
than one equipped with internal coils, so its use can be
justified only when the advantages outweigh the extra
cost.

The common glass-lined reactor is an example of a
jacketed vessel. Glass-lined steel resists most corrosive
mixtures, and its smoothness reduces or eliminates fouling due to the buildup of tars, polymers or solids. The
use of glass-lined steel also eliminates the problem of
yield losses due to product decomposition on hot coils,
mechanical loss, or cross-contamination between
batches.

Jacket designs available to meet various process
heat-transfer needs include, in approximate order of
increasing cost:
- Simple jacket.
- Jacket with agitation nozzles.
- Spirally baffled jacket.
- Dimple jacket.
- Partial-pipe coil jacket.
- Integral plate- or panel-type coil jacket.

In selecting a design, the engineer looks for the best
match of heat-transfer capabilities with process needs,
at minimum cost. If the heating or cooling demand is
low, and water or steam can be used, a simple jacket
may be adequate. However, if such a jacket cannot be
used, the designer needs a way to predict heat-transfer
coefficients for more-complex designs.

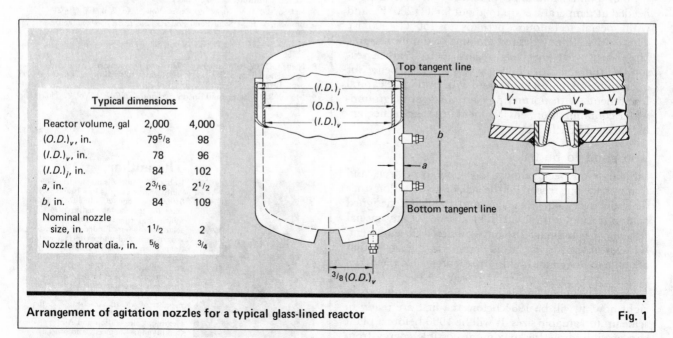

Typical dimensions		
Reactor volume, gal	2,000	4,000
$(O.D.)_v$, in.	$79^5/_8$	98
$(I.D.)_v$, in.	78	96
$(I.D.)_j$, in.	84	102
a, in.	$2^3/_{16}$	$2^1/_2$
b, in.	84	109
Nominal nozzle size, in.	$1^1/_2$	2
Nozzle throat dia., in.	$^5/_8$	$^3/_4$

Arrangement of agitation nozzles for a typical glass-lined reactor

Fig. 1

Originally published September 20, 1982

Typical heat-transfer coefficients

Heat-transfer step	Coefficient (h), Btu/(h)(ft²)(°F)
Between water (inside vessel) and vessel wall	600-730
Between organics (inside vessel) and vessel wall	40-210
Through clad vessel wall, 3/16-in. SS on 5/8-in. steel	290
Through 0.06 (± 0.02)-in. glass lining on 13/16-in. steel	81 (64-111)

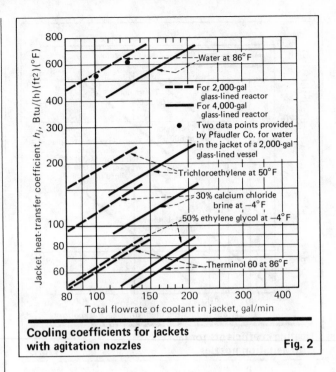

Cooling coefficients for jackets with agitation nozzles   **Fig. 2**

Heat transfer in a simple jacket

Simple jackets have a 2- to 3-in.-wide annular space, in which the flow is low—about 0.1 ft/s—and natural convection is the principal mode of heat transfer.

Uhl and Gray recommend the following equation for simple-jacket heat transfer [1]:

$$h_j/k = 0.8K''(c\mu/k)^{1/3}(\rho^2 g\beta(\delta t)/\mu^2)^{1/3} \qquad (1)$$

where h_j is the heat-transfer coefficient for flow of heat between the jacket fluid and the vessel wall.

Eq. (1) without the 0.8 constant was derived for vertical channels; Uhl and Gray recommend the 0.8 factor for jackets to compensate for the nearly horizontal surfaces at the bottom of a fully-jacketed vessel. K'' has a value of 0.15 for upward flow of a heating fluid or downward flow of a cooling fluid, and a value of 0.128 for flows in the opposite directions.

Jacket heat-transfer coefficients, h_j, calculated from Eq. (1) for brines and organic fluids fall in the range of 30–70 Btu/(h)(ft²)(°F), and a coefficient of 320 Btu/(h)(ft²)(°F) is calculated for water when the temperature difference between the fluids in the vessel and the jacket (δt) is 114°F.

The table shows typical values of heat-transfer coefficients other than h_j. The overall process of heat transfer

Nomenclature

A_f	Cross-sectional flow area in jacket, $a \times b$, ft²
A_o	Annular area for leakage in spirally baffled jacket, gap times baffle circumference, ft²
a	Width of jacket annulus, ½[(I.D.)$_j$ − (O.D.)$_v$], ft
b	Straight-side length of jacket with agitation nozzles, or spiral channel height for spirally baffled jacket, ft
c	Specific heat of fluid, Btu/(lb)(°F)
D	Duct dia., ft
D_e	Hydraulic dia., 4 × flow area/wetted perimeter, ft; for spiral flow, $D_e = 2ab/(a + b)$
F_t	Frictional flow resistance per spiral baffle turn, ft-lb$_f$/lb$_m$
f	Friction factor, dimensionless
g	Acceleration due to gravity, 32.17 ft/s² or 4.17 × 10⁸ ft/h²
g_c	Gravitational constant, 32.17 ft-lb$_m$/(lb$_f$)(s²) or 4.17 × 10⁸ ft-lb$_m$/(lb$_f$)(h²)
h	Individual heat-transfer coefficient, Btu/(h)(ft²)(°F)
h_j	Heat-transfer coefficient for flow of heat between jacket fluid and the vessel wall, Btu/(h)(ft²)(°F)
(I.D.)$_j$	Inner dia. of jacket, ft or in.
(I.D.)$_v$	Inner dia. of vessel, ft or in.
K''	Constant, dimensionless; $K'' = 0.15$ for upward flow of heating fluid or downward flow of cooling fluid; $K'' = 0.128$ for downward flow of heating fluid or upward flow of cooling fluid
k	Thermal conductivity of fluid, Btu/(h)(ft)(°F)
L	Vessel circumference, ft
N_{Pr}	Prandtl number, $c\mu/k$, dimensionless
N_{Re}	Reynolds number, $DV\rho/\mu$, dimensionless
(O.D.)$_j$	Outer dia. of jacket, ft or in.
(O.D.)$_p$	Outer dia. of pipe coil, ft or in.
(O.D.)$_v$	Outer dia. of vessel, ft or in.
t	Temperature of jacket fluid, °F
δt	Log-mean-temperature-difference between process and jacket fluids, °F
U	Overall heat-transfer coefficient, Btu/(h)(ft²)(°F)
V	Velocity of fluid in channel, ft/s
V_j	Swirl velocity of fluid downstream of nozzle, ft/s
V_n	Velocity of fluid leaving nozzle, ft/s
V_1	Velocity of fluid upstream of nozzle, ft/s
w_l	Mass flowrate of fluid leaking between baffle and jacket wall, lb/s
w_p	Mass flowrate of fluid leaving nozzle, lb/s
β	Coefficient of thermal expansion of fluid, 1/°F
μ	Fluid viscosity, lb/(ft)(h) or lb/(ft)(s)
ρ	Fluid density, lb/ft³

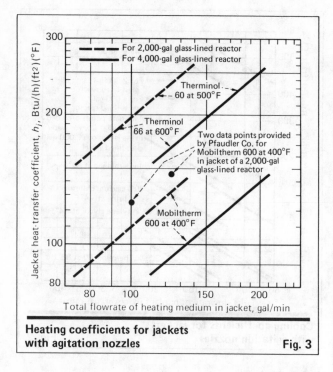

Heating coefficients for jackets with agitation nozzles   **Fig. 3**

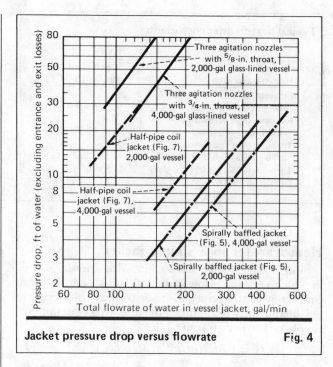

Jacket pressure drop versus flowrate   **Fig. 4**

between the vessel contents and the jacket contents involves a series of three steps—convective heat transfer between the liquid in the vessel and the vessel wall, conduction through the vessel wall, and convective heat transfer between the vessel wall and the fluid in the jacket—and the step having the lowest rate, or the lowest individual heat-transfer coefficient, h, limits the overall heat-transfer rate.

Jacket heat-transfer coefficients in the range of 30–320 Btu/(h)(ft²)(°F) often limit the overall heat-transfer coefficient, U, depending upon where in its respective range each individual heat-transfer coefficient falls.

Agitation nozzles

Agitation nozzles are used to improve jacket heat transfer for glass-lined vessels. Fig. 1 shows a typical agitation-nozzle-jacket installation. Three nozzles, all pointing in the same direction, are recommended. They produce some localized turbulence, but their main effect is to impose a spiral flow pattern tangential to the jacket wall by momentum exchange between the high-velocity stream leaving the nozzle and the jacket fluid. This momentum exchange results in "swirl velocities" in the range of 1–4 ft/s, which is high enough to cause turbulent flow and to permit the use of conventional (e.g., Sieder-Tate) equations for calculating jacket heat-transfer coefficients for forced convection.

Determining such coefficients for jackets having agitation nozzles begins with the calculation of the swirl velocity—which is eventually used, as part of the Reynolds number, to obtain h_j.

The swirl velocity, V_j, is determined by trial-and-error solution of the following equation:

$$w_p(V_n - V_j) = (4fL/D_e)(V_j^2/2)\rho A_f \qquad (2)$$

A value for the friction factor, f, is assumed and used to calculate a V_j; V_j is then used to calculate a Reynolds number, $N_{Re} = D_e V_j \rho/\mu$; finally, that N_{Re} is used to cal-

culate a new f. The process is repeated until successive values for f agree to the designer's satisfaction.

Eq. (2) is based on the following assumptions:

■ Jacket fluid velocities upstream and downstream of the agitation nozzles are equal.

■ Fluid flow area in the jacket is essentially equal to the area bounded by the straight sides of the jacket and the vessel.

■ The nozzle in the bottom head of the jacket provides the same momentum contribution as the nozzles in the straight side.

■ Local turbulence downstream of the nozzles has little effect on the average heat-transfer coefficient.

Once reasonable agreement has been achieved for Eq. (2), the jacket heat-transfer coefficient can be calculated using one of the following two equations. Eq. (3), which is an adaptation of an equation in Perry [2], applies to water, and Eq. (4), which is an adaptation of the Sieder-Tate equation given by Kern [3], applies to brines and organics.

$$h_j = 91(1 + 0.011t)V_j^{0.8}/D_e^{0.2} \qquad (3)$$

$$h_j D_e/k = 0.027(N_{Re})^{0.8}(N_{Pr})^{1/3} \qquad (4)$$

The results of solving Eq. (2)–(4) for various heat-transfer fluids are plotted in Fig. 2 and 3. Fig. 2 relates h_j to total jacket flowrate for cooling fluids; Fig. 3, for heating fluids. Both figures are based on the use of three agitation nozzles installed as shown in Fig. 1. In all of the calculations, V_n is the total flow at one nozzle, which handles one-third of the total jacket flow.

This method agrees reasonably well—±15%—with experimental data obtained from a manufacturer of glass-lined reactors for water (Fig. 2, Curve 1) and Mobiltherm 600 (Fig. 3, Curve 2).

As can be seen in Fig. 2 and 3, agitation nozzles result in jacket heat-transfer coefficients two or three times higher than those in simple jackets, which range from 30–320 Btu/(h)(ft²)(°F). However, the price paid for

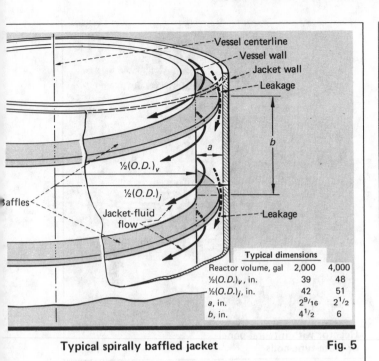

Typical dimensions		
Reactor volume, gal	2,000	4,000
½(O.D.)$_v$, in.	39	48
½(O.D.)$_j$, in.	42	51
a, in.	2⁹⁄₁₆	2¹⁄₂
b, in.	4¹⁄₂	6

Typical spirally baffled jacket   Fig. 5

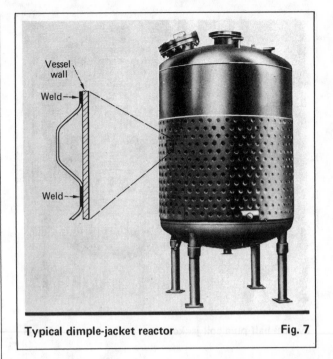

Typical dimple-jacket reactor   Fig. 7

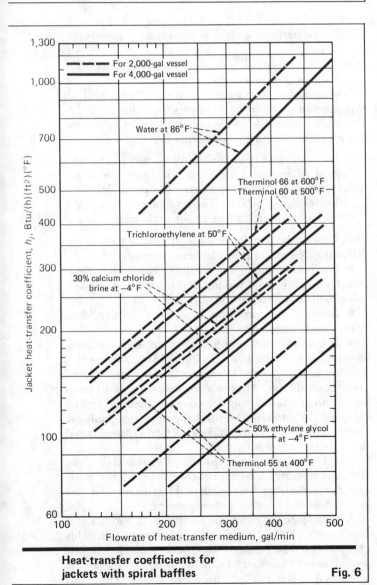

Heat-transfer coefficients for jackets with spiral baffles   Fig. 6

this increased jacket heat-transfer is more pumping energy required to overcome nozzle pressure drop. Fig. 4 relates nozzle pressure drop to flow through the nozzle, and is based on a manufacturer's data [4]. (Fig. 4 also shows pressure drop vs. flowrate for other jacket configurations that will be discussed later in this article.)

Because agitation nozzles inject fluid at several points in a jacket, it is probable that the jacket fluid is relatively well-mixed. For this reason, the outlet temperature of the jacket fluid should be used in calculating the heat-transfer driving force.

Spirally baffled jackets

A spiral baffle consists of a metal strip spirally wound around a vessel wall from jacket entrance to exit. This strip directs flow in a spiral path to obtain fluid velocities in the range of 1–4 ft/s. Fig. 5 illustrates one full turn of a spiral baffle.

After the baffle strip is welded to the vessel, the jacket is slipped over the assembly, drawn tight with banding or temporary lugs, and welded longitudinally and circumferentially. This fabrication method leaves a gap between the baffle and jacket wall. The gap is a second path for fluid flow from the jacket inlet to outlet, perpendicular to the desired path along the baffle. The fluid that bypasses the spiral baffle through the gap (leakage) does not contribute directly to heat transfer at the vessel wall.

The area for leakage is relatively large. For a 2,000-gal reactor, a ¹⁄₃₂-in. gap between the baffle and jacket wall is equivalent to a flow area of 8.3 in.², compared to an 11-in.² flow area along the desired spiral path.

Design calculations for a spirally baffled jacket begin with an assumed flow velocity, V, in the range of 1–4 ft/s. This is used to determine the jacket-side heat-transfer coefficient, using Eq. (3) or (4).

Once heat-transfer requirements have been satisfied, the pressure drop due to frictional resistance for flow

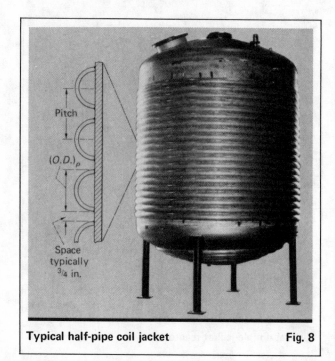

Typical half-pipe coil jacket   Fig. 8

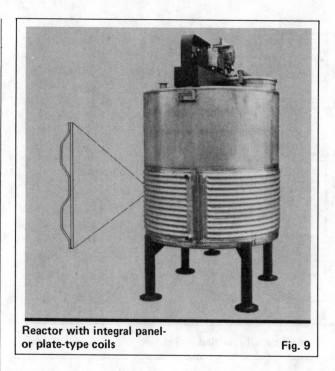

**Reactor with integral panel-
or plate-type coils**   Fig. 9

around one turn of the spiral, which is equal to the pressure drop causing leakage between adjacent turns, is calculated from:

$$F_t = (4fL/D_e)(V^2/2g_c) \qquad (5)$$

This pressure drop is then used to determine leakage flow by the following conservative adaptation of the standard annular-orifice-flow equation:

$$w_l = 0.65A_o(g_c F_t \rho^2)^{0.5} \qquad (6)$$

The total required flow is the sum of the leakage rate plus the channel flow (expressed in consistent units).

In general, leakage flow amounts to one-third to one-half of the total flow circulated to a spirally baffled jacket. The total jacket pressure drop is the pressure drop per turn, F_t, multiplied by the number of turns, plus jacket nozzle entrance and exit losses.

Fig. 6 relates heat-transfer coefficients to total jacket flowrate for several heat-transfer media in spirally baffled jackets. Data on pressure drop vs. water flowrate are plotted in Fig. 4.

In comparison to agitation nozzles, spiral baffles require up to twice the flow, but result in only one-fifth the pressure drop, for equivalent heat-transfer rates. Spirally baffled jackets, therefore, require about 40% of the energy needed by jackets with agitation nozzles.

The flow pattern in spirally baffled jackets approximates plug flow, so the log-mean-temperature-difference should be used in calculating the heat-transfer driving force.

Other jacket designs

The jacket designs described thus far—the simple jacket, the jacket with agitation nozzles, and the spirally baffled jacket—are structurally weak because the jacket pressure is applied over a large area of unsupported metal. For higher jacket-fluid pressures (over about 150 psig), one should consider an alternative design—the

dimple, the partial-pipe coil, or the plate- or panel-type coil jacket:

Dimple jackets can withstand pressures up to 300 psig. Fig. 7 shows a dimple-jacket reactor.

Unfortunately, few data on pressure drop, and no experimental data on heat transfer, in a dimple jacket are readily available. A representative of a fabricator has made the following comments [5]:

"... the velocity in the dimple jacket has to be limited to around 2 ft/s, due to higher pressure drop per foot (the dimples create a great deal of turbulence) ... [which] is approximately 10 to 12 times higher than that of an open channel."

In view of the scarcity of data on dimple jackets, the following design approach is recommended:

■ For preliminary vessel sizing, assume that heat-transfer coefficients at the low end of the range for spirally baffled jackets can be obtained.

■ Request pressure-drop and heat-transfer data from vessel fabricators.

■ If manufacturer-supplied data are not available, assume a maximum flow velocity of 2 ft/s, and use this velocity in Eq. (3) or (4) to calculate the jacket heat-transfer coefficient. Assume a pressure drop of 10 times that of an annulus having the same cross-section and no dimples. To estimate the total flow volume, multiply the smaller of the horizontal or vertical cross-sectional areas at the dimples by 2 ft/s.

This is a conservative approach. Reliable data from the manufacturer would probably show higher heat-transfer coefficients than calculated by this method, but might also show higher pressure drops.

A *partial-pipe coil jacket*, shown in Fig. 8, consists of a continuous channel welded to the vessel wall. This channel could be made of structural channel steel, but would more commonly be made from a pipe whose cross-section was cut into halves or thirds. Thus, this type of jacket is called a half-pipe or third-pipe coil jacket, depending on the amount of the circumference of the pipe welded to the vessel wall.

Typical flowrates in a half-pipe coil are in the range of 2½ to 5 ft/s. Heat-transfer coefficients can be calculated by assuming various velocities in this range and using Eq. (3) or (4). The lowest velocity that results in a suitable transfer of heat (in turbulent flow) should be selected.

Coil pressure drop (excluding entrance and exit losses) can be obtained for water from Fig. 4; the pressure drop for other fluids or other jacket arrangements can be calculated using Eq. (5). A safe estimate for total losses at one entrance and one exit is 1.5 velocity heads—$1.5V^2/2g_c$, in ft of fluid— where V is the fluid velocity in the partial-pipe coil channel.

At the same flowrates, pressure drops for half-pipe coil jackets are less than those for agitation nozzles, but more than those for spirally baffled jackets.

Panel-type coils or *plate-type coils* are fabricated from two metal plates. For reactors, one plate is smooth and forms the vessel wall. The outer plate is embossed to form a series of flowpaths between the plates. Fig. 9 illustrates a panel-coil jacket.

A vessel made from panel- or plate-type coil can have a thinner inner wall than a conventional vessel because the jacket pressure does not affect the design of the inner wall. This advantage may be offset by the need to make the outer plate of the same corrosion-resistant material as the inner plate. (This could occur if the vessel wall were made of stainless steel; if the outer plate were made of less-expensive carbon steel, the impurities in the carbon steel could migrate into the stainless and cause corrosion of the inner plate. This situation can also occur in dimple and partial-pipe coil jackets where the jacket and vessel walls are in direct contact.)

Single-embossed panel- or plate-coils have a relatively high pressure drop per square foot of heat-transfer area. For example, one 29x143-in. panel (28.8 ft²) has a 10-psi pressure drop for a flow of 14 gal/min (4.3 ft/s), or 20 psi for 20 gal/min (6.2 ft/s). A 4,000-gal vessel would require eight such panels to allow a flow of 112 gal/min at a pressure drop of 9–15 psi. This 8-panel arrangement would need 16 piping connections, as compared with 2 to 4 for a conventional reactor.

As with partial-pipe coils, typical flows are in the range of 2½ to 5 ft/s, and Eq. (3) or (4) can be used to calculate heat-transfer coefficients for several assumed velocities in this range. The lowest rate that provides suitable heat transfer should then be selected.

Vendor catalogs provide coil dimensions and data on pressure drop per panel for water at various flowrates; pressure drops for other fluids can be estimated by multiplying the pressure drop for water by the specific gravity of the other fluid. (Vendor-supplied pressure ratings generally include entrance and exit losses.) The total pressure drop across the jacket is the pressure drop for one coil times the number of coils in series in the jacket arrangement.

Selection of a jacket design

When overall heat transfer is limited by simple-jacket heat transfer, h_j can be increased by using one of the more complex jacket designs. The selection of a jacket design depends on several factors:

- Cost. (The jacket designs were discussed above in approximate order of increasing cost.)
- Heat-transfer requirements.
- The heat-transfer medium.
- Pressure limitations. Simple, spirally baffled, agitation-nozzle, and panel- or plate-type coil jackets ordinarily would not be used at pressures greater than 150 psi; dimple jackets, 300 psi; and half-pipe coil jackets, depending on the design, 1,000 psi. For vessels where the internal pressure is more than twice the jacket pressure, the spirally baffled jacket will generally be the most economical [5].
- Temperature limitations. High temperatures and large differences between the vessel- and jacket-fluid temperatures require consideration of (a) thermal-expansion differentials between the materials used for the vessel and jacket, and (b) the difference in thickness between the vessel and jacket walls. In some cases, special materials may be required; or the vessel and jacket may have to be made of the same material, or of materials having the same coefficient of thermal expansion; or special closures that can withstand the stresses of expansion may be needed at the top and bottom of the jacket [5].
- Suitability. For example, glass-lined vessels can use only agitation nozzles or partial-pipe coils.
- Other design considerations. One such is the need for a high heat-transfer-fluid flowrate for "hot" reactions, or for rapid heatup or cooldown for batch reactors. Heatup or cooldown time for batch reactors is proportional to the ratio:

$$\frac{\text{weight} \times \text{specific heat of batch liquid}}{\text{mass flowrate} \times \text{specific heat of heat-transfer media}}$$

If a high heat-transfer-fluid flowrate is needed, then a spirally baffled or enlarged partial-pipe coil jacket may be necessary.

References

1. Uhl, V. W., and Gray, J. B., "Mixing: Theory and Practice," Vol. I, p. 314, Academic Press, New York, 1966.
2. Perry, J. H., ed., "Chemical Engineers' Handbook," 4th ed., p. 10–14, McGraw-Hill, New York, 1963.
3. Kern, D. Q., "Process Heat Transfer," p. 103, McGraw-Hill, New York, 1950.
4. The Pfaudler Co., "The Pfaudler Agitating Nozzle," Bulletin 950, 1957.
5. Markovitz, R. E., "Picking the Best Vessel Jacket," *Chem. Eng.*, Nov. 15, 1971, pp. 156–162.

The author

Donald H. Bolliger is Manager, Process Chemicals R & D, Monsanto Chemical Intermediates Co., 800 N. Lindbergh Blvd., St. Louis, MO 63166, (314) 694-4951. He received a B.S. in chemical engineering from the U. of Michigan, an M.S. in chemical engineering and a Master's in engineering administration from Washington U. Employed by Monsanto since 1950, he has done plant engineering and project supervision, process design, project management, and research and development work. He is a licensed professional engineer in Missouri, and a member of AIChE and the Engineers' Club of St. Louis.

Compressor intercoolers and aftercoolers: Predicting off-performance

Here is a simplified method for estimating how the operation of heat exchangers used with air and gas compressors changes when service conditions differ from the original design.

Peter Y. Burke, Sundstrand Fluid Handling

☐ Intercoolers and aftercoolers (IC/AC), which are heat exchangers in gas compression systems,* control the temperature of the compressed gas flowing to a downstream process or to another compression step. In the case of using an intercooler before a second compressor, lowering the gas temperature before it enters the unit reduces mechanical deterioration, and more-efficient compression can take place.

An example of an aftercooler application is the liquefaction of chlorine. Predried chlorine gas, from electrolytic cells, enters a compressor (shown in Fig. 1) and is precooled in a heat exchanger before being liquefied in a Freon-cooled condenser.

There are three basic IC/AC configurations:

In the *counterflow* type, the gas and liquid coolant flow essentially in opposite directions. This type is used in most commercial IC/AC designs because it generally requires less surface area for a given heat flowrate. Fig. 2 shows this type of IC/AC.

In the *parallel-flow* type, the two fluids move essentially in the same direction.

In the *cross-flow* type, the fluids move at right angles with respect to each other. The design and application of these configurations, which vary extensively, are covered in most heat-transfer texts.

*The above photo shows a vertical, inline compressor that might be used in such systems.

Heat exchanger manufacturers who design and build intercoolers and aftercoolers for the compressed air and gas industry use proprietary procedures and data to arrive at optimum designs. However, like many engineering calculations, the optimum is based on a single set of operating conditions specified by the compressor manufacturer or end-user. Unfortunately, the actual operating conditions for the IC/AC often do not correspond with the "design point." The reasons for this are many—e.g., environmental changes, incorrect initial specifications, system changes, and performance efficiency reductions.

Dealing with performance changes

The plant engineer must often react quickly when faced with unexplainable IC/AC performance shifts, or when changes (flowrate or temperature) are proposed in the plant cooling-water system. A precise estimate would necessitate calculation of the overall heat-transfer coefficient, U. And this requires evaluation of parameters such as: (a) gas-side film coefficient, (b) water-side film coefficient, (c) fin efficiency, (d) geometric factors, and (e) tube conductivity.

While IC/AC manufacturers are well equipped to predict performance changes, in many cases quick and less-sophisticated estimates produce acceptable results. The suggested evaluation method is based on several

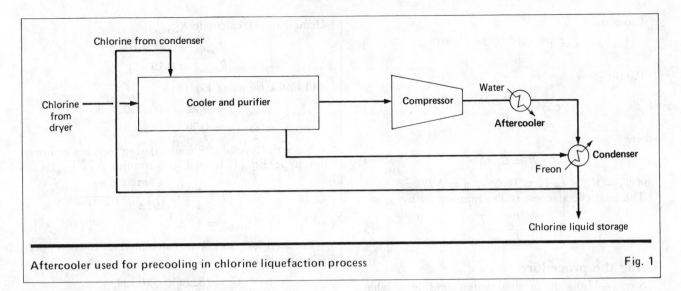

Chlorine from condenser

Chlorine from dryer → Cooler and purifier → Compressor → Water → Aftercooler

Freon → Condenser

Chlorine liquid storage

Aftercooler used for precooling in chlorine liquefaction process   Fig. 1

assumptions regarding the IC/AC applications, and assumes no ready access to their design intricacies. The assumptions are:

■ The performance of the IC/AC at design conditions is known from the manufacture's predictions or experimentation.

■ The overall heat-transfer coefficient, U, remains constant for the original exchanger design and new operating conditions.

■ The off-design conditions result from reasonable deviations in flowrates for both the hot gas and coolant (i.e., changes of not more than 25%).

■ The fluids entering the IC/AC are unmixed, traveling separate paths by tubes or channels.

Basis of the method

This method uses NTU (number of transfer units) and heat-exchanger-effectiveness relationships to predict performance at an alternate operating point by comparing the actual IC/AC performance to that of an ideal model [design] performance, and determining how the ideal model performance is affected by changing parameters. The parameters of concern generally are the gas and liquid flowrates, and entering and exiting temperature of both gas and coolant.

In instances where other parameters, such as relative humidity and operating pressure, deviate from design conditions, their effect on performance should be evaluated by other known methods. However, these parameters will not have a significant impact on the procedure discussed here.

The standard NTU and heat-exchanger relationships used by exchanger designers are as follows for the three types of exchangers [1]:

Counterflow

$$\epsilon = \frac{1 - e^{-NTU(1-R)}}{1 - Re^{-NTU(1-R)}} \qquad (1)$$

Counterflow is one of three exchanger types   Fig. 2

T_{g1}

T_{l2}

T_{l1}

T_{g2}

Nomenclature

A	Heat-transfer area, ft^2
C	Flowstream capacity rate, Btu/(h)(°F)
c	Specific heat, Btu/(lb)(°F)
NTU	Number of transfer units, dimensionless, $\left(NTU = \dfrac{UA}{C_g}\right)$
R	Capacity rate ratio, dimensionless
T	Temperature, °F
U	Overall heat-transfer coefficient, Btu/(h)(ft^2)(°F)
W	Mass flowrate, lb/h
ϵ	Heat exchanger effectiveness, dimensionless
η	$NTU^{-.22}$

Subscripts

1	Entering condition
2	Exiting condition
g	gas
l	liquid

Crossflow

$$\epsilon = 1 - e^{[[e^{(-R\eta NTU)} - 1]\, 1/\eta R]} \qquad (2)$$

Parallel-flow

$$\epsilon = \frac{1 - e^{-NTU(1+R)}}{1 + R} \qquad (3)$$

where:

$$R = C_g/C_l \qquad (4)$$

and $C_g = W_g \times c_g$, $C_l = W_l \times c_l$, $\eta = NTU^{-.22}$
The heat effectiveness is also represented by:

$$\epsilon = \frac{T_{g1} - T_{g2}}{T_{g1} - T_{l1}} \qquad (5)$$

Using the procedure

Step 1—Using the original design conditions, calculate C_g, C_l, R (Eq. 4), and ϵ (Eq. 5).

Step 2—Now using the appropriate heat-exchanger-effectiveness relationship—Eq. (1), (2) or (3)—calculate the design NTU.

Step 3—After calculating C_g and C_l for the new operating conditions, determine the NTU from the following relationship:

$$NTU_{new} = NTU_{design} \times \frac{C_{g\,design}}{C_{g\,new}} \qquad (6)$$

which is obtained from $NTU = UA/C_g$, when UA is assumed constant.

Step 4—After recalculating R for the new conditions, calculate the effectiveness, using the same equation used in Step 2.

Step 5—The new effectiveness can be used in Eq. (5), which is then solved for the exiting-gas temperature, T_{g2}.

An example

To illustrate the technique, let us calculate the change in performance (T_{g2}) for a water-cooled counterflow aftercooler used on an air compressor. The original design and new conditions are:

	Design	New
W_g	53.5 lb/min	62.06 lb/min
W_l	8.0 gal/min	6.0 gal/min
T_{g1}	250°F	275°F
T_{g2}	100°F	?
T_{l1}	60°F	80°F

and $c_g = 0.241$ Btu/(lb)(F) for air; $c_l = 0.999$ Btu/(lb)(F) for water at 60°F.

Following Step 1, the design parameters C_g and C_l are determined:

$$C_g = (53.5)(60)(0.241)$$
$$C_g = 774 \text{ Btu/(h)(°F)}$$

and

$$C_l = (8.0)(60)(62.4)(0.134)(0.999)$$
$$C_l = 4014 \text{ Btu/(h)(°F)}$$

Using Eq. (4), calculate R_{design}:

$$R_{design} = 774/4014$$
$$R_{design} = 0.193$$

and find ϵ by using Eq. (5)

$$\epsilon_{design} = 250 - 100/250 - 60$$
$$\epsilon_{design} = 0.79$$

Since the type of aftercooler design used is a counterflow type, Eq. (1) is used to determine NTU_{design}:

$$0.79 = \frac{1 - e^{-NTU(1-0.193)}}{1 - 0.193\, e^{-NTU(1-0.193)}}$$

$$NTU_{design} = 1.722$$

For Step 3, C_g and C_l are calculated for the new operating conditions:

$$C_g = (62.06)(60)(0.241)$$
$$C_g = 897 \text{ Btu/(h)(°F)}$$

and

$$C_l = (6.0)(62.4)(60)(0.134)(0.999)$$
$$C_l = 3010 \text{ Btu/(h)(°F)}$$

NTU_{new} can now be calculated from Eq. (6):

$$NTU_{new} = 1.73 \times (774/897)$$
$$NTU_{new} = 1.49$$

The new R is found using Eq. (4) and is then used in Eq. (1) to determine the new ϵ:

$$R_{new} = 897/3010$$
$$R_{new} = 0.298$$

and

$$\epsilon_{new} = \frac{1 - e^{-1.49(1-0.298)}}{1 - 0.298 e^{-1.49(1-0.298)}}$$

$$\epsilon_{new} = 0.725$$

Finally, the new ϵ is used in Eq. (5), which is solved for T_{g2}:

$$0.725 = (275 - T_{g2}/275 - 80)$$
$$T_{g2} = 134°F$$

References

1. Desmond, R. M., and Karlekar, B. V., "Engineering Heat Transfer," West Publishing Co., St. Paul, Minn., 1977.

The author

Peter Y. Burke is director of engineering at Sundstrand Fluid Handling, a unit of Sundstrand Corp., P.O. Box FH, Arvada, CO 80004; telephone: 303-425-0800. A professional engineer in New York and Maryland, he previously worked as a product manager at Worthington Engineered Pump, a division of the Worthington Pump Co. He holds a B.S. in mechanical engineering from Virginia Polytechnic Institute and an M.S. in the same subject from Rensselaer Polytechnic Institute. He is a member of the National Management Assn.

Are liquid thermal-relief valves needed?

Thermal stresses, or pressures from hydrogen evolution due to corrosion, can rupture a pipe or tube. Here is how to find such forces to see whether a relief valve is needed.

Sudhir R. Brahmbhatt, *MG Industries*

☐ In protecting process pipelines, engineers seldom perform calculations to see whether thermal stresses or pressure buildups (as a result of H_2 evolution from corrosion) exist. Such neglect could lead to rupture, especially if relief valves have not been employed.

Here, we present a calculation method that will determine whether or not a relief valve is required. An example illustrates the method.

Thermal expansion and corrosion

In the case of thermal expansion, when pipes, tubes or process equipment are full of liquid and are blocked in, rupture can result from heat from several sources:

■ Solar radiation.

■ Heat-tracing coils.

■ Heat transfer by radiation and conduction from nearby process equipment.

A typical example would be a heat exchanger blocked-in on the cold side, with flow continuing on the hot side. Such a situation could occur during normal operation, e.g., when a hot product-stream was being pumped through an exchanger to a flash drum and the stream was being used to preheat the feed to a reactor. If a level controller that operated a control valve between the exchanger and flash drum were to close the control valve, the feed would be blocked in.

Then, of course, the system can be subject to thermal expansion. Stresses result from the difference in coefficients of thermal expansion between the liquid and the metal.

On the other hand, corrosion, which generates hydrogen, can also give rise to severe stresses caused by gas pressure. Such a case may occur when, for example, the flow of sulfuric acid is stopped in a steel pipeline and it remains full. After some time, enough hydrogen may be generated to create pressures that will burst the pipe.

To see whether a relief valve is needed, the stresses due to thermal expansion and corrosion must first be found. Here is a method to do this:

Determining thermal stresses

The net volume of liquid that expands is found by subtracting the pipe expansion due to both temperature rise and pressure rise from the liquid volume expansion due to temperature rise:

$$\Delta V = (V_t - V_{to}) - \Delta V_p - \Delta V_{Pr} \qquad (1)$$

The volumetric expansion of the liquid at temperature t above a reference temperature is given by:

$$V_t = V_{tr}(1 + at + bt^2 + ct^3) \qquad (2)$$

The reference temperature tr is 0°C. Values of a, b, and c are given for some liquids in Table I.

When a liquid expands from temperature to to temperature t, its volume at t is:

$$V_t = V_{to} + V_{tr}[a(t - to) + b(t^2 - to^2) + c(t^3 - to^3)] \qquad (3)$$

The pipe expansion due to temperature is given by:

$$\Delta V_p = \pi/4[(D + CD)^2(L + CL) - D^2L] \qquad (4)$$

In Eq. (4), L is the total pipe length including fittings, i.e., it must account for the total restricted fluid volume. Typical values of C appear in Table II.

The increase in pressure of the liquid—due either to heat or to gas buildup—will expand the pipe further [1]. Such an increase is usually minimal when compared with the effects of heat alone, and can be ignored in making calculations. However, an equation for this effect (pipe expansion due to pressure) is presented:

$$\Delta V_{Pr} = [\pi (\Delta R)^2] L + \pi R^2 [\Delta L]$$
$$= \pi L \left[\frac{R}{E} \left(\frac{PR}{T_h} - \gamma \frac{PR}{2T_h} \right) \right]^2 + \pi R^2 \left[P \frac{L}{E} \frac{A_{ID}}{A_{metal}} \right] \qquad (5)$$

Poisson's ratio, γ, is 0.3 for steel and Alloy 20, as given in the American Natl. Standards Institute (ANSI), New York, ANSI Standard 31.3. Liquid compressibility is defined as:

$$\beta = - \left(\frac{\text{Final volume} - \text{Initial volume}}{\text{Final pressure} - \text{Initial pressure}} \right) \times \qquad (6)$$
$$\left(\frac{1}{\text{Initial volume}} \right)$$

For a liquid of volume V_t in a pipe of fixed volume V_{to}, β is expressed as:

$$\beta = - \left(\frac{V_{to} - V_t}{\Delta P} \frac{1}{V_t} \right) \qquad (7)$$

Originally published May 14, 1984

Coefficients of cubical expansion for some typical liquids   Table I

Liquid	Range, °C	$a*$, 10^{-3}	b, 10^{-6}	c, 10^{-8}
Benzene	11- 81	1.17626	1.27755	0.80646
Toluene	0-100	1.028	1.779	—
Methyl alcohol	−38-+70	1.18557	1.56493	0.91113
n-Butyl alcohol	6-108	0.83751	2.8634	-0.12415
iso-Propyl alcohol	0- 83	1.04345	0.44303	2.7274
Chloroform	0- 63	1.10715	4.66473	-1.74328
Carbon tetrachloride	0- 76	1.18384	0.89881	1.35135
Ethyl acetate	−36-+72	1.2585	2.95688	0.14922
Hydrochloric acid, 33.2%	0- 33	0.4460	0.215	—
Sulfuric acid, conc.	0- 60	0.5758	-0.864	—
n-Pentane	−190-+30	1.50697	3.435	0.975

The coefficients are for $t_r = 0°C$

*For example, for benzene: $V = V_{tr}[1 + 1.17626 \times 10^{-3} t + 1.2775 \times 10^{-6} t^2 + 0.80646 \times 10^{-8} t^3]$

From Ref. [2].

Coefficients of linear thermal expansion for a few typical alloys   Table II

Temperature, °F	Carbon steel	18-8 stainless steel	25-Cr 20-Ni	Wrought iron
50	−0.14	−0.21	−0.16	−0.16
70	0	0	0	0
100	0.23	0.34	0.28	0.26
125	0.42	0.62	0.51	0.48
200	0.99	1.46	1.21	1.14
250	1.40	2.03	1.70	1.60
300	1.82	2.61	2.18	2.06
400	2.70	3.80	3.20	3.01
500	3.62	5.01	4.24	3.99
700	5.63	7.50	6.44	6.06
900	7.81	10.12	8.78	8.26

From Ref. [5].

Compressibilities of some selected liquids   Table III

Liquid	Temperature, °C	Pressure, atm	β, 10^{-6}, atm
Ethyl alcohol	20	1-50	112
	20	200-300	86
Methyl alcohol	0	1-500	79.4
n-Butyl alcohol	17.4	8	90
iso-Butyl alcohol	17.95	8	98
n-Propyl alcohol	0	1-500	69
iso-Propyl alcohol	5.65	8	95
	17.85	8	103
Chloroform	20	0-98.7	94.9
	100	8-9	211
Ethyl acetate	13.3	8.1-37.4	104
	99.6	8.13-37.15	250
Benzene	20	1-2	95.3
	20	98.7-197.4	58.4
Toluene	10	1-5.25	79
	100	1-5.25	150
Sulfuric acid, 70%, wt.	60.3	(wide range)	50-100
Water	0	1-25	52.5
	100	100-200	46.8

From Ref. [2, 4].

Nomenclature

a, b, c	Coefficients for Eq. (2)
A	Surface area of pipe in contact with liquid, ft²
A_{ID}	Pipe cross-sectional area, ft²
A_{metal}	Cross-sectional area of pipe metal (of the "ring" of a section of pipe), ft²
C	Coefficient of linear thermal expansion of pipe material, ft/ft
C_R	Corrosion rate, mils/yr
D	Pipe inside dia., ft
E	Modulus of elasticity, lb/in²
H	Henry's law constant, atm
L	Pipe length between block valves, ft
L_m	Gas evolved, moles H_2/wk
M_p	Hydrogen equivalent weight of pipe material, moles
P	Pressure, psi or atm
ΔP	Pressure change due to thermal stresses, psi or atm
ΔP_c	Pressure change due to corrosion, psi or atm
R	Pipe radius, ft
T_h	Pipe thickness, ft
t	Temperature, °F or °C
t_{avg}	Average temperature, °F or °C
t_r	Reference temperature, here 0°C
V_t	Volume of liquid at t, ft³
V_{to}	Volume of liquid at to, ft³
V_{tr}	Volume of liquid at tr, ft³
ΔV	Net volume change, ft³
ΔV_p	Pipe volume change due to temperature rise, ft³
ΔV_{pr}	Pipe volume change due to pressure rise, ft³
$\Delta V_{thermal}$	Liquid volume expansion due to temperature rise, ft³
X_L	Amount of liquid in pipe, moles
Greek letters	
β	Liquid compressibility, atm^{-1}
γ	Poisson's ratio
ρ_p	Pipe material density, lb/ft³

Eq. (7) can be rearranged and expressed in terms of the coefficients of liquid volume expansion to yield the following expression:

$$\Delta P = \left(\frac{1}{\beta}\right) \left(\frac{a(t - to) + b(t^2 - to^2) + c(t^3 - to^3)}{1 + at + bt^2 + ct^3}\right) \quad (8)$$

Usually, $(V_t - V_{to})$ is much greater than ΔV_p and ΔV_{Pr}. Thus, Eq. (7) is generally sufficient for calculating the pressure rise due to thermal stresses.

Determining stresses due to corrosion

In an aqueous medium, the corrosion reaction of a metal may be represented by:

$$\text{Metal} + H_2O \rightarrow H_2 + \text{Metal oxide} \quad (9)$$

One mole of hydrogen is released for every mole of

bivalent metal that reacts. If L_m is the moles of hydrogen evolved/wk, then:

$$L_m = \left(C_R, \frac{mils}{yr}\right)\left(\frac{in.}{1,000\ mils}\right)\left(\frac{ft}{12\ in.}\right)(A)(\rho_p) \times$$
$$\left(\frac{mole}{M_p}\right)\left(\frac{yr}{52\ wk}\right) =$$
$$\left(\frac{1}{624\times10^3}\right)\left(\frac{C_R\,A\,\rho_p}{M_p}\right),\ moles/wk \qquad (10)$$

The number of moles of H_2 evolved will vary if the valence of the metal is different. Such an effect must be accounted for in calculations.

Using Henry's law (i.e., the solubility of a gas in a liquid—its mole fraction—is directly proportional to the pressure of the gas above the liquid), the pressure rise is:

$$\Delta P_c = H\left(\frac{L_M}{X_L + L_M}\right) \qquad (11)$$

H, Henry's law constant, is available in the literature. Eq. (11) is based on the following assumptions: (1) H is not a function of pressure; (2) the hydrogen gas-phase is ideal; (3) the pipe is full of liquid and the liquid is incompressible; and (4) the pipe is rigid.

Eq. (11) gives an approximate pressure rise for one week with no flow in the pipe.

Example

A 4-in. Alloy 20, Sch. 10 pipeline is used to transport 70% H_2SO_4. The pipe's pressure rating is 500 psi and its I.D. is 4.26 in. The temperature is expected to rise occasionally to 130°F (54.8°C) from its normal value of 100°F (38.1°C). The pipeline could be out of service for up to one week. The distance between block valves is 9,900 ft. Is a thermal-relief valve needed?

Consider the worst case—130°F and 1 wk. From Table I, for 70% H_2SO_4, use the volume for concentrated sulfuric acid, which is expressed as:

$$V_t = V_{to} + V_{tr}[0.5758 \times 10^{-3}$$
$$(t - to) - 0.864 \times 10^{-6}(t^2 - to^2)] \qquad (12)$$

Eq. (12) is written in °C. The coefficient of linear thermal expansion for Alloy 20 (25Cr, 20Ni) is, from Table II, 0.556 in./100 ft, or 46.3×10^{-3} ft/100 ft.

Calculate the liquid volume expansion due to the temperature rise. In Eq. (12), substituting the following:
$$t - to = 16.7\ °C,\ and\ to = 38.1\ °C$$
and rearranging, yields:

$$\left(\frac{V_t - V_{to}}{V_{to}}\right) =$$
$$\left(\frac{0.5758\times10^{-3}(16.7)-0.864\times10^{-6}(54.8^2-38.1^2)}{1+0.5758\times10^{-3}(38.1)-0.864\times10^{-6}(38.1)^2}\right) =$$
$$\frac{0.008275}{1.020684} = 0.008107 \qquad (13)$$

The initial volume of acid in the pipe is:

$$V_{to} = (\pi/4)D^2L =$$
$$(\pi/4)(4.26/12)^2(9,900) = 979.9\ ft^3 \qquad (14)$$

And the change in volume is:

$$\Delta V_t = 979.9 \times 0.008107 = 7.94\ ft^2 \qquad (15)$$

Calculate the pressure rise due to the liquid volume expansion. Rewriting Eq. (7) and substituting values from Eq. (13) into it:

$$\Delta P = -\left(\frac{V_{to} - V_t}{\beta}\right)\left(\frac{1}{V_t}\right) =$$
$$-\left(\frac{V_{to} - (V_{to} + 0.008107\ V_{to})}{V_{to}(1.008107)}\right)\left(\frac{1}{\beta}\right) =$$
$$\left(\frac{0.008042}{\beta}\right) \qquad (16)$$

From Table III, β for 70% H_2SO_4 is: $50 \times 10^{-6} < \beta < 100 \times 10^{-6}$. Thus, substituting into Eq. (16) yields: 80 atm $< \Delta P <$ 160 atm. Calculations for ΔV_{Pr} and ΔV_p are omitted, since they are assumed to be much smaller than the above-calculated pressure rise. Now, the pressure rise due to corrosion is calculated.

The corrosion rate for Alloy 20 in H_2SO_4 at 54.8°C is about 5 mils/yr. Assume that this value is accurate enough. The density of Alloy 20 is 499 lb/ft³, and M_p is 50.8. M_p is calculated by multiplying the moles × valence/2 for each element in an alloy and adding up the total of such terms. Substituting into Eq. (10):

$$L_m = \left(\frac{1}{624 \times 10^3}\right)\left(\frac{5\pi\,(4.26/12)\,9,900 \times 499}{50.8}\right) =$$
$$0.87\ moles\ of\ H_2\ generated/wk \qquad (17)$$

The moles of H_2SO_4 in the pipe are:

979.9 ft³ × 115.19 lb/ft³ × (1/98) mole/lb = 1,152 moles; where 115.19 lb/ft³ is the density of sulfuric acid and 98 is its molecular weight.

Henry's law constant for H_2 in water at 20°C is 6.83×10^4 atm. Thus:
$\Delta P_c = (6.83 \times 10^4)[0.87/(1,152 + 0.87)] = 51.5$ atm, or about 760 psi in one week.

A relief valve is needed, since the pressure calculated due to either heat or corrosion (or certainly from both) exceeds the design pressure of the piping.

References

1. Roark, R. J., "Formulas for Stress and Strain," 3rd ed., 1954, McGraw-Hill Book Co., New York, p. 268.
2. Lange, N. A., ed., "Handbook of Chemistry," 10th ed., 1961, McGraw-Hill Book Co., New York, pp. 1670-1679.
3. Mukerji, A., How to Size Relief Valves, Chem. Eng., June 2, 1980, Vol. 87, No. 11, p. 79.
4. Hodgman, P. C., ed., "Handbook of Chemistry and Physics," 32nd ed., 1950, Chemical Rubber Publishing Co., Cleveland, pp. 1803-1805.
5. Weaver, R., "Process Piping Design," Vol. II, Gulf Pub. Co., Houston, 1973, p. 155.

The author

Sudhir R. Brahmbhatt is an applications engineer for the Gas Products div. of MG Industries, P.O. Box 945, Valley Forge, PA 19482. Tel: (215) 630-5400. His work includes developing process alternatives, selecting processes, and designing energy-saving systems. Previously, he was a senior process engineer with Air Products & Chemicals, Inc. He holds a bachelor's degree in chemical engineering from Nadiad Inst. of Technology (India), an M.S. in chemical engineering from Stevens Inst. of Technology, and an M.B.A. from Fairleigh Dickinson University. He is a member of AIChE.

INDEX